HOMES WITHOUT HANDS.

BEING A DESCRIPTION OF

THE HABITATIONS OF ANIMALS, CLASSED ACCORDING TO

THEIR PRINCIPLE OF CONSTRUCTION.

BY

THE REV. J. G. WOOD, M.A. F.L.S. ETC.

AUTHOR OF

THE "ILLUSTRATED NATURAL HISTORY," "COMMON OBJECTS OF THE SEASHORE AND COUNTRY,"

ETC., ETC. ETC.

WITH NEW DESIGNS BY W. F. KEYL, AND E. SMITH.

ENGRAVED BY G. PEARSON.

NEW YORK:

D. APPLETON & CO., 445 BROADWAY.

1866.

LONDON
PRINTED BY SPOTTISWOODE AND CO.
NEW-STREET SQUARE

OSPREY AND GRAKLES.

PREFACE.

THE object of this work is so fully given in the title-page, that little more remains to be said in the preface.

Beginning with the simplest and most natural form of habitation, namely, a burrow in the ground, the work proceeds in the following order:—2d, those creatures that suspend their homes in the air; 3d, those that are real builders, forming their domiciles of mud, stones, sticks, and similar materials; 4th, those which make their habitations beneath the surface of the water, whether salt or fresh; 5th, those that live socially in communities; 6th, those which are parasitic upon animals or plants; 7th, those which build on branches. The last chapter treats of miscellanea, or those habitations which could not be well classed in either of the preceding groups.

In all these classes a definite order has been preserved, the Mammalia having precedence, and being followed in regular order by the other members of the group. Thus, in the first few chapters, which treat of the Burrowers, the following system has been observed:—First comes Man, the chief of all the mammalia, and in due zoological order follow the Moles and Shrews, the Foxes, the Weasels, the Rodents, and the Edentates. The White Bear alone is removed from its legitimate place, on account of its singular habitation in the snow. The Burrowing Birds come next in order, those which burrow in the earth taking precedence of those which make holes in wood. Burrowing Reptiles follow next in order; and then come the Burrowing Invertebrates, headed by the Crustacea. The same system is followed throughout, so as to give the reader a clear and definite idea of the subject. For this reason, a table of contents is appended to the work, as well as an alphabetical index; the one to enable the reader to form a general conception of the subject, and the other to enable him to find out any particular creature.

On perusing the work, the attentive reader will probably discover that various animals are placed in one class when they might very well be in another. The reason is, that many creatures, such as the wasp, the ant, the squirrel, &c., might with equal propriety find a place in several of these classes, and I have therefore placed them in that class of which some peculiarity in nest-making renders them fit illustrators.

I must now return my thanks to the many friends who have assisted me in the work, by the loan or gift of specimens, or by affording valuable information. Among them I must especially mention J. GOULD, Esq., who kindly took an interest in the ornithological portion of the work; F. SMITH, Esq., of the British Museum; and the late CHARLES WATERTON, Esq., who permitted me the use of his museum, and gave me much interesting and useful information.

CONTENTS.

CHAPTER I.

BURROWING MAMMALIA.

CHAPTER II.

BURROWING BIRDS.

CHAPTER III.

BURROWING REPTILES.

CHAPTER IV.

BURROWING INVERTEBRATES

CRUSTACEA.

CHAPTER V.

BURROWING MOLLUSCS.

CHAPTER VI

BURROWING SPIDERS.

CHAPTER VII.

BURROWING INSECTS.

HYMENOPTERA.

CHAPTER VIII.

BURROWING BEETLES.

CHAPTER IX.

WOOD-BORING INSECTS.

CHAPTER XIII.

PENSILE BIRDS (CONTINUED).

CHAPTER XIV.

PENSILE INSECTS.

CHAPTER XV.

BUILDERS.

CHAPTER XVI.

BUILDING BIRDS.

CHAPTER XVII.

BUILDING BIRDS (CONTINUED).

CHAPTER XVIII.

BUILDING INSECTS.

CHAPTER XIX.

SUB-AQUATIC NESTS. VERTEBRATES.

CHAPTER XX.

SUB-AQUATIC NESTS.

INVERTEBRATES.

CHAPTER XXI.

SOCIAL HABITATIONS.

SOCIAL MAMMALIA.

CHAPTER XXII.

SOCIAL BIRDS.

CHAPTER XXIII.

SOCIAL INSECTS.

CHAPTER XXIV.

SOCIAL INSECTS (CONTINUED).

CHAPTER XXV.

PARASITIC NESTS.

CHAPTER XXVI.

PARASITIC NESTS (CONCLUDED).

CHAPTER XXVII.

BRANCH-BUILDING MAMMALIA.

CHAPTER XXVIII.

FEATHERED BRANCH-BUILDERS.

CHAPTER XXIX.

FEATHERED BRANCH-BUILDERS (CONCLUDED).

CHAPTER XXX.

BRANCH BUILDERS.

SPIDERS AND INSECTS.

CHAPTER XXXI.

MISCELLANEA.

LIST OF ILLUSTRATIONS.

FULL PAGE ILLUSTRATIONS.

ILLUSTRATIONS IN THE TEXT.

HOMES WITHOUT HANDS.

CHAPTER I.

BURROWING MAMMALIA.

Introduction—MAN as a Burrower—The MOLE and its Dwelling—Difficulty of observing its Habits—Complicated structure of its Fortress, and its Uses—Character of the Mole—Adaptation of its Form to its mode of Life—Common Objects—The SHREW MOLE, ELEPHANT SHREW, and MUSK RAT—The ARCTIC FOX—Structure of its Limbs—Form of its Burrow—Its Character, Fur, and Flesh—The common FOX—Mode of Burrowing and economy of Labour—The young Family—The WEASEL, and some of its Habits—The BADGER and its Burrow—The PRAIRIE DOG, or WISH-TON-WISH—Dog-towns—Unpleasant Intruders—The RABBIT, and the Warren—Self-sacrifice—Study of animal Life—The CHIPPING SQUIRREL—Curious form of its Dwelling—Its subterranean Treasures—The WOODCHUCK, the POUCHED RAT, the CAMAS RAT, the MOLE RAT, and the SAND MOLE—The WHITE BEAR—Its curious Dwelling—Snow as a Shelter—The PICHICIAGO—Its Form, Armour, and Burrow—The ARMADILLOS and their Habits—The MANIS—The AARD VARK, its Food and Dwelling—The MALLANGONG—Its strange Habits and its Burrow—The PORCUPINE ANT-EATER—Its burrowing Powers.

AT some period of their existence, many of the higher animals require a Home, either as a shelter from the weather, or a defence against their enemies. Of all forms of habitation, the simplest is a burrow, whether beneath the surface of the ground, or into stone, wood, or any other substance.

The lowest grades of human beings are found to adopt this easy and simple substitute for a home, and the Bosjesman of the Cape, and the "Digger" Indian of America, alike resort to so obvious an expedient. If the country be craggy and mountainous, a casual cleft or hollow affords a habitation exactly suited to a race of mankind who have never undergone any training in industry, who never exert themselves until forced to do so by some imperative demand of nature, and who reduce such exertion to the minimum of labour which some present emergency requires.

Such debased tribes of humanity will occasionally adapt to their current circumstances the hole or crevice in which they take up their residence, and which can scarcely be called a home. No domestic associations hang around the habitation of the earth-dweller. The cave in which he dwells, or rather, in which he sleeps and shelters himself from inclement weather, possesses none of the thousand little amenities which constitute the home of man when even partially civilized. It is hallowed by no domestic joys, sanctified by no domestic trials, and those who take casual shelter therein know nothing of those "homely" feelings which in ancient times made the hearth an inviolable sanctuary, and which were outwardly symbolized by the Lares and Penates that surrounded the sacred spot. The inhabitants may adapt for the present, but they make no arrangements for the morrow, and, indeed, their memories seem to be as forgetful of the past and its lessons as their minds are incapable of fore-thought for the future.

They may possibly remove a stone which incommodes them while they seek repose, if, indeed, they cannot contrive to arrange themselves so as to save the trouble of removing it; and if the labour should not be severe, *i.e.* if the whole tribe need not do more work in a day than an English workman will perform in an hour, they may possibly enlarge or slightly alter their subterranean home.

Civilized man may, and does frequently, employ the rocky cavern as his dwelling-place; but with this difference, that he converts the rude cavern into a permanent home. Some of my readers have probably seen those curious rock-houses in Derbyshire, which have been hollowed out of the solid sandstone, and present to the astonished traveller a view of windows and doors cut into the face of the rock, and of a chimney just projecting out of the level ground above. Local traditions report that this peculiar construction was intended for the purpose of affording gratuitous nourishment to the inhabitants, who were supposed to feed on the hares, rabbits, lambs, and other creatures that stumbled over the chimney top and fell into the fire below.

Except, however, that the walls of these houses are carved from the living rock, instead of being built up by successive series of stones or bricks, there is nothing in them which differs from the ordinary dwellings raised by builders, so that in reality they have little in common with the rock habitations of savage tribes.

If the country in which the earth-dweller is placed should not be of a rocky or stony character, affording no caverns already excavated by the hand of nature, the savage is obliged to do violence to his temperament, and to set to work. Furnished as he is with the most miserable of tools—his usual implement, a stick with a sharpened end charred in the fire to make it harder—he can make but little progress, humble though the task may be. The sandy nature of the soil in which he is generally placed offers but little resistance to the rude tool with which he labours, and as the savage is content with a mere apology for a dwelling-place, his task is soon accomplished. If he desires to be peculiarly comfortable, he may stick a few dried bushes on the windward side of the hole, and hang a skin on them; but it is only on very wet and windy days that he will take so much trouble.

All subterranean dwellings are not of this simple nature. The underground palaces of India are wonderful examples of workmanship; but then they are nothing more or less than buildings placed below the level of the ground, and inhabited in the hot season by the luxurious. Even in such cases, however, the inherent defects of an underground dwelling make themselves painfully apparent. The rooms, though cool, are close and depressing in the extreme. Ventilation cannot be properly accomplished—the coolness is but the damp chilliness of a cellar, and brings no invigorating freshness to the languid frame, so that the edifice is only inhabited occasionally for the sake of grandeur, and the owner gladly retreats to the upper air, where he seeks the needed coolness by means of fans and evaporating water.

Human habitations, however, do not come within the scope of the present work, which is restricted to those homes that are constructed without the aid of hands, and are planned, not by reason, but by instinct. We pass, therefore, from the handiwork of man to those dwellings which are constructed with feet or jaws or beaks, and which are never marred by incompetence or improved by practice.

FORTRESS OF THE MOLE.

Of all the mammalia, the Mole is entitled to take the first place in our list of burrowers.

This extraordinary animal does not merely dig tunnels in the ground and sit at the end of them, but forms a complicated subterranean dwelling-place, with chambers, passages, and other arrangements of wonderful completeness. It has regular roads leading to its feeding-grounds; establishes a system of communication as elaborate as that of a modern railway, or to be more correct, as that of the subterranean network of metropolitan sewers; and is an animal of varied accomplishments.

It can run tolerably fast, it can fight like a bulldog, it can capture prey under or above ground, it can swim fearlessly, and it can sink wells for the purpose of quenching its thirst. It is, indeed, a most interesting animal, and our comparatively small knowledge of its habits gives promise of much that is yet to be made known.

Take the Mole out of its proper sphere, and it is as awkward and clumsy as the sloth when placed on level ground, or the seal when brought ashore. Replace it in the familiar earth, and it becomes a different being,—full of life and energy, and actuated by a fiery activity which seems quite inconsistent with its dull aspect and seemingly inert form. The absence of any external indication of eyes communicates a peculiar dulness to the creature's look, and the peculiar formation of the fore limbs gives an indescribable awkwardness to its gait.

I have always taken much interest in this animal, and have watched many of its habits, as far as can be done under the very untoward circumstances that always must exist when the animal to be watched is essentially subterranean in its habits. The Mole cannot develop its nature unless it is buried below the surface of the ground, and when it is there, we cannot see it. Many marine and aquatic animals can be tolerably watched by placing them in the aquarium; but when they take to burrowing, they put an effectual stop to investigation.

To catch a living Mole without injuring it is not an easy task, and when it is caught, the duty of supplying it with food entails so severe a labour, and necessitates such very early rising, that no one can hope for success who does not combine perseverance, patience, and resolution.

Dull and sombre as the Mole appears to be, it is by far the fiercest and most active mammal within the British Isles. Indeed, so remarkable is it for both those qualities, that I doubt whether the great feræ of tropical climates can equal it either in ferocity, activity, or voracity. We need not pity the Mole for the dull life which we suppose it to lead below the ground. There the Mole is happy, and there only can it develop its various capabilities. We must not judge other beings by ourselves. We are apt to envy the swallow for its sunny flight through the air in chase of flies, and to pity the Mole for its darkling passage through the earth in chase of worms. Yet, there is no doubt but that both beings receive equal pleasure in carrying out the object of their existence, and that the Mole feels no less gratification in the capture of a worm, than the swallow in the capture of a fly. Such, at all events, is the inference which is to be drawn from the manner in which the Mole acts when it has seized a worm; for no one can witness the active eagerness with which it flings itself upon its prey, and the evident enjoyment with which it consumes its hapless victim, without perceiving that the creature is exultantly happy.

The notion that worms must be miserable is a very natural one. A very little boy of my acquaintance was lately excusing some contemplated acts of cruelty towards worms, by saying that they were already in misery under the stones, and therefore that a little more pain would not be of much consequence to them.

We all know that the Mole burrows under the ground, and that it raises those little hillocks with which we are so familiar; but we do not generally know the extent or variety of its tunnels, or that the animal works upon a regular system, and does not burrow here and there at random. How it manages to form its burrows in such admirably straight lines is not an easy problem, because it is always in black darkness, and we know of nothing which can act as a guide to the animal. As for ourselves and other eye-possessing animals, to walk in a straight line with closed eyelids is almost an impossibility, and every swimmer knows the difficulty of keeping a straight course under water, even with the use of his eyes.

The ordinary mole-hills, which are so plentiful in our fields, present nothing particularly worthy of notice. They are the shafts through which the quadrupedal miner ejects the materials which it has scooped out, as it drives its many tunnels through the soil, and if they be carefully opened after the rain has consolidated the heap of loose material, nothing more will be discovered than a simple hole leading into the tunnel. But let us strike into one of the large tunnels, as any mole-catcher will teach us, and follow it up until we come to the real abode of the animal.

A section of this extraordinary habitation is given in the illustration. The hill under which this domicile is hidden is of considerable size, but is not very conspicuous, because it is always placed under the shelter of a tree, a shrub, or a suitable bank, and would not be discovered but by a practised eye. The subterraneous abode within the hillock is so remarkable that it involuntarily reminds the observer of the well-known maze, with which the earliest years of youth have been puzzled throughout many successive generations.

The central apartment, or keep, if we so term it, is a nearly spherical chamber, the roof of which is nearly on a level with the earth around the hill, and therefore situated at a considerable depth from the apex of the heap. Around this keep are driven two circular passages, or galleries, one just level with the ceiling and the other at some height above. The upper circle is much smaller than the lower. Five short descending passages connect the galleries with each other, but the only entrance into the keep is from the upper gallery, out of which three passages lead into the ceiling of the keep. It will be seen, therefore, that when a Mole enters the house from one of his tunnels, he has first to get into the lower gallery, to ascend thence to the upper gallery, and so descend into the keep.

There is, however, another entrance into the keep from below. A passage dips downwards from the centre of the chamber, and then, taking a curve upwards, opens into one of the larger tunnels, or high roads, as they have been appropriately termed. It is a noteworthy fact, that the high roads, of which there are seven or eight, radiating in different directions, never open into the gallery opposite one of the entrances into the upper gallery. The Mole, therefore, is obliged to turn to the right or left as soon as it enters the domicile, before it can find a passage to the upper gallery.

By continual pressure of the Mole's fur, the walls of the passages and the roof of the central chamber become quite smooth, hard and polished, so that the earth will not fall in even after the severest storm.

The use of so complicated a series of cells and passages is extremely doubtful, and our total ignorance on the subject affords another reason why the habits of this wonderful animal should be better studied. The only object that can at present be surmised is, that the rightful owner of such a stronghold may rest safely in his middle chamber, tasting the reward of repose which sweetens labour, and that in case of alarm, he might escape through either of the many passages which surround his home. I do not know, however, whether the Mole always retires to his fortress in order to rest, but rather imagine that he contents himself with lying in the high road. Such, at all events, seems to be the opinion of professional mole-catchers, who tell me that the Mole works and rests at regular intervals of three hours, making no distinction whatever between day and night.

Wonderful as is this subterranean habitation, it is not the only one which is constructed by the animal. It may be well adapted to a solitary individual, but it is not at all suited for a family, for whom a more extended nursery must be provided. The nursery is much simpler than the habitation, consisting merely of a large chamber, in which is laid a considerable mass of dried grass, the young blades of corn being sometimes employed for that purpose. The Mole chooses for this purpose the spot where two or more passages intersect each other, so that in case of alarm, the mother and young may escape in the direction which seems farthest removed from danger. This nursery is almost invariably placed at some distance from the fortress.

About the middle of June, or commencement of July, the Moles begin to fall in love, and are as furious in their attachments as in all other phases of their nature. At that time, two male Moles cannot meet without a mortal jealousy, and they straightway begin to fight, scratching, tearing, and biting with such insane fury, that they seem to be unconscious of everything but the heat of battle. Not content with fighting in their burrows, they often emerge into the open air, and may then be caught without the least difficulty. A few days before writing this account, I heard that a pair of Moles were thus taken in the fields near Erith, and one of my friends made a similar capture on Shooter's Hill.

Indeed, the whole life of the Mole is one of fury, and he eats like a starving tiger, tearing and rending his prey with claws and teeth, and crunching audibly the body of the worm between the sharp points. Some writers say that the Mole eats snails and other molluscs, but I am disposed to doubt that assertion. I have kept several Moles and never saw them eat anything but worms. They even rejected the julus millipede, kicking it aside with utter contempt.

It is also asserted that the Mole skins the worm before he eats it, "stripping the skin from end to end, and squeezing out the contents of the body." To prove a negative is proverbially a difficult task, and therefore I will not venture to say that the

Mole does not trouble himself about stripping off the skin of the worm. I do not see how he could do so, for even with the assistance of knives, scissors and forceps, such a task presents many difficulties, and how the Mole is to succeed in such an undertaking with no tools but his teeth and claws, I cannot comprehend. No Mole that I have ever seen, gave the slightest indication of skinning or emptying the worm, but proceeded without the least ceremony to devour the writhing prey, and then looked out for another victim.

It is hardly possible to conceive, and quite impossible to describe the fury with which the Mole eats. It hunches its back in a most curious manner, retracts the head between the shoulders, and uses its fore paws to assist it in pushing the worm into its jaws. In this respect there is a singular resemblance between the Mole and the carnivorous chelodines of America. I have kept several of them, and have always noticed that they ate exactly after the fashion employed by the Mole, seizing their food in their jaws, and tearing it to pieces by the aid of the armed fore paws—one foot being applied at each side of the mouth, so as to push the food forwards, while the head draws it back.

How the Mole assumes this peculiar attitude I cannot conceive. I have often seen it engaged in eating, and have sketched the creature while so employed; but, when the Mole has been dead, I have been unable to place it in the proper attitude, though anxious to do so in order that the artist might be able to make his drawing properly.

From seeing the animal eat, I can readily conceive the fury with which it must be animated when it fights, and can perfectly appreciate the truth of the assertion, that it has been observed to fling itself upon a small bird, to tear its body open, and to devour it while still palpitating with life.

Nothing short of this fiery energy could sustain an animal in the lifelong task of forcing itself through the solid earth; and it may well be imagined that when two male Moles of equal strength happen to meet, the combat must be of the most furious kind.

To those who are accustomed only to look at animals from their own stand-point, these battles may appear too insignificant to attract attention; but to the eye of a naturalist, who instinctively identifies himself with the nature of the animals which he is observing, these combats lose all their insignificance, and even partake in some degree of the sublime. Size is only of relative importance; and, in point of fact, a battle between two Moles is as tremendous as one between two lions, if not more so, because the Mole is more courageous than the lion, and, relatively speaking, is far more powerful and armed with weapons more destructive.

Magnify the Mole to the size of the lion, and you will have a beast more terrible than the world has yet seen. Though nearly blind, and therefore incapable of following prey by sight, it would be active beyond conception, springing this way and that way as it goes along, so as to cover a large amount of space, leaping with lightning quickness upon any animal which it meets, rending it to pieces in a moment, thrusting its blood-thirsty snout into the body of its victim, eating the still warm and bleeding flesh, and instantly searching for fresh prey.

Such a creature would, without the least hesitation, devour a serpent twenty feet in length, and so terrible would be its voracity, that it would eat twenty or thirty of such snakes in the course of a day. With one grasp of its teeth and one stroke of its claws it could tear an ox asunder; and if it should happen to enter a fold of sheep or an inclosure of cattle, it would kill them all for the mere lust of slaughter. Let, then, two of such animals meet in combat, and how terrific would be the battle. Fear is a feeling of which a Mole seems to be unconscious; and when fighting with one of his own species, he gives his whole energies to the destruction of his opponent, without seeming to heed the injuries which are inflicted upon himself.

From the foregoing sketch the reader will be able to estimate the extraordinary energies of this animal, as well as the wonderful instincts with which it is endowed.

What a surprising effort of intuitive skill is shown in the fortress, with its central chamber and the circular galleries that surround and defend it. How enormous is the space of ground which a single Mole covers with its network of roads and galleries, driven in every direction from the fortress, and sunk at various depths, according to the state of the ground and the position of the worms. Sometimes it burrows along just at the surface of the ground, its back visible as it passes along, and so making a shallow trench rather than a tunnel. Sometimes, as in very dry weather, it is obliged to dive deeply into the earth before it can find the worms, which detest drought, and cannot exist but in damp situations.

How marvellous is the amount of muscular power that is concentrated into so small a space. Every one who has worked at digging a pit is well aware of the labour involved in his undertaking, even with the aid of crowbar, pickaxe, and spade. If the reader should happen to have excavated a cubic yard of earth, he will know by experience the amount of muscular exertion that is required for the task, and will be the better able to appreciate the tremendous powers of the Mole, which is able to drive its tunnels so rapidly through the solid earth, and to throw up at short intervals those well-known mole-hills, which contain as much earth as would make a heap twelve feet in height and twenty feet in diameter, were a man to be the workman instead of the Mole.

On looking over the list of burrowing mammalia, the observer cannot but be struck with the wonderful manner in which they emerge from the earth with unsoiled fur. This capability is the more remarkable in the animal now under consideration, because it is continually engaged in making new tunnels, and is not content merely to pass up and down a passage already excavated. The sides of the passages, which are popularly known as the high roads, are by degrees worn quite smooth by the attrition of the Mole's body, so that in them there is little danger of injury accruing to the fur. But that an animal should be able to pass unsoiled through earth of all textures is a really remarkable phenomenon, which is partly to be explained by the character of the hair, and partly by that of the skin.

The hair of the Mole is notable for its velvety aspect, and its want of "set." The tips of the hairs do not point in any particular direction, but may be pressed equally forwards or backwards or to either side. The microscope reveals the cause of this peculiarity. The hair is extremely fine at its exit from the skin, and gradually increases in thickness. When it has reached its full width, it again diminishes. This alternation of tenuity and thickness occurs several times in each hair, and gives the peculiar velvet-like texture with which we are all so familiar. There is scarcely any colouring matter in the slender portions of the hair, and the characteristic changeability of the blackish-brown hues is owing to this structure.

Perhaps the reader may not have noticed that when the fur of the Mole has been thoroughly cleansed, it has a strong iridescence in certain lights, assuming various beautiful tints, among which a ruddy copper is the most prevalent.

Another reason for the cleanliness of the fur is the strong, though membranous muscle beneath the skin. While the Mole is engaged in tunnelling, particularly in loose earth, the soil falls upon the fur, and for a time clings to it. But, at tolerably regular intervals, the creature gives the skin a sharp and powerful shake, which throws off at once the whole of the mould that has collected upon the fur. Some amount of dust still retains its hold, for, however clean the fur of a Mole may seem to be, if the creature be placed for an hour in water, a considerable quantity of earth will be dissolved away, and fall to the bottom of the vessel. The improvement in the fur after being well washed with soft tepid water and soap, is almost incredible.

Many persons have been struck with such admiration of the fur of the Mole, that they have been desirous of having a number of the skins collected and made into a waistcoat. This certainly can be done, but is not a commendable plan, for the garments thus made are very hot, so that they can only be worn in winter; they are very expensive, costing from two to three pounds, and they possess but little lasting powers. There is also a wonderfully strong smell about the Mole; so strong, indeed, that dogs will sometimes point at Moles instead of game, to the great disgust of their masters. This odour adheres obstinately to the skin, and even in furs, which have been dried for more than ten years, I have noticed this unpleasant savour.

The Mole is one of those animals which, like the sloth, are formed expressly for the condition of life in which they are placed. There are many burrowing animals, but the Mole is emphatically *the* burrower, the very type of a creature which is intended to pass the whole of an active existence under ground. I say an *active* existence, because there are many other creatures which lead a subterranean life, but which are comparatively quiet and listless during their sojourn underground. But the Mole absolutely riots in the exuberance of animal spirits and muscular activity, passing through the earth almost like a fish through the water, and giving to its strange, and apparently sombre life, a poetry and an interest which we fail to find in the lives of many creatures more richly endowed with external beauty.

Let me recommend the reader to procure, or at all events to examine, the skeleton of a Mole, and to note the wonderful structure by which such effects are produced. The enormous shoulder-blades, projecting far above the spine, the short, bowed, and powerful bones of the fore limbs, the wide, flattened palms, and the strong, sharp, and curved claws, look almost like a miniature model of some machine invented for the purpose of tearing the stubborn earth.

See how all the power is thrown into the fore quarters, while the hind quarters are feeble, and, in comparison, common-place. See how enormously strong must be the muscles of the neck, where the ligament (popularly called the paxwax) is hardened into bone. The nose, too, is furnished with an accessory bone, which projects into the snout, and gives it that combined strength and mobility which distinguishes the creature. Immediately after death the snout of the Mole is flexible and elastic, springing back when bent, as if cut from solid India-rubber. But in a very few hours it becomes stiff, unsightly, and shrivelled, and loses all its plump rotundity. Once lost, this is never restored. You may immerse the Mole in water as long as you like, but as the shrivelling is more from within than without, the moisture fails to penetrate the tissues, and to enable them to regain their pristine contour. As to stuffed specimens, I never yet saw one in a museum that gave much idea of the animal, and the snout in particular, is always crumpled, black, and withered.

In order to give greater spread and power to the fore paws, there is an accessory bone shaped something like a sickle, projecting from the carpus, and it will be found that in this extraordinary animal still exist certain remarkable peculiarities of structure, which are seen in no other living form, but have been discovered in the fossil skeletons of animals long extinct.

I have given much space to the Mole on account of its many claims to our notice. Had the creature been a rare and costly inhabitant of the tropics, how deep would have been the interest which it excited. How the scientific world would have crowded to see the marvellous structure of a skeleton wherein are several accessory bones, and which exhibits peculiarities hitherto found only in fossil remains. How great would have been the admiration evoked by its soft, velvet-like fur, its tiny eyes deeply hidden in the fur, so as to be sheltered from the earth through which the animal is continually making its way, the strange mixture of strength and softness in the palms of its fore feet, and the elastic springiness of its nose.

But, because it is a native of our own country, and to be found in every field, there are but few who care to examine a creature so common, or who experience any feelings save those of contempt or disgust, when they see a Mole making its way over the ground in search of a soft spot in which to burrow, or pass by the place where the mole-catcher has strung up his victims on the trees as Louis XI. was accustomed to suspend the bodies of those who had committed the crime of trespassing on the royal domains. For my own part, I am but too glad that such wonderful beings *are* common, and am thankful for so many opportunities of studying the works of Him who has made the lowly Mole as carefully as the lordly man.

THERE are many other burrowing animals allied to the mole; and although it will be impossible to give illustrations of their burrows, they ought not to be passed by without a casual notice.

The Shrews, for example, are among the burrowers, and although their eyes are full and round, their fore quarters of ordinary proportions, and their fore feet of the usual shape, there is something about the head, with its long mobile snout, which strongly reminds the observer of the same member in the mole. These pretty little creatures reside within their burrows during the day, and are therefore seldom seen in a living state, except by those who are in the habit of traversing the country by night in search of specimens. Dead Shrews are common enough, having probably been killed by predatory animals, but left uneaten on the ground, in consequence of the powerful odour which they evolve.

At the end of the burrow the Shrew makes its nest, which is composed of dry grasses and other herbage, and is of a partly globular form.

The SHREW MOLE of North America (*Scalops aquaticus*), is one of the best burrowers among this family, scarcely yielding to the mole itself in the extent of the tunnels which it excavates. Like the mole, it drives its burrows below the surface of the ground, throws up hillocks at intervals, and feeds chiefly on earthworms. The eyes of this creature are very minute, and deeply hidden in the soft fur. Unlike the mole, however, it is in the constant habit of coming to the surface of the ground, and passing into the full blaze of the noontide sun. At that time of day the animal may be caught by driving a spade under it, so as to cut off its retreat, and by flinging it to some distance from its tunnel.

Mr. Peale mentions that a Shrew Mole in his possession was able to bend the snout to such an extent as to force food into its mouth. The European mole, flexible as is its mobile snout, possesses no such power, but is obliged to perform that task with its fore paws.

Then, there is the ELEPHANT SHREW of Southern Africa (*Macroscelides typicus*), a thick-furred, long-snouted, short-eared burrower, which has a rather remarkable method of sinking its tunnels, first boring a nearly perpendicular shaft, and then driving its burrow at an angle. It is not so devoted to a subterranean existence as either of the preceding animals, and loves to come out of its burrow and bask in the genial sunbeams. It is, however, as wary as the rest of its kindred, and at the least alarm darts off to its subterranean fastnesses. While basking in the warm rays, it generally sits erect, facing the sun, so as to receive every ray.

Our last example of the Shrews is the remarkable animal which is popularly called the MUSK RAT (*Myogalea moschata*), though it is an insectivorous animal, and far removed from the rodents. The river Wolga is the favourite resort of this curious quadruped, which seems to hate dry land as much as the beaver, and to spend the greater part of its time in the water. The Musk Rat is an admirable burrower, making its tunnels of considerable length, some of them extending to a distance of twenty feet. There is only one entrance, which is always below the water; and the burrow rises gradually upwards, so that at the extremity the animal is lodged on dry ground. It is instinctively careful to avoid too close a proximity to the surface of the earth, lest the roof of its home might fall, and disclose the interior to the unwelcome light.

The odour which has already been mentioned as belonging to

all the members of this family, is in the Musk Rat so powerful as to raise the animal into an article of commerce, the musky scent being nearly as powerful as that produced by the musk deer, and the perfume obtained at a much cheaper rate. The well-known Musk Rat of India belongs also to this odoriferous family.

"EARTH" OF THE FOX.

THE FOX is a well-known burrower, its 'earth' being familiar to many by sight, and to all by name.

Few persons, who do not know the history of the Fox, would believe it to be capable of forming excavations of such extent. The fore feet of the mole are clearly formed for digging, their sharp claws penetrating the earth, their broad palms acting as shovels, and their powerful muscles giving the needful force. These limbs are essentially used for digging, and are but little employed as means of locomotion. But the Fox is an admirable runner, as any hunter can avouch, and its fore limbs are formed for speed and endurance, their length enduing them with the one quality, and their muscular lightness with the other. Yet, just as the digging limbs of the mole are used for locomotion, and enable the animal to proceed at no contemptible speed; so the running limbs of the Fox are used for digging, and enable the creature to excavate burrows of no contemptible dimensions.

The ARCTIC FOX (*Vulpes lagopus*), an animal which dwells in the polar regions, is notable for the extent and structure of the burrow. In order to shield itself from the inclemency of the climate, it digs to a considerable depth; and it is rather remarkable that a solitary burrow is seldom found, twenty or thirty Foxes generally sinking their tunnels in close proximity to each other.

Perhaps this semi-sociality may be accounted for in a very simple manner, namely, the suitability of some particular piece of ground, to which the Foxes flock by instinct, and in which they drive as many burrows as the ground will accommodate. This conjecture is the more likely to be true, because sandy spots are always chosen for this purpose, where twenty or thirty burrows are often sunk in close proximity to each other. Such spots would be peculiarly suitable to the Fox, because the sandy soil is not so likely to be hardened by the frost as that of a more compact and watery nature, and would be easily thrown out by the small though powerful feet of the animal.

If one of these little colonies could be laid open, a very curious sight would present itself. The earth would be seen to be pierced with multitudinous tunnels, each complete and independent in itself, and never interfering with burrows belonging to other owners. Each burrow, too, is of a very complex character, and by no means consists of a single tunnel, with a rude nest at the extremity. There are three or four distinct passages, each of which opens into the common chamber, which is of considerable dimensions, and serves as a starting-place whence the inhabitant can seek refuge in either of its passages, according to the direction in which it apprehends danger.

This chamber is not, however, the nursery for the young, a second cavity being used for that purpose. The nursery is not of great dimensions, and communicates by a passage with the chamber already mentioned. The reader will see, therefore, that in some respects the habitation of the Arctic Fox corresponds with that of the mole, both having a kind of fortress from which a number of passages lead in different directions, and the nursery being in both instances separate from the general habitation.

Five or six young ones are mostly bred in these subterranean nurseries; and in the outer chamber, and in several of the passages that lead to it, are placed good stores of food. In one such nest were found many bodies of two species of lemming, and several stoats; and the abundance of bones belonging to hares, fishes, and ducks, showed that the wants of the young Foxes had been amply supplied.

This Fox is a very intelligent animal, though the accounts of the earlier Arctic voyagers would lead the reader to imagine that it was peculiarly stupid. It could be caught without the least difficulty; any rough contrivance answered for a trap, and any number of Foxes might be taken in it. The hunter might make a trap and bait it while the animal looked on, and as soon as he retired, the Fox would walk into the snare. Fifteen have thus been taken in four hours in a single trap.

Other voyagers tell very different tales of the same animal. Their note-books abound in anecdotes of its exceeding craft, and the difficulty of catching it; and they tell us how the Arctic Foxes learned to remove the baits without falling into the traps, or being shot by the spring-guns. The reason is evident. Those who commemorate its stupidity are the earlier voyagers—those who describe its craft are the later travellers. Before the advent of the former, the Foxes knew nothing of the European, his traps and his guns, and had no suspicion that a bait was meant for any other purpose than to be eaten. Before the European came into these northern regions, the Fox was very little molested; but when men found that its skin was a little fortune, and that its flesh was generally eatable, the Fox was subjected to a merciless persecution. For, even in its ordinary state, the skin of the Arctic Fox is in great favour as a fur; but when it is bleached by the dread cold of the regions in which the animal resides, and is of a pure snowy whiteness down to the very roots of the hair, it is so exceedingly costly, that a mantle made of that fur is only to be purchased by millionaires, or placed on imperial shoulders. I am afraid to say how many thousand pounds have been paid for a mantle of white fox-skins. In consequence of the value of the fur, scarcely a Fox can show his sharp nose without being tempted by baits or followed by riflemen; and so many have fallen victims, that the survivors have learned wisdom.

All persecuted animals learn wisdom. Try to catch an old rat, and see how long you will have to wait before you see him in the trap. Try to snare an old raven, or even to hook an old trout, and you will find that your best energies will be taxed and all your ingenuity tested, before you will succeed. So it has been with the Foxes. They like the bait as well as ever, but they have acquired a rooted distrust of wires, or sticks, or strings, or indeed of anything to which they are not accustomed in their everyday life, and therefore keep carefully aloof from everything that conveys suspicion to their eyes or nostrils.

The flesh of the young Fox is very good eating, but that of the old animal is almost valueless except to starving men, being hard and stringy, and having a very unpleasant flavour. Even the water in which it has been boiled is acrid, and apt to blister the mouth and gums. But, although the flesh is valueless, the skin is almost beyond price, and the fur of a fine old Fox in perfect condition is worth many times its weight in gold.

THE habitation of the common Fox of this country is by no means so complicated as that of the Arctic species.

Whenever it can, the Fox avoids the labour of burrowing, and avails itself of the deserted home of a badger, or even a rabbit. In the former case there is very little to be done to the burrow, and in the latter the cunning animal finds its labour greatly diminished; for though the Fox is a much larger animal than the rabbit, and needs a rather larger tunnel, it finds that the task of enlarging a ready-made burrow is very much less than if it had to drive a passage through solid ground. Every one who has worked with carpenters' tools knows that a large gimlet passes easily through wood, if it follows the track of a smaller one, and on the same principle, the Fox passes easily through the earth on the track of the rabbit. The burrow of the latter animal is moreover much larger than is absolutely required for its passage, while the former is quite satisfied if he can pass through the tunnel with tolerable rapidity.

Sometimes, however, the animal is not fortunate enough to find any ready-made habitation, and in such cases sets determinately to work, and scoops out a burrow on its own account. Herein it lies asleep all day, as is the custom with most predaceous animals, and only sallies forth at night. Herein the mother produces and nurtures her young, and sometimes on a summer's evening, the whole family, the father, mother, and cubs, come out to enjoy the fresh air. They never wander far from the mouth of the burrow, and as the young are gamesome little creatures, as playful as puppies, and much prettier, and the mother helps her young ones in their sports as a good mother ought to do, the group presents a very pretty sight. When young the cubs are certainly not prepossessing, and scarcely any one would take the sprawling grey-coated, broad-muzzled creatures, with their little short pointed tails and stumpy ears, for the young of the Fox, with its ruddy fur, its active limbs, its narrow muzzle, its fine bushy tail, and its erect, intelligent-looking ears.

Though there is but one burrow for the nursery, the Fox generally has access to "earths" as they are called, at considerable distances apart, and, as all huntsmen know, when he finds that one of his earths is stopped, will straightway start off for another which may probably be at a distance of several miles, not to mention his accurate knowledge of drains and similar places of refuge. Therefore to keep an old experienced Fox above ground is a task which needs great skill and considerable endurance, for he is by no means above availing himself of clefts in rocks, should the country be of a mountainous nature, or using holes in decayed trees ; and, indeed, if within a radius of some ten or twelve miles there is a cavity which is capable of concealing a Fox, the cunning animal is sure to know it.

THE Weasels have been said to be great burrowers, but I am inclined to think that very few of them are in the habit of tunnelling below the ground. The Otter is generally thought to be a burrower because it has certain caves in the river banks, to which it flies for refuge when pursued, and in which it produces and rears its young. But I believe that in every instance the animal takes advantage of some ready-made cavity, mostly contenting itself with accepting the retreat, and at the best, merely scraping and adapting the spot to suit its own purposes.

The Weasel is certainly no excavator. It takes up its habitation in rocky crevices, under the gnarled roots of old trees, in the interstices between stones, and similar localities, stone-heaps being always favourite spots. Ladies who build their picturesque rockeries for the culture of ferns, would be very much surprised if they knew how often the Weasel takes possession of the stones, and how the interior of the mimic rock is tenanted by these snake-necked, red-bodied, bright-eyed little creatures. Should they perchance see a Weasel poking its intelligent little head out of a crevice, they should not be alarmed, but do their best to encourage an animal so useful, a little ally that will do more towards clearing the garden of mice and other nocturnal depredators than all the ratcatchers in the neighbourhood.

One of the Weasel tribe is, however, a most powerful and industrious excavator. This is the BADGER (*Meles taxus*), an animal which was formerly considered as our only surviving British representative of the bear tribe, but is now found to belong to the weasels.

The Badger makes a most gloomy, dark, and tortuous burrow, generally excavated in some retired and shadowy spot, such as dense thickets, or the recesses of thickly-wooded forests. As is the case with several burrowing animals, there are several chambers in its domicile, one of which is appropriated as a nursery, and is warmly padded with dry mosses and grass.

The Badger is a creature that cannot live in close proximity to human beings, and has, in consequence, been gradually banished from the greater part of England. Forest after forest falls before the woodman's axe, mile upon mile of barren bog-land is drained and converted into fertile, food-producing soil; and so, to the very great satisfaction of the political economist, and the very great discomfiture of the naturalist, all our large carnivora, whether furred or feathered, are gradually ousted from the soil wherein they formerly exercised unquestioned sway. The Badger has long ago been driven out of the land; the otter is but seldom seen in the rivers where it was once so plentiful; the polecat and martens have retired into the deepest recesses of the few forests which are still left to us, but over which the demon of bricks and mortar already casts an evil eye; and the stoat and weasel only hold their own on account of their diminutive size, and the comparative ease with which they obtain a supply of food. They are among the animals which are gradually eliminated out of existence by the encroachments of man, and it may be that in a few years a stoat or weasel may be as rare in England as a Badger is at the present day.

The fossorial limbs of the Badger are useful in various ways; for not only do they enable their owner to dig a domicile which none dare invade without the help of man, but they aid him in obtaining a kind of food to which he is particularly partial. The Badger is tolerably omnivorous, but has a special liking for insects in their immature state, and will dig up the nests of wasps and other subterranean hymenoptera, for the sake of devouring the larvæ. Some writers say that the Badger scrapes out the wasp-combs on account of the *honey* contained in them, but as no British wasp makes a cell that can hold honey, or is capable of gathering and storing that sweet substance, the Badger might scrape for a very long time before it earned a meal.

In like manner the Mink, the Vison, and other weasels of Northern America are in the habit of retiring to holes and crevices, but do not appear to form burrows for themselves. The Honey Ratel (*Mellivora Ratel*) does, it is true, scrape deep holes in the ground with very great rapidity, but then the creature cannot be ranked among the true weasels, and by many authors is thought to approach closely to the bears.

THE exact classification of animal habitations involves a task not easily accomplished, inasmuch as so many of them partake of characteristics which might entitle them to be placed under various categories. The rabbit, for example, might be considered either as a social or a burrowing animal, and the same may be said of the common wasp, the humble bee, and many other insects.

The PRAIRIE DOG (*Spermophilus Ludovicianus*) may, like the rabbit, be considered equally as a burrower or a social animal, and we will therefore place it in the former of these categories.

This animal is sometimes called the WISH-TON-WISH, but it is usually known by the name of Prairie Dog, though it is a rodent and not a carnivorous animal. The reason of its popular name

A PRAIRIE DOG TOWN.

lies in the short yelping sound which it is fond of uttering, and which bears some resemblance to the bark of a young puppy. Even in captivity it utters this short, impatient yelp, which may generally be extorted from the little animal by placing the hand near the cage. Though so gentle and affectionate to its keeper, it dislikes strangers; and if their fingers approach the bars of its house too closely, it barks at the intruders like an angry squirrel, and scratches smartly at their hands with its sharp and powerful claws.

It is a pretty, and rather curious animal, measuring about sixteen inches in total length. Its general shape is round and flattish, and the head is peculiarly flat, giving to the animal a very remarkable aspect. The fur is greyish red, with a grizzled effect, produced by the alternate chestnut and grey colour of each hair. The disposition of the Prairie Dog is pleasant and sociable, and the little creature is very susceptible of domestication. There are at the present time (July, 1863) two fine specimens of this animal in the Zoological Gardens, and both of them are notable for their tameness. The male (called Charley by the keeper), seems remarkably fond of his master, and loves to be taken up and nestle in his breast. The female is also a very tame animal, and was a great favourite of its late owner, following him about like a dog, and residing chiefly in his coat pocket.

In spite of the formidable foes by which it is attacked, and which take up their residence in the very centre of its habitations, the Prairie Dog is an exceedingly prolific animal, multiplying rapidly, and extending its excavations to vast distances. Indeed, when once the Prairie Dogs settle themselves in a convenient spot, their increase seems to have no bounds, and the little heaps of earth which stand near the mouth of their burrows extend as far as the eye can reach.

The burrows are of considerable dimensions, and evidently run to no small depth, as one of them has been known to absorb five barrels of water without being filled. It is not impossible, however, that there might have been a communication with some other burrow, or that the soil might have been loose and porous, and suffered the water to soak through its substance. They are dug in a sloping direction, forming an angle of about forty-five degrees with the horizon, and after descending for five or six feet, they take a sudden turn, and rise gradually upwards.

Thousands upon thousands of these burrows are dug in close proximity to each other, and honeycomb the ground to such an extent that it is rendered quite unsafe for horses.

The scene presented by one of these "dog towns" or "villages," as the assemblages of burrows are called, is most curious, and well repays the trouble of approaching without alarming the cautious little animals. Fortunately for the traveller, the Prairie Dog is as inquisitive as it is wary, and the indulgence of its curiosity often costs the little creature its life. Perched on the hillocks which have already been mentioned, the Prairie Dog is able to survey a wide extent of horizon, and as soon as it sees an intruder, it gives a sharp yelp of alarm, and dives into its burrow, its little feet knocking together with a ludicrous flourish as it disappears. In every direction a similar scene is enacted. Warned by the well-known cry, all the Prairie Dogs within reach repeat the call, and leap into their burrows. Their curiosity, however, is irrepressible, and scarcely have their feet vanished from sight, than their heads are seen cautiously protruded from the burrow, and their inquisitive brown eyes sparkle as they examine the cause of the disturbance.

A good marksman will take advantage of this peculiarity, and, by aiming at the eye, will make sure of killing the animal on the spot. It is marvellously tenacious of life, and unless its head be almost knocked to pieces, is sure to escape into its home. A pea-rifle is almost useless in shooting Prairie Dogs, a large bullet being needed to produce instantaneous death.

The Prairie Dog has not the privilege of possessing a home exclusively devoted to its own use, for the Burrowing Owl, sometimes called the Coquimbo Owl (*Athene cunicularia*), and the terrible rattlesnake, take forcible possession of the burrows, and devour the inmates, thus procuring board and lodging at very easy rates. The rattlesnake at all events does so, the bodies of young Prairie Dogs having been found in its stomach.

On the discovery of owls and rattlesnakes within the burrows of the Prairie Dog, it was generally thought that these incongruous beings associated together in perfect harmony, forming in fact a "Happy Family" below the surface of the ground. The ruthless scalpel of the naturalist, however, effectually dissipated all such romantic notions, and proved that the snake was by no means a welcome guest, but an intruder on the premises, self-billeted on the inmates, like soldiers on obnoxious householders, procuring lodging without permission, and eating the inhabitants by way of board.

The reason for the presence of the owls is not so evident, though it is not impossible that they may also snap up an occasional Prairie Dog in its earliest infancy, while it is very young, small, and tender. These winged and scaled intruders are not found in all the burrows, though many of the habitations are infested by them.

The general aspect of the Prairie Dog is not unlike that of its near relative, the Alpine Marmot, so familiar in this country through the mediumship of Savoyard boys, who carry the animal about in a box, and exhibit it for halfpence.

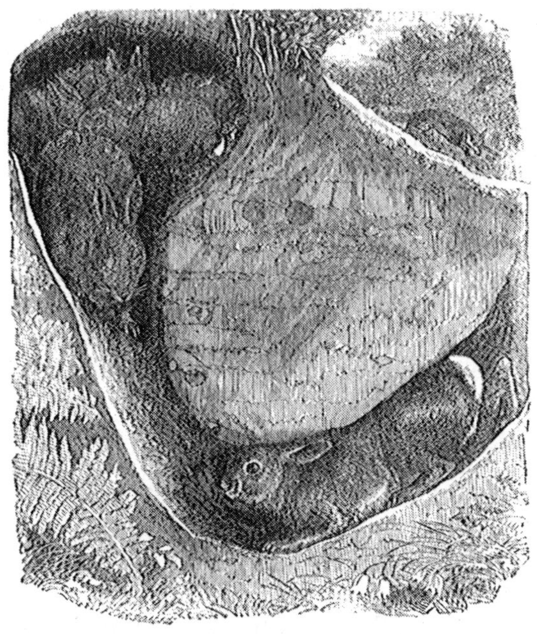

THE RABBIT WARREN.

One of the most familiar of the British burrowing rodents, is the common Rabbit (*Lepus cuniculus*), an animal notable for "sporting," as gardeners would say, into a vast number of varieties, some of which are so unlike the original stock, that they seem to be species and not varieties, and indeed might have taken rank as species, did they not invariably display a tendency to recede to the ancestral short brown fur and upright ears of the wild Rabbit.

This animal lives, as we all know, in burrows, and is mostly of a social nature, a considerable number of burrows being gathered together and known by the name of a Warren. Whenever the Rabbits find an undisturbed spot, which combines the advantages of a sandy situation with the vicinity of food, they establish themselves forthwith, and sink their multitudinous tunnels into the ground. The favourite locality for the Rabbit is a loose, sandy, or gravelly soil, covered with patches of furze bushes; for the soil is easily excavated, and is very dry, and the young shoots of the furze yield a food equally grateful and nutritious. Moreover, the tangled roots of the furze afford an excellent protection to the burrows, and the overhanging branches, with their prickly verdure, serve admirably to shelter the entrances.

When once they have established themselves, the Rabbits increase with almost incredible rapidity, nearly rivalling the rats and mice in fecundity, and converting the land into a very honeycomb of burrows. Indeed, were not the flesh of the Rabbit marketable, and its fur valuable—were not the stoat, the weasel, the hawk, and other furred and feathered depredators extremely fond of young Rabbits, the animals would spread so fast as to become a positive nuisance. In some places they have increased to such an extent, that the safety of buildings has been greatly endangered by the deep and ramifying tunnels which they have sunk beside the foundations; and I know of a case where they have multiplied so inordinately, that the proprietor of the ground, albeit a most staunch conservator of animal life, has been obliged in self-defence to have them exterminated.

It is not a very easy matter to drive the creatures from any place of which they have already taken possession, and even after employing all the paraphernalia of ferrets, nets, and guns, two or three isolated individuals are apt to escape, and if they should chance to be of opposite sexes, the Rabbit host is marvellously soon reproduced. The creature becomes a parent at a very early age, and by the time that a Rabbit is a year old, it may have attained the dignity of a grandparent

As is the case with most animals, the Rabbit seeks a quiet and retired spot for her little nursery. She does not produce her young in any of the burrows to which the general Rabbit colony has access, but prepares an isolated tunnel, at the end of which she forms her nest. The bed on which the young recline is beautifully soft and fine, being composed chiefly of the downy fur which grows on the mother's breast, and which she plucks off with her teeth in tufts of considerable size. Any one who keeps tame Rabbits may see the female preparing her cradle with this soft fur, and note how perseveringly she denudes her breast of its covering.

Much stress has been laid on the self-sacrifice exhibited by the animal, but I cannot believe that there is any sacrifice in the case. We know that if we were to pull out handfuls of our hair, we should suffer exceeding pain, and should, moreover, feel very uncomfortable for want of the locks which we had torn away. But the case of the Rabbit has no real analogy with such a proceeding, for the fur of the animal is at that time so loosely attached to the skin that it falls off as easily as the hairs of a cat in summer, and its evulsion produces no such disastrous effects as would follow the forcible plucking out of human hair. No raw and bleeding surface is seen when the fur of the Rabbit is removed, and scarcely a sign of inflammation is visible upon the skin. The act is a purely instinctive one, involving no more self-sacrifice than is occasioned by any other instinctive act; and perhaps we should not be very far wrong if we were to say that the animal would experience more self-sacrifice in omitting than in committing the act in question.

The incubation of the eider duck presents similar features, the parent birds stripping themselves of their down in order to form a warm bed for their young. Yet, I do not think that there is any self-sacrifice in the case, and fully believe that the creature experiences a feeling o gratification rather than of inconvenience, when it plucks away the down and arranges it for the reception of the eggs. And, even if we grant that the Rabbit or eider duck did find that they were less comfortable after they had denuded themselves than before, and moreover that they knew beforehand what would be the effect of the operation, we are met by the fact that they are obeying an impulse which they cannot resist, and that the idea of self-sacrifice is therefore untenable. A sacrifice it may be, but not a self-sacrifice as we understand that phrase at the present day.

In like manner, also, the beautiful gold-tail moth (*Porthesia chrysorrhœa*), so well known for the soft downy plumage of its wings, and the large round tuft of golden hairs upon the end of its body, robs itself of the latter adornment after it has laid its eggs, and shelters the future brood with an elegant thatch, composed of those downy hairs. Yet there is no self-sacrifice involved in the act, which is as purely instinctive as that of laying the eggs; and I do not believe that the insect has the least idea of the future prospects of the eggs, or possesses any foreknowledge of the snow and bitter frost of the coming winter. Even if she did know that she was depriving herself of a natural adornment for the sake of her offspring, the act would lose little of its instinctive character, and may find a parallel in humanity, when a fond mother devotes the once-cherished robes of her bridehood to her babe, and feels the keenest enjoyment in wrapping the costly furs and sheeny satins around its little limbs.

So with the ordinary incubation of birds. Didactic writers are in the habit of holding up for our admiration the conduct of the bird who leaves all her accustomed pleasures, and submits to a voluntary imprisonment in her nest until her eggs are hatched. These writers are entirely wrong, inasmuch as they assign to the lower animals certain attributes which belong only to man. I do not intend to depreciate in any way the faculties of the animal creation. On the contrary, I believe that the lower animals are endowed with gifts more extended than we generally suppose. But, as has already been remarked, we must not judge them by our own standard; and, instead of elevating them to our position, we should try to lower ourselves to theirs. By doing so there is nothing derogatory to the preeminence of human nature. We know that the best schoolmaster is not merely he who is the best scholar, but he who possesses the faculty of descending from his lofty elevation, of identifying himself with his pupils, and, for the time, sharing their ignorance, and so placing himself in their position. In like manner, the best naturalist is not necessarily he who has read the most about animals, nor he who attributes to them the feelings which he himself would experience in similar situations; but he who can divest himself, for the nonce, of his superiority, and conjecture the thoughts which would enter the limited minds of the creatures with which he is brought in contact.

Suppose, for example, that I am training a dog, which happens to be the case at present. If I were to judge the motives of that dog by my own ideas, I should fall into grievous errors, and fail entirely in my object. At present the animal commits various misdemeanours; but, as he does not know them to be such, I should be very wrong were I to punish him for them. He is at present obedient to instinct alone, and, until his reasoning powers have been brought into play, I should not only have no right to punish him for any instinctive act, but should greatly retard the duration of his training.

Being unused to human society, he had no idea that he might not jump on the table and help himself to meat; and his only idea of shoes, shawls, and other articles of clothing was, that they were charming playthings, which could be bitten and shaken without hurting his teeth. So, when I see him standing on the table, busily at work on a joint, or come upon him in the act of worrying my favourite shoe, I do not fly in a passion with him and beat him, but quietly put a stop to his proceedings, and tell him that he is not to do so again. Not being terrified by the fear of ill-treatment, he perfectly understands the sense though not the *ipsissima verba* of the remonstrance, and proves his intelligence by his acts. For example, if he takes another shoe and is discovered, he immediately drops his tail and ears, and looks like the culprit that he is. Being a delinquent, and knowing that he is so, he receives the punishment due to the offence, and ever afterwards the very word "shoe" will make him look utterly ashamed of himself.

Again, if he scampers over my newly-dug flower-borders, and lies down on my best bed of mignonette, I reflect that, in his position, I should have done just the same thing, not knowing that there was any harm in it. So I call him off the flowers, and explain to him that he is to restrict his gambols to the gravel and lawn. If, after the explanation, he persists in running over the borders, he becomes a conscious delinquent, and is punished accordingly.

Moreover, I manage so that he fancies the punishment to be inherent in the offence. For example, the animal at present in training is a Skye puppy of the purest breed, and as such is liable to be carried off by the numerous vagabonds who turn a dishonest penny by dog-stealing. His orders are, that he is not to go into the road on any pretext whatever, and, as a necessary consequence, he is always burning to do so. Therefore, I keep a watch upon him, and as soon as he pushes his way under the gate, he gets such a shower of swan-shot about his ears that he yelps in sudden terror, and dives back again. On such occasions he always comes up to me, as if to report himself, and I always pat him and pretend to know nothing about his intended escapade.

It is through the neglect of this simple principle that so many animals are made savage, or sulky, or stupid, by the clowns—whether in fustian or broadcloth—who have the charge of them. They lay down certain arbitrary rules of their own, neglect to teach these rules to the poor animals, and then ill-treat them because they break the laws which have never been taught to them. Farm labourers are, as a body, continually committing this error, and it is to their mingled ignorance and obstinacy that the savage nature of so many animals is due. A horse, for example, strays into some spot where he has no business, and is immediately chased, and shouted at, and pelted with stones, and struck with the first weapon that can be seized. Or he turns his head to the right instead of the left, and straightway is "chucked" by the bridle, and kicked, and anathematized with that copious flow of foul language with which our unsophisticated rural swains are so amply supplied.

Now the horse does not in the least know *why* he is thus maltreated. He is unconscious of error, and can only attribute the pain which he suffers to the arbitrary and inscrutable will of the being whom he hates, but is mysteriously forced to obey. He knows not that he has broken any law, or disobeyed any command, and, in consequence, will probably repeat the offence, and so earn the character of an obstinate and disobedient beast. If he be a horse of some spirit (and such an animal is always the best servant when properly treated), he will resent the injustice of which he is the victim, and bring hoofs or teeth to bear upon his tormentor, thus earning the character of a savage as well as an obstinate brute. So it happens, that a fine animal, which would have cheerfully laboured with all his vast strength in the service of man, is prevented from assuming his rightful place as a humble friend and servant, and is converted into a trembling slave or a dangerous foe. Those who trained him have not identified themselves with him.

So it is with the study of zoology; and the only method of gaining a true insight into the motives of animals, and of identi fying ourselves with them, is to put ourselves, as far as possible in their condition, and to think how we should act under the circumstances, were our instincts as powerful and our reason as weak as theirs.

North America is peculiarly rich in burrowing animals belonging to this order—so rich, indeed, that many curious species must be omitted for lack of space.

BURROW AND STOREHOUSES OF CHIPPING SQUIRREL

Among these burrowers, the Chipping Squirrel, or Hackee, or Chipmuck (*Tamias Lysteri*), is peculiarly conspicuous. It is a very pretty little creature, brownish grey in colour, with five stripes of black and two of pale yellow drawn along the back; so that it cannot be mistaken for any other animal. Below, and on the throat, it is a pure snowy white. These are the normal hues of the fur; but it is somewhat variable in point of colour, the grey and yellow being sometimes quite superseded by the black.

The burrow of the Chipping Squirrel is rather complicated in structure, and is always made under the shelter of a wall, an old tree, or a bank. The hole descends almost perpendicularly for nearly a yard, and then makes several devious windings in a slightly ascending direction. Two or three supplementary galleries are driven from the principal burrow, and by means of them the animal is able to escape almost any foe. The stoat, however, cannot be deceived by this complicated arrangement of tunnels, but winds its lithe body through all the deviating passages, and kills every Chipping Squirrel which it finds. One of these bloodthirsty weasels has been known to enter the burrow of a Chipping Squirrel, and in a short time to leave it, having in the space of a very few minutes killed six victims, a mother and five young, whose lifeless bodies were found in the nursery when the burrow was opened.

The nest is made of dried leaves of various kinds, and in it the mother and her offspring can rest in security from all ordinary foes. Owing to the complexity of the burrow, no little skill is required to trace its various windings, and much exertion is needed before they can all be laid bare.

Within this burrow, the Chipping Squirrel lays up a wonderfully large store of food. It is, in fact, a perfect miser in respect of provisions, gathering and secreting much more than it can possibly eat, and never seeming to be satisfied so long as another acorn or nut can be obtained. The common beaked nut (*Corylus rostrata*) is a favourite article of food, and lest the sharp beak should hurt its mouth, the Chipping Squirrel takes care to bite off the beak before putting the nut into the ample cheek-pouches with which it is furnished. It is able to carry four nuts at a time, three being held in the mouth and the fourth between the teeth. When thus laden, it presents a most absurd aspect, its cheeks being so puffed out with its treasure, that it reminds the classical observer of Alcmæon in the treasury of Crœsus. Unclassical observers may be reminded of a man suffering from a severe attack of mumps.

The provisions are stored in the lateral galleries, and are of a very mixed nature, so as to afford variety of diet to the animal. In one burrow were discovered two quarts of buckwheat, some grass seeds, nearly a peck of acorns, some Indian corn, and a quart of the beaked nuts already mentioned.

The popular names of Chipping Squirrel and Chipmuck are given to this little animal on account of its cry, which sounds like the chirping of young chickens. It is a quick, lively little creature, ever on the move, and delighting to dart in and out of the dense underwood.

THE WOODCHUCK (*Arctomys monax*) is another well-known burrower of America. On account of its whistling cry, it is called SIFFLEUR by the French; and its peculiar form has earned for it the somewhat inappropriate title of GROUND HOG.

The burrow of the Woodchuck is rather long, running to a distance of twenty or thirty feet from the entrance, which is almost invariably placed under the shelter of a projecting rock, or on the side of a hill. It descends obliquely for a few feet, and then rises gradually towards the surface of the ground. At the extremity of the burrow is placed the nest, which is a rounded chamber of considerable dimensions. Here the young Woodchucks are born, and here they remain until they are about five months old, when they separate from each other, and begin an independent existence, each digging a small hole about a yard in depth. Many of these little burrows are made and left unoccupied. Digging out Woodchucks is a favourite amusement with boys.

OUR next example is the CANADA POUCHED RAT (*Pseudostoma bursarius*), sometimes called the GOPHER, or MULO.

This remarkable animal drives burrows of very great extent, and whenever it gains admission into a garden, it works much damage to the roots of the plants. Every root that crosses the tunnel the Pouched Rat will eat; and not only herbs and flowers, but even fruit trees of many years' growth have been killed by this destructive animal. In such cases, the extremity of the burrow is always to be found among the roots of some tree, which act at the same time as a defence and a larder; for the Rat hides itself under their protection, and eats away their tender shoots.

Like the mole, the Gopher throws up little hillocks at irregular intervals, sometimes twenty or thirty feet apart, and sometimes crowded closely together. The nest of the Gopher is made in a burrow constructed expressly for the purpose, and is placed in a small globular chamber about eight inches in diameter. The bed on which the mother and her young repose

CANADA POUCHED RAT.
(Plan of Burrow.)

is made of dried herbage and fur plucked from the body. This chamber is the point from which a great number of passages radiate, and from these other tunnels are driven. These radiating burrows evidently serve two purposes, enabling the animal to escape in any direction when alarmed, and serving to conduct it to its feeding grounds.

In two instances where the Gophers had entered a garden, their tunnels were traced throughout the greater part of their extent, and were found to be driven at an average depth of a foot or eighteen inches below the surface, except when they crossed a path, in which case they dipped to a much lower level. Two of these burrows were opened, but in neither case was the animal captured, as the lateral galleries ran on every side into the fields and woods, so as to afford an easy mode of escape.

The Canada Pouched Rat is nearly a foot in length, and is notable for the great development of its incisor teeth, which project beyond the lip; and for the dimensions of its cheek-pouches, which measure about three inches in length, and extend as far as the shoulders. It was formerly thought that the animal employed these pouches for the conveyance of earth out of its burrow, but it is now known that it does not make any such use of those natural pockets.

Another species belonging to the same genus, the Camas Rat (*Pseudostoma borealis*), is an indefatigable burrower, making tunnels at a very little distance below the surface of the ground, and throwing up hillocks as it passes along. Like the rabbit, it is a partially social animal, living in small communities, and probably drawn together by no personal affection, but by the convenience of the locality. The name of Camas Rat is derived from its food, which consists chiefly of the quamash root (*Scilla esculenta*).

One or two other burrowing rodents deserve a passing notice. There is the Slepez, or Mole Rat of Asia (*Spalax typhlus*), an animal which looks very like a mole, though it belongs to the rodents. It is even less capable of sight than the mole, for its eyes are not only as small as those of that animal, but are actually covered by the thickly-furred skin. The burrow which it makes is very like that of the Gopher, and consists of a series of tunnels driven through the earth at no great depth from the surface, and from which radiate a vast number of lateral passages.

The Sand Mole of Southern Africa (*Bathyergus maritimus*) is another of these shallow burrowers, and by the abundance of its tunnels, and the very trifling depth at which they are driven, renders the ground absolutely dangerous to horsemen. It is rather a large animal, nearly equalling our wild rabbit in dimensions. The fur of the Sand Mole is remarkable for its woolly texture.

Owing to the peculiar nature of the substance in which the White Bear (*Thalarctos maritimus*) makes its curious burrows, I have placed it after, instead of before, the earth-burrowing rodents.

POLAR BEAR

We are told by experienced travellers in northern climes that nobody need be frozen to death in the snow. They look upon such a misfortune with a species of contemptuous pity, compassionating the victim of cold for his sufferings, but despising him for his ignorance. The aboriginal Australian cannot comprehend how a white man can be so foolish as to die of thirst while there are so many water-bearing vegetables around him; the aboriginal American is at a loss to understand how a European can perish of hunger while in the midst of plenty; and those who have passed much of their lives amid the snow can hardly conceive an act of such supreme folly as to be frozen to death while the means of warmth are at hand.

There is no need of a constitution especially organized or sedulously acclimatized to the snow; the benighted traveller who loses himself in the white expanse, with the heavy flakes falling thickly round him, need not possess the hardihood of the highland "reiver," who cares for no covering save his plaid, and looks upon a snow-pillow as an effeminate luxury. He who finds himself in such a position, and knows how to avail himself of the means around him, will welcome every flake that falls, and instead of looking upon the snow as an enemy whose white arms are ready to enclose him in a fatal embrace, he hails the soft masses as a means of affording him warmth and safety.

Choosing some spot where the snow lies deepest, such as the side of a bank or a tree, or a large stone, he scoops out with his hands a hollow in which he can lie, and wherein he is sheltered from the freezing blasts that scud over the land. Wrapping himself in his garments, he burrows his way as deeply as he can, and then lies quietly, allowing the snow to fall upon him unheeded. The extemporised cell in which he reclines soon begins to show its virtues. The substance in which it is hollowed is a very imperfect conductor of heat, so that the traveller finds that the caloric exhaled from his body is no longer swept off by the wind, but is conserved around him, and restores warmth and sensation to his limbs. The hollow enlarges slightly as the body becomes warm, and allows its temporary inhabitant to sink deeper into the snow, while the fast falling flakes rapidly cover him, and obliterate the traces of his presence.

There is no fear that he should be stifled for want of air, for the warmth of his breath always keeps a small passage open, and the snow, instead of becoming a thick uniform sheet of white substance, is broken by a little hole round which is collected a mass of glittering hoarfrost, caused by the congelation of the breath. There is no fear now of perishing by frost, for the snow-cell is rather too hot than too cold, and the traveller can sleep as warmly, if not as composedly, as in his bed at home.

The reader may possibly remember that, even in the British Islands, the snow-bed is almost annually brought into requisition.

The use of snow as a warm mantle to protect the young crops from the frost is familiar to all. Some of us have seen, and we have nearly all read of, the wonderful scenes that take place among the Scottish mountains, where the snow-drifts are heaped like white hills by the wayward tempest, taking all kinds of fantastic forms, and scooped into bays, and precipices, and craggy mountains, with outlines as bold and sharp as if cut in unyielding granite. After such storms as raise these strange mockeries of rugged landscape, whole flocks of sheep are missing, and must be sought by the shepherd and his faithful dog.

As the two allies press onwards in their quest, they walk at random, for the snow masses have swept over hill and dale, have obliterated all the well-known landmarks, raised hills where hollows had been, and have changed the face of nature. Left to himself the shepherd would scarcely discover a single sheep, and in all probability would find himself in the very predicament from which he seeks to rescue his woolly charge. Were it not for the fine instincts of the quadruped many a flock would be lost, for the dog sniffs and runs about, and raises his nose in the air as the well-known odour salutes his nostrils, and finally dashes forward and comes to a stand-still over a little hole in the snow, around which is gathered a slight incrustation of hoarfrost. This is a sure indication that the sheep are below and still living, and then the shepherd breaks through the roof of the snowy cell with his pole, and rescues the starving animals from their perilous position.

The sheep which are thus preserved from the effects of the cold do not voluntarily burrow into the snow. They are not intended to pass a large portion of their lives in a subnivean abode, and their presence under the snow is quite accidental. Striving to avoid the chilling blasts of the wind, they crowd towards any object that may shelter them from the cruel tempest, and while huddled together, the snow-drifts are heaped around them, and cover them effectually. Under such circumstances they often die of starvation if they are left undiscovered for too long a period, after having nibbled all the wool from each others' backs.

But the White Bear intentionally places herself in such a position, and towards the month of December retreats to the side of a rock, where, by dint of scraping, and allowing the snow to fall upon her, she forms a cell in which to reside during the period of her *accouchement.* Within this strange nursery she produces her young, and remains with them beneath the snow until the month of March, when she emerges into the outer air, bringing with her the baby bears, who are then about as large as ordinary rabbits. As the time passes on, the breath of the family, together with the warmth exhaled from their bodies, serves to enlarge the cell, so that in proportion with their increasing dimensions, the accommodation is increased to suit them. As is the case with the snow-covered sheep, the hidden Bear may be discovered by means of the little hole which is made by the warm breath, and is rendered more distinguishable by the hoarfrost which collects around it.

This curious abode is not sought by every Polar Bear. None of the males trouble themselves to spend so much time in a state of seclusion; and as the only use of the retreat is to shelter the young, the unmarried females roam freely about during the winter months. The habit of partial hibernation is common to most, if not to all true Bears, and we find that the White Bear of the Polar regions, the Brown Bear of Europe, and the Black Bear of Northern America, agree in this curious habit. Before retiring into winter quarters, the Bear eats enormously, and, driven by an unfailing instinct, resorts to the most nutritious diet, so that it becomes prodigiously fat. In this condition it is in the best state for killing, as the fur partakes of the general fulness of the body, and becomes thick and sleek, as is needful when we consider the task which it has to perform.

During the three months of her seclusion, the Polar Bear takes no food, but exists upon the store of fat which has been accumulated before retiring to her winter home. A similar phenomenon may be observed in many of the hibernating animals, but in the Bear it is more remarkable from the fact that she has not only to support her own existence, but to impart nourishment to her offspring. It is true, that in order to enable them to find sufficient food, they are of wonderfully small dimensions when compared with the parent; but the fact remains, that the animal is able to lay up within itself so large a store of nutriment that it can maintain its own life and suckle its young for a space of three months without taking a morsel of food.

It is worthy of notice, also, that in the Bears of the Old as well as the New World, is found the curious phenomenon of the "tappen," a hard concreted substance, which plugs up the intestine, and seems to be of service in retaining the animal in condition. In Scandinavia, where the Bears of both sexes retire to winter quarters, and remain in their hidden recesses for five full months, the tappen is very seldom cast until the animal leaves its den. In the rare instances where such an event has happened, the Bear is said to have become miserably thin and weak. Full particulars of the tappen and the hibernating habits of the Brown Bear may be found in Lloyd's "Field Sports of Northern Europe."

THERE are other animals which burrow under the snow, though they do so for the sake of finding food, and not of forming a habitation. Several of the Arvicolæ, or field-mice of North America, are in the habit of driving long tunnels under the snow in search of food, and are so successful in this curious mode of foraging that they in general become quite fat during the winter months, when every green leaf has fallen, and every herb is covered with a thick mantle of snow.

FROM a work of this character, so remarkable an animal as the PICHICIAGO ought not to be omitted. Its scientific name is *Chlamyphorus truncatus,* and is very happily chosen, as will presently be seen.

The Pichiciago is not larger than an ordinary mole, and in its general habits somewhat resembles that animal. The shape of its body sufficiently indicates its burrowing propensities, and the view of the skeleton confirms the aspect of its outward form. The bones of the fore legs are short, thick, and arched in that manner which is so indicative of great muscular power, and even those of the hind legs are remarkably strong in proportion to the size of the animal. The fore paws are enormously large, palm-shaped, and furnished with five strong, curved, and compressed claws, so as to form admirable digging instruments. The snout is rather long and pointed, and, as in the mole, the eyes are very small, and hidden under the soft dense fur.

It is a native of Chili, and seems to be of rare occurrence, though it may probably be more plentiful than is generally imagined, its subterranean habits and timid nature seldom permitting it to be seen. Like the mole, it lives beneath the earth, scooping out long galleries in the soil, and probably feeding upon insects like the rest of the edentate animals.

PICHICIAGO

The chief point of interest which strikes an observer when looking at a Pichiciago, is the cuirass with which its body is defended. It is made and arranged in a very peculiar manner. The cuirass looks as if a number of squared plates of horn had been sewn upon short lengths of tape, and then the tape bands laid side by side and fastened to each other. It is not fixed to the animal throughout its whole extent, as might be supposed, but is only attached along the spine, and on the top of the head. It does not merely protect the back, but when it reaches the insertion of the tail, turns suddenly downwards as if on hinges, and forms a kind of flap over the hind-quarters, which are short and square, as if abruptly cut off by a perpendicular blow with a sharp instrument. This arrangement affords a perfect protection to the hind-quarters while the animal is burrowing, and effectually repels any attack that might be made from the rear, reminding the observer of the shell with which the testacella is furnished.

This coat of mail is as flexible as the chain or scale armour of the olden times, and accommodates itself to every movement of the animal. The rest of the body is covered with a coat of soft, yellowish fur, nearly as fine as that of the mole, and much longer, but not so dense. The scientific name of the Pichiciago relates to the mail-clad body and the peculiar form of the hind-quarters, the generic title signifying "mantle-bearer," and the specific name, "abruptly shortened."

THE ARMADILLO can run with considerable speed, and some species are said to be able to outstrip a man. This may possibly be the case in the native country of the animal, but those specimens which have been brought to England in a living state would certainly be overtaken by a man of ordinary powers. They get over the ground at a sharp pace, using a queer little jog-trot kind of movement, and display an easy flexibility of body and agility of limbs, which never fail to astonish those who have only seen the stiffened specimens in a museum.

They are mostly nocturnal animals, concealing themselves in their burrows by day, and coming out at night to search for food. The burrows in which they live are generally about thirteen or fourteen feet in length, descending in an abruptly sloping direction for some three or four feet, and then taking a sudden bend, and inclining slightly upwards. In these subterranean homes the mother Armadillo produces and nurtures her young, which are on an average about four or five in number.

All the Armadillos are natives of the tropical and temperate regions of Southern America.

The various species of Armadillo, all mailed animals, are mighty burrowers, residing in holes which they have dug with their powerful fore limbs, and obtaining much of their food below the surface of the earth. They are carnivorous beings, and feed upon insects and all kinds of animal substances. One species, the GIANT ARMADILLO (*Priodonta gigas*), is so determined a burrower, that it has often been known to dig up dead bodies for the purpose of feeding on them. All these creatures, however, are fond of animal substances, and many of them may

THE GIANT ARMADILLO.

be found upon the savannahs of South America, feasting greedily upon the bodies of the cattle which are slaughtered so recklessly for the sake of their hides. In all these animals the coat of mail is exceedingly hard, so hard indeed, that it is used for sharpening the long Spanish knife, which is universally carried by the Gauchos.

Digging these animals out of their retreat is no easy business. According to Mr. Waterton, the method adopted is simple, though laborious. As the Armadillos burrow like rabbits in a warren, the first point is to ascertain whether the inhabitant is at home. This is done by pushing a stick into each hole, and watching for mosquitos. If any of these troublesome flies emerge, the inhabitant is at home ; if not, there is no use in searching further. When the presence of an Armadillo is satisfactorily ascertained, a long rod is thrust into the burrow in order to learn its direction, and a hole is dug in the ground so as to meet the end of the stick. A fresh departure is taken from that point, the rod is again introduced, and by dint of laborious digging the animal is at last captured.

Meanwhile, the Armadillo is not idle, but continues to burrow in the sand, in hopes of escaping its persecutors. It cannot, however, dig so fast as they can, and is at last obliged to yield. Mr. Waterton mentions that he has been obliged to work for three-quarters of a day, and to sink half a dozen pits before a single specimen could be secured.

If an Armadillo should be surprised, and its retreat to the burrow intercepted, it at once sets to work at sinking a fresh tunnel. So fast, indeed, does it excavate, that if a horseman sees one of these animals, he must almost tumble from his steed if he wishes to capture the active creature. And, when he has grasped it, he must be careful about his hands, or he will suffer severe wounds from the powerful claws of the Armadillo. As with the pichiciago, the coat of mail, which appears so hard and stiff in the stuffed specimen, is perfectly flexible during life, enabling the limbs of the animal to enjoy their full play, and even permitting the owner to roll itself into a ball when it is threatened with danger.

THE different species of Manis deserve a passing notice. They are all burrowers, and are furnished with armour even better calculated for defence than that of the armadillo, inasmuch as it assumes somewhat of an offensive as well as a defensive character. All these animals are covered with large, sharp-edged scales, of a stout horny consistence, which overlap each other like the tiles of a house. They are of wonderful hardness, and form a buckler which is impenetrable to any weapons possessed by the carnivorous animals of the regions wherein it resides. A specimen of the BAJJERKEIT, or SHORT-TAILED MANIS of India (*Manis pentadactyla*), now before me, affords a good example of this weapon-resisting power. Edwin Arnold, Esq. to whom I am indebted for this specimen, possessed it in a living state for a considerable time, and, when he was about to leave India, determined to kill the animal and take the skin with him. Accordingly, he fired three barrels of a Colt's revolver pistol at the Manis, but without the slightest effect, and was at last obliged to introduce the point of a dagger under the scales, and drive the weapon into the heart. On examining the interior of the skin, the wound caused by the double-edged dagger is plainly perceptible, but I cannot find the slightest trace of the bullets. One of the balls, indeed, recoiled upon the intending destroyer.

When the Manis is alarmed, it rolls itself up, wraps its tail over the body, and lies in conscious security, the horny scales acting as a buckler, and their sharp edges deterring enemies from the attack as much as the quills of the porcupine or the spines of the hedgehog.

THE curious AARD VARK of Southern Africa (*Orycteropus Capensis*) is another of the earth-burrowers, residing, for the most part, in great holes which it scoops in the ground.

The name Aard Vark is Dutch, signifying Earth-hog, and is given to the animal on account of its extraordinary powers of excavation and the swine-like contour of its head. The claws with which this animal works are enormous, as, indeed, is needful for the task which they are intended to perform. They are by no means intended merely to excavate burrows in soft or sandy soil, though they are frequently employed for that purpose; but they are designed for labours far more arduous. By means of these implements, the Aard Vark tears to pieces the enormous ant-hills which stud the plains of Southern Africa—edifices so strongly made as to resemble stone rather than mud, and capable of bearing the weight of many men on their summits. These marvellous dwellings (of which we shall see something in a future page) are absolutely swarming with inmates; and it is for the purpose of feeding upon the tiny builders that the Aard Vark plies its destructive labours.

AARD VARK.

Towards evening the Aard Vark issues from the burrow wherein it has lain asleep during the day, proceeds to the plains, and searches for an ant-hill in full operation. With its powerful claws it tears a hole in the side of the hill, breaking up the stony walls with perfect ease, and scattering dismay among the inmates. As the ants run hither and thither, in consternation, their dwelling falling like a city shaken by an earthquake, the author of all this misery flings its slimy tongue among them, and sweeps them into its mouth by hundreds. Perhaps the ants have no conception of their great enemy as a fellow-creature, but look upon the Aard Vark as we look upon the earthquake, the plague, or any other disturbance of the usual routine of nature. Be this as it may, the Aard Vark tears to pieces many a goodly edifice, and depopulates many a swarming colony, leaving a mere shell of irregular stony wall in the place of the complicated and marvellous structure which had sheltered so vast a population.

The ant-hills thus destroyed are metamorphosed into caverns, which form hiding-places for the jackals and other predaceous beasts, and are resorted to by various serpents. The Kaffir tribes often use them as extemporised burial vaults, and thrust into them the dead bodies of their comrades. Owing to the great burrowing powers of the Aard Vark, the capture of a living specimen is a task of enormous difficulty, the claws being instruments of excavation that conquer the spade of civilized man. Unless disturbed, however, and forced to dig deeper through fear of capture, the Aard Vark makes but a shallow burrow, and lies at a short distance from the surface of the earth. The excavations are, however, deep enough and plentiful enough to be dangerous to the traveller, causing the wheels of wagons to sink into them, so that the machines capsize. Horses, too, frequently fall into these treacherous pitfalls while the hunter is in full chase; and severe injuries are sometimes the consequence of such a mishap.

THERE are two large islands, one large enough to take rank as a continent, which are pre-eminent for the strange character of the creatures which inhabit them. Whenever an animal of more than usual oddity is brought to England, we may safely conjecture that it was taken either in Madagascar or Australia. The creatures which we are now about to examine are natives of the latter country.

Perhaps there never was a more extraordinary and unique being than the well-known animal which is so familiar to us under many titles. Some call it the DUCKBILL, on account of its mandibles, which are ludicrously like those of the bird from which it derives its name. Others call it the WATER MOLE, on account of its aquatic habits and mole-like fur.

Some scientific naturalists have called it the *Ornithorhynchus paradoxus*; others have given it the name of *Platypus anatinus* —the former title being to my mind by far the more appropriate and expressive of the two. The natives of Australia have several names for this remarkable animal; some calling it Mallangong, others Tambreet, and others Tohunbuck—the second of these titles being most generally in use.

MALLANGONG OR DUCKBILL.

Until Dr. Bennett prosecuted his well-known researches in Australia, no European knew precisely whether the Duckbill was a burrower, or, indeed, whether it had a home of any kind. The natives were well aware of the fact that the animal dug tunnels into the ground, and showed great address in discovering the burrows and unearthing the inmates. There, however, their knowledge seemed to end. The only value of an animal to a native Australian lies in its capability of being eaten, and the only lore which an Australian troubles himself to acquire is the knowledge of the habits of the animal with reference to catching it. When he knows where to look for an animal, and how to kill it, he has reached the limit of his education, and never troubles his brain about any branch of learning which does not assist him in procuring something to eat.

In accordance with this principle, Dr. Bennett found that the native Australians were admirable assistants and safe guides up to a certain extent. They could discover the hidden burrows of the Duckbill with instinctive certainty, and could then dig out the animal with their pointed sticks faster than the Europeans with their spades. They knew, moreover, that the burrow had a very evil savour, as is the case with many burrows, and cautioned Dr. Bennett not to thrust his hand into the tunnel, because "he make smell hand." But in all points of abstract natural history, they were totally at fault.

They did not agree as to the domestic economy of the Duckbill, and were not at all sure whether the young were born alive or hatched from eggs. Some advocated the latter opinion, and said that "old woman have eggs, there in so many days;" but their ideas of the eggs in question were exceedingly vague, oval and spherical eggs being equally declared to belong to the Duckbill. Others rejected the egg theory. "Bel cambango (no egg)," said they, "tumble down; pickaninny tumble down." And, as if to show the value of inductive reasoning, it more than once happened that when the natives positively averred that the Duckbill could not possibly be hidden in certain localities, Dr. Bennett conjectured that she was very likely to be there, and succeeded in finding her in the very spot which he had pointed out.

On looking at a living Duckbill, few would set it down as an excavator of the soil; yet it is a burrower, and makes tunnels of great length and some complexity. The soft broad membrane that extends beyond the claws while the animal is walking or swimming, and in the latter case forms a paddle by which the creature can propel itself swiftly through the water, falls back when the foot is employed for digging, and aids the animal in flinging back the soil which its claws have scraped away. The rotund body is admirably adapted for traversing the burrows, though the stuffed specimens which generally are seen in museums give but little idea of such capability. As a general rule, these stuffed specimens are much too long, too stiff, too straight, too flat, and too shrivelled. During life, the body is round, and the skin hangs in loose folds around it, having a very curious aspect when the creature is walking upon the land. The Duckbill is, in fact, so very odd a being, that dogs who see it for the first time, as it scrambles along with its peculiar waddling gait, will sit and prick up their ears, and bark at the strange animal, but will not dare to meddle with it; while cats fairly turn tail, and scamper away from so uncanny a beast. The hair with which the body is so densely covered is admirably suited to an animal which passes its time in the water or underground. Next the skin there is a thick, close coating of woolly fur, through which penetrates a second coat of long hairs, which are very slender at their bases, and can therefore turn in any direction, like those of the mole. The eyes are fuller and rounder than might be expected in an animal that passes so much of its time underground; but they are defended from the earth by a remarkable leathery flap, which surrounds the base of the mandibles, and looks very like the leathern guard of a foil. This curious appendage has probably another use, and is intended to prevent the bill from being thrust too deeply into the mud when the animal is engaged in searching for food.

The wonderful duck-like mandibles into which the head is prolonged are sadly misrepresented in the stuffed specimens which we generally see, and are black, flat, stiff, and shrivelled, as if cut from shoeleather. The dark colour is unavoidable, at all events in the present state of taxidermy. Bare skin invariably becomes blackish brown by lapse of time, no matter what the previous colour may have been, so that the delicate tints of an English maiden's cheek and the sable hue of the blackest negro would, in a few years, assume the same dingy colour and become quite undistinguishable from each other. But, there is no excuse now-a-days for allowing the bare skin to become shrivelled. The colours we cannot preserve, the form we can and ought to reproduce. No one would conceive, after inspecting a dried specimen, how round, full and pouting were once those black and wrinkled mandibles, and how delicately they had been coloured while the animal retained life. Their natural hue is rather curious, the outer surface of the upper mandible being very dark grey, spotted profusely with black, and its lower surface pale flesh-colour. In the lower mandible the inner surface is flesh-coloured, and the outer surface pinky white, sometimes nearly pure white.

Having now glanced at the general form of the Duckbill as it is in life, and not as it is in museums, we will pass to the habitation which it constructs.

Being a peculiarly aquatic animal, the Duckbill always makes its home in the bank of some stream, almost invariably at those wider and stiller parts of the river, which are popularly called ponds. There are always two entrances to the burrow, one below the surface of the water and the other above, so that the animal may be able to regain its home either by diving, or by slipping into the entrance which is above the surface. This latter entrance is always hidden most carefully under overshadowing weeds and drooping plants, and is so carefully concealed that the unaccustomed eyes of an European can very seldom find it.

When the grasses, &c. are put aside, there is seen a hole of moderate size, on the sides of which are imprinted the footmarks of the animal. By the dampness and sharpness of these impressions, the natives can form a tolerably accurate opinion whether the creature is likely to be at home or not, as in the former case, the footmarks which point upwards are fresher and wetter than those which point downwards. While digging out the Duckbill, they occasionally pull out a handful of the clay, inspect the marks, and then fall to work afresh. From this hole the burrow passes upwards, winding a sinuous course, and often running to a considerable length. From twenty to thirty feet is the usual average, but burrows have been opened where the length was full fifty feet, and where the course was most annoyingly variable, bending and twisting about so as to tire the excavators, and make them quite disgusted with their work. The natives never dig out the entire burrow, but push sticks along it, and sink shafts upon the sticks; just, in fact, as a boy digs out a humble bee's nest, by inserting twigs into the hole, and digging down upon them.

This serpentine form of burrow is in all probability attributable in a great degree to the peculiar instincts of the animal. As, however, the course of the tunnel is extremely variable, and no two burrows have precisely the same curves and windings, it is likely that various obstacles, such as roots and stones, may turn the animal out of its course while engaged in digging its subterranean home, and therefore that the shape of the burrow may in some measure depend upon the character of the ground.

At the upper extremity of the burrow is placed the nest, an excavation of a somewhat oval form, much broader than the width of the burrow, and well supplied with dry weeds and grasses, upon which the young may rest. They appear to remain in these burrows until they have attained half their full growth, for Dr. Bennett captured a pair of young Duckbills, ten inches in length, which seemed not to have left the burrow. Sometimes there are four young in one nest, and sometimes there is only one, but the usual number is two.

The Duckbill is a far more active animal than could be conceived merely by looking at its form. It is very powerful in proportion to its size; so strong, indeed, that it cannot be held in the hands without great difficulty, slipping through the grasp almost as if it were oiled. The loosely-hanging integuments aid it in this method of escape, and under them may be felt the powerful subcutaneous muscles working with vast energy. It is an admirable climber, not only in its wild state, but among civilized objects. Dr. Bennett found that a pair of tame Duckbills, which he kept for some time, were in the habit of clambering to the tops of bookshelves and other articles of furniture, achieving this feat by a process similar to that which is employed by chimney-sweepers, and those whose business calls upon them to ascend narrow perpendicular passages, namely, by placing their backs against the wall and their feet against the bookcase, and so working their way upwards, in a strictly vermicular fashion.

I remember that during my childhood I was frequently found upon the roof of a stable by my parents, who could not conceive the method by which so small a boy could have reached so great an altitude. The fact was, that a summer-house had been built within a foot or two of the stable-wall, so that by means of placing my feet against one wall and my back against the other, I was soon deposited on the roof of the summer-house. A jump and a clutch then transferred me to the stable-roof, *where the grapes grew;* whereby was manifested the practical advantage of watching the climbing boys ascend the chimney.

This mode of climbing may probably be called into operation while the animal is engaged in ascending the almost perpendicular part of the burrow, just above the water's edge.

The whole life of this animal is very similar to that of the musk-rat already mentioned, and is an alternation between the water and the burrow. While swimming, the animal hardly looks like a living and breathing creature, but bears a great resemblance to the loose bundles of weeds that float vaguely in the water. Though decidedly aquatic in its habits, the Duckbill cannot withstand a very long immersion in the water; and Dr. Bennett found that few of them could endure an immersion of fifteen, or at the most twenty minutes, without being much fatigued by the exertions which they made in order to keep themselves afloat. Several persons who have procured living specimens have drowned them by placing them in water from which there was no mode of escape.

THERE is another strange Australian animal, also remarkable for its power of burrowing. This is the creature which is known as the PORCUPINE ANT-EATER (*Echidna hystrix*), and is called by the very erroneous names of Porcupine, or Hedgehog. The natives have several names for it, some calling it Nicobejan, others Jannocumbine, and others Cojera.

Here I may mention the curious circumstance that in three orders of animals, widely separated from each other in structure, habits, and locality, the hairs of the back are greatly enlarged, and developed into stout and sharp spines. There are several species of hedgehog, natives of Europe, and the Tanrecs of Madagascar, all insectivorous animals; there are the various porcupines which inhabit Africa, Asia, Southern Europe, and America, and belong to the rodent group; and there is the Echidna, one of the monotrematous animals, which is found only in Australia.

The Echidna is a wonderful burrower, and, in spite of its small size, can make its way through very hard ground. It can pull up stones of great size if it can only contrive to insert its paws and find a convenient crevice for them, and is so quick at this task that to confine the animal is by no means an easy matter, even a paved yard affording but a poor safeguard against its escape. In the open country it digs with such extreme rapidity that it can hardly be captured, gathering its back into an arched form, collecting the legs under the body, scratching away with the feet, and sinking like a stone in a cup of treacle.

These paws are not only potent in digging, but in clinging to any object, and their hold is so wonderfully firm that they cannot be disengaged even from smooth boards without very great trouble. To grasp the creature is impossible, because the sharp points of the projecting spines are capable of inflicting painful wounds, and its feet are so completely hidden under the body that they cannot be separately detached. Dr. Bennett gives a very graphic account of its clinging powers:—"When one of these animals was given to me, and placed in the box of the gig to bring home, on arriving there I could not by any effort remove it, from its adhering to the boards like a limpet to the rocks (the head and snout being drawn in). Only a formidable array of prickles was visible, so sharp that on the least touch they left a very painful feeling on the hands. So firmly was the animal fixed, that it was impossible to stir it from that position. At last, the method of removing limpets and chitons from the rocks was resorted to, and a spade being inserted gradually at one extremity of the animal, it was scraped from its position with some difficulty, and even then it was some length of time before we succeeded in grasping the hind legs, and conveying the troublesome creature to the place of confinement allotted to it."

Grasping it by the hind leg is the only method of conveying this animal with safety, for it kicks so hard with its powerful and armed feet, that the hands and clothes will suffer severely from the strokes; while the violent plunges of the body are sure to bring the pointed prickles into unpleasant contact with the fingers. In spite of the difficulty of procuring living specimens, and the interest which attaches itself to an animal of whose habits so little is known, Dr. Bennett was not very sorry when his specimen—which we cannot call a tame one—was one day found dead; for its burrowing propensities were so destructive, and its prickles so annoying, that it made itself into a positive nuisance.

If attacked when on ground into which it cannot burrow rapidly, the Porcupine Ant-eater immediately curls itself into a ball, hedgehog-wise, and sets its foes at defiance. The large perforated spur with which the hind feet of the male are armed, and through which is poured a liquid secreted by a gland of considerable size, is a very formidable-looking weapon, but to all appearances is really harmless. Dr. Bennett often handled the animal, but never saw it attempt to use the spur, and found that the duckbill, which is armed in a similar manner, was equally innocuous.

At the present date, January, 1864, the living animal may be seen in the collection at the Zoological Gardens.

CHAPTER II.

BURROWING BIRDS.

The SAND MARTIN—Mode of burrowing and shape of the tunnel—Enemies of the Sand Martin—Midges and Martins—The KINGFISHER and its habits—Its burrow and peculiar nest—Number of the eggs—The PUFFIN a feathered usurper—The Feroe Islands and the Puffins—Pro aris et focis—The MUTTON BIRD and its burrows—Snakes and birds—The JACKDAW, STOCKDOVE, and SHELDRAKE—Nest of the Sheldrake—The BEE-EATER and its habits—Its burrow and nest—The STORMY PETREL—Its mode of nesting and shallow tunnels—mode of feeding its young—Evil odour of its burrow—The WOODPECKER—Its uses and misunderstood character—Method of burrowing—The Fungus and the Woodpecker—American Woodpeckers—The WRYNECK—Its popular names and locality of its nest—The STARLING—Its social character—Locality of its habitation—The TREE CREEPER—The NUTHATCH and the HOOPOE—Curious nest of the Hoopoe—The COLE-TIT and its habits—A Cole-tit's nest at Walton Hall—The TOUCAN—The enormous beak and its uses—Nest of the Toucan—The SWIFT—Its nest and eggs—Its curious feet and their structure.

WE now take leave of the furred burrowers, and proceed to those which wear feathers instead of hair.

One of the best examples of Bird Burrowers is the well-known SAND MARTIN (*Cotile riparia*), so plentiful in this country. The powers of this pretty little bird seem to be quite inadequate to the arduous labours which it performs so easily, and few would suppose, after contemplating its tiny bill, that it was capable of boring tunnels into tolerably hard sandstone. Such, however, is the case, for the Sand Martin is familiarly known to drive its tunnels into sandstone that is hard enough to destroy all the edge of a knife.

The bird does not prefer a laborious to an easy task, and if it can find a spot where the soil is quite loose, and yet where the sides of the burrow will not collapse, it will always take advantage of such a locality. I have frequently seen such instances of judgment, where the birds had selected the sandy intervals between strata of stone, and so saved themselves from any trouble except scraping and throwing out the loose sand.

When, however, the Sand Martin is unable to find such a situation, it sets to work in a very systematic fashion, trying several successive spots with its beak, until it discovers a suitable locality. It then works in a circular direction, using its legs as

SAND MARTIN.

a pivot, and by dint of turning round and round, and pecking away as it proceeds, soon chips out a tolerably circular hole. After the bird has lived for some time in the tunnel, the shape of the entrance is much damaged by incessant passing to and fro of the inmates, but while the burrow is still new and untenanted, its form is almost cylindrical. In all cases the tunnel slopes gently upwards, so as to prevent the lodgement of rain, and its depth is exceedingly variable. About two feet and a half is a fair average length. Generally, the direction of the burrow is quite straight, but sometimes it takes a curve, where an obstacle, such as a stone or a root has interrupted the progress of the bird. Should the stone be a large one, the Sand Martin usually abandons the burrow, and resumes its labours elsewhere, and in a piece of hard sandstone rock many of these incomplete excavations may be seen.

At the furthest extremity of the burrow, which is always rather larger than the shaft, is placed the nest—a very simple structure, being little more than a mass of dry herbage and soft feathers, pressed together by the weight of the bird's body. Upon this primitive nest are laid the eggs, which are very small, and of a delicate pinky whiteness.

Few foes can work harm to the Sand Martin during the task of incubation. Rats would find the soft sandy soil crumble away from their grasp; and even the lithe weasel would experience some difficulty in gaining admission to the nest. After the young Sand Martins are hatched, many foes are on the watch for them. The magpie and crow wait about the entrance of the holes, in order to snap up the inexperienced birds while making their first essays at flight; and the kestrel and sparrow-hawk come sweeping suddenly among them, and carry off some helpless victim in their talons.

Man is perhaps the worst foe of the Sand Martin, for there is a mixture of adventure and danger in taking the eggs, which is irresistible to the British schoolboy. To climb up a perpendicular rock, to cling with one hand, while the other is thrust into the burrow, and to know that a chance slip will certainly snap the invading arm like a tobacco-pipe stem, is a combination of joys which no well-conditioned boy can withstand.

Fortunately for the Sand Martins, many of their nests are placed in situations which no boy can reach, and there are happily some instances where the services which they render to mankind are properly appreciated. Mr. C. Simeon, in his "Stray Notes on Fishing and Natural History," gives an interesting account of some Sand Martins which were thus gratefully protected:—

"Whilst waiting for the train one afternoon at Weybridge, I amused myself with watching the Sand Martins, who have there a large establishment on either side of the cutting, and got into conversation with one of the porters about them. On my saying, I supposed that the boys robbed a good many of the nests, he answered, 'Oh, sir, they would if they were allowed, but the birds are such good friends to us, that we won't let anybody meddle with them.' I fancied at first that he spoke of them as friends in the way of company only, but he explained his meaning to be, that the flies about the station would be quite intolerable if they were not cleared off by the martins, which are always hawking up and down in front of it; adding, that even during the few hot days which occurred in the spring before their arrival, the flies were becoming very troublesome. 'Now,' he said, 'we may now and then see one, but that is all.'

"It was a bright sunny day in July, and the scene was a very lively and interesting one. The mouths of the holes on both sides of the cutting were crowded with young martins—as many perhaps as four or five in each—sunning their barred white breasts, and waiting to be fed: the telegraph wires formed perches, of which advantage was taken by scores of others more advanced in growth, and of old ones reposing after their exertions; while the air was filled with others employed in catering for their families. All of a sudden the young ones retreated into their holes; the wires were deserted, and only a few remained, describing distant circles. I thought that a hawk must have made his appearance, but it turned out that the alarm had been caused by two men walking over the heath above, and approaching the holes. The young ones in the holes had, no doubt, felt the jar caused by their tread, and those on the wing, who saw them, had probably given warning, by note, to the others perched on the wires, who could not have seen, nor, I should think, heard their approach."

ALTHOUGH the KINGFISHER (*Alcedo ispida*) does not excavate the whole of the burrow in which it resides, it does, at all events, alter and arrange a ready-made burrow to suit its own necessities.

This lovely bird, which is one of the few indigenous British species that can vie with the bright-feathered denizens of the tropics, is happily very plentiful in England, scarcely any stream or lake being without its Kingfishers.

All who are fond of angling, or of walking by the side of streams, must have noticed the Kingfisher as it sits motionless on a stone or overhanging branch, peering eagerly into the water beneath, and watching the fish as they pass and repass its place of vantage. Brilliant in colour though the bird may be, its azure back and red belly seldom betray it except to a practised

KINGFISHER.

eye, so immovable is its attitude. Suddenly, down it drops into the water, splashes furiously for a few seconds, emerges with a small fish in its mouth, and then returns to land. Sometimes it seeks again the perch from which it descended, and then, throwing in the air the fish, which has all the while been held across the beak, catches it dexterously head downwards, swallows it with a few eager gulps, and then looks out for another victim. Sometimes it darts with lightning speed along the bank, and, with a quick flash of the azure plumage, settles for a moment upon the bank, looks cautiously round, and then pops swiftly into a little hole. Into this hole we will follow the bird. She always chooses her residence by the water side, and selects for nidification the deserted hole of some quadrupedal burrower. I have even seen a Kingfisher's nest made in the side of a tiny rivulet across which a child could step, and which served to conduct the drainings of an upper to a lower field.

Generally, the nest is placed in the deserted burrow of a water-vole, but in this instance it had been made in the empty tunnel of a water-shrew, so that the hole was of comparatively small dimensions, and would not admit my hand and arm without some artificial enlargement. In all cases, the bird takes care to increase the size of the burrow at the spot where the nest is made, and to choose a burrow that slopes upwards, so that however high the water may rise, the nest will be perfectly dry.

That the eggs are laid upon dry fish-bones is a fact that has long been known, but for an accurate account of the nest we are indebted to Mr. Gould, the eminent ornithologist.

Until he succeeded in removing the nest entire, no one had been able to perform such a feat, and so well known to all bird-nesters is the difficulty of the task, that a legend was, and perhaps is still, current in various parts of England, that the authorities of the British Museum had offered a reward of 100*l.* to any one who would deposit in their collection a perfect nest of the Kingfisher. This feat has been admirably accomplished by Mr. Gould.

Having discovered the retreat of a Kingfisher, and ascertained by digging down upon the nest that the bird was laying, he replaced the earth, and waited for three weeks before attempting any further operations. The chief difficulty was, of course, to prevent the earth from falling into the nest, and becoming mixed with the delicate bones of which it was composed. In order to obviate such a mishap, Mr. Gould introduced a quantity of cotton wool into the burrow, pushing it to the extremity with a fishing-rod. He then dug down upon the nest, and captured the female, who was sitting upon eight eggs. With very great care he removed the fragile nest, and transferred it to the British Museum, where it may be seen by any one who will look for it in the room devoted to such objects.

The nest is composed wholly of fish-bones, minnows furnishing the greater portion. These bones are ejected by the bird when the fles his digested, just as an owl ejects the pellets on which her eggs are laid. The walls of the nest are about half an inch in thickness, and its form is very flat. The circular shape and slight hollow show that the bird really forms the mass of bones into a nest, and does not merely lay her eggs at random upon the ejector. The whole of these bones were deposited and arranged in the short space of three weeks.

It may possibly be owing to these bones and the partial decomposition which must take place during the time occupied in drying, that the burrow possesses so exceedingly evil an odour. This unpleasant effluvium, which may indeed be called by the stronger name of stench, is wonderfully enduring, and clings to the bird as well as to its dwelling. The feathers of the Kingfisher are most lovely to the eye, but the proximity of the bird is by no means agreeable to the nostrils, the "ancient and fish-like smell" being extremely penetrating. I have now before me a stuffed and perfectly dry skin of a Kingfisher, which has been washed and soaked in water for many hours, and yet retains the peculiar odour, which is so strong that after I had prepared it, many and copious ablutions were required to divest my hands of the horrible emanation.

To those who collect eggs, and care for numbers, the discovery of a Kingfisher's nest is a singular boon. Not only does the bird lay a great multitude of eggs, the aggregate mass of which exceeds her own dimensions, but she is a fearless and indefatigable layer, and if the eggs are removed with proper care, she will produce an enormous number in the course of a season.

THE comical little PUFFIN (*Fratercula arctica*) may be reckoned among the true burrowers, possessing both the will and the power of excavation, but exercising neither unless pressed by necessity.

As is the custom with most diving birds, the Puffin lays only one egg, and always deposits it in some deep burrow. If possible, the bird takes advantage of a tunnel already excavated, such as that of the rabbit, and "squats" upon another's territory, just as the Coquimbo owl takes possession of the excavations made by the prairie dog. The rabbit does not allow its dominion to be usurped without remonstrance, and accordingly the bird and the beast engage in fierce conflict before the matter is settled. Almost invariably the Puffin wins the day, its powerful beak and determined courage being more than a match for the superior size of its antagonist.

When it is unable to obtain a ready-made habitation, it sets to work on its own account, and excavates tunnels of considerable dimensions.

The Feroe Islands are notable haunts of the Puffin, because the soil, which is in many places soft and easily worked, is

PUFFIN.

favourable for its excavations. The male is the principal excavator, though he is assisted by the female; and so intent is the bird upon its work, that it may be captured by hand by thrusting the arm into the burrow. The average length of the tunnel is about three feet, and it is seldom straight, taking a more or less curved form, and being furnished with a second entrance.

No nest of any kind is used, but the egg is laid on the earth, at the end of the burrow, so that, although it is at first beautifully white, it becomes in a short time stained so deeply that it can seldom be restored to its primitive purity.

So deeply do the burrows run, that when a passenger is walking near the edge of the precipice upon which the Puffins breed, he can hear the old birds grunting below his feet, angry because they are disturbed by the footsteps above them.

The young Puffin has many foes, who endeavour to seize it before its bill has attained its full proportions and its muscles have gained their full powers. The parent birds, however, bravely defend their young, and have been known, as a last resource, to grasp the invader in the beak, and hurl themselves and the foe into the sea. Once among the waves, the Puffin is in its natural element, for it is an admirable swimmer and practised diver, being able to catch the swift-finned fishes and bear them home to its nest. The foe, therefore, must either remain on dry land or lose the victory, if not its life, for there are few enemies for which the Puffin is not more than a match when in the water.

MR. M'GILLIVRAY, in his "Voyage of the Rattlesnake," gives a curious account of the nesting-place of an allied bird, which burrows in Goose Island, off the North Australian coast.

"The rock is a coarse syenite, forming detached bare masses and ridges, but none of considerable height. In the hollows the soil appears rich, dark, and pulverulent, with much admixture of unformed bird guano. The scanty vegetation is apparently limited to a grass growing in tussocks, and a few maritime plants. The ground resembles a rabbit warren, being everywhere undermined by the burrows of the MUTTON BIRD (*Puffinus brevicaudus*) the size of a pigeon. A person in walking across the island can scarcely avoid frequently stumbling among these burrows, from the earth giving way under his feet; and I was told by the residents that snakes are very numerous in these holes, living upon the Mutton Birds. I myself trod upon one which, fortunately, was too sluggish to escape before I had time to shoot it, and ascertain it to be the well-known 'black snake' of the Australian colonists (*Acanthophis tortor*), a very poisonous species.

. . . About dusk, clouds of Mutton Birds came in from the sea, and we amused ourselves with chasing them over the ground among their burrows, and as many specimens as I required were speedily provided by knocking them down with a stick. As usual with the petrel tribe, they bite severely if incautiously handled, and disgorge a quantity of offensive oily matter, the smell of which pervades the whole island, and which the clothes I then wore retained for a long time afterwards." The curious association of the burrowing bird and the venomous snake is very remarkable, and reminds the observer of the burrowing owl and the rattlesnake which inhabit the tunnels of the prairie dog.

THERE are many other birds which pass a semi-burrowing life, making their nests in hollows already excavated, and either using them without adaptation or altering them very slightly for the purpose of depositing their eggs and rearing their young. The JACKDAW, for example (*Corvus monedula*), is frequently one of the semi-burrowers, making its nest within deserted rabbit burrows, when it can find no more congenial locality. The STOCKDOVE (*Columba œnas*) is frequently found in similar situations, placing its rude platform of sticks within the burrow ; and the common SHELDRAKE (*Tadorna Vulpanser*) possesses the same habit.

The nest of the last-mentioned bird is always placed close to the water, so that the young may be fed with marine crustacea. The female is accustomed to cover the eggs with down plucked from her own breast. Rabbit warrens upon sea-edged cliffs, are favourite resorts of the Sheldrake. In default, however, of rabbit burrows, the Sheldrake is well content with any moderately deep holes in the shore, and therein lays her enormous deposit of eggs, which are from ten to fifteen in number, and of a white colour. Burrows thus tenanted may be found in many situations, especially on the banks of estuaries, localities which are always sheltered, and almost always produce an abundant supply of food for the bird and its young brood.

THERE is a bird which sometimes visits our island, and which is more lovely in colour than the kingfisher, and far more elegant in shape. This is the BEE-EATER (*Merops apiaster*), so common in the warmer parts of the Old World. Not only is it rich in colour and delicately shaped, but it is most graceful in its movements, sweeping through the air with the ease and rapidity of the swift, and much resembling that bird in the character of its flight.

It is not a large bird, being about the size of the common thrush, and formed in a more slender manner. Its feathers are coloured in the most exquisite manner, green, azure, yellow, orange, and chocolate-brown, being mixed in a singularly harmonious manner, and relieved by a little white on the forehead, and a narrow band of deep blue-black under the throat. When the sunbeams fall on these gorgeously decorated feathers, the effect is magnificent in the extreme, for there is not only the light azure hue which gives to our kingfisher so brilliant an aspect, as it darts along in its meteor-like flight, but with every movement of the bird the colours change like those of "shot" silk.

The peculiarly graceful flight of these birds is calculated to display their lovely colours to the best advantage, and as they are partially gregarious in their habits, and love to assemble in little flocks, they afford a magnificent spectacle as they sweep through the air in devious flight, crossing and recrossing each other's track, rising and sinking on facile wing, wheeling swiftly, as some more active insect strives to escape from the lovely destroyers, and ever and anon shooting with arrowy speed as their ruby-cinctured eyes catch a glance of some distant prey. Were it not for their harsh, screaming notes, they would seem almost too beautiful to belong to this world. The splendour of their plumage is quite tropical, and though the Bee-eater may not possess the metallic radiance of the humming-bird, the extreme beauty of the silken plumage, which shines in the sunbeams like spun glass, cannot be surpassed, and is far too subtle to be approached by human art. The greatest master of the brush, aided by all the resources of the chemist, can do no more than indicate the wondrous beauty of this bird.

The bird does not restrict itself to insects. My friend, E. Arnold, Esq. tells me that in India, he has seen the Bee-eater catch and devour small fish equal in dimensions to the well-kncwn minnow of England. These fish are eaten by human beings as well as by birds, and seem to be intermediate between the smelt and the whitebait. Among the residents they are known by the quaint name of "Havildar-and-ten," because a dish consists of eleven or twelve fishes. In the Indian army the Havildar is equivalent to our sergeant, and has ten men under him, so that a "Havildar-and-ten," signifies a sergeant and his guard, making altogether eleven men. The external resemblance between the Bee-eater and the kingfisher is patent to all, in spite of the short, clumsy-looking body, and stunted tail of the one, and the slight, elegant form, and lengthened tail of the other; and it is sufficiently curious, that a light and airy bird, like the Bee-eater, which feeds upon the most active insects, and catches them on the wing, should invade the realms of the kingfisher, and procure a meal from the water.

Near the spot where the Bee-eater hovers about in search of its daily food, the nest may be found.

As the bird is generally as gregarious in its nesting as in its flight, there is little difficulty in finding the locality in which it has formed its temporary home. The Bee-eater is one of the true burrowers, excavating a hole in some bank, and depositing its eggs therein. The burrow is not a deep one, seldom exceeding a foot in length, so that the sitting bird is plainly visible from the exterior. The extremity of the hole is floored and partially lined with moss, upon which are placed five or six eggs, of pearly whiteness. Whenever the bank happens to be a convenient one, it is pierced with holes as numerous as those of the sand martin of our own country; and if the observer can manage to conceal himself in close proximity to the nest, and will remain perfectly quiet, he will witness a scene which is unsurpassable for beauty.

Although gifted with the rapid wing of the swift, the Bee-eater does not possess the untiring flight of that bird, and, therefore, is accustomed to repose at short intervals, whenever it has caught a butterfly, or some large and active insect. Numbers of these lovely birds may be seen perched in rows upon the branches of neighbouring trees, exhibiting masses of colour that have a peculiarly magnificent effect to the eye. Here they sit for the purpose of eating the prey which they have captured, and the ground beneath these favoured branches is thickly strewn with wings of butterflies and other insects which they have devoured.

I do not think that the Bee-eater ever makes its nest in England, for it is only an occasional visitor, and is generally shot before it has passed many days, or even hours, in this country. But its nest may be plentifully found in the warmer latitudes of Europe, and many parts of Asia and northern Africa.

WE often find burrowers where we least expect them.

Who would think, on inspecting a specimen of the well-known STORMY PETREL (*Thalassidroma pelagica*), that it was able to

PETREL

dig into the ground, and form the burrow in which it makes its nest? Such, however, is the case, and the pretty little traverser of the ocean shows itself to be as accomplished in excavating the ground as it is in flitting over the waves with its curious mixture of flight and running. If the Stormy Petrel can find a burrow already dug, it will make use of it, and accordingly is fond of haunting rocky coasts, and of depositing its eggs in some suitable cleft. It also will settle in a deserted rabbit-burrow, if it can find one sufficiently near the sea, and is found breeding in many places which would equally suit the puffin.

Failing, however, all natural or ready-made cavities, the Stormy Petrel is obliged to excavate a tunnel for itself, and even on sandy ground is able to make its own domicile. Off Cape Sable, in Nova Scotia, there are many low-lying islands, the upper parts of which are of a sandy nature, and the lower composed chiefly of mud. Not a hope is there in such localities of already existing cavities, and yet to those islands the Petrels resort by thousands, for the purpose of breeding. The birds set resolutely to work, and delve little burrows into the sandy soil, seldom digging deeper than a foot, and, in fact, only making the cavity sufficiently large to conceal themselves and their treasure.

Each bird lays a single egg, which is white, and of small dimensions. The young are funny-looking objects, and resemble puffs of white down rather than nestlings. The parent attends to its young with great assiduity, feeding it with the oleaginous fluid which is secreted in such quantities by the digestive organs of this bird. So large indeed is the amount of oil, that in some parts of the world the natives make the Stormy Petrel into a lamp, by the simple process of drawing a wick through its body. The oil soon rises into the wick, and burns as freely as in any of the really rude and primitive, though ornamental lamps of the ancients.

Many shore-going birds are equally notable for their capacities of producing fat in large quantities when they procure a plentiful supply of food. The common sand-piper, for example, which haunts the muddy shores of tidal rivers, and feeds upon the various living creatures with which the mud is peopled, accumulates fat to such an extent that to skin the bird properly is almost impossible, the thick coating of fat between the skin and the flesh melting by the heat of the fingers, and running like oil over the feathers. I could never bring myself to believe in the petrel-lamp until I had opened the sand-piper when in full condition.

The Petrel only feeds its young by night, remaining on the wing during the day, and flying to vast distances from the land.

Owing to this habit, and its custom of taking to the sea during the fiercest storms, it has long been an object of dread to sailors, whose illogical minds are unable to discriminate between cause and effect, and fancy that the Petrel, or Mother Carey's Chicken, as they call the bird, is the being which, by the exercise of some magic art, calls the storm into existence. They even fancy that the Petrel never goes ashore nor rests; and will tell you that it does not lay its egg in the ground, but holds it under one wing, and hatches it while engaged in flight. To the vulgar mind, everything incomprehensible is fraught with terrors, and so the harmless, and even useful Petrel, is hated with strange virulence.

The bird is essentially a storm-lover, for by the violence of the wind upon the waves the substances on which the bird feeds are thrown to the surface, and can be snapped up before they sink again. The Petrel perceives by some innate faculty the approach of a storm, and its appearance is the signal for the careful mariner to reduce his sails. The ignorant sailors, who know, from long experience, that the Petrel is the forerunner of a storm, salute it by the title of Devil's bird, together with sundry other epithets, all very forcible, but on that very account not to be printed.

Throughout the breeding season, the Petrel is indefatigable in search of food, and will follow ships for considerable distances, in hopes of obtaining some of the offal that is thrown overboard by the cook. Even if a cupful of oil be emptied into the water, the Petrel will scoop it up in its bill, and take it home to its young. During the night it mostly remains with its offspring, feeding it, and making a curious grunting noise, something like the croaking of frogs. This noise is continued throughout the night, and those who have visited the great nesting places of the Petrel, unite in mentioning it as a loud and peculiar sound. The ordinary cry is low and short, something like the quacking of a young duck. By day, however, the birds are silent, and only those who keep nightly watch on the ship's deck, can have an opportunity of hearing their chattering cry.

The burrow in which the young Petrel is hatched is extremely odoriferous, the oily food on which the bird lives having itself a very rancid and unsavoury scent; and in consequence of feeding upon this substance, both the habitation and the inmates are extremely offensive to the nostrils. The young bird is at first very helpless, and remains in its excavated home until it is several weeks of age. One of these birds was seen on the Thames in the month of December, 1823, where it attracted some attention, its peculiar mode of pattering over the water causing it to be taken for a wounded land bird, and inducing many persons to go in vain pursuit of the supposed cripple.

WOODPECKER.

THE birds that have hitherto been mentioned are either burrowers into the earth, or adopters of burrows which have been made and deserted by fossorial mammalia. Those which now come before us are burrowers into wood, and either form their tunnels with their own beaks, or adapt to their purposes the excavations made by other creatures, and the hollows formed by natural decay.

The first in order of these birds are necessarily the WOODPECKERS, examples of which are found in most parts of the world. They are easily distinguished from any other birds by the peculiar construction of the beak, the feet, and the tail; the beak enabling them to chip away the bark and wood, the feet giving them the power of clinging to the tree-trunk, and the tail helping to support them in the attitude which gives to their strokes the greatest force. Their beaks are long, powerful, straight and pointed; their feet are formed for grasping, and are set far back upon the body; and their tails are short and stiff, and act as props when pressed against the rough bark.

From England, the Woodpeckers are fast disappearing, and except in the few forests and woods that still remain, a Woodpecker is now seldom seen in this country. The birds, however, possess that remarkable instinct which tells them where they will be safe; and any one who possesses a sufficiency of trees surrounded by a wall, and who will not permit a gun to be fired within those precincts, will not have to wait very long before his eyes are gladdened by the bright colours of the Woodpecker's plumage as it darts from tree to tree, and his ears gratified by the rapid tattoo of its beak upon the wood.

We should probably have possessed many more specimens of the Woodpecker, had it not been subjected to such persecution. It was universally thought to be very hurtful to trees, and its reiterated blows were considered as so many direct injuries. If the observer could quietly make his way to a tree on which the Woodpecker was at work, he would find great flakes of bark lying on the ground, as marks of the bird's industry, and might be led to suppose that, in separating them from the trunk, the bird was inflicting a positive injury upon it. If, however, he should examine the flakes of bark, he would find that they had already been separated from the tree in the course of nature, and that they were mere useless excrescences upon its surface. Under these bark-flakes whole tribes of insects find a shelter, and it is in order to obtain the insects that the Woodpecker removes the flakes.

As is well known, this bird makes its nest in a tunnel which it hollows in the tree, and to a superficial observer might easily be reckoned among the enemies of the forest. If it were to burrow into sound timber, as is often supposed to be the case, it would certainly rank among the deadliest foes of our trees; for in the spots where it still resides, its burrows may be seen in plenty, perforating the trunks and branches of the finest and most picturesque trees. But, in point of fact, none of the British Woodpeckers are able to cut so deep a tunnel into sound and growing wood, and are perforce obliged to choose timber which is already dead, and which has begun to decay.

Sometimes the bird selects a spot where a branch has been blown down, leaving a hollow in which the rain has lodged and eaten its way deeply into the stem. In such places the wood is so soft that it can be broken away with the fingers, or scraped out with a stick; and in many a noble tree, which seems to the eye to be perfectly sound, the very heart-wood is being slowly dissolved by the action of water, which has gained access through some unsuspected hole. Water, when thus admitted to the interior of a tree, fills its centre with decay; and if a perforation be made through the trunk, so as to let out the contained fluid, gallon after gallon of dark brown water will gush forth, mixed with fragments of decayed wood, and betray, by its volume and consistency, the extent of the damage which it has occasioned.

Oftentimes a large fungus will start from a tree, and in some mysterious manner will sap the life-power of the spot on which it grows. When the fungus falls in the autumn, it leaves scarcely a trace of its presence, the tree being apparently as healthy as before the advent of the parasite. But the whole character of the wood has been changed by the strange power of the fungus, being soft and cork-like to the touch. Although the eye of man cannot readily perceive the injury, the instinct of the Woodpecker soon leads the bird to the spot, and it is in this dead, soft, and spongy wood that the burrow is made. Mr. Waterton, who, I believe, was the first to point out this fact, has shown me many examples of the fungus and its ravages among the trees, several fine ash-trees and sycamores having been reduced to mere stumps by the silent operation of the vegetable parasite.

It is, by the way, a rather remarkable fact, that a tree which has been penetrated by water, can be relieved of its burden by the hand of man. An auger, or long-shanked centre-bit, judiciously inserted, will draw off the water, and if the aperture through which it gained admission be carefully trimmed and covered, so as to prevent any further lodgment of moisture, the bark will roll over the orifice, and soon obliterate it. The same process of self-repair will heal the smaller aperture made by the auger. Sometimes, when a semi-burrowing bird, such as the titmouse, enters a hollow thus formed, and builds its nest therein, the bark grows over the entrance, and so buries the nest in the hollow of the tree. Sawyers not unfrequently find various objects in the trunks of trees, which have been embedded by the curative powers of the tree.

Should, however, a fungus show itself, the tree is doomed. Perhaps the parasite may fall in the autumn, and the tree may show no symptoms of decay; but at the first tempest that it may have to encounter, the trunk snaps off at the spot where the fungus has been, and the extent of the injury is at once disclosed. As long as any portion of that tree retains life, it will continue to throw out these destructive fungi, and even when a mere stump is left in the ground, the fungi will push themselves out in profusion.

The pickaxe-like beak of the Woodpecker finds no difficulty in making its way through the decayed wood, and thus the bird is enabled to excavate its burrow without very much trouble. The nest itself can scarcely be called by that name, being nothing more than a collection of the smaller chips which have fallen to the extremity of the tunnel while the bird was engaged in the task of excavating. The burrow of the Woodpecker is as unpleasantly odorous as that of the kingfisher. The eggs are pure white.

In the British Museum may be seen samples of the burrows made by this bird, a portion of the tree having been cut off, and a section made, so as to show the shape, direction, and interior aspect of the hole. The specimens were obtained by Mr. Gould, the celebrated ornithologist, and the illustration was drawn from them.

According to Wilson and Audubon, some of the Woodpeckers of North America are able to excavate tunnels in the sound and still undecayed wood. They do not however select the hard wood in preference to that which is decayed, but always give the precedence to the latter. Still, they are often obliged to bore through several inches of solid wood, in order to reach the decayed portion in the centre.

The burrowing powers of the great Ivory-billed Woodpecker are marvellous, its chisel-like beak having been known to chip splinters from a mahogany table, and to cut a hole fifteen inches in width through a lath-and-plaster partition. Even the small Downy Woodpecker is able to bore its way through solid wood, and to make a most ingenious nest, the burrow sloping for some six or eight inches, and then being driven perpendicularly down the tree. The bird takes care to make the sloping tunnel only just large enough to admit the passage of its body; but the perpendicular hole, in which it resides, is quite large and roomy, so as to deserve the name of a chamber. When first made, the hole through which the bird enters its nest is as truly circular as if cut by a centre-bit; but it loses the sharpness of its outline after it has been in use for any length of time. Both the male and female Woodpecker work at this task of excavation, labouring alternately, relieving each other in regular rotation, and pecking continually until the burrow is finished, even though they should occupy several days in completing their home. They are so intent upon their labour, that they work all day and far into the evening, hammering away like carpenters at the bench.

The nest is not unfrequently stormed and seized by an usurper, diminutive in size, but unconquerable in spirit. This is the common house-wren, which is fond of building in holes and crannies, and is by no means particular as to its domicile. Kind-hearted persons are in the habit of nailing boxes to poles in their gardens for the use of this little bird, which is sure to take immediate possession, and to repay them for their benevolence by ridding their plants of noxious insects. Empty cocoa-nuts, gourds, earthen pots, and similar objects, are eagerly appropriated by the wren; and, in default of better premises, it will build in old hats, or in the sleeves of coats that have been hanging undisturbed on their pegs.

The burrow of the Woodpecker is far too comfortable a dwelling to be neglected by the wren, who allows the Woodpecker to proceed with its labours until he thinks that the hole is large enough for his purpose, and then assaults the unfortunate burrowers, driving them off to seek another and a less disturbed locality. In one case, a pair of Woodpeckers began to make their tunnel in an apple-tree, and were driven from the spot by the house-wren. They then pitched upon a pear-tree, completed their burrow, and had laid one egg, when they were again attacked by the fiery little bird, and obliged to abandon the locality altogether.

Safe as the habitation of the Woodpecker may seem, it is in America exposed to many and strange dangers. One of the perils which environ the burrows of this bird has already been mentioned, but a worse remains to be told. The black snake espies the parent birds as they enter and leave their nest, presses its sombre body against the tree, glides slowly up the trunk, and enters the apartment of the Woodpecker. Eggs or young are equally acceptable to the snake, which, finding itself in a comfortable and sheltered spot, coils itself round and abandons itself to repose. "The eager schoolboy," writes Wilson, "after hazarding his neck to reach the Woodpecker's hole, at the triumphant moment when he thinks the nestlings his own, and strips his arm, launching it down into the cavity, and grasping what he conceives to be the callow young, starts with horror at the sight of a hideous snake, and almost drops from his giddy pinnacle, retreating down the tree with terror and precipitation.

"Several adventures of this kind have come to my knowledge, and one of them that was attended with serious consequences, where both snake and boy fell to the ground, and a broken thigh and long confinement cured the adventurer completely of his ambition for robbing woodpeckers' nests." The unlucky bird-nester might have saved himself a fall, had he been anything of a naturalist. The black snake which is mentioned in the anecdote (*Coryphodon constrictor*) is as harmless as the common snake of England, though it is a fierce-looking reptile, and very irascible of temper, darting with open mouth at the hand of any one who annoys it, and making believe to bite. It is sometimes called the racer snake, on account of its swiftness.

THERE are many birds which make use of holes in trees for the deposition of their eggs, but which seldom, if ever, excavate the burrow by their own exertions. One of the best known examples of these birds is the WRYNECK (*Yunx torquilla*) or EMMET-EATER, a pretty though not a gorgeously-decorated creature. In Wales it is known by the name of Gwas-y-gog, or Cuckoo's knave, because it is said to follow the cuckoo as a servant follows a master. It is a rather elegantly-shaped bird, with plumage beautifully mottled with various shades of brown, and with a kind of low crest on the head, movable at pleasure. This bird selects for its home some hollow in a tree, sometimes taking the deserted burrow of a woodpecker, when it is fortunate enough to secure so convenient a residence. Any holes in trees, however, are used by the Wryneck; and it is very fond of those hollowed places where a branch has been broken away, and the interior of the tree has in consequence begun to decay. In such cavities the bird makes its nest, or rather, lays its eggs, for it is quite satisfied with chips of decayed wood for a bed whereon to place its numerous white eggs; and, like all those birds which build in burrows, it cares little about an elaborate nest. It will, however, take advantage of the nest which has been made by some other bird, and has been known to deposit its eggs in the deserted habitation of a redstart.

Though a timid bird, it is never slow in defence of its home and its numerous young; and though its slight beak can inflict but very trifling damage upon ordinary foes, it often frightens away the novice who has approached its domicile, by writhing its neck, darting its head forward, ruffling its crest, and hissing like an angry viper. On account of this habit, it is known in many parts of England by the popular name of Snake-bird. The head and neck of the bird have quite a formidable aspect as they thus present themselves at the orifice of the burrow, and at a little distance the writhing neck and angry hiss give to the creature a very sanguinary character.

AMONG the semi-burrowers we may rank the STARLING (*Sturnus vulgaris*), as this bird invariably lays its eggs in a hollow of some kind.

Its instinct teaches it to select spots wherein it can be hidden; and the deeper the burrow, and the narrower the entrance, the better does the Starling seem to be pleased. In all kinds of places the Starling makes its home, and its pale blue eggs may be found wherever there is a hole that will contain them. Under the eaves of houses the Starling contrives to creep, and finds some retired spot where it can sit in security and tend its young. Oftentimes when it resides in towns it displays so much ingenuity in concealing the locality of its habitation, that the impatient cry of the young birds affords the first indication of a nest. Frequently it gets into pigeon-cotes, and associates quite amicably with the rightful inmates. It has a great liking for the same domicile as the jackdaw, and there is often a trial of skill between the birds, each trying to gain the disputed nesting-place.

In towers, old trees, and similar places, the jackdaws endeavour to supplant the owls, and the Starlings endeavour to extirpate the jackdaws. On the ground under such places may be seen whole heaps of sticks, dropped by the jackdaw while endeavouring to make its nest, and showing how strangely circumscribed are the reasoning powers of lower beings.

In many things the jackdaw is a wonderfully clever bird, displaying such an amount of ingenuity in its actions, that its rational capacities are evidently very great. But, however clever a bird may be, and however admirably it may adapt its actions to surrounding circumstances, it is sure to break down suddenly in an unexpected manner, and to fail in the easiest part of its task. A jackdaw, for example, will go afield in search of sticks, and will spend some time in selecting a branch that will serve its purpose. It lifts, and drops, and turns, and weighs the branch, displaying great acumen in its task, and occupying much time in making a proper selection. When it has chosen a suitable branch, it flies away to the spot it has chosen for its nest.

When taking up the branch, the jackdaw mostly carries it by the middle, because it can be easily balanced when so held. But the bird forgets that a branch held crosswise will not enter a small aperture, and accordingly it finds itself checked when endeavouring to gain admission to its domicile. It flutters about in great dismay, and tries with all its powers to force the branch into the hole; but it never thinks of the simple expedient of taking the branch by the end, and pushing it lengthwise through the entrance; and after it has wearied itself out in vain attempts, it drops the branch and goes off for another. Beneath the many nesting-places at Walton Hall may be seen a wonderful number of sticks which have been thus dropped. Mr. Waterton lately drew my attention to a rapidly-accumulating pile of dead branches, which had been dropped by a jackdaw which was making its nest in a small window in the stable front. In this favoured place the birds know that they will not be injured, and so they permit their proceedings to be watched without exhibiting any shyness.

Old towers are very favourite haunts of the Starling, who

builds in close proximity to the owl and jackdaw, neither of the three appearing to be disturbed by the presence of the others. In forest lands the Starling lays its eggs in old trees; and I have frequently looked into a little hole high on the trunk, and seen the eggs lying far below, out of the reach of any foe except the rat, the weasel, and the British schoolboy, with his fertile invention and ready limbs. Starlings which choose such situations are strangely indifferent to observation, and are so noisy in all their conversation, that they may be heard at a distance of several hundred yards.

ONE or two other British birds must be mentioned, because they lay their eggs in excavations either natural or artificial. There is the elegant little TREE CREEPER (*Certhia familiaris*), so well known for its delicate form, its slender and slightly-curved beak, and the great agility with which it traverses the trunks of trees. The nest of this bird is mostly placed in the hollow of some decaying tree, and is of rather more ambitious a character than is generally found with birds which lay their eggs in similar situations, being formed of moss, grass, and other soft vegetable substances, and lined with downy feathers. There are about seven or eight eggs, which are small and of a light grey, variegated with brownish dots.

THE short-bodied, stout-beaked, strong-limbed NUTHATCH (*Sitta Europæa*) is another example of the semi-burrowers, inasmuch as it always chooses the hollow of a decaying tree for its nursery. The general habits of this curious little bird are very well known, and as they bear but very slightly on the principle of nesting, there is no need to mention them in this place.

The cavity which the bird selects is usually one which has but a very small entrance; and it is said that when the orifice is too large, the mother bird lessens it by kneading clay into the sides. It has already been mentioned that the wryneck defends its nest by the simulation of offensive powers, though it is, in truth, a very harmless bird, without the means to work an injury to an enemy. The Nuthatch defends its home with equal success, but not by the same deception; for whenever an enemy approaches too closely to the nest, out dashes the bird in a state of wild excitement, darts at the intruder, and pecks so fiercely with its powerful beak, that it can drive away any ordinary foe. The bite which the Nuthatch can inflict is of no trifling force; for the beak is strong enough to crack the shell of any nut, and when employed on softer substances, is very apt to leave behind it a tangible mark of its powers.

The nest of the Nuthatch is hardly deserving of the name, for it merely consists of a few dried leaves, intermingled with little bits of decaying wood.

ANOTHER of these semi-burrowing birds is the HOOPOE (*Upupa epops*), one of the handsomest, though not the most brilliant, of English birds. It is now very rare in this country, and, from all appearances, is unlikely to become plentiful.

The Hoopoe makes its nest in some decaying tree, and often prepares the hollow for its nest, though without intending to do so. The food of the bird consists chiefly of insects, in various stages of existence, most of which are dug by the long bill from the decayed wood wherein they burrow. The larvæ of many beetles exist in such localities, and as they are mostly fat and plump, they afford abundant nourishment to their destroyer. In dislodging these larvæ from their strongholds, the Hoopoe not only enlarges the hollow, but flings a quantity of small chips of the spongy wood to the bottom of the cavity. The nest is made of grass, feathers, and similar materials, and in many cases is placed upon the layer of dried fragments.

The cavity in which the Hoopoe makes its nest is notable for a most horrible stench, which, in countries where the bird is plentiful, has become proverbial. The odour which emanates from the kingfisher is most unsavoury, but it does not possess the pungent offensiveness which distinguishes that of the Hoopoe. The food of the Hoopoe was long considered to be the cause of this unpleasant peculiarity; but as the bird lives entirely on insects, it is evident that some other cause must be sought. This is found in certain glands near the tail, which secrete a substance that certainly must be useful to the bird in some mysterious way, just as the odorous secretion of the musk-deer must be beneficial to the animal; but it possesses a singularly offensive smell, and renders the nest unendurable to human nostrils.

ONE or two of the Titmice are in the habit of making their nests in similar situations. The COLE TIT (*Parus ater*) will always take advantage of hollow places, though it is perfectly capable of building a nest among the dense underwood, and its habitation may be mostly found in such localities. Young fir plantations are favourite resorts of this bird, which finds a congenial resting-place among the low, horizontal branches.

In Mudie's "Feathered Tribes of the British Islands," there is a brief and valuable summary of the bird-attracting powers of the fir in its different stages of development. "In a fir plantation, which is neither so low as to partake of the mushroom growth of pines (especially *Pinus sylvestris*) upon too rich soils, nor too inland and upland, there is a succession of birds. Linnets and other brake-birds come to them as long as they are mere bushes; but the note of the cuckoo is not heard in them. After a while the Cole Tit becomes one of their most plentiful inhabitants; and by that time the cuckoo perches and sings on the margin. A few years longer, and the ringdove moans in the tops of the trees, which have then begun to open towards the surface of the ground, and the covers for the brake-birds, and resting-places for all birds that build hideling and near the earth, are gone. The cuckoo is then heard less frequently, unless there are coppices of deciduous trees, or young pines come up in succession, in the vicinity. If the trees form a belt between rich grounds, the magpie, though he loves the 'home' trees better, will sometimes come, a little after the woodpigeon; and if the plantation is deep and secluded, the jay will, perhaps, come a little earlier. To all these succeeds the rook, which nestles in the mature trees, with the long boles clear of branches, and he quits them not until they are cut down or perish in the lapse of time."

In my note-book there is a sketch of a curious habitation occupied by a Cole Tit. One of the large trees at Walton Hall had been infested by the fungus, which has already been mentioned, and had broken asunder some eighteen or twenty feet from the ground. Several spots where these fungi had softened the wood were excavated by Mr. Waterton, in order to make nesting-places for various birds. In such spots the owls come and breed, and so do the jackdaws, starlings, and other birds. To one of these cavities Mr. Waterton fitted a door, composed of bark, and in the upper part of the door, he cut a little circular hole. The Cole Tit soon found out the hollow, and discovering that the cavity would make a good dwelling-place, and that the hole afforded an easy mode of entrance and egress, she proceeded to make her nest therein. When I saw the tree, the Cole Tit had not begun to build, but the relics of the old nest were there and could easily be seen by opening the door.

TOUCAN.

ANOTHER example of birds that make their nests within the hollow of trees is the strangely-formed TOUCAN (*Ramphastos Ariel*).

There are many species of Toucan, all of which are easily recognisable on account of the colours of the beak, for in all these birds the enormous bill is decorated with strangely brilliant tints. In one species the beak is rich orange and black, in another it is scarlet and yellow, and in another it is green and red; and in all it is of enormous dimensions when compared with the body, and is of great strength, though very light. Indeed, it is but a mere shell of horny substance, in some places not thicker than writing paper, and coloured by means of certain membranes in the interior, which shine through the semi-transparent horn.

It has long been known that the Toucan nested in hollow trees, and that it preferred those cavities which could only be entered by a small aperture, the reason for this predilection being rather absurd. It was supposed that the young of the Toucan were liable to the attacks of monkeys and large birds of prey, and that whenever the parent bird was alarmed, all she had to do was to poke her beak out of the aperture. The assailant, on seeing such a huge bill, fancied that an animal of corresponding size must be behind it, and therefore fled from so doughty a foe. One writer puts this idea in a very quaint manner. The monkeys, he says, are very noisome to young birds, and try to pull the unfledged Toucans out of their nests. But the mother bird, when she sees a monkey approaching, "so settles herself in her nest as to put her bill out at the hole, and gives the monkeys such a welcome therewith, that they presently pack away, and glad they escape so."

According to some writers, the Toucan makes the burrow for itself, using the huge beak as the tool wherewith it excavates its work. I very much doubt, however, whether the bird has the power of doing so, and think that, at the most, it only adapts and slightly alters the interior of the hollow, in order to suit its own purposes.

The Toucan is always a tree-loving bird, and does not wander from the forests. It is a native of South America, and may generally be seen perched on the topmost boughs of the lofty mora-tree, far beyond the reach of the shot-gun, and requiring a single bullet, or the Indian's tiny poisoned arrow, to bring it from its lofty elevation. It flies only by jerks, takes no long aerial journeys, and its body always seems overweighted by the enormous beak, which makes the head bow downwards as the bird passes through the air.

PERHAPS the SWIFT (*Cypselus apus*) may take rank among the semi-burrowing birds.

It always lays its long white eggs and makes its simple nest in holes, and in some cases is able to form the tunnel in which it breeds. When it takes up its habitation far from human abodes, it contents itself with crevices in rocks, hollow trees, and similar localities. But, when it resides near the habitations of man, it attaches itself to him like the swallow and the martin. Slates and tiles have, however, driven the Swift away from many a spot wherein it was once plentiful, for it loves to penetrate into thatch and therein to rear its hungry brood.

I can well remember the gradual ejection of the Swifts from a country town, on account of architectural improvements. Formerly, when all the less pretending houses were covered with thatch, the Swifts had their nests in every roof, and the "Jacky Screamers," as the peasants called the birds, used to hunt for flies in the streets, and boldly carry their prey to their young. The houses were so low that a man could touch the eaves merely by standing on a chair, and the habits of the birds were easily watched. Their nests were frequently robbed, but the birds seemed to care little for their bereavement, and when the eggs were removed, would quietly lay another couple or so. I seldom found more than three eggs in a nest.

By watching the Swift enter the tunnel leading to its nest, the object of the oddly-formed feet is clearly ascertained. The legs are very short but strongly made, and the toes are all furnished with strong curved claws, and directed forward, so that the bird is unable to clasp a branch with its feet. This structure enables it to scramble through its tunnel with great rapidity, and it is most interesting to see the Swift wheel round in the air with a piercing cry, answered by a little complacent chirrup from its mate within the nest, dart into the hole as if shot from a bow, closing its wings as it enters the tunnel, and then scramble away with a quick and certain gait that never fails to excite admiration.

The nest itself is a very simple affair. Any soft material seems to suit the Swift, which brings hay, flakes of wool, bits of rag, feathers, paper, string, and many other substances into the burrow. With these materials it makes a tolerably compact nest, which is generally to be found at a distance of eighteen inches or two feet from the entrance of the burrow. The holes which have been made by rats are mostly used for this purpose, but if the bird cannot find a hollow already existing, it is quite capable of forming one for itself, and by dint of pulling out the straws in some weak spot, and pushing aside those which it cannot extract, it soon makes a burrow large enough for its purpose.

CHAPTER III.

BURROWING REPTILES.

The Reptiles and their hibernation—The Land Tortoise and its winter dwelling—The Crocodiles—Snakes—The Yellow Snake of Jamaica—Its general habits — Its burrowing powers discovered — Presumed method of removing the earth.

The Reptiles are, as a body, not remarkable for the burrows which they make.

Many of them bore their way into the ground, pass a few months in a state of torpidity, and then push their way out again. But the hole which they make in the earth is scarcely to be called a home, inasmuch as the inhabitant merely enters it as a convenient place wherein it may become torpid, and abandons it as soon as the ordinary functions of the system are restored by the warmth of the succeeding year.

The common Land Tortoise, for example (*Testudo Græca*), is in the habit of slowly digging a burrow with almost painful deliberation, and then concealing itself below the surface of the earth during the cold months of winter. Many Tortoises which have lived in this country have been noticed to perform this act, and I have lately seen a very good example of a burrow which had been sunk amid some strawberry plants, and from which the inmate had just emerged.

Many other reptiles follow a similar course of action. The crocodiles, for example, sink themselves deeply in the mud, and have more than once caused much alarm by awakening out of their hibernation, and protruding their unwelcome snouts from the mud close to the feet of the astonished spectator.

Snakes are accustomed, in like manner, to conceal themselves during the period of their hibernation, resorting to hollow trees, holes in the ground, and similar localities. Labourers while engaged in digging, especially in breaking down banks, frequently unearth a goodly assemblage of snakes, all coiled up in an unsuspected cavity, which they must have entered through the deserted burrow of a mouse or some other little animal. But that a snake should be able to form its own burrow is a feature so remarkable in herpetology, that a single accredited example must not be passed without notice.

In his very interesting work on the natural history of Jamaica, Mr. Gosse gives a curious account of a burrow made by the YELLOW SNAKE (*Chilabothrus inornatus*). This snake is very plentiful in Jamaica, and is perfectly harmless to man, being destitute of poison-fangs, and not reaching a size which would render it formidable to human beings. Its average length, when full-grown, is eight feet. So far, indeed, from being obnoxious to man, it may rank among his best friends, as being a determined foe to rats, feeding largely upon them, and even entering houses in search of its prey. Like the weasel, indeed, of our own country, which feeds mostly on mice and other destructive animals, but occasionally makes a raid upon the fowl-house, the Yellow Snake enters the farmyard, and, instead of eating rats as it ought to do, proceeds to the hen-roosts, and robs them. No less than seven eggs have been found inside a single Yellow Snake, and not a single egg was broken.

There is now (1863) a good specimen in the Reptile-room of the Zoological Gardens of London.

One of these snakes was seen to crawl out of a hole in the side of a yam-hill—*i. e.* a bank of mould prepared for the purpose of growing yams—and when the earth was carefully removed, a large chamber was discovered in the middle of the hill, nicely lined with strips of half-dried plantain leaves, technically called "trash," and containing six eggs, all fastened together. Just outside the hole was a heap of loose mould, which had evidently been thrown out when the excavation was made.

The Yellow Snake generally makes its home in the deep spaces between the spurs of the fig or the buttresses of the cotton-tree, and always lines it with "trash;" but that the creature should be able to excavate a burrow, and throw out the earth, seems almost incredible. How did the snake remove the earth? As the reptile was not seen in the act of excavating, this question could not be precisely answered. Mr. Hill, however, to whom the subject was referred, gave as his opinion that the snake loosened the earth with its snout, and then worked the loose soil out of the hollow by successive contractions of the segments of the abdomen, which would thus "deliver" the soil after the manner of the Archimedean screw.

The eggs which were found in the chamber were removed, and from one of them, which was opened, was taken a young snake, about seven inches in length.

CHAPTER IV.

BURROWING INVERTEBRATES.

CRUSTACEA.

The LAND CRABS and their habits—The Violet Land Crab—Its burrows, its combativeness, and its pedestrian powers—The FIGHTING CRAB, why so called —The RACER CRAB of Ceylon—Its burrows and mode of carrying off the soil— The ROBBER CRAB—Its form and general habits—Food of the Robber Crab—A soft bed, and well-stocked larder—The CHELURA, and its ravages among timber —The GRIBBLE and its kin.

THE reader will doubtless perceive that among such a multitude of mammals and birds, each of which has some habitation, it is impossible to give more than a selection of some of the more remarkable examples. Although, therefore, there are many other burrowing and semi-burrowing vertebrates, we must leave the furred, feathered, and scaled tribes, and pass to those which occupy a lower place in the animal kingdom.

Among the Crustacea, there are very many species which form burrows, and which conceal themselves under the sand or mud. As, however, these creatures cannot be said to form their habitations, and the burrows are mostly obliterated by the return of the water, they can scarcely be reckoned among those which make "homes without hands." Some, however, there are which are as fully entitled to be ranked among the true burrowers, as any creature which we have mentioned, digging a regular burrow in the earth, residing in their subterranean home, issuing forth to procure food, and retiring to it when alarmed. These are the creatures so widely famous as LAND CRABS (*Gecarcinus*), respecting which so many wonderful tales are told, some true, some false, and many exaggerated. The Land Crabs are found in various parts of the world, and are notable for very similar habits. They all burrow in the ground, run with very great speed, bite with marvellous severity, and associate in considerable numbers. As a general fact, they are considered as great dainties, and when properly prepared, may be ranked among the standing luxuries of their country.

LAND CRAB.

As the Violet Land Crab of Jamaica (*Gecarcinus ruricola*), is the most familiar of these creatures, we will take it as our first example of the burrowing crustacea. This species, which is sometimes called the Black Crab, and sometimes the Toulourou, is exceedingly variable in its colouring, sometimes black, sometimes blue, and sometimes spotted. Whatever may be the colour, some tinge of blue is always to be found, so that the name of Violet Crab is the most appropriate of the three. Wherever the Land Crab makes its home, the ground is filled with its burrows, which are as thickly sown as those of a rabbit warren, and within these habitations the crabs remain for the greater part of the day, coming out at night to feed, but being always ready to scuttle back at the least alarm.

Should, however, their retreat be intercepted, they are as ready to fight as to run, and have a curious habit of seizing the foe with one of the large claws, and then shaking off the limb at its junction with the body. As the muscles of the claw retain their tension for some little time after the connexion with the body has been severed, the enemy feels as much pain as if the crab were still living; and in the momentary confusion caused by the bite, the crab takes the opportunity to conceal itself in some crevice. As is the case with all crustaceans, it suffers but a temporary loss, a new limb soon sprouting out, and taking the place of the discarded member.

Although these warrens are seldom less than a mile from the sea, and are often made at a distance of two or even three miles, the Land Crabs are obliged to travel to the shore for the purpose of depositing their eggs, which are attached to the lower surface of the abdomen, and are washed off by the surf. Large numbers of the crabs may be seen upon their journey, which they prosecute so eagerly that they suffer no opposition to deter them from their purpose. This custom has probably given rise to the greatly exaggerated tales that have been narrated respecting these crabs, and their custom of scaling perpendicular walls rather than turn aside from the direct line of their route.

Twice in the year the Land Crabs become very fat and heavy and are then in the best condition for the table, their flesh being peculiarly rich and loaded with fat. No one seems to be tired of the Land Crab, and new comers are apt to indulge in the novel dainty to such an extent that their internal economy is sadly deranged for some little time after the banquet.

About the month of August, the Land Crab is obliged to cast its shell, and for that purpose retires to the burrow, which has been well stocked with grass, leaves, and similar materials. It then closes the entrance, and remains hidden until it has thrown off its old shell, and indued its new suit, which is then very soft, being little but a membranous skin, traversed by multitudinous vessels. At this time the crab is thought to be in the best condition for the table. Calcareous matter is rapidly deposited upon the membrane, and in process of time the new shell becomes even harder and stronger than that which has been rejected.

Many species of Land Crab are known, some of which possess rather curious habits. The FIGHTING CRAB (*Gelasimus bellator*), is a good example of them. This species possesses one very large and one very little claw, so that it looks as if a small man were gifted with one arm of Hercules and the other of Tom Thumb. As it runs along, with the wonderful speed which belongs to all its kin, it holds the large claw in the air, and nods it continually, as beckoning to its pursuer. While so engaged it has so absurd an aspect that it has earned the generic title of Gelasimus, *i.e.* laughable. As may be conjectured from its popular name, it is a very combative species, holding its fighting claw across his body, just as an accomplished boxer holds his arm, and biting with equal quickness and force. It is also a burrower, and lives in pairs, the female being within, and the male remaining on guard at the mouth of the hole, his great fighting claw across the entrance.

Another Land Crab, which has earned the generic title of Ocypode, or Swift-footed, and is popularly called the RACER, from its astonishing speed, is a native of Ceylon, where it exists in such numbers that it becomes a terrible nuisance to the residents. Having no respect for the improvements of civilization, this crab persists in burrowing into the sandy roads, and is so industrious at its excavations, that a staff of labourers is constantly employed in filling up the burrows which these crabs have made. Were not this precaution taken, there would be many accidents to horsemen.

The mode of excavation employed by this creature is rather peculiar. It " burrows in the dry soil, making deep excavations, bringing up literally armfuls of sand, which, with a spring in the air, and employing its other limbs, it jerks far from its burrows, distributing it in a circle to the distance of many feet."

THERE is a very remarkable burrowing crustacean, called the ROBBER CRAB (*Birgus latro*). This creature is of a strange, weird-like shape, difficult to explain, but easily to be comprehended by reference to the illustration. The reader can, however, form some notion of its general form, by removing a common hermit

THE ROBBER CRAB.

crab from its residence, and laying it flat before him. The Robber Crab, however, does not live in a shell, and its abdomen is consequently defended by hard plates, instead of being soft and unprotected like that of the hermit crab, to which it is closely allied.

The Robber Crab inhabits the islands of the Indian ocean, and is one of those crustacea which are able to exist for a long time without visiting the water, the gills being kept moist by means of a reservoir on each side of the cephalothorax, in which the organs of respiration lie. Only once in twenty-four hours does this remarkable crab visit the ocean, and in all probability enters the water for the purpose of receiving the supply which preserves the gills in working order.

It is a quick walker, though not gifted with such marvellous speed as that which is the property of the racer and other land crabs, and is rather awkward in its gait, impeded probably by the enormous claws. While walking, it presents a curious aspect, being lifted nearly a foot above the ground on its two central pairs of legs, and if it be intercepted in its retreat, it brandishes its formidable weapons, clattering them loudly, and always keeping its face towards the enemy. Some writers aver that it is capable of climbing up the stems of the palm-trees, in order to get at the fruit, but this assertion seems to require very strong corroboration before it can be believed.

The food of the Robber Crab is of a very peculiar nature, consisting chiefly, if not entirely, of the cocoa-nut. Most of my readers have seen this enormous fruit as it appears when taken from the tree, surrounded with a thick massy envelope of fibrous substance, which, when stripped from the nut itself, is employed for many useful purposes. How the creature is to feed on the kernel seems quite a mystery; and, *primâ facie,* for a crab to extract the cocoa-nut from its envelope, to pierce the thick and stubborn shell, and to feed upon the enclosed kernel, seems an utterly impossible task. Indeed, had not the feat been watched by credible witnesses, no one who was acquainted with the habits and powers of the crustacea would have credited such an assertion. Yet Mr. Darwin, Messrs. Tyerman and Bennett, and other observant men, have watched the habits of the creature, and all agree in their accounts.

According to Mr. Darwin, the crab seizes upon the fallen

cocoa-nuts, and with its enormous pincers tears away the outer covering, reducing it to a mass of ravelled threads. This substance is carried by the crabs into their holes, for the purpose of forming a bed whereon they can rest when they change their shells, and the Malays are in the habit of robbing the burrows of these stored fibres, which are ready picked for them, and which they use as "junk," *i.e.* a rough kind of oakum, which is employed for caulking the seams of vessels, making mats, and similar purposes. When the crab has freed the nut from the husk, it introduces the small end of a claw into one of the little holes which are found at one end of the cocoa-nut, and by turning the claw backwards and forwards, as if it were a bradawl, the crab contrives to scoop out the soft substance of the nut.

According to the observations of Messrs. Tyerman and Bennett, the well-known missionaries to the South Seas, the Robber Crab has another method of getting at the cocoa-nut, and displays an instinctive knowledge of political economy which is very remarkable.

"These animals live under the cocoa-nut trees, and subsist upon the fruit which they find upon the ground. With their powerful front claws they tear off the fibrous husk; afterwards, inserting one of the sharp points of the same into a hole at the end of the nut, they beat it with violence against a stone until it cracks; the shell is then easily pulled to pieces, and the precious fruit within devoured at leisure. Sometimes, by widening the hole with one of their round, gimlet claws, or enlarging the breach with their forceps, they effect sufficient entrance to enable them to scoop out the kernel, without the trouble of breaking the unwieldy nut.

"These crabs burrow in the earth, under the roots of the trees that furnish them with provisions—prudently storing up in their holes large quantities of cocoa-nuts, stripped of their husk, at those times when the fruits are most abundant, against the recurring intervals when they are scarce. We are informed that if the long and delicate antennæ of these robust creatures be touched with oil, they instantly die. They are not found on any of these islands except the small coral ones, of which they are the principal occupants. The people here account them delicious food."

The palm-climbing habits of the Robber Crab are mentioned

by Mr. T. H. Hood, in his "Notes of a Cruise in H.M.S. *Fawn*, in the Western Pacific." In the Samoan group of islands, the crab is called "Ou-ou," and is a favourite article of food. While the vessel remained off Samoa, Mr. Hood asked about these crabs; and though he did not see any of them performing so strange a feat, he shows that there are very good grounds for believing the possibility of such an action.

"I inquired of them about the habits of the Ou-ou, or great cocoa-nut-eating crab, common here, and found the reports previously received from the natives corroborated. Mr. Darwin mentions that, in the Seychelles and elsewhere, there is a species which is in the habit of husking the nuts on the ground, and then tapping one of the eyes with its great claw, in order to reach the kernel. Its congener here ascends the cocoa-trees, and having thrown the nuts down, husks them on the ground; this operation performed, again ascends with the nuts, which he throws down, generally breaking them at the first attempt, but, if not successful, repeating it till the object is attained.

"Before leaving, an old Savage Island man at the mission brought in three or four immense Ou-ous, which evinced in their efforts to escape, bursting coils of cocoa-nut sinnet, a strength quite sufficient to husk the toughest cocoa-nut. As to the method of obtaining the contents afterwards, every native (both Samoans and Niuans) confirms the account mentioned before. The Niuans understand their habits best. The old man who brought them to-day dug them out of the holes in which they remain many weeks torpid. The female differs from the male in having three flippers, well furnished with strong borers, on the right side of the sac."

When full grown, this crab is more than two feet in length, and, as may be seen by the illustration, is stoutly made in proportion to its length. The colour of the creature is very pale brown, with a decided tinge of yellow.

PASSING by many other species of crustacea which burrow in the earth, or mud, or sand, we come to a very remarkable being, which makes its habitation in solid wood. This is the WOOD-BORING SHRIMP (*Chelura terebrans*), one of the sessile-eyed crustacea, nearly related to the well-known sand-hopper, which is so plentiful on our coasts.

Although very small, it is terribly destructive, and does no small damage to wooden piles driven into the bed of the sea. It is furnished with a peculiar rasping instrument, by means of which it is enabled to scrape away the wood and form a little burrow, in which it resides, and which supplies it with nourishment as well as with a residence. The tunnels which it makes are mostly driven in an oblique direction; so that when a large number of these creatures have been at work upon a piece of timber, the effect of their united labours is to loosen a flake of variable dimensions. As long as the weather is calm, the loosened flake keeps its position; but no sooner does a tempest arise, than the flake is washed away, and a new surface is exposed to the action of the Chelura.

When the Chelura is placed on dry land, it is able to leap nearly as well as the sand-hopper, and performs the feat in a similar manner.

THIS is not the only wood-boring crustacean with which our coasts are pestered; for the GRIBBLE (*Limnoria terebrans*) makes deeper tunnels than the preceding creature, though it is not so rapidly destructive, owing to the direction of its burrows, which are driven straight into the wood, and do not cause it to flake off so quickly as is the case when the Chelura excavates it. Still, it works very great harm to the submerged timber, boring to a depth of two inches, and nearly always tunnelling in a straight line, unless forced to deviate by a nail, a knot, or similar obstacle. The Gribble is a very tiny creature, hardly larger than a grain of rice, and yet, by dint of swarming numbers, it is able to consume the wooden piles on which certain piers and jetties are supported; and in the short space of three years these destructive crustacea have been known to eat away a thick fir plank, and to reduce it to a mere honeycomb. Sometimes these two wood-boring shrimps attack the same piece of wood, and, in such cases, the mischief which they perpetrate is almost incredible, considering their small dimensions and the nature of the substance into which they bore. The common fresh-water shrimp, so plentiful in our brooks and rivulets, is closely allied to the Gribble, and will convey a very good idea of its appearance. In some parts of our coasts the ravages of these animals are so destructive, that the substitution of iron or stone for wood has become a necessity.

CHAPTER V.

BURROWING MOLLUSCS.

The Boring Snail of the Bois des Roches—Opinions as to its method of burrowing—Shape of the tunnels—Solitary habits of the Snail—The Piddock, its habits and appearance—Structure of the Shell, and its probable use—Method of burrowing—Use of the Piddock and other marine burrowers—The balance of Nature preserved—The Wood-borer and its habits—The Date Shell—Its extraordinary powers of tunnelling—The Razor Shell—Its localities and mode of life—The Flask Shell and the Watering-Pot Shell—The Shipworm—Its appearance when young and adult—Its curious development—Its ravages, and the best method of checking them—Its value to engineers—The Giant Teredo—Form, dimensions, and structure of the shell—How and where discovered.

Ill fitted as the Molluscs seem to be for the task of burrowing, there are several species which are able not only to make their way through soft mud, or into the sandy bed of the sea, but to bore deep permanent tunnels into stone and wood. Even the hard limestone and sound heart-of-oak timber cannot defy these indefatigable labourers, and, as the sailor or the dweller on the coast knows full well, the rocks and the timber are often found reduced to a mere honey-combed or spongy texture by the innumerable burrows of these molluscs.

There is now before me a piece of very hard calcareous rock, in which are bored several deep holes, large enough to admit a man's thumb, and remarkably smooth in the interior, the extremity being always rounded. Indeed, if a hole were made in a large lump of putty by putting the thumb into it and turning it until the sides of the hole became smooth, a very good imitation of these miniature tunnels would be produced. This fragment of stone was taken from a little wood in Picardy, called Le Bois des Roches, on account of the rocky masses that protrude through its soil, and was brought to England by Mr. H. J. B. Hancock, who kindly presented it to me.

In the winter time, each of these holes is occupied by a specimen of the Helix saxicava, a small snail, closely resembling the common banded snail of our hedges (*Helix nemoralis*), and it is thought that the holes are excavated by the snail which inhabits them. Mr. Hancock, who has lately re-opened in the columns of the *Field* newspaper a controversy respecting these snails, which was initiated in 1839, is of opinion that the snails really form the hole, and that they burrow at the average rate of half an inch per annum. The late Dean Buckland was of the same opinion. Other naturalists, however, think that the holes were originally excavated by pholades and other marine molluscs when the rocks in question formed part of the ocean bed, and that the snails merely inhabit the ready-formed holes. Mr. Pinkerton upholds this opinion, and states that at least three other species of helix possess similar habits, the garden and the banded snail being among the number.

I have compared the burrows of the mollusc, which we will call the Boring Snail, with those of the pholas and lithodomus, both of which will be presently described, and find that there is no resemblance in their forms, the shape and direction of the holes being evidently caused by an animal of no great length in proportion to its width. In my own specimen, every hole is contracted at irregular intervals, forming a succession of rounded hollows. If we return to our lump of putty, we may form the holes made by the thumb into a very good imitation of those in which the Boring Snail lives. After the thumb has been pushed into the putty and well twisted round, put in the forefinger as far as the first joint and turn it round so as to make a rounded hollow. Push the finger into the hole as far as the second joint, and repeat the process. Now introduce the whole of the finger, enlarge the extremity of the hole and round it carefully, when there will be a very correct representation of the tunnel formed in the rock.

Granting that the snail really does form the burrow, we have still to discover the mode of working. Mr. Hancock says that it must do so by means of an acid secretion proceeding from the foot, which corrodes the rock and renders it easy to be washed away. If the snail be removed and placed on litmus paper, the ruddy violet colour which at once tinges the paper shows that there is acid of some kind, and if the paper be applied to the spot whence the snail has been taken, the same results follow. It is a remarkable fact that although the snail leaves the usual slimy marks of its progress when crawling in the summer time, no mucus is perceptible on the approach of winter. When the cold months come round, the Boring Snail leaves its food and attaches itself to the rock, remaining in the same spot until summer approaches. During this time, the portion of rock to which it clings is worked away, and the stone around the excavation is impregnated with a greasy matter which soon dries up after the admission of the atmosphere. In a letter to me, dated October 14th, 1863, Mr. Hancock remarks that the rock at Monte Pellegrino in Sicily, which is crystalline and hard as marble, is perforated by the same snail and in the same manner. I may here mention that the stone of the Bois de Roches is that of which the column at Boulogne is built, which has retained its sharpness of outline after exposure to wind and weather for nearly sixty years. It is therefore called *marbre Napoléon.* Mr. Hancock proceeds to say, " The following are a few of the peculiarities which I have not mentioned in my letter in the *Field :*

1st. There is no instance at Bois de Roches of a tunnel being formed on the horizontal surface of a rock, or on the sides facing the south and south-east. They are always on the sides facing the north or north-east.

2nd. The snail forms no epiphragm.

[The "epiphragm" is the barrier of hardened mucus with which snails mostly close the entrance of their shells. There are generally several epiphragms in each shell.]

3rd. Though during the summer it leaves behind it the usual slimy mucus track; in the winter on returning to the rocks no track is perceptible except the corrosion of the rock by frequent passage. This would seem to point to a system of secreting organs for the acid, separate from that for the mucus.

4th. Contrary to the usual habits of burrowing molluscs, who generally have a bed of muddy matter between their shells and the walls of their dwelling, the Helix saxicava keeps his tunnel perfectly clean and neat.

5th. When the liquor alluded to as forming a fatty aureole round the tunnel penetrates into pre-existing clefts in the rock, it provokes the growth of a microscopic lichen, which also grows in the tunnels in places after the liquor has evaporated.

6th. The tunnels of the Helix saxicava are always irregular, bearing no relation to the size or shape of the excavators, whereas, in other excavating molluscs, the shape of the hole always bears some relation to its occupant, and also the excavations are alike for all animals of the same species."

There is an opinion that the gastric juice secreted in the stomach may be the means through which the tunnelling is conducted, and that instead of being employed as food within the body it is poured out upon the stone, so as to dissolve it, the softened substance being then removed by the foot. The Boring Snails do not congregate together during hibernation, as is the well-known custom of the garden species, but are always solitary. Sometimes two or even three are found in the same burrow, but then they are always at some distance from each other, and form supplementary tunnels of their own. In my own specimen there is a curious example of this peculiarity, where the snail has contrived to bore completely through the barrier that separates it from a neighbouring tunnel, and has made a hole as large as the keyhole of an ordinary writing-desk, and nearly of the same shape.

THERE are many marine boring molluscs, some of which excavate mud, others stone, and others timber. Of the mud-borers I have little to say, few of them possessing points worthy of notice. Perhaps the most noteworthy of these is the common GAPER SHELL (*Mya arenaria*), so called, because one end of the shell gapes widely, in order to permit the passage of a long and stout tube. In a specimen now before me, the tube is between three and four inches in length, and at the base is large enough to admit the thumb. As, however, it gradually tapers to the extremity, the aperture at the other end is scarcely capable of receiving the little finger. The walls of this tube are very thin and membranous, and it is more or less retractile, carrying within it the siphons through which the mollusc respires and takes nourishment.

The Gaper Shell inhabits sandy and muddy shores, and to an inexperienced eye is quite invisible. The shell itself, together with the actual body of the mollusc, is hidden deeply in the mud, seldom less than three inches, and generally eleven or twelve inches from its surface. In this position it would be unable to respire, were it not for the elongating tube, which projects through the mud into the water, and just permits the extremities of the siphons to show themselves, surrounded by the little radiating tentacles which betray them to the experienced shell-hunter. These tentacles or fringes are never seen in the dried specimens, and can only be partially preserved by plunging the animal into spirits of wine, glycerine, or other antiseptic liquid. The Gaper Shell is esteemed as an article of food by man, beast and bird; for not only do human beings dig it up with tools, cook it, and eat it, but the wolves and the arctic fox scratch it out of the mud and eat it raw, and the various sea birds peck it out with their beaks, prize the shell open, and devour the contents.

THE well-known LIMPET is a kind of borer, though the holes which it excavates are of very trifling depth, and are probably made by the mechanical friction of the shell and foot against the rock, without any intention on the part of the animal. Those who have been accustomed to wander along the sea-shore must have noticed that the Limpet shells always sink more or less into the rocks on which they cling, and that in very old specimens which are covered with algæ and barnacles, the shells are often sunk fully half their depth into the solid rock. Grooves, too, of various depths may be seen in the same rock, showing the slow and tedious track which the Limpets have made over its surface, until they finally settled down into some convenient situation.

OUR next example of the burrowing molluscs is the well-known PHOLAS, popularly called the PIDDOCK (*Pholas dactylus*), the shells of which are extremely plentiful upon our coasts, whether empty and thrown upon the beach, or still adhering to the living animal and deeply sunken in the rock. Almost in every part of our shores the Piddock is to be found wherever there is rock, and its dimensions and general appearance vary together with the locality. The chalk cliffs, which bound so many miles of our coast, are thickly studded with the burrows of the Piddock, which takes up its residence as high as the mid-water zone of the coast, and in some places is so plentiful, that the hand can scarcely be laid upon the rock without covering one or two of the holes.

The shell itself is extremely fragile, and of a rather soft texture, and its outer surface is covered with ridges, that sweep in the most graceful curves from the hinge to the edge, and bear

PHOLAS IN WOOD. LITHODOMUS. RAZOR SHELL. HOLES OF PHOLAS IN ROCK.

some resemblance to the projections upon a file. Yet practical naturalists have proved that, by means of these tiny points and ridges, the Pholas is able to work its way into the rock; for not only can a similar hole be bored by using the shell as a bradawl is used to pierce wood, but the creature has actually been watched while in the act of insinuating itself into the chalk rock, a feat which was performed by gently turning the shell from right to left, and back again.

The Pholas burrows to a considerable depth, and if a piece of the rock be detached and broken to pieces by the hammer, it will be seen to be completely riddled with the perforations. Chalk-rock is mostly the richest in specimens, but even the hard limestone formations are penetrable by the fragile shell of the Pholas. It has been well remarked, that the size of the Pholas and the sharpness of its markings vary in inverse ratio to the hardness of the rock in which it burrows. From the softest sea-beds are taken the largest and most perfect shells, while those specimens which are obtained from the hard limestone rocks, are comparatively small, and the surfaces are rubbed nearly smooth. The very worst examples, however, are those which are found in gritty rocks, interspersed with pebbles. The shells that have burrowed into such substances are dwarfed, abraded, and often misshapen, and are valueless except to the physiologist.

We naturally ask ourselves why the Pholas should bury itself in the rocks instead of passing its life in the open sea, like the generality of bivalves. The creature does not feed upon the substances in which it forms its curious tunnel, and to all appearance would obtain as much food without as within the burrow. One obvious answer to this question is, that the creature buries itself in the rock for the sake of safety, its shell being, as has already been remarked, of a soft and fragile texture. This opinion is further corroborated by the fact that one of the British species, the Paper Pholas (*Pholas papyracea*), has a peculiarly thin and delicate shell, so as to earn for it the name by which it is popularly known. Yet, although this may be a reason, it is not *the* reason; for there are many well-known shells which are far more fragile than those of the Pholas, yet which need no such protection, and instead of concealing themselves in any way, roam the ocean freely.

In my own opinion, the burrowing instincts of this and many other marine creatures of similar habits, are implanted in them for other than mere individual purposes. Judging by the effects which these animals must have produced upon the line of coast throughout a succession of centuries, I cannot but think that they are, at all events partially, intended as instruments which aid in producing those mighty changes that are continually taking place over the whole face of the globe.

We know that although the general proportions of sea and land are maintained, a continual change is being worked in their relative positions. Even within the memory of man, fields now blooming with corn were once covered with the salt water, and buildings that were once a mile from the shore are now in hourly danger of falling into the sea. And I have very little doubt but that the Pholas plays a very important part in these changes. If the reader will examine any of our chalk-bound coasts, and will walk from the foot of the existing cliff to the extreme of low-water mark, he will see that the Pholas has everywhere left the tokens of its industry. Even at the verge of low water, the spot whereon he stands was once the base of the cliff which has now receded so far from the waves, and continues to recede yearly. And as he looks out to sea, and watches the breakers fling their white foam over the sunken rocks, and notes the dark masses of tangle lifting with the waves, he sees the remains of former cliffs, that have long since crumbled away piecemeal and fallen into the sea.

It is true that the continual dashing of the billows will, in time, destroy the hardest rocks, and that even a granite cliff cannot withstand the action of water. But the process of destruction—if we may use such a word—is greatly accelerated by the multitudinous holes bored by the Pholas and other burrowers, and the rock is rapidly undermined by the joint action of the molluscs and the waves. The upper portion of the cliff is thus left without support, and when a few heavy rains have loosened the earth, down comes a mixed mass of rock, soil, and herbage, and a fresh face is thus made to the cliff. The loose earth is soon washed away by the waves, the submerged portion of the fallen rock is eagerly seized upon by the burrowers, and in process of time, the whole mass crumbles away, is broken up by the storms, and the fragments are gradually rubbed to atoms by the action of the waves. Thus it is that the cliffs recede on one side, while the soil advances on another, and so the whole face of the world is gradually changed.

PERHAPS the DATE SHELLS are even more powerful as burrowers than the molluscs which have just been mentioned. One species, the Fork-tailed Date Shell (*Lithodomus caudigera*), is able to bore into substances which the pholas cannot penetrate. It is truly a wonderful little shell. Some of the hardest stones and stoutest shells are found pierced by hundreds of these curious beings, which seem to have one prevailing instinct, namely, to bore their way through everything. Onwards, ever onwards, seems to be the law of their existence, and most thoroughly do they carry it out. They care little for obstacles, and if one of their own kind happens to cross their path, they quietly proceed with their work, and drive their tunnel completely through the body of their companion.

The precise method employed in excavation is at present unknown, for the shape of the shell, and the exactitude with which it fits the burrow, prove that the mollusc does not form its tunnel by means of the protuberances on the surface of the shell, and no other method of boring has at present been discovered.

THERE is another notable burrower among the bivalve marine shells, which is remarkable for the depth to which it bores, and the hard nature of the substances through which it makes its way. This is the shell called *Saxicava rugosa;* one of the most variable of the molluscs, so variable indeed, that no less than fifteen different names have been given to it, each being supposed to be a separate species. Not only species, but even genera have been formed from the varieties of this curious shell. It is a flattish bivalve, of no very great size, symmetrical in shape when young, but oblong when old.

This creature burrows as rapidly as the species which has just been described, and the process by which the feat is accomplished is quite as enigmatical. Several conchologists have expressed an opinion that the animal must secrete some liquid solvent, which softens the rock, and permits the shell to pass. But, although it is possible for the boring snail to excavate by such means, the Saxicava can hardly do so. For the boring snail is a terrestrial species, and would be able to employ a solvent without interruption; but as the Saxicava works below the surface of the sea, any solvent which it could employ would be either washed away, or so diluted by the water that it would have no effect upon the stone.

Still, that the creature must employ some means not yet

known to naturalists, is evident from the shape of the hole and the comparative hardness of the shell and the substances in which it is embedded. The shell is of ordinary hardness, while the rock in which it is found is often of adamantine density. Sometimes it bores into corals, frequently into limestone, and often into shells, which it penetrates as deeply as the date shell. On every rocky shore which the Saxicava inhabits, its burrows may be found, no matter what may be the hardness or composition of the stone. The clay ironstone which is found about Harwich, and is popularly called cement-stone, is filled with the burrows of the Saxicava. Its tunnels are found in the Kentish rag, while even the well-known Portland stone, of which the Plymouth breakwater is constructed, is often honeycombed by the multitudes of these bivalves that inhabit it. Some of the enormous stones which were employed in building the breakwater are now much wasted by the holes made in them by the Saxicava.

As is the case with the burrows of the date shell, those of the Saxicava do not run parallel with each other, but are driven into the stone at any angle. In consequence of this custom, it is not of unfrequent occurrence that one of the creatures hits upon the burrow of another, and if it does so, it will not, and in fact cannot, alter the direction of its tunnel. Neither is it able to wait until the other shell has burrowed further, but eats its way silently and unrelentingly along, cutting through the shell and body of its luckless companion, and thus bringing on it a violent death, which its rocky home seems especially intended to avert. The hole is on the average about five or six inches in depth, and the animal does not lie free in its burrow, but attaches itself to the side by means of a byssus or cable, like that of the mussel, save that it is smaller, because the strain upon it is not so great as when the shell is anchored in the open sea. This shell has a very wide range of locality, and is sometimes found at a very great depth, specimens having been procured at a depth of nearly 900 feet. It attains its largest dimensions in the colder seas.

Another member of this family (*Xylophaga dorsalis*), burrows, as its name implies, into wood and not into stone. It is a small species, and of a very globular form, and never burrows to any great depth, an inch being the ordinary length of its tunnel. The shells of this creature are often found in floating wood, or in the sea-covered portions of wooden piles, and it is a notable fact that the burrows are always made acróss the grain of the timber in which it lives.

THOSE who are fond of wandering on the sea-shore, will often have experienced tangible proofs of the existence of another burrowing mollusc, the RAZOR SHELL (*Solen ensis*).

In some parts of our coast it is impossible to walk on the mixed rock and sand, when the tide has receded, without noticing innumerable jets of water, which start from the ground without any perceptible cause, leap for a foot or so in the air and then disappear. On watching one of these miniature fountains, and looking at the exact spot whence it proceeds, two little round holes are generally seen in the sand, so close to each other as to resemble a keyhole, and large enough to receive an ordinary goosequill. If the finger be placed on the spot, or even if the foot descends heavily on the ground, the curious object vanishes far out of the reach of a probing finger. The jets are thrown up by the Solen, and the two little holes are the open extremities of the siphon, that wonderful instrument through which the creature obtains its nourishment.

If a Razor Shell should be required for any ordinary purpose such as baiting fishing-hooks, it can easily be procured by pushing into the hole an iron rod turned up at the end, and twisting the Solen out of its burrow. If, however, a perfect specimen of the animal or shell be required for scientific purposes, it can be obtained by the simple process of dropping a spoonful of salt down the hole. The Solen has a strange hatred of salt, and as soon as the obnoxious substance is felt, up comes the Razor Shell in a hurry, thrusting itself out of the hole, and enabling the operator to seize it before it can again withdraw to its shelter.

THE curious group of molluscs called Gastrochænidæ, deserve a passing notice. All the species of this family are burrowers and some of them are capable of making their way through substances of considerable hardness. A common British species, the FLASK SHELL (*Gastrochæna modiolina*), is notable for its habit of burrowing through various shells, those of the oyster being often perforated and fixed to the creature by some natural cement. In such cases, the animal constructs a flask-shaped case from any loose materials that are within reach, and has by this means earned its specific name of *modiolina*.

In two remarkable genera the shell is very small and is cemented to the base of a long shelly tube, through which the siphon runs. The Clavagella is remarkable for the successive frills that decorate the tube; from three to six of these curious appendages being seen in various specimens. These frills are formed by the orifice of the siphon when the tube is elongated. It is a remarkable fact, that the left valve of the shell is always cemented to the side of the burrow, so that the animal possesses no locomotive power, and in all cases the shell is very small, and sometimes scarcely visible.

The WATERING-POT SHELL (*Aspergillum*), is well known to conchologists. In this creature the shell is exceedingly small, and so deeply sunk into the tube that only the umbo of each valve is visible. The base of the tube is expanded into a rounded and perforated disc, which bears a remarkable resemblance to the "rose" of a watering-pot, and its opposite extremity is mostly decorated with frills, from one to eight in number.

THE reader will remember that the wood-boring pholas always makes its burrow *across* the grain of the timber which it is commissioned to destroy. The SHIPWORM (*Teredo navalis*), on the contrary, always burrows *with* the grain, and never makes a transverse tunnel, unless turned from its course by some obstacle, such as a nail, or the burrow of another Teredo.

At first sight, few would perceive that the Shipworm belongs to the same class as the oyster and the snail, for it is long, slender, and worm-like in shape, from six to eight lines in diameter, and nearly a foot in length. One end is rather larger than the shaft, if we may use the term, and is furnished with a pair of curved and very narrow shell-valves, while the other is divided into a forked apparatus containing the siphon. The colour is greyish-white.

Such is the aspect of the Shipworm when adult, but in its early stages of existence it possesses a totally different form. When it first issues from the sheltering mantle of its parent, it is a little, round, lively object, covered with cilia, like a very minute hedgehog, and, by the continual movement of these appendages, passing rapidly through the water. It does not, however, retain this form for more than six and thirty hours, but undergoes a further process of development, and is then furnished with a distinct apparatus for swimming and crawling. It also possesses rudimentary eyes, and in that portion of the body which may be considered the head, there are organs of hearing resembling those of certain molluscs. When it has passed its full time in this stage of development, it fixes upon some favourable locality, and then undergoes its last change, which transforms it into the worm-like mollusc with which naturalists are so familiar.

SHIPWORM.

The ravages committed by this creature are almost incredible. Wood of every description is devoured by the Shipworm, whose tunnels are frequently placed so closely together that the partition between them is not thicker than the paper on which this account is printed. As the Teredo bores, it lines the tunnel with a thin shell of calcareous matter, thus presenting a remarkable resemblance to the habits of the white ant. When the Teredos have taken entire possession of a piece of timber, they destroy it so completely, that if the shelly lining were removed from the wood, and each weighed separately, the mineral substance would equal the vegetable in weight.

The Shipworm has been the cause of numerous wrecks, for it

silently and unsuspectedly reduces the plankings and timbers to such a state of fragility, that when struck by the side of a vessel, or even by an ordinary boat, large fragments will be broken off. I have now before me two specimens of "worm-eaten" timber, one of which is so honeycombed by this destructive mollusc, that a rough grasp of the hand would easily crush it. Yet this fragment formed part of a pier on which might have depended a hundred lives, and which was so stealthily sapped by the submarine miners, that its unsound state was only discovered by an accident.

The copper sheathing, with which the bottoms of ships are covered, is placed upon them for the express purpose of baffling the Shipworm, and though so expensive a process, is cheaper than permitting the destructive creature to work its own will on the vessel. It is possible, however, that an equally effectual, and very much cheaper method of protecting ships and submerged timber may soon be brought into active operation. M. de Quatrefages has discovered that mercurial salts of any kind are instantaneously fatal to the Shipworm, and that, by their use, not only the existing animals may be killed, but their eggs destroyed also. A vessel that has been attacked by these pests may be rid of them by throwing a few pounds of corrosive sublimate into the dock where she lies, and it would not be very difficult to keep a special dock for the purpose.

The most effectual method, however, of checking the ravages of the Shipworm is, by saturating the timber with corrosive sublimate; a process which is effected by exposing the timber for a long period of time, so as to allow the sap to escape, and then by forcing a solution of the metallic poison into the minute interstices of the wood. This is done in a curiously simple manner, namely, by laying the logs of timber on the ground, introducing a tube into one end, carrying the tube to a height of forty or fifty feet, and then connecting it with a tank filled with the solution. It is, of course, necessary that the timber should be thoroughly seasoned before it is thus treated. M. de Quatrefages suggests that the prepared wood might be sawn into thin planks, which could then be used in the same manner as the copper sheathing now in use.

Another species of the same genus, *Teredo corniformis*, is remarkable for the locality in which it is found. This curious mollusc burrows into the husks of cocoa-nuts, and other thick woody fruits which may be found floating in the tropical seas. In consequence of the locality which it selects for its habitation, it cannot proceed in one direction for any great distance, and is obliged to make its burrows in a crooked form, which has earned for the creature the specific title of corniformis, or horn-shaped. Fossil woods are often found perforated with these burrows.

Destructive as it may be, the Shipworm will ever be an object of interest to Englishmen, inasmuch as its shell-lined burrow gave to Sir I. Brunel the idea which was afterwards so efficiently carried out in the Thames Tunnel. And, though from the alteration of surrounding circumstances, that wonderful monument of engineering skill has not been so practically useful as was anticipated, it has proved of incalculable value as pioneer to the numerous railway tunnels of this and other countries.

The largest species of this curious genus is the GIANT TEREDO (*Teredo gigantea*), which produces a shell more than five feet in length, and three inches in diameter. The substance of the shell is of very great strength, being about half an inch in thickness, radiated in structure, and so hard that when the first specimen was brought to England many naturalists took it for a hollow stalactite.

This creature is a burrower into mud, and was discovered in a very curious manner. In the year 1797, a violent shock of earthquake took place in Sumatra, and caused great upheavals of earth and corresponding floods of water. When the sea receded from one of the bays, certain unknown objects were seen protruding from its muddy bed, and were pulled out with tolerable ease. They projected about eight or ten inches from the mud, and as the projecting portions were beset by serpulæ, bivalves, and other marine parasites, it was evident that they were not forced out of the mud by the shock, but had been in that position for a considerable time. All, however, were damaged, one or both ends being broken off. Their colour was pure white on the exterior, and yellowish within. None of them were perfectly straight, and the greater number more or less contorted.

CHAPTER VI.

BURROWING SPIDERS.

The SCORPION and its habits—The burrow of the Scorpion—How detected—Suicide among the Scorpions—Spiders and their burrows—The Atypus—Madame Merian and her book of the BIRD SPIDER—Mr. Bates' Discoveries—Hair of the Bird Spider—The TARANTULA—Its ferocity and courage—The TRAP-DOOR SPIDER—Its tunnel and the lining thereof—Its appearance under the microscope—The "Trap-door" itself, and its structure—Curious example of instinct—Activity of the Spider—Specimen in the British Museum—Strength and obstinacy of the Trap-door Spider—An Australian Trap-door Spider.

AMONG the burrowers belonging to this order may be reckoned the well-known SCORPION, of which there are several species, resembling each other in their general appearance, their structure and their habits.

Scorpions are found in all the warmer portions of the globe, and under the tropics they may be said to swarm. They are, as a general rule, intolerant of light, creeping by day into every cranny that can shelter them from the unwelcome sunbeams, and often causing very great annoyance by this custom. Old travellers, who have learned by experience the habits of these creatures, do not retire to rest before they have carefully examined the bed and surrounding furniture, especially taking up the pillow, and seeing that no enemy has lodged within the folds of the bedding. The left hand is generally employed in lifting the clothes, while the right is armed with a boot-jack, or stout shoe, or some other convenient weapon, with which the Scorpion may be immolated to the just wrath of its discoverer, before it can run off and hide itself afresh. Shoes, boots, and gloves are also favourite resorts of the Scorpion, which has caused many an inexperienced traveller to buy future caution at rather a dear rate.

Scorpions may be found everywhere, under every stone, and in every crevice; and it not unfrequently happens that when a pedestrian is passing over a sandy bank, and happens to break away a portion of it with his feet, a great black scorpion comes tumbling down, rolling over and over among the sandy avalanche, disengaging itself with an angry snap of its claws and a savage whisk of its tail, and showing fight as if it expected immediate attack from some present enemy. In such cases, the Scorpion has been a true burrower, excavating a temporary dwelling in the sandy soil, and living therein during the day.

The burrows of the Scorpion can always be detected by the peculiar shape of the entrance, which is of a semilunar form, exactly fitting the outline of the animal which digs it. The shape of the aperture is not unlike that of the hole which is cut in the seats of wooden stools for the purpose of introducing the hand when they are lifted. Wherever the soil is suitable for their purpose, the Scorpions take every advantage of it, so that a great number of these venomous creatures may be found in a comparatively small space of ground. Captain Pasley, R.A., tells me that, while in India, he has often destroyed, in the space of an hour or so, more than forty Scorpions, which had dug their sandy burrows in his garden.

The semilunar shape of the entrance is an infallible indication of the inhabitant, and in order to find out whether the Scorpion is at home, a jug full of water is poured into the burrow. Scorpions detest water, and when they feel the stream pouring upon them, they issue from their holes in high dudgeon, their pincers preceding them and snapping wildly at the enemy. A fork or spade is then driven under the Scorpion, and its retreat being thus cut off, it is easily killed.

The same officer also mentioned, that he had repeatedly tried the experiment of surrounding the Scorpion with a ring of fire, and that it had invariably stung itself to death. The fiery circle was about fifteen inches in diameter, and composed of smouldering ashes. In every instance the Scorpion ran about for some minutes, trying to escape, and then deliberately bent its tail over its back, inserted the point of its sting between two of the segments of the body and speedily died. This experiment was repeated seven or eight times, and always with the same results, so that a further repetition would have been but a useless cruelty. The heat given out by the ashes was very trifling, and not equal to that which is caused by the noontide sun, a temperature which the

Scorpion certainly does not like, but which it can endure without suffering much inconvenience. Generally, the Scorpion was dead in a few minutes after the wound was inflicted.

MANY of the true spiders are among the burrowers, and, even in our own country, it is possible to see a sandy bank studded with their silk-lined tunnels.

There is such a bank that skirts a fir-wood near my house, the material being the loosest possible sandstone, scarcely hard enough in any place to resist a pinch between the fingers and thumb. About an inch or two above the soil, this sandstone is quite excavated by the spiders, and as the sandy sides of their tunnels would fall in were they not supported in some manner, every tunnel is carefully lined by a coating of tough webbing, very strong, very elastic, very porous, and yet not suffering one particle of sand to pass through its interstices. From the opening of each burrow a web is spread, looking very much like a casting net, with a hole through its middle. From this again, radiate a number of separate threads, which extend to a considerable distance from the entrance.

At the very bottom of its silken tunnel the living architect lies concealed, its sensitive feet resting on the web, so that it is enabled to perceive the approach of the smallest insect that crosses the spot which it has so elaborately fortified. It is curious to watch the various insects that are caught by different species of spiders. The common garden spider (*Epeira diadema*) enjoys the greatest variety of diet, and the water spider, of which we shall see something in a future page, is also capable of varying its food to a considerable extent. The Burrowing Spiders, however, of which there are several species, are much restricted in their diet, the chief food that is found in their webs consisting of small beetles and midges. These spiders belong to the family Agelenidæ.

ONE of the best, if not indeed the very best, examples of the British burrowing Arachnida is the remarkable species, *Atypus Sulzeri*, a creature which is so rare as to have received no English name. It is a small species, not half an inch in length, but it is a curiously-constructed being; and were it made on a larger scale, would be a really formidable species. Its jaws are long, sharply pointed, and remarkably stout at their bases—so stout, indeed, that, but for a remarkable adaptation of structure, it would not be able to see anything in front.

None of these spiders have a separate head, that part of the body and the thorax being fused together, and forming what is called by naturalists a "cephalothorax," *i.e.* a head-thorax. The same structure may be observed in the scorpion, and also in the common lobster, the shrimp, and other crustacea. The eyes, as in all spiders, are rather close together, and are placed upon the upper part of this cephalothorax; but so large are the bases of the jaws, that they rise far above the level of the cephalothorax: and if the eyes were placed in the ordinary manner would act like the "blind" that is hung over the eyes of a bad-tempered bull. In order, however, to enable the spider to see objects in its front, a sort of little turret rises from the cephalothorax, and on its summit are placed the eyes. Naturalists familiarly call this projection the "watch-tower."

This spider inhabits moist situations, and burrows into the banks, the direction of the burrow being at first horizontal and then sloping downwards. It is lined with a remarkably compact silken tube, beautifully white, and about half an inch in diameter. The upper part of the tube is rather larger than the lower, and projects from the earth, falling forward so as to form a flap, which protects the mouth of the burrow. Specimens of this remarkable spider have been obtained from several parts of England.

NEARLY one hundred and fifty years ago, Sibilla Merian published her famous account of the insects of Surinam, wherein are several statements that were first received without scruple, afterwards doubted, and finally disbelieved. The most important of these controverted statements was that in which she mentioned that the gigantic spiders of Surinam catch the humming-birds, kill them, and suck all the juices out of their bodies. This statement appeared to be of so wild a character, that naturalists might well be pardoned for refusing credit to it, especially as Madame Merian did not offer herself as an eye-witness, but merely related the story on the authority of the natives.

There is certainly nothing in the comparative sizes of the two creatures which would render such a feat impossible, for the spider has a body nearly as large as that of a sparrow, and its expanse of limb is seven or eight inches, while the humming-bird is scarcely bigger than the common humble bee of our gardens and fields. Still, it did seem so strange that a spider should attack a bird, that, failing a credible eye-witness, the story was not believed. That want, however, has been recently supplied, for Mr. H. W. Bates, who spent eleven years upon the banks of the Amazon River, has been an eye-witness to the murder of a small bird by a great spider, and the question is now finally set at rest.

"In the course of our walk, I chanced to verify a fact relating to the habit of a large hairy spider, belonging to the genus Mygale, in a manner worth recording. The species was *M. avicularia,* or one very closely allied to it; the individual was nearly two inches in length of body, but the legs expanded seven inches, and the entire body and hair were covered with coarse grey and reddish hairs. I was attracted by a movement of the monster on a tree-trunk; it was close beneath a deep crevice in the tree, across which was stretched a dense white web. The lower part of the web was broken, and two small birds, finches, were entangled in the pieces. They were about the size of the English siskin, and I judged the two to be male and female. One of them was quite dead, and the other lay under the body of the spider, not quite dead, and was smeared with the filthy liquor or saliva exuded by the monster. I drove away the spider, and took the birds, but the second one soon died. . . . I found the circumstance to be quite a novelty to the residents hereabout."

One of these spiders, kindly presented to me by Mr. Bates, is now before me, and after examining the terrible fangs as they lie folded under the head, and the enormous power of the long, clinging legs, I believe that a small bird would stand a very poor chance of life if once entangled in the fatal clutch. There are several species of Mygale, some of which are great burrowers, making holes of considerable depth. One species, *Mygale Blondii,* which is easily known by the yellow stripes which run down its limbs, is an admirable burrower, digging tunnels of two feet in depth, and rather wide, and lining them with a silken coating, so as to prevent the earth from falling in. In the evening, the spider may be seen at the mouth of its hole, evidently watching surrounding events, but as soon as it perceives an approaching footstep, it pops back into the dark recesses of the tunnel, and will not make its appearance for some time afterwards. Others live under stones; and others, again, make their dens in the thatched roofs of houses. The natives do not seem to entertain any feelings of abhorrence towards these creatures, which to an European mind are so repulsive; for Mr. Bates once saw a group of children amusing themselves with a gigantic Mygale, which they had secured by tying a string round its waist, and were leading about as if it had been a dog.

While living, the Mygale sheds its hairs very easily; and as these hairs penetrate the skin and are of a painfully irritant character, like those of the palmer-worm and other British caterpillars, the incautious naturalist is apt to buy his experience of the Mygale rather dearly. The natives call these creatures "Aranhas carangueijeiras," or Crab-spiders, because they are so strong and so large.

Several large spiders that live mostly upon the ground are confounded together under the general name of Tarantula. There is scarcely a part of the world where is not found some great Lycosa, or Wolf-spider, that is popularly called by the dreaded name of Tarantula, and feared lest its bite should produce the disease which was once so rife through Europe, and called Tarantismus. These are all more or less burrowers, and line their tunnels with a silken coating, so as to prevent the earth from falling in upon them. Some of them hunt about after prey, while others sit at the entrance of the den and wait for the approach of any passing insect, which they may seize and devour at their leisure in the safe retreat of the neighbouring burrow. In this tunnel their young are hatched, and, as soon as they can struggle themselves free from the egg, they clamber upon their mother's back, and there cling in heavy clusters, often hiding her shape by their numbers.

One species of spider that goes by the name of Tarantula is resident in Siberia, and hides in holes in the ground. The peasantry are greatly afraid of it, fancying that it will bite them, and that its bite will cause great injury. For their terrors there are really some grounds, inasmuch as the spider is a savage kind of creature; and if a knife be pushed into its den, it will rush out in a fury, and try to bite the blade. In all probability, how ever, it is not very venomous, for it is actually eaten by sheep as they graze.

Of all the burrowing spiders, however, none is so admirable an excavator as the TRAP-DOOR SPIDER of Jamaica, and none displays so much ingenuity in the arrangement of its burrow. Specimens of both the tunnel and the spider are now before me, and it is impossible to inspect them without admiration. When removed from the earth which surrounded it, the silken tube is seen to be double, the outer portion being thick, deeply stained of a ruddy brown, and separated into a great number of flakes, lying loosely upon each other. This outer covering is so thick, harsh, and crumpled, that it looks more like the rough bark of a tree than a spider's web, and its true nature would hardly be recognised even by the touch. The exterior of a common wasp's nest bears some resemblance to this part of the tube. Beneath this covering is an inner layer of a very different character. This is uniformly smooth to the eye, and of a silken softness to the touch. It is but slightly adherent in places to the outer tube, and can be separated without any difficulty and without injuring the one or the other.

The texture of the interior surface is quite unlike that of the inner or outer tube, being nearly white and of a smoothness and consistency much resembling the rough and unsized paper on which continental books are usually printed. It is curiously stiff also, and is so formed that no one who saw it for the first time would be likely to guess at its real character. The microscope, however, reveals its true character at once. If the interior of the tube be submitted to a moderately low power, say from thirty to forty diameters, a curious sight is presented to the observer. The surface looks like very rough felt, covered with little prominences, and composed of threads twisted together without the least apparent order. The threads are very coarse, in comparison to ordinary spider-web, and seem to be stiff, as if covered with size or gum.

The entrance of the tube is guarded by the "trap-door," from which the spider takes its name. This is a flap of the same substance as the tube, circular in shape, so as to fit the orifice with perfect accuracy, and attached to the tube by a tolerably wide hinge, so that when it closes it does not fall to either side,

THE TRAPDOOR SPIDER AND NEST.

but comes true and fair upon the opening which it defends. The inner surface of the trap-door is white and felt-like, and exactly resembles the interior of the tube, but its outer surface is covered with earth, taken from the soil in which the hole is dug. As the trap-door is flush with the surface of the ground, it is evident that, when it is closed, all traces of the burrow and its inhabitant are lost.

The spider is urged by a curious instinct to make its tunnel in some sloping spot, and to keep the hinge uppermost, so that when the inhabitant leaves its home, or retreats to the extremity of its burrow, the door closes of its own accord, and effectually conceals it. New-comers into the country which the Trap-door Spider inhabits are often surprised by seeing the ground open, a little lid lifted up, and a rather formidable spider peer about, as if to reconnoitre the position before leaving its fortress. At the least movement on the part of the spectator, back pops the spider, like the cuckoo on a clock, clapping its little door after it quite as smartly as the wooden bird, and in most cases succeeds in evading the search of the astonished observer, the soil being apparently unbroken, without a trace of the curious little door that had been so quickly shut.

In the British Museum there is one of these tubes, which tells a curious story, and shows that the spider which made it had chosen cultivated ground for its residence. About three inches from the mouth of the tube, there is a tough, leathery flap, the object of which is not very apparent. A closer examination shows that this flap was once a trap-door, and that the spider had lengthened its cell and made a second door at the new entrance. This fact proved that the ground had increased in thickness since the spider completed its habitation, and that the addition to the surface was very rapid, for the spider is not remarkable for longevity, and yet, in its short life, three inches of soil had covered the entrance of its silken cell. Evidently the creature had burrowed in cultivated soil, most probably in a garden, and in process of tilling the ground, a spadeful or two of earth had been thrown over the trap-door. Being thus imprisoned, the spider had no other resource but to push its way through the earth, lengthen its tube, and make another door level with the new surface.

The spider itself is an odd-looking creature, with rather short, but very powerful legs, and a most formidable pair of fangs. These fangs are notable for the fact that their bases are furnished with a series of sharply-pointed barbs. This peculiarity has given to the spider the generic name of Cteniza, this title being derived from a Greek word signifying a comb. The abdomen is very large, round, and firm, and from its tip project the spinnerets, by means of which the silken tunnel is made. Altogether, it has so crustacean an aspect, that, in common with many other species, it is called by the French the Crab-spider. The length of the specimen now before me is about an inch and a quarter, exclusive of the legs.

It is nocturnal in its habits, and during the night it leaves its burrow and hunts for prey. Insects of various kinds fall victims to this spider, and at the bottom of its tunnel may be found the relics of its feast, often including the remains of tolerably large beetles. If, when it is within its home, the lid be lifted gently, the spider hastens to the entrance, hooks its hind legs to the silken lining of the lid, and the fore legs to the side of the tube, and resists with all its might. Some writers have averred that it employs the curved fangs for this purpose, and that the comb-like array of barbs is useful in giving it a stronger hold ; but a very slight examination of the spider will show that such an action would be impossible, and that even if the fangs were hitched into the silk, the barbs would have no effect whatever on the firmness of the hold.

Nothing short of actual violence will induce the Trap-door Spider to vacate the premises which it so courageously defends. It will permit the earth to be excavated around its burrow, and the whole nest to be removed, without deserting its home ; and in this manner specimens have been removed and placed in positions where their proceedings could be watched. Some few months ago, several examples of the Trap-door Spider and its nest were to be seen in the reptile-room of the Zoological Gardens. Boldly as the spider guards its home, and energetic as it is while engaged in defence, it is no sooner removed from the burrow than it loses all its activity, remains fixed to the spot as if stupified, or, at the best, walks languidly about without appearing to have any definite object in view.

Trap-door Spiders inhabit many parts of the world. In the British Museum is a curious specimen of a nest, which is furnished with two doors, one at each end. The door of one end is rather loosely and irregularly made, as is, indeed, the whole end of the nest; but, at the other extremity, the door is beautifully rounded, very smooth, and fitting with astonishing neatness into the aperture. This curious specimen was discovered in Albania, and presented by W. Wilson Saunders, Esq.

The gem of the collection, however, for accuracy and finish, is one that is the work of an Australian spider, and was found at Adelaide. Only the upper part of the tube is preserved, so as to show the valve which closes it; but no one who really takes an interest in natural history can pass this nest without pausing in admiration. The workmanship is wonderful, and the hole, with its cover, looks as if it had been made in clay, by means of the potter's wheel, so regular and true are its outlines. The hole itself is circular, but the door is semi-circular, the hinge extending across the middle of the aperture.

Two points in this door are specially worthy of notice, the one being that its edge, as well as that of the aperture, is bevelled off inwards, so that the accurate closure of the entrance is rendered a matter of absolute certainty. The second point is, that the outer surface of the door, together with the surrounding earth, is ingeniously covered with little projections, so that when the door is closed, the line which, on smooth ground, would have marked its presence is totally hidden. The shape of the door, too, is remarkable. Towards its hinge it is comparatively thin, but upon the edge it is very thick, solid, and heavy, so that its own weight is sufficient to keep it firmly closed. The "hinge," to which allusion has frequently been made, is not a separate piece of workmanship, but is a continuation of the silken tube which lines the tunnel. An exact imitation of its principle may be made by taking the cover of a book, and cutting it across from the inside, until all its substance except the cloth or leather is severed, and then bending the two portions back. The cloth or leather will then form a hinge precisely similar to that of the Trap-door Spider, the pasteboard taking the place of the earthen door.

CHAPTER VII.

BURROWING INSECTS.

HYMENOPTERA.

The SAÜBA ANT and its habitation—Use of the "parasol" leaves—Mr. Bates account of the insect—Enormous extent of the Dwelling—The DUSKY ANT—Its Strength and Perseverance—Man and insect Contrasted—The BROWN ANT—Form of its Habitation—Regulation of Temperature—Necessity of Moisture—How the Ant constructs Ceilings—Mining Bees—The ANDRENA and its burrowing Powers—The EUCERA—Its Habitation and curious method of liberating the Antennæ—The SCOLIA, its Burrows and its Prey—The INDIAN SPHEX and its Ingenuity—The MELLINUS and OXYBELUS—Curious method of Catching Prey—The PHILANTHUS, its Burrow, and the Food of its Young—The HUMBLE BEE—Its general Habits—Locality of its Dwelling—Development of the Young—The LAPIDARY BEE, its Colours, Disposition and Habits—The WASP—Its Food and Habitation—Materials and Architecture of the Nest—Disposition, Form, and Number of the Cells—Biography of a Queen Wasp, and History of her Nest—Other British Wasps and their Homes—The MONEDULA and its Prey—Boldness of the insect, and its Uses to Travellers—The BEMBEX—Its energetic Habits, its Food, and Mode of Storing the Nest.

THE burrowing INSECTS now come before our notice.

Of these creatures there is much store, for, indeed, the greater number of insects are wholly or entirely burrowers at some period of their existence. It frequently happens that the very insects which we most admire, which are decorated with the most brilliant colours, and which soar on the most ethereal wings, have passed the greater portion of their lives as burrowers beneath the surface of the earth.

Take, for example, the well-known Mayfly, or Ephemera, so called because its existence was once thought to be comprised within the limits of a single day. How delicate are its gauzy wings; how wonderful are the iridescent tints which play over their surface with a changeful radiance, like that of the opal or the pigeon's neck; and how marvellous is the muscular power which enables the new-born being to disport itself in the air for a period which, in comparison with our own lives, is equal to at least forty years! It never seems to weary. It wavers up and down, up and down in the air, together with myriads of its companions, and for the greater portion of its terrestrial existence is an inhabitant of the air; yet its life has not altogether been spent in amusing itself, for it has passed an existence of some three years or more hidden from human gaze.

To-day it is a bright denizen of the sunbeams, exulting in its beauty, and dancing in very rapture in the air; yesterday, it was a denizen of the mud, a slimy, crawling, repulsive creature, breathing through the medium of the water, and feeding greedily upon any prey that might come within its reach. Yesterday, had it been removed from the water and laid in the sunbeams, it would have died as with poison, and in an hour would have been reduced to a dry and withered semblance of its former self; to-day, were it to be plunged beneath the waters, it would quickly perish, and be shortly eaten by its former companions. For it is fitted for a higher position and a purer atmosphere, so that the element which but a few hours ago was its very life, has now become a present death, and the food in which it so lately revelled can no longer be received into that etherealized form.

So is it with many other insects. Some of our most tender and downy-plumed moths, whose exquisitely delicate raiment is destroyed by a touch, have entered upon their winged state while in the bowels of the earth, and have made their way through the soil without losing a single feather of the myriad plumes with which their bodies and wings are covered. Flies, too, whose slender bodies and light gauzy wings always excite our wonder, that a thing so light should contend with the world, have passed the greater part of their lives in some dark hole, where the fresh air never entered, and into which the sunbeams never cast a ray.

WERE this work to be arranged according to the rigid systems of zoological schoolmen, the list of burrowing insects must have been headed by the beetles; but, as the subject of the book is to describe the peculiar dwellings which are needful for the welfare of various animals, a different arrangement is necessary, so that a well-built home takes precedence over a well-developed animal. If we wish to select an order of insects which surpasses every other in the variety and excellence of their burrows, we turn at once to the Hymenoptera, a large and important group of insects, which includes the wasps, bees, ants, sawflies, ichneumons, and one or two other families. The greater number of these insects burrow in the ground; but others are remarkable for their wonderful powers of excavating the hardest wood, and of forming therein a series of beautifully made cells, for the protection of the future brood.

Turn we first to some exotic Ants which inhabit tropical America.

I HAVE felt considerable doubt whether the SAÜBA, or COUSHIE ANT (*Œcodoma cephalotes*), ought to be reckoned among the burrowers or the builders, inasmuch as it makes large excavations below the ground, and raises dome-like edifices on its surface. As, however, the burrows are very much larger than the buildings, I shall place it with the former class, reserving for the corresponding example of the building-insects the Termites, whose edifices are more important than their burrows. It must first be mentioned that, although this species has often been described as the Visiting Ant, it is in reality a distinct species, as will be seen in the course of a few pages.

The Saüba Ant is restricted to tropical America, where it exists in such vast profusion, that it oftentimes takes forcible possession of the land, and drives out the human inhabitants who have cultivated and planted it. Broad columns of these ants may be seen marching along, each individual carrying in its jaws a circular piece of leaf, about the size of a sixpence, which is held vertically by one of its edges. In the British Museum there is a specimen of a Saüba Ant, with the leaf still grasped in its jaws, the ruling passion strong in death. From this curious habit the creature is sometimes called the Parasol Ant, and many persons have thought that the leaves are carried in order to protect the insect against the hot sunbeams. The real reason, however, has been discovered by Mr. H. W. Bates, who has studied with great care the habits of this remarkable insect, and has disentangled its history from many doubts and difficulties.

There are, as is usual with all ants, three distinct ranks —namely, the winged, the large-headed, or soldiers, as they are popularly called, and the ordinary workers. The large-headed individuals are sub-divided into two classes, namely, the smooth-heads and rough-heads, the former wearing a polished, horny, translucent helmet, and the head of the latter being opaque and covered with hair. The large-headed ants do no ostensible work, all the labour falling to the lot of the workers. These creatures make raids upon the trees, always giving the preference to cultivated trees, such as the orange and the coffee, and cut away the leaves so fast that the growth is stopped, and the entire plant sometimes dies.

The use of the leaves is to thatch the domes of their curious edifices, and to prevent the loose earth from falling in. Some of these domes are of gigantic dimensions, measuring two feet in height and forty feet in diameter, the mightiest efforts of man appearing small and insignificant when the comparative dimensions of the builders are taken into consideration. Division of labour is carried out to a wonderful extent in these buildings, for the labourers which gather and fetch the leaves do not place them, but merely fling them down on the ground, and leave them to a relay of workers, who lay them in their proper order. As soon as they have been properly arranged, they are covered with little globules of earth, and in a very short time they are quite hidden by their earthy covering.

The functions performed by the large-headed ants are not very evident. Those with smooth fronts seem to do nothing but walk about. They do not fight like the soldier-termites, nor do they appear to exercise any rule over the workers. Moreover, they have no sting, and even when assaulted they scarcely ever resent the insult. The hairy-headed variety is still more enigmatical in its duties. "If the top of a small, fresh hillock, one in which the thatching process is going on, be taken off, a broad cylindrical shaft is disclosed, at a depth of about two feet from the surface. If this be probed with a stick, which may be done to the extent of three or four feet without touching the bottom, a small number of colossal fellows will slowly begin to make their way up the smooth sides of the mine. Their heads are of the same size as the class No. 2, but the front is clothed with hairs instead of being polished, and they have in the middle of the forehead a twin ocellus, or simple eye, of quite different structure from the ordinary compound eyes on the sides of the head. This frontal eye is totally wanting in the other workers, and is not known in any other kind of ant. The apparition of these strange creatures from the enormous depths of the mine reminded me, when I first observed them, of the Cyclopes of Homeric fable. They were not very pugnacious, as I feared they would be, and I had no difficulty in securing a few with my fingers. I never saw them under any other circumstances than those here related, and what their special functions may be I cannot divine."

The subterranean galleries which these creatures form are of almost incredible extent—so vast, indeed, and so complicated, that they have never been fully investigated. A conjecture as to their size may be formed from the fact, that when sulphur smoke was blown into a nest, one of the outlets was detected at a distance of seventy yards. The Saüba has often done considerable damage to property, having pierced the embankment of a large reservoir, and let out all the water before the damage could be detected.

The winged class is composed of the perfect male and female, which take their departure from the nest in January and February. They are quite unlike the other workers and soldiers, being larger and darker, with rounder bodies and a more bee-like aspect. The female is a really large insect, measuring more than two inches in expanse of wing, and the body being equal in size to a hornet; but the male is much smaller, as is generally the custom with the insect race. Of the hosts which pour out of the nests, only a few individuals remain after a space of twelve hours, the nest having been devoured by birds and other insect-eating creatures. Those which survive address themselves to the founding of new colonies; and so prolific are these insects, that, in spite of the vast destruction wrought among the winged individuals, to whom alone the task of reproduction belongs, man often has to retire before them, and even his art cannot conquer them.

The Saüba is one of the very few ants that does not attack other creatures. The real DRIVER, or VISITING, or FORAGING ANT, of which there are several species, belongs to another genus, Eciton, which will be described among the building-insects.

MOST of the British ants are among the burrowers, hollowing out subterranean abodes of great extent, and constructing them upon some intricate plan, the principle of which is not very evident. The DUSKY ANT (*Formica fusca*) generally prefers banks with a southern aspect, in which it forms its elaborate dwelling. Like many other ants, it is somewhat of a builder as well as a miner, and can raise story upon story, as well as add them by excavation. This task is achieved by covering the former roof with a layer of fresh and moist clay, and converting it into a floor for the next story. Dry weather has the effect of retarding the ants in their labours, because they find a difficulty in procuring sufficient moisture wherewith to mix the clay.

The muscular power and the energy and endurance of the ant are truly wonderful; and if a human being, even if aided by tools, could perform such a day's work as was achieved by a single ant without them, he would be a wonder of the world. M. Huber had the curiosity and good sense to devote the whole of a rainy day to watching the proceedings of a single Dusky Ant. The insect began by scooping out a groove in the earth, about a quarter of an inch in depth, kneading the earth, which it removed into little pellets, and placing them on each side of the groove, so as to form a kind of wall. The interior of the groove was beautifully smooth and regular, and when completed it looked very like a railway cutting, and performed a similar office. After completing this task, it looked about and found that there was another opening in the nest to which a road must be made, and straightway set to work upon a second sunken path of a similar character, parallel to the first, and being separated from it merely by a wall of a third of an inch in height.

Compare the size of an ant with that of a man, and then see how vast are the powers of so small a creature. Taking all the calculations in round numbers, and very much to the disadvantage of the ant, we find that a single man, who would have achieved a similar work in a single day, must have acted as follows:—

He must have excavated two parallel trenches, each of seventy-two feet in length and four feet six inches in depth; he must have made bricks from the clay he dug out, and with them built a wall along each side of the trenches, from two to three feet in height and fourteen or fifteen inches in thickness; and lastly, he must have gone over the whole of his work again, and smoothed the interior until it was exactly true, straight, and level. All this work must also have been done without the least assistance, and the ground must be supposed to be filled with huge boulders, and covered with tree trunks, broken logs, and other impediments.

The most admirable subterranean architecture is perhaps that of the BROWN ANT (*Formica brunnea*), a species which is not very commonly known in this country, and is probably confined to certain localities. Its habitation and the mode of its construction have been carefully noted by M. Huber.

This ant works mostly at night, and during light, misty rain, the sunbeams being obnoxious, and heavy showers causing much inconvenience. The nest is a most complicated structure, composed of a series of stories, often reaching thirty or forty in number, and generally being built in a sloping direction. These stories are not composed of regular cells, like those of the bee, wasp, and hornet, but of chambers and galleries of very irregular form and dimensions, beautifully smoothed in the interior, and about one-fifth of an inch in height. The walls are about the twenty-fourth of an inch in thickness. The object of so many stories is to be able to regulate the heat and moisture of their establishments. If, for example, the sun is not very powerful, and the instinct of the little insects tells them that more heat is required in order to hatch the pupæ which are undergoing their metamorphosis, they take up the white burdens and carry them into the upper chambers, where the heat is greater than below.

Again, if there should be a heavy rain, which floods all the lower stories, nothing is easier for the inhabitants than to remove themselves and brood into the upper sets of chambers, where they will be secure from the inundation. On those days when the sun is peculiarly hot, the ants secure a more equable temperature, by removing the young brood to the central flats, if they can be so called, while they themselves can obtain the needful moisture from the lower parts of the nest, to which the sunbeams cannot penetrate. Were it not for this provision which they instinctively make, all building operations would be stopped during a drought, whereas, by descending to the cellars or crypts of the mansion, the ants can obtain sufficient clay for ordinary work.

In order to watch the ants closer, Huber constructed a kind of vivarium in which they could work, and supplied them with earth, sand, and other necessaries. As, in this artificial state of existence, the insects could not procure moisture from the depths of the earth, moisture from other sources was necessary. Whenever the insects had ceased to work, they could almost always be induced to renew their labours by dipping a stiff brush in water, and striking the hand upon it in such a manner that the water descended like very fine rain upon the earth. As soon as the formerly quiescent ants felt the grateful shower, they regained their activity, ran about with renewed energy, and set to work upon the soil, moulding it into little pellets, and testing each tiny ball with their antennæ before they applied it to the purposes for which it was made.

While some of the ants were engaged in this task, which must be considered analogous to brickmaking as practised by mankind, others were scooping out shallow hollows in the clay floor, the little ridges that were left standing being the foundation of the new walls. On these were dabbed the earthen pellets, and adjusted by means of the mandibles or by pressure of the fore feet, thus receiving compactness and uniformity. The most difficult part of such a task is the formation of the ceiling, but the ants do not appear to be at all embarrassed by so formidable an undertaking, but can lay ceilings of two inches in diameter with perfect certainty. The method of constructing the ceiling is by moulding the clay pellets into each angle of the chamber and also to the top of the pillars. As fast as one row of pellets becomes dry, a second is added; and the insects perform this delicate duty with such accuracy, that although so many centres are employed, the parts always coincide in the proper spots. The peculiar kneading and biting to which the clay pellets are subjected makes them exceedingly tenacious, so that they adhere strongly on the slightest contact.

When once these walls are completed, they are of very great strength, and are only the more consolidated by rain and heat. Before their completion, however, they do not appear to endure the extreme either of heat or moisture, and are taken to pieces by the little architects if a drought should check the supply of that moisture, without which the work cannot be properly compacted.

Mr. Rennie, who followed up the observations of Huber, makes the following remarks on the nest of the ant:—"On digging cautiously into a natural ant-hill, established upon the edge of a garden walk, we were enabled to obtain a pretty complete view of the interior structure. There were two stories, composed of large chambers, irregularly oval, communicating with each other by arched galleries, the walls of all which were as smooth as if they had been passed over by a plasterer's trowel.

MYRMELEON. AMPULEX. SCOLIA.

The floors of the chambers, we remarked, were by no means either horizontal or level, but all more or less sloped, and exhibiting in each chamber at least two slight depressions of an irregular shape. We left the under story of this nest untouched, with the notion that the ant might repair the upper galleries, of which we had made a vertical section; but instead of doing so, they migrated during the day to a large crack, formed by the dryness of the weather, about a yard from their old nest."

This description is accompanied by a sketch of a portion of the dwelling. Five chambers are shown, two large and three small, communicating with the gallery by very short corridors.

It is a noteworthy fact, that the ant will always avail itself of any accidental circumstances that may assist it in building. For example, one of these industrious little beings has been observed to take advantage of some straws that happened to cross one another, and to convert them into beams, wherewith the ceiling could be supported. It began the work by depositing the little clay pellets in the angles formed by the straws, and then laid several rows of the pellets along the sides of each straw. The ceiling rapidly grew under the jaws and feet of the ant, and on account of the extemporized beams, was necessarily of much greater strength than those which were constructed in the usual manner.

The common YELLOW ANT (*Formica flava*), so abundant in marshes and gardens, is also a good burrower, though its habitation is not so large or so elaborate as that of the Brown Ant. This species is very fond of making its subterranean houses under stones or similar substances, and I have found hundreds of the nests under flat stone tiles that had once been employed in edging the walks of a large kitchen garden, and had been pressed aside or sunk flat upon the earth. It is a curiously sociable species, for it is often found occupying one side of a little hillock, while another species of ant, *Myrmica scabrinodis*, has possession of the other. This latter species is sometimes extremely abundant, and it is a rather remarkable fact, that some of our rarest British beetles are only to be found in the nests of the ants.

AS is well known, the ants do not retain their wings for any lengthened period, and after these members have served the purpose for which they were intended, they are broken off by the insect by means of a transverse seam near the base. There are, however, many of the permanently winged hymenoptera which possess very great powers of burrowing, and are able to excavate soil so hard that a knife can scarcely make its way through the solidly impacted mass of earth and stones.

The mining bees, which belong to the genus Andrena, are admirable burrowers, and in spite of their small size, drive their little tunnels into the earth with astonishing ease. I once came on a whole colony of the Andrena, in a peculiarly hard and stony path near Dieppe. The ground was full of little holes, from which the bees were continually issuing, and into which others were as continually passing; their bodies yellow with the pollen of the flowers which they had been rifling, and which was intended to serve as a provision for the future brood.

An ordinary pocket-knife could make no impression on the ground, mixed as it was with stones, trodden by daily traffic, and baked by the heat of summer, into a mass nearly as hard as brick, harder perhaps than the bricks that are employed for modern houses. I was obliged, therefore, to return to my room and fetch a great, rude, thick-bladed clasp-knife that was reserved for rough work, and with much labour succeeded in tracing several of the burrows. They were sunk, on an average, about eight inches into the ground, and near the end they took a sudden turn, and were ended by a rounded chamber, in which was almost invariably a ball of pollen about as large as a pea. No larva was found in any of the burrows. The whole of the labour falls upon the female, the fore-legs of the male being unable to dig, and the hind-legs unable to carry the pollen.

The genus Andrena is of enormous extent, for in 1855, not less than sixty-eight acknowledged species had been discovered in England, and the number is probably increased after a space of nearly ten years.

One of the most interesting members of this family is the pretty insect known by the name of *Eucera longicornis,* and believed to be the only British representative of its genus. The name longicornis, or long-horned, is derived from the very long antennæ of the male, which is also remarkable for a notch on the first joint of the fore-legs. The use of this notch we shall presently see. Like the bees which have just been mentioned, the Eucera digs a rather deep burrow, but prefers a clay soil. The extremity of the burrow is widened into an oval cell, the walls of which are beaten and pressed by the insect until they are quite hard. The reason for this precaution is, that the cell is stored with a mixture of honey and pollen, which is of a semi-fluid consistency, and would be absorbed by the earth if the walls of the cell were not "puddled," as engineers call the operation.

Within the cell is placed the egg, and in due time the larva is hatched, and feeds on the soft sweet mixture with which it is surrounded. It then changes into the pupal condition, and is remark able for being enveloped in a very thin pellicle, something like the slough of a snake. Even the antennæ are enveloped separately in the pellicle, and the male would find great difficulty in divesting itself from the membrane, were it not for the notch in the fore-legs. As soon, however, as it is partially free, the insect bends down its head, lodges successively each antenna in the notch, closes the joint upon it, and then, by drawing the antenna through the notch, strips off the pellicle with perfect ease.

Among these insects, the females are treated much as the wives of savages are treated. All the work falls to their lot, and the males do nothing but amuse themselves, circling about the nests in graceful undulations, while the females are hard at work, digging the burrows and fetching home the food. Still, there is no doubt but that this disparity is only in appearance, and that the one sex feels as much enjoyment in following the instinct which teaches her to dig, as does the other in following the instinct which teaches him to fly about.

At the right-hand side of the illustration on page 128 may be seen a figure of a remarkable burrowing bee, called *Scolia flavifrons*, a native of Europe, but not as yet proved to be British. In common with other fossorial bees, this insect is carnivorous in its larval state, and is supplied by its mother with the creatures on which it feeds.

Some bees feed upon larvæ, others upon full-grown insects. Some eat beetles, some devour bees, some prefer spiders, and others flies, while a very great number of species are caterpillar-eaters, and are in consequence extremely useful to the gardener and farmer. This particular insect has a curious predilection, and stocks its nest with the grub or larva of a beetle, belonging to the genus Oryctes. At the bottom of the cell may be seen certain grubs, the smaller of which is the larva of the Scolia, and the larger that of the beetle. As may be seen from the illustration, the grub of the beetle is very much larger than that of the creature which feeds upon it. The species which is here represented is a large and remarkably striking one, the four conspicuous spots at once distinguishing it from any other insect.

In the middle of the illustration another example of a bee-burrower is given, in order to show the manner in which the insect takes its prey into the nest. The technical name of this species is *Ampulex compressa,* and its nest is stocked with cockroaches, one of which is being dragged into the hole, wherein it will be shortly eaten by the inhabitant.

As space is valuable, I will merely give the names of our most conspicuous burrowing bees, together with a brief notice of their habits.

All the species belonging to the genus Pompilus are burrowers, and stock their nests with spiders. Sandy soils are favoured spots with these bees, some species preferring the dry, hard sand-banks, while others choose soft and loamy sand for the site of their habitation. The Sand Wasps, belonging to the genus Ammophila, are always mighty burrowers, and set about their task with a fiery zeal that never fails to excite the admiration of the spectator, their antennæ quivering and their wings flirting with excitement. When the burrow is completed, the mother insect flies off in search of a caterpillar or spider, according to the species, and conveys it to the bottom of the tunnel, where a small chamber is excavated.

She always enters the burrow backwards, grasping her prey in her jaws and dragging it after her. It is so large that she can scarcely force it along the tunnel, and were it not for the comparatively wide chamber, she would not be able to make her way out again. When she has fairly lodged it in the chamber, she creeps round it, deposits an egg upon it and crawls out again, taking care to stop up the entrance with some small pebbles. She then flies away in search of a fresh victim, and after some four or five caterpillars have been placed in the nest, she closes the entrance carefully, flies off and dies, the great duty of her life being then at an end.

There is an allied insect residing in India, which measures about three-quarters of an inch in length, and is of a fine polished green colour. Scientifically, it is known as Sphex scutigera. The habits of this insect have been carefully watched by Sir J. Hearsey, K.C.B., who gave me much information respecting the method by which it prepares a habitation for its young. This species preys upon large spiders and cockroaches, and sometimes displays a wonderful amount of ingenuity in achieving its object.

One of these insects had captured a spider, which was too heavy to be carried through the air. The Sphex then dragged it to a little bank, dropped it into the water, and perched upon it, sitting there until it had been carried some distance down the stream. Finding that the spider was sinking, the Sphex left it, and sat on a straw, which was floating down the stream, still, however, keeping company with its prey. After a while, the spider struck against the shore, and the Sphex then grasped it afresh, and tried to drag it along. The steep bank, however, baffled all its endeavours, and at last the industrious creature was obliged to leave the spider on the ground, and to go off in search of another.

The dark and sombre little bee called *Mellinus arvensis* is an excellent example of the burrowers. This insect preys on various flies, and packs away a large number of its victims in the burrow. The flies which it chooses are all swift of wing, whereas the Mellinus is rather a slow flier, so that it cannot take its prey by open assault, but is obliged to trust to craft. In order, therefore, to obtain its victims, the Mellinus watches some spot where flies most love to congregate, and walks to and fro as if it were quite unconcerned. It continues to run about in this manner until it comes close to a fly, when it springs upon the luckless insect, trounces it in its claws, and carries it off like a falcon with a partridge.

Six or seven flies are generally taken by the Mellinus, and as soon as the larva is hatched, it begins to devour the fly which is nearest to the bottom of the cell. It eats them in succession, usually devouring six of the victims, consuming the softer parts only, and leaving the head, shell of the abdomen, part of the thorax, and the limbs. Ten days suffice for the completion of its feeding, and it then spins a tough, dark-coloured cocoon, wherein it remains during the winter and part of the spring, changes into the pupal condition in the summer, and attains its perfect state at the beginning of autumn.

Another species of burrowers, *Oxybelus uniglumis*, has similar habits. Mr. F. Smith writes of it as follows: "I once observed several females running amongst the blades of grass which shot up from the surface of a little hillock upon which the sun shone and tempted various diptera, occasionally to alight. The Oxybeli continued to run about, apparently unheedful of the flies, until, at length, the latter became somewhat accustomed to their presence; but when the Oxybelus came within five or six inches, it darted upon the luckless fly in the same manner as a cat springs upon its prey." The burrows of this species are generally made in hard white sand.

The boldest of the British bee-burrowers, is undoubtedly the insect which is called *Philanthus triangulum*, inasmuch as it provisions its nest with the common hive-bee, seizing the luckless honey-makers, and carrying them off to its nest. It is a very fierce-looking creature, with a large head and wide jaws, and has a suspiciously waspish look, owing to its yellow abdomen and black dots. It does not confine itself to the hive-bees, but seizes also the andrenæ, and similar insects.

The members of the genus Cerceris are remarkable for the variable colouring of the species, and for the widely different insects with which they store their nests. Generally they prefer beetles, and, strangely enough, they often select those species which are not only small in body, but are furnished with very hard shells, so that the larva would seem to experience some difficulty in making a meal. Some beetles which Mr. Smith found in the cells of the Cerceris, were so hard that he could with difficulty pass a pin through their bodies. Fortunately for agriculturists, the Cerceris generally selects the very beetles which are most injurious to vegetation, such as the various weevils and the turnip-fleas. Mr. Smith is of opinion that the shells of the beetles are softened by the dampness of the ground in which they lie.

In the accompanying illustration are shown the nests of two common species of British Humble Bee.

Both these species are burrowers, and sometimes make their nests at a considerable depth beneath the surface. The common Humble Bee (*Bombus terrestris*) generally makes its subterranean house in the side of some bank, and the nest is usually found at a depth of a foot or eighteen inches. Sometimes, however, in places where the soil is light and friable, the nest has been found at a very great depth from the surface, so that a perpendicular shaft of five feet in length has been required before the nest could be reached. In all probability the bee has been aided by the burrow of a field mouse, when the gallery has been of such a length.

The history of the nest is really a curious one.

At the end of autumn, nearly all the Humble Bees die. The males invariably perish, but one or two of the females survive, and pass the winter in a state of hibernation. They do not select the nest for this purpose, convenient though the locality may seem, but hide themselves away singly in sheltered spots, such as the eaves of thatched barns, hollow trees, haystacks, or old ruins. When the sunbeams of spring gain warmth and strength, the sleepers awaken from their torpor, and immediately search for a spot wherein the new home may be excavated.

BOMBUS TERRESTRIS. BOMBUS LAPIDARIUS

These bees, which are the Methuselahs of their short-lived race, may be seen in any warm spring day, flying about in all directions, prowling over every spare yard of ground, and settling here and there, as if to test the quality of the soil.

They are very jealous of observation at this time, and if they think that they are being watched, will take instant offence and fly off with a quick, eager sound, very different from the steady, monotonous hum with which they accompany their researches. To watch one of these insects in hopes of seeing her begin her labours, is an endless task, for she will never dig an inch of soil as long as she sees any suspicious object, and will often make her way under a thick tuft of herbage, and remain quietly in the retired nook until she fancies that the danger has passed away.

When, however, she has suited herself with a locality, she scrapes away the ground quickly, and when she has dug to a sufficient depth, she scoops out a small cavity or chamber, and therein constructs her first nest. There are but few cells at the beginning of the year, and these contain the first workers, who are intended to assist in constructing the enlarged nest. The larvæ are large, fat, white, round-bodied creatures, with little horny heads, and their bodies always slightly curved. When they have completed their feeding, each spins for itself an oval cocoon of coarse silk, rather irregular in shape, very soft, tough, and thick in consistency.

Herein they remain until they have attained their perfect state, when they gnaw a round piece from one end of the cocoon, just as a chicken chips off the top of the egg, and emerge into the nest. They do not venture out into the air for several days, the thick hair with which they are covered being all matted together, their wings soft and crumpled, and their limbs scarcely able to bear them. Two or three days are generally passed in the nest, and not until having gained their full strength do they venture out into the wide world. None but worker bees are developed for the first part of the year, the females and males not making their appearance until the summer weather has set in.

As may be seen from the illustration, the cells of the Humble Bee are not arranged in regular rows, like those of the hive bee, but are set carelessly side by side, mostly fixed together in groups of greater or lesser dimensions. Now and then a very little group of two or three cells is found, and single cells are occasionally to be seen, detached from the general mass.

This species is more prolific than any other, and the nests contain more individuals. Mr. Smith mentions that in one nest he found one hundred and seven males, fifty-six females, and one hundred and eighty workers, making a total of three hundred and forty-three inhabitants. Compared with the numbers that inhabit the hive, this may seem to be but a small amount; but when the reader takes into consideration the fact that the insects are very much larger than the hive bee, and that the cells in which they are hatched and nurtured are not only of corresponding size but are also set very irregularly, so as to occupy a large amount of space, he will see that the cells which produced nearly three hundred and fifty humble bees must have formed very large groups, and that the cavity which contained them must have been exceedingly large in proportion to the size of the excavators.

As far as my own experience goes, there is little danger in unearthing and exploring the nest of the Humble Bee. Opinions differ greatly on this point, some practical observers saying that the bees are dangerous when irritated, and that they execute stinged vengeance on the disturbers of the home; while others report that the insects do not attack at all; and others, again, say that they attack, but that their stings cause so little pain as to excite no fear. I have opened many a nest of the Humble Bee, and never been stung by the inhabitants, though the bare hand might easily have suffered, as it was thrust into the chamber wherein the cells reposed.

The honey of the Humble Bee is peculiarly sweet and fragrant, but not suitable for general consumption, as many persons, myself among the number, are always attacked with severe headache after eating the contents even of a single cell. The colour of all Humble Bees is most variable, especially in the male sex—so variable, indeed, that Mr. Kirby, in his monograph of the British bees, has divided one insect into seven distinct species. As if to add to the difficulty of identifying the different varieties, bees of more than one species will sometimes be found in one nest.

The colours, however, of the present species are generally as follows:—The female is nearly an inch in length, and the general colour is black. The collar is orange-yellow; a band of the same hue is drawn near the second segment of the abdomen; the hinder edge of the fourth segment and the whole of the fifth segment are pale yellow, and the tip of the abdomen is naked.

The worker is barely half the size of the queen, and is marked in a similar manner, except that white hairs are mixed with the yellow. The male is intermediate in size between the female and worker, being about three-fourths of an inch in length. The yellow is brighter than in the female, and the tip of the abdomen is covered with light tawny hairs.

THE right-hand nest in the illustration is that of the Red-tipped Humble Bee of Shakspere, known as the LAPIDARY BEE (*Bombus lapidarius*), which derives its specific name from its habit of making its nest within heaps of stone. This beautiful insect is plentiful in most parts of England, and may be known by the bright orange-red hue which decorates the last three segments of the abdomen. The female and worker of this species are precisely alike, except in their size; the former, which is popularly called the queen bee, measuring nearly an inch from the head to the tip of the tail, while the worker is scarcely half that length. The male is very variable in colour, but is generally black, with thick yellowish hairs upon the face, the fore part of the thorax, and the first segment of the abdomen.

I have always found this species to be fiercer than the preceding, and have more than once been driven away from the neighbourhood of the nest by its rapid and incessant attacks. The sting with which this bee is armed is a very formidable weapon, and the poison which it conveys into the wound is extremely virulent, causing much pain, and leaving a dull, aching sensation for several days afterwards. These symptoms, however, vary according to the individual who is stung, and those which are mentioned are described according to personal experience.

Generally, the Lapidary Bee makes its nest in heaps of stone, sometimes choosing those hillocks of rough stones which are heaped on the sides of roads, awaiting the stone-breaker and his hammer. Sometimes the fallen *débris* of limestone rocks affords a residence for this bee, and, in many instances, it burrows into the ground, and there makes its nest, just like that of the common humble bee.

Eighteen species of the true British Humble Bees are now recognised, all of them social in their habits, but varying much in the localities and form of their dwellings. Several of them will be described in a future page. Meanwhile, it will suffice to mention that the use of scientific terms, when speaking of these insects, is absolutely necessary, the popular mind not having recognised the different species, which are, in consequence, without popular names. Any Humble Bee, no matter what species, is known as a Bumble Bee, a Foggie, a Dumbledore, or a Hummel Bee, according to the peculiar dialect of the locality; and very few persons seem to have any idea that there can be more than one species.

THERE is one well-known and very handsome insect, which is equally disliked by the bee-keeper, the gardener, and the grocer, as it annoys them greatly in their respective callings. This is the common Wasp (*Vespa vulgaris*), which is equally fond of honey, fruit, and sugar; and as it is armed with a potent weapon, is not merely a hateful marauder, but a formidable enemy. The gardener, however, is the least injured of the three, for the Wasp confers upon him some slight benefits, which counteract in some degree the inroads which it makes upon his treasures. It is true that the Wasp is very fond of ripe fruit, and that with an unfailing instinct it prefers the choicest fruits, exactly when they are in their best condition, gnawing holes in them, and spoiling them for the market. Still it is more of a predacious than a vegetable-feeding insect, and kills so many flies that it relieves the gardener of other foes, which, in the end, would be more injurious than itself, inasmuch as the larva endangers not only the fruit but the very life of the plant. It is a strangely bold insect, and has recourse to singular methods of procuring food. In the farming department at Walton Hall, I have seen the pigs lying in the warm sunshine, the flies clustering thickly on their bodies, and the Wasps pouncing on the flies and carrying them off. It was a curious sight to watch the total indifference of the pigs, the busy clustering of the flies, with which the hide was absolutely blackened in some places, and then to see the yellow-bodied Wasp, just clear the wall, dart into the dark mass, and retreat again with a fly in its fatal grasp. On the average, one Wasp arrived every ten seconds, so that the pigsty must have been a well-known storehouse for these insects.

As is well known to every boy who has participated in the delight of taking a Wasp's nest, the habitation of the insect is mostly under ground, and is a marvel of ingenious industry. The shape is more or less globular, and the material of which it is composed is very much like coarse brown paper, though not so tough. If it be opened, a wonderful scene is disclosed; terrace upon terrace of hexagonal cells being arranged in regular rows, and enclosed in a shell of papery substance, some half-an-inch in thickness, which is evidently intended to prevent the earth from falling among the combs, as these cell-terraces are called.

We will now suppose ourselves to be present at the construction of the nest, and, Prospero-like, will see without being seen.

In the early days of spring, a Wasp issues from the place in which it has passed the winter, and anxiously surveys the country. She does not fly fast nor high, but passes slowly and carefully along, examining every earth-bank, and entering every crevice to which she comes. At last she finds a burrow made by a field mouse, or perhaps strikes upon the deserted tunnel of some large burrowing insect, enters it, stays a long while within, comes out again and fusses about outside, enters again, and seems to make up her mind. In fact, she is house-hunting, and all her movements are very like those of a careful matron selecting a new home.

Having thus settled upon a convenient spot, she proceeds to form a chamber, at some depth from the surface, breaking away the soil, and carrying it out piece by piece. When she has thus fashioned the chamber to her mind—for she has a mind—she flies off again, and makes her way to an old wooden fence which has stood for many years, and which, although not rotten, is perfectly seasoned. On this she settles, and, after running up and down for a little time, she fixes upon some spot, and begins to gnaw away the fibres, working with all her might, so eagerly engaged that even were we not invisible we might stand by and watch her proceedings. At last, she has gathered a little bundle of flies, which she gnaws and works about until she reduces them to a kind of pulp, and then flies back to the burrow.

She now runs up the side of the chamber, and clings to its roof with the two last pairs of legs, while with the first pair, aided by her jaws, she fixes the woody pulp on the roof, kneading it until it forms a kind of little pillar. Another and another supply is brought, until this pillar, which is pendent from the roof, like a *paper-maché* stalactite, is completed. The Wasp now begins to form the comb, and at the end of the pillar she places three very shallow cells, of a cup-like shape, not hexagonal, as are the completed cells. In each of these little cups she deposits an egg, and then constructs a roof over them, made from the same material as the cells, but laid in a different manner, the length of the fibres being nearly at right angles to the centre of the proposed comb. More cells are then added, eggs are laid in them, and the roof extended over them.

The eggs that were laid in the first three cells are now hatched, and have produced very tiny grubs, which are always hungry and require much attention. They grow rapidly, and, in proportion to their growth, the parent Wasp adds to the walls of their cells, so that the young grubs are suspended, with their heads downwards, as, indeed, is the custom with very many hymenopterous larvæ. The Wasp proceeds in her task, having all the cares of the nest upon her—the enlargement of the chamber, the building of the nest, the transport of materials, the deposition of the eggs, and the feeding of the ever-hungry grubs.

In due time, however, the oldest grubs cease to feed, spin a silken cover over their cells, and release their parent from further attendance upon them. In the cells they undergo the change to the perfect state, and, after they have passed a short season in retirement, they tear away the silken cover with their jaws, and come forth as perfect Wasps. As soon as they have gained strength to use their limbs, they take the heavy labours upon them, and the work goes merrily on, the mother Wasp having little to do but to deposit eggs in the cells as fast as they are made.

Before very long, the first cell-terrace is completely full, and more accommodation is needed. This is supplied in a very curious manner. Taking the junction point of these cells as the foundation, the Wasps construct several pendent pillars, exactly like the one which has already been described, and, by dint of adding cells to each, they all unite, and form a second terrace, below the first, the distance between them being just sufficiently large to permit the Wasps to cross each other. In this, as in the former terrace, all the mouths of the cell are downwards and their bases upwards, so that the bases of the second terrace form a floor on which the Wasps can walk while feeding the young contained in the first. A third, fourth, and fifth terrace are added in this manner, all alike, the cells being so small that the mother Wasp cannot even put her head into them.

It will be seen, therefore, that, as insects never grow after they have assumed the perfect form, the Wasps which have been bred in these cells must be very much smaller than their parent. They are, in fact, the worker wasps, or neuters, as they are sometimes called, whose entire life is devoted to labour, and who, in fact, are undeveloped females.

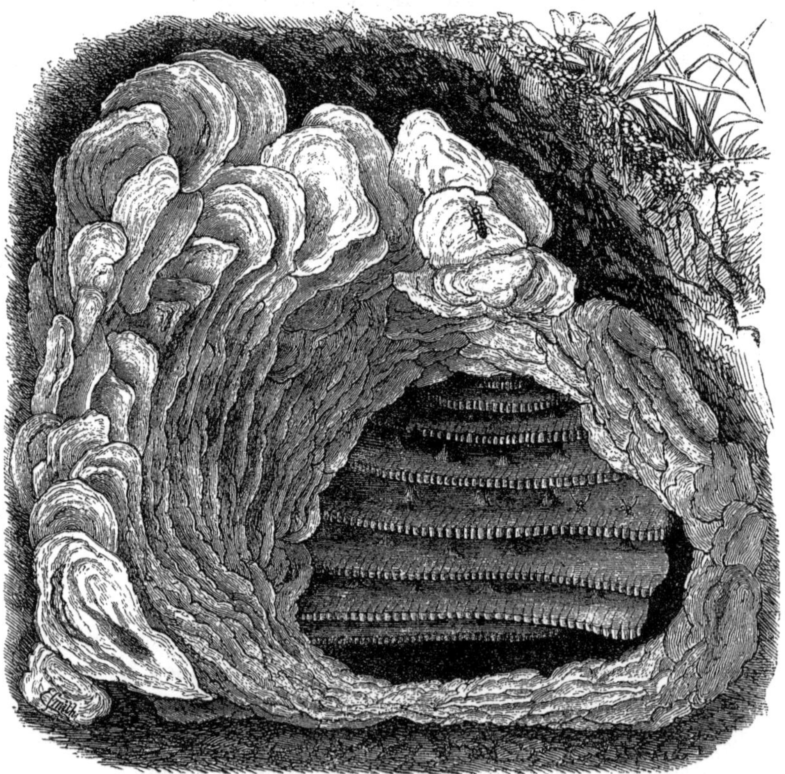

WASP NEST.

Now, however, a change takes place. The cells of which the next few terraces are composed are of very much larger dimensions than the others, and are intended for the purpose of hatching the grubs which will afterwards become perfect male and female wasps. It will be seen, therefore, that the workers are hatched in the earlier part of the year, and that the male and female do not make their appearance until the end of the season. The cell-terraces increase gradually in diameter until the fourth or fifth, when they usually decrease slightly, and in exact accordance with their enlargement the covering is extended over them. A large nest will contain about seven or eight thousand cells; and, on the average, each cell is the birthplace of three generations. As all the young grubs have to be fed with animal substances, usually flies, the reader can easily imagine the havoc which Wasps make in the insect world. Mr. F. Smith, however, who has given so much time and labour to this subject, remarks that as he has never found the cells of the males and females to contain the remains of more than one lining, these cells only accommodate a single brood.

The silken cover is always convex, and draws the mouth of the cell into a rounded form, so that if one of the cells is removed from the comb while the pupa is still within it, the two ends are of very similar form. The Wasps do not break through the cover in the same manner. Sometimes they burst their way through the centre, leaving a rude and rugged opening; sometimes they bite out a circular hole, and push their way through it, tearing the edges as they pass through its substance; and sometimes they cut it neatly round the edge, so that the entire covering can be lifted like the lid of an ancient tankard, and the imprisoned insect is able to emerge without any trouble, the lid closing again as soon as the inmate has escaped.

The covers of the cells are not precisely perpendicular, but radiate slightly from the centre of each comb or terrace. Nor is the flooring precisely flat, for the edges of each comb are slightly raised, so as to form a trifling concavity in the centre. At their mouths, the cells are perfectly hexagonal—those, at all events, which occupy the centre of each comb; but their bases are always cup-shaped, the walls changing gradually into hexagons as the cells increase in height, or, to be more accurate, in depth. When viewed from above, the forms of the bases are plainly perceptible, and they look very like the mosaic teeth in a skate's jaw.

The successive layers of which the cell-walls are composed can be easily seen when the comb is held to a strong side light;

and it sometimes happens that the Wasp finds pieces of paper lying near the nest, bites them to pieces in the same manner that it bites the wood-fibres, and then uses them for its nest. I have seen a nest which was made almost entirely of the blue and white paper used for cartridges, the Wasps having taken advantage of the expended papers, and used them instead of taking the trouble to gnaw hard wood. The covering of the nest is of much rougher texture than the cells themselves, and looks like a number of tiny oyster-shells piled on each other like the "grotto" of metropolitan children. It is made very simply by laying a lump of the fibrous paste upon the nest, and sweeping it backwards and forwards to flatten it, just as a bricklayer spreads a lump of mortar with his trowel. No attempt is made to smooth the surface, and the impression of the little architect's head can be seen upon each successive patch, or tile, if we may so call it.

This woody fibre seems but a flimsy substance for the materials of a nest which can contain so many individuals. In a large nest there are always from two to three thousand inhabitants, more than half that number being the fat and weighty grubs. If the insects were removed from the nest, and placed in a pair of scales, their united mass would be so heavy as to cause a feeling of wonder that so slight a habitation could endure their weight. The walls, however, are stronger than they seem to be, and the hexagonal shape of the cells affords such mutual support that the walls can not only bear the weight of the insects within them, but, as has already been mentioned, are strong enough to uphold a series of cells that are suspended to them.

At the end of the season, after successive bands of worker-wasps have passed through the cells, and the single generation of the males and females has come to maturity, the nest shows symptoms of dissolution. If there are any grubs still left in the comb, the workers at once change their behaviour. Instead of feeding and tending them with jealous care, instead of defending them at the risk of their own lives, they pull these helpless white things out of their cradles, carry them far out of the nest, and abandon them. It seems a cruelty, and so it is; but it is a cruel mercy, substituting a quick death by exposure, or, perchance, being eaten by birds, for a slow and lingering death by starvation within the nest. For the instinct of the workers tells them that their labour is over, and their course is run, and that in a short time they will all die of old age, so that the helpless nurslings in the cells would find no food, and must perish by starvation.

At last, the entire population deserts the nest, the workers die, and so do all the males, none of them surviving their brief wedlock for more than a few hours; and the majority of the females die also, some from exposure to cold, and others by a violent death. Those, however, that are fortunate enough to find a crevice in which they can lie dormant during the long months of winter, creep into it, and there remain until the following spring, when they emerge to be the queens and mothers of future colonies. It is a remarkable fact that the Wasp never passes the winter in the nest, convenient as that spot may seem, but always seeks some other place of refuge. The reader will now comprehend, that whenever a Wasp is seen in the spring tide, it is one of the females which have survived the winter, and is about to found a new colony. Those, therefore, who pride themselves on their wall-fruit will do well to kill such Wasps, inasmuch as a single queen Wasp in spring is equivalent to many thousand Wasps in autumn.

Three species of burrowing Wasps inhabit England; namely, the common Wasp, which has just been described, the German Wasp (*Vespa Germanica*), distinguished from the preceding species by having three black spots at the root of the first ring of the abdomen; and the Red Wasp (*Vespa rufa*), known by the black anchor on the top of the head, and the reddish-yellow limbs.

Mr. F. Smith mentions a very remarkable circumstance connected with the removal of Wasp grubs by the worker. "There is a point which I have more than once witnessed in the history of Wasps, which does not appear to have been recorded. In the spring I have found as many as three nests in a bank, not more than two hundred yards apart; and on visiting the spot a month or so later, I observed no Wasps issuing from the first nest, and, on digging into the bank, discovered that it was deserted, a single empty comb alone remaining. I then passed on to the second nest, and was surprised to observe a few Wasps come out, each carrying something away: at length I captured one, and found it was conveying larvæ from the nest. I traced them in their flight, and was astonished to see a Wasp enter the third nest, with a larva from the nest No. 2.

"Here there is a mystery, which time will unravel, but which I am unable to determine. Were the larvæ carried off to be nursed, and to add to the swarm, or were they destined to become the food of the larvæ of the nest in the freebooter community? I have several times observed that swarms had deserted their nests, but have only once witnessed anything which would in any way account for such a proceeding."

Sometimes the Wasp dispenses with a burrow and becomes a builder, placing its nest on a beam or under a thatched roof. In this case the outer shell of the nest is much more handsome than that which surrounds the subterranean combs, being of a yellowish-brown colour, and the individual flakes of which it is composed being sharply defined. They are more porous than the ordinary grey flakes of the underground nest, and are less capable of resisting moisture. Before concluding this account of the Wasp and its nest, I may mention that the character of the insect has been generally misunderstood. The popular impression is, that the Wasp derives some especial gratification from the act of stinging, is of a savage and malicious disposition, and lives wholly upon the proceeds of theft. Now, in fact, the Wasp never stings, until it is compelled to do so, either by alarm or when it retaliates upon an adversary. It seldom survives the act of stinging, because the secreted point of its weapon is held in the wound, and in many cases the entire poison bag and gland are torn out of the body together with the sting itself. In defence of its home it can be fierce enough, as indeed it ought to be, and cares nothing for its own life, provided that it can only inflict a wound upon the enemy.

The reader may perhaps be surprised to hear that Wasps can be kept as easily as bees, and that, like those insects, they never injure those with whom they are familiar. Indeed, they are even less likely to sting than the hive bees, whose olfactory nerves are so sensitive that they assault any passenger who happens to have been recently smoking, or who has used perfume of any kind. Bees usually treat me very well; but during the last summer, as I was looking at a neighbouring hive, the bees began to dash past me with that peculiar menacing sound which always heralds an attack. Taking warning by the sound, I retired quietly to the further end of the garden, but was followed even there by one pertinacious enemy, who at last made a dash at my face, and passed on, leaving its sting as a memorial of its anger. I afterwards discovered that a handkerchief in my breast-coat-pocket retained a faint scent of eau-de-cologne. Now, if the assailants had been Wasps instead of bees, a bystander would assuredly have considered the attack as a proof of the malignant nature of the Wasp.

BEFORE taking leave of the earth-burrowing hymenoptera, it will be necessary to mention two very remarkable insects which are described by Mr. Bates, in his well-known "Naturalist in the River Amazon." Neither of the insects have any popular name.

The first is called *Monedula signata*, and is a handsome-looking insect, very much resembling an ordinary wasp, and ornamented, like that insect, with bold black marks on the thorax and abdomen. The antennæ, however, are twisted, and at once prove that the creature is but distantly related to the true wasp. The burrows of this insect seem to be made only in the sandbanks which project above the surface of the river; so that they would not be discovered by ordinary travellers. Fortunately for the residents in that part of the country, the Monedula stores its nest with one of the most obnoxious insects that haunt the Amazon river. This is the Motuca fly (*Hadaus lepidotus*), an insignificant-looking creature, smaller than an ordinary house fly, and of a bronze-black colour, with the wings of an ashy brown, except a whitish spot near the tips.

This fly belongs to the well-known family of Tabanidæ, and, like them, is furnished with a very formidable apparatus, by means of which it obtains its food. Whenever the Motuca can attack a human being, it dashes at him, settles, and in a moment drives a broad, sharp-edged lancet through the skin, cutting quite a gash, and causing the blood to flow fast. Fortunately the wound is not very painful, and it is possible that the flowing blood may be useful in washing out any poison that has been injected. It is of a sluggish nature, and can be easily taken with the fingers—a very happy circumstance, inasmuch as a dozen Motuca flies may often be seen clustering upon the ankle, just above the shoe.

The Monedula destroys multitudes of the Motuca flies and will travel for half a mile in order to procure its prey, always taking care to close the entrance of its burrow, and to reopen it on its return. Mr. Bates mentions that he has been frequently indebted to the Monedula for saving him from sundry gashes from the Motuca; and that the Monedula would charge straight towards his face or neck, pick up a Motuca as it was about to settle, and fly off with it. The fly was not captured with the jaws, but seized in the first and second pairs of feet.

The other burrower is that which is known to entomologists as *Bembex ciliata,* and is remarkable for the eager assiduity with which it plies its labour of love. In colour, it is shining green; and when it has fixed upon a suitable spot for its burrow, it scratches away the sandy soil with such furious haste, that a nearly continuous fountain of sand is thrown up behind it. Even after it has penetrated for two or three inches into the ground, the sandy stream issues from the orifice, propelled as if by a miniature engine, and being flung under the body by means of the powerful fore-feet with their bristly armature.

When it has completed its tunnel, which is always driven in a slanting direction, and from two to three inches in depth, it emerges from the orifice, walks about for a time, as if to take bearings of the locality, and then darts off and is lost to sight. After a while it returns, bringing in its grasp a fly, which is destined to be the food of the young Bembex. Only one fly is placed in each tunnel, and then the entrance is carefully stopped up with sand, so that it cannot be distinguished from the surrounding soil. It is a remarkable fact that, however many nests may be made in a sand-bank, and however closely they may be set, the insect which dug them never mistakes another dwelling for its own, and always flies directly to the spot which it has selected as the cradle of its posthumous offspring.

Although the ants have been postponed to a future page, a few words must be given to the insect whose nest will be seen on the right hand of the plate.

This insect is the *Formica compressa* of India. Now and then, the nest of this species is above ground, and is made of mud.

fastened to trees and leaves; but as in ninety-nine cases out of a hundred its home is subterranean, it will here be considered as one of the burrowers.

The section of the nest which is presented to the reader will give a very good idea of its general structure. There are generally some five or six entrances to the nest, but they are so ingeniously hidden under stones, clods of earth, and any object which can shelter them, that they would not be detected by a casual passenger. The few upper passages or galleries are extremely irregular, often having a zigzag direction, and being of no very great length. Those at a greater depth, however, are much more regular in their structure, and when they are driven at some three or four feet from the surface, they are large in diameter, cylindrical, and extend to a considerable distance. In the nest of the British species, *Formica fusca*, there is a somewhat similar structure; and although the ant is so small, these tunnels are sometimes an inch in diameter, and five feet or even more in length. Into these deep-set galleries, the tropical ants retire during the rainy season, and in our own country the insects may be found in them throughout the cold months of winter.

Near the surface of the ground, the reader may observe several enlargements of the galleries, forming spacious chambers. In these chambers the ants are accustomed to lay the white pupæ as well as the eggs, in order that they may be warmed by the sun, without enduring the full fury of his beams. At night, if rain should come on, the vigilant workers take up their helpless charges, and convey them to hiding-places far beneath the surface. If, during the months of April or May, the nest of the Dusky Ant be opened, a very curious state of things will be disclosed. Within the chamber may be seen a vast mass of pupæ and their attendant ants; and, what is still more remarkable, specimens of certain beetles may also be found in company with the ants.

There are several species of British beetles which are never seen in any other localities, and, until their singular mode of life was discovered, were ranked among the rarest of our insects. No less than thirty-seven species of ant's-nest-beetles have already been acknowledged, besides the larvæ of three other species. One very rare species of the Staphylinidæ, or Cocktail-beetle (*Atemeles emarginatus*), has now become quite common, so frequently is it found in the nest of the ant which is now under consideration. The locality of this beetle was discovered by a collector, who saw an ant carrying one of the beetles into its nest. As to the beetles themselves, they seem to be quite as much at home as the ants, and when the nest is laid open, their first attempt is to escape into the farthest galleries, or to hide themselves in the nearest crevice. The ants, however, watch them carefully, run after them, seize them in their jaws, and carry them back again into the nests.

CHAPTER VIII.

BURROWING BEETLES.

The TIGER BEETLE, and its habits—Beauty of the Insect, its Larva, and mode of life—Curious form of its Burrow—The SEXTON BEETLE and its power of digging in the ground—The DOR BEETLE—Its polished surface, and the substances into which it Burrows—Use of the Dor Beetle—The SCARABÆUS of Egypt and its wonderful Instincts—The Egg, the Grub, and the Cocoon—Cocoon in the British Museum—The MOLE CRICKET, its form and elaborate Dwelling—Its general Habits, and wide distribution. The FIELD CRICKET, and its Tunnels—Structure of the Ovipositor—The MIGRATORY LOCUST and its development—The ANT LION, its form, food, and mode of life—The Pitfall and its structure—Mode of catching Prey—Perfect form of the Ant Lion.

WE now come to the Burrowing Beetles, of which there are no few species. As is the case with the generality of insects, the subterranean habitations which they excavate are seldom intended for their own use—at all events, after they have attained their perfect form; but are either formed by the parent while preparing a home for the young brood which it will never see, or by the larva itself while feeding, or while forming a cell in which it can lie dormant in the pupal state.

FIRST among the British coleoptera comes the lovely TIGER BEETLE (*Cicindela Campestris*), an insect which, though small, can challenge comparison with the most beautiful exotic specimens. It is the fiercest, handsomest, and most active of all the British coleoptera, using legs and wings with equal agility, running or flying with such speed that its form cannot be clearly defined, and settling on the ground or taking to wing with equal ease. As it darts through the air, the burnished surface of the abdomen flashes in the sunbeams as if a living gem had passed by, earning for its owner the popular title of Sparkler Beetle.

This insect is, or rather has been, a mighty burrower, exhibiting, even in its larval condition, something of that fiery energy which actuates it when it has reached its perfect condition. Sandy banks are the chief resorts of the Tiger Beetle, which in this country seems seldom or never to alight upon trees, restricting itself to bare and sandy soil. It even avoids those spots which are covered with grass and herbage, cares nothing for shade, and delights to settle upon banks with a southern aspect, and to run about upon soil that has been rendered so hot by the sun that the bare hand can hardly endure contact with its surface. In America, however, the Tiger Beetles possess different habits, preferring trees to the ground, and either running about on the trunk or darting from leaf to leaf in search of their prey. The English entomologist, however, who wishes to find this beetle, must look for it on the ground; and near the spots which the adult beetles traverse so rapidly may be found the larva in its burrow.

These larvæ are most remarkable beings. They are whitish in colour, and strangely moulded in form, the head being of enormous size, and of a horny consistency, and the eighth segment developed into a hump-like projection, carrying upon its upper surface a pair of bent hooks. The larva never is seen above the surface of the ground, and, indeed, never exhibits more than the smooth horny head and mandibles. It lives in perpendicular burrows, about a foot in depth, which it is able to traverse with great rapidity, and which are only just of sufficient diameter to permit the inhabitant to pass up and down.

It is a carnivorous being, feeding chiefly on insects, which it is able to capture, in spite of the apparent disadvantage under which it labours of being confined to one spot. The mode by which it obtains its daily food is as follows. Ascending to the upper portion of its burrow, it fixes itself firmly by means of its hooks, and then lays its jaws level with the soil. While in this attitude, it is almost invisible, and as soon as an insect passes by the ambushed larva, the sickle-like jaws grasp it, and it is dragged to the bottom of the tunnel, where it is devoured. Not only is the larva carnivorous, but it is combative in proportion to its voracity, and if a straw be thrust into its burrow, the angry grub will fasten upon it with the tenacious gripe of a bull-dog, and suffer itself to be dragged out of its home rather than release its supposed enemy.

The burrow is made by the larva, and not by the parent, and is a work of some little time, the earth being loosened by means of the feet and jaws, and then carried to the surface on the flattened head.

OTHER beetles are in the habit of driving deep tunnels into the ground, wherein may be deposited the eggs which are destined to produce a fresh brood in the ensuing season. Our own country can boast of possessing many such beetles, but in the hotter parts of the world their number is quite wonderful.

Our first example will be the well-known SEXTON, or BURYING BEETLES, some of which may be seen at work at the left hand of the plate, busily engaged in burying the dead bird. There are several species of Burying Beetles; but as their habits are very similar, they need not be separately described. Any one who wishes to see them at work may do so by taking a dead mouse, bird, or piece of meat, and laying it on a soft spot of ground. I was about to add the frog to the number of objects for sepulture, but have omitted that creature because the porous nature of its skin causes it to dry up so rapidly, that the beetle will seldom take the trouble of burying it.

Sometimes, but very rarely, a pair of the beetles will come to the bait by daylight, their wide wings bearing them along with great speed; but in general they prefer night as the time to begin their work. If the bird be visited in the early morning, it will be no longer upon the surface of the ground, but will be half sunken below it, as though the earth had given way, just as a piece of dark cloth sinks into snow. If, however, the bird be removed, the cause of its gradual disappearance will be seen in the form of one or two beetles, sometimes black, and sometimes beautifully barred with orange. Then let the bird be replaced, and a trowel carefully introduced under it, so that the bird and beetles can be gently transferred to a vessel of earth and covered with a glass shade.

During the day, the beetles will mostly remain quiet; but in the evening they begin to be active. To dig a hole, and then to drag the bird into it, would be a task far beyond their powers, and they therefore employ another plan. They entirely burrow beneath the bird, emerging every now and then to scrape out the loose soil, walk round the bird, mount it as if to see how the work is proceeding, and then disappear afresh and renew their labours. Sometimes they dig rather too much on one side, and then they appear sadly puzzled, running round and round the bird, getting on it as if to press it down with their weight, pulling it this way and that way ; and at last they **do** what they ought to have done at first, namely, disappear under the bird and scrape away the earth until the hole is large enough to allow the bird to sink into the required position.

The time occupied in the transaction necessarily varies, according to the size of the buried object and the condition of the beetle; but on the average an ordinary finch, or a mouse, can be buried in the course of a day. When the task is completed, a number of eggs are laid upon the buried animal, and then the beetles emerge, cover it with earth, and then fly away. In some cases they will bury a whole series of corpses; and in the well-known experiments of M. Gleiditsch, four beetles buried, in a small piece of earth, four frogs, three birds, two fishes, one mole, two grasshoppers, the entrails of a fish, and two pieces of meat. And so strong and persevering are these insects, that a single beetle succeeded in burying a mole in two days. Now the mole is at least forty times as large as the beetle, so that we can estimate the strength and perseverance of the beetle by calculating the labour which would be necessary for a man to inter, in two days, an animal forty times as large as himself.

Perhaps the reader may remember a curious analogy between the mode of sepulture employed by these beetles, and the mode of sinking wells in sandy soil. Instead of digging a hole, and then building a brick-lining to it, a circular tower is first built, and then, by scraping away the sand from within, the workmen cause it to sink into the ground. When it has sunk sufficiently, some twelve or fourteen feet are added, and the sand again scraped out; and in this manner the brick tube sinks gradually down, and becomes the lining of the well.

THE beetle just mentioned conveys into its burrow the whole of the substance on which the grub is intended to feed; but those which we shall now examine select only a portion for that purpose. There is a very large tribe of beetles, of which the British type is the common DOR BEETLE (*Geotrupes vulgaris*),

SCARABÆUS.

BURYING BEETLE. ANT-NEST.

sometimes called the Watchman, or Clock, whose heavy hum drones upon the ear in the evening, as the

"Beetle wheels his drowsy flight,"

and whose hard and notched head occasionally strikes against the face with a violence less agreeable to the man than to the insect, the latter being quite undisturbed by the shock.

Catch one of these beetles, and examine the wondrous beauty of its colour, how its polished surface gleams as if made of burnished steel, pure and bright as armour just out of the smith's hands. Yet this creature has, in all probability, been burrowing deeply into the ground, has been meddling with the most noxious substances, and still retains no trace of its past labours. Save for the round-bodied yellow parasites that cling to its body, and insert their beaks between the joints of its armour, it is brilliantly clean. Not a speck of mould remains upon its surface, not a stain defiles its limbs, neither does it retain the least odour which would betray its occupation. Other beetles are not so fortunate. The burying beetles just mentioned are mightily ill-savoured insects, and so are many others with similar habits. But the Dor beetle is free from such noisomeness, and both the eye and the nostrils pronounce it pure.

Let us now watch this beautiful insect, as it wheels through the air. Either by the development of the sense of smell, or by some sixth sense with which humanity is practically unacquainted, the beetle is made aware that the object of its search is at hand. The dull, monotonous buzz is immediately exchanged for a triumphant hum, the circling flight ceases, and the beetle darts through the air, with arrow-like rapidity, to the spot which it seeks. A few more circles, lessening at every round, and down it settles, on an object uninviting to Europeans, but in great favour with Hindoos, Kaffirs, and scarabæi, namely, a patch of cow-dung.

No sooner has it settled, than it dives downwards until it reaches the earth, and then bores a perpendicular hole, some eight inches in depth, and large enough to admit a man's finger. I have often watched the beetles at their work, and seen them thus engaged, and have turned many a Dor beetle out of the burrow which it had been so industriously excavating. Having ascended to the surface, it carries a quantity of the cow-dung to the bottom of the burrow, deposits an egg, and ascends, repeating this process as long as its powers endure. There are several other British beetles which prepare the cradle for their offspring in a similar manner.

Merely to dig a hole, to place at the bottom of it the food which the young are intended to eat, and to fill it in with earth, is a process of great simplicity, and makes but few calls on the industry or ingenuity of the labourer. Some allied beetles there are, however, which feed their young on similar substances, and in like manner bury them in the earth, but which exercise extraordinary industry in the performance of the task. All the world has heard of the famous SCARABÆUS of the Egyptians, (*Scarabæus sacer,*) an insect which is found in many parts of the globe, and very much resembles the Dor beetle of our own country. This insect sets to work in a curiously systematic manner.

As soon as the sensitive organs of the Scarabæus announce to it that the desired substance is at hand, it proceeds to the spot, alights, and sets at once to work. First, it sinks a tolerably deep and perpendicular hole in the ground, and, having returned to the cow-dung, it separates a sufficient quantity for its purpose, lays an egg in it, and forms it into a rude ball. She, for the female insect is the worker, then begins a curious and laborious task. Seizing the ball between her hind feet, she begins to roll it about in the hot sunshine, not taking it direct to the shaft which she has sunk, but remaining near the spot. Should rain come on she ceases to roll, or should the ball be made just before sunset, she waits for the morning before recommencing her labour. The consequence of all this curious rolling about, is twofold; it accelerates the hatching of the enclosed egg by the exposure to the sunbeams, and it forms a thin, hard, clay-like crust round the soft material in which the egg reposes.

When the ball is sufficiently rolled, it is taken to the hole, dropped down and the earth filled in. The egg is very soon hatched, and from it proceeds a little white grub, which finds itself at once in the midst of food, and begins to eat vigorously. By the time it has devoured the whole of the contents of its cocoon—if the mere empty shell may be so called—it is ready for its change into the pupal form, and there lies in the earth until it again changes its form and becomes a perfect beetle.

If the reader will refer to the plate, he will there see two of these beetles at work upon a ball, for it is not an unusual circumstance that two insects should propel the same ball. And, upon the accompanying illustration may be seen the completed cocoon.

Several good examples of this cocoon are in the British Museum, as well as those belonging to allied insects.

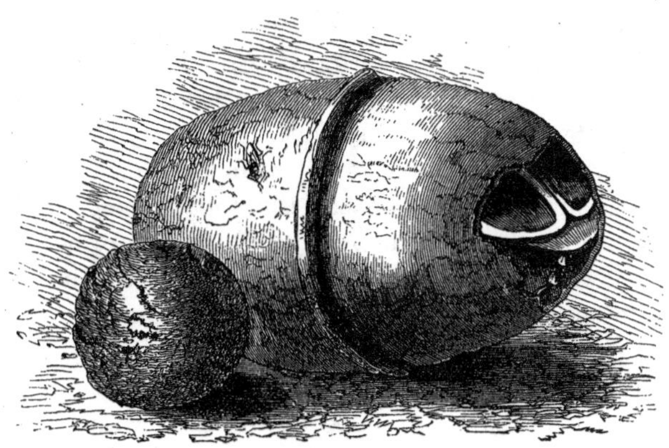

COCOONS OF SCARABÆUS AND GOLIATH.

Frequently the beetle is very much puzzled to discover a place wherein it may dig a hole for the reception of the ball, especially where the ground is uniformly hard. The material which it desires is generally to be found plentifully upon roads, but as roads are usually too hard to be penetrated by the beetle's limbs, the unfortunate insects may be seen rolling their pellets with a patient and hopeless industry only to be equalled by that of Sisyphus. I may perhaps mention in this place that an allied species living in America is popularly called the tumble-bug—the latter inappropriate monosyllable being indiscriminately used for every insect that even looks like a beetle.

There is also in the British Museum one enormous cocoon made of clay. It is almost the size of a six-pounder cannon-ball, with walls of such thickness, that the hollow in its centre is barely the size of a crab-apple. The weight of this cocoon is enormous, when the size of its inmate is considered, and that so comparatively small a beetle should construct and roll so large a ball, seems almost incredible. The beetle belongs to the genus Copris. There is the cocoon of another species of Copris, but in this case, the walls are very thin, and the entire ball would go into the hollow of that which has just been described. A British beetle, *Geotrupes vernalis,* also makes a cocoon about the size of that which has just been mentioned. It is made of mixed clay and cow-dung, and specimens may be seen in the Museum.

Perhaps the most extraordinary of these cocoons is that which is represented in the illustration. This is made by one of the gigantic beetles of the tropics. The insect which made it has no English name, but is scientifically called *Goliathus Drurii.* This wonderful cocoon is as large as a swan's egg, and, as may be seen by reference to the illustration, has very thin walls in proportion to its size. It is strengthened by a remarkable belt, which runs around its centre, exactly like that of the bullet which is used for the two-grooved rifle. How the belt is formed is perfectly unknown, as is its use, unless the strengthening of the walls be its only object. I have carefully examined the cocoon itself, and specimens of the insect which made it, and can find nothing which affords the least clue to the difficulty.

There is no doubt as to the species of insect which made it, for the creature lies inside, a small portion of the ends of the elytra and part of one leg being visible through the fracture. The colour of the beetle is peculiarly beautiful, being rich dark chocolate, soft and deep as made of velvet, and upon the thorax and round the elytra are drawn broad streaks of creamy white. On account of the large dimensions of the cocoon, it has necessarily been reduced in size, but a common house-fly is introduced into the drawing, in order to show the comparative size of the cocoon and the insect.

Many of the Orthopterous insects are burrowers, either digging holes wherein they themselves reside, or preparing a subterranean habitation for their young.

The best-known and most important of these insects is the MOLE CRICKET (*Gryllotalpa vulgaris*), called in some places the CROAKER, or CHURR-WORM, on account of the peculiar sound which it produces. It is a truly wonderful insect, one of those beings, which for the sake of force, we may perhaps call the anomalies of nature, though, in fact, nature is perfectly harmonious, and can have no real anomalies. A cursory glance at the insect will at once point out its habits, for the general shape, as well as the strange development of the fore-limbs, and the peculiar formation of the first pair of feet, are so similar to the corresponding members of the mole that the identity of their pursuits is at once evident.

Like the mole, the insect passes nearly the whole of its life underground, digging out long passages by means of its spade-like limbs, and traversing them with some swiftness. Like the mole, it is fierce and quarrelsome, is even ready to fight with its kind, and if victorious, always tears to pieces its vanquished opponent. Like the mole, it is exceedingly voracious, and requires so much food, that if several of them be confined in the same cage and kept only for a short time without food, the strongest will fall upon the weakest, kill and devour them. Like the mole also, it is useful enough in the fields, where its tunnels form a kind of subsoil drainage, but it is equally destructive in the garden, working great havoc among young plants and flowers. One species that inhabits Jamaica has done great damage to the young sugar-canes soon after they had begun to shoot up.

Though spread over the face of the earth, and though almost every portion of the globe can boast its Mole Cricket, it is ever a local insect, being very fastidious in its choice of soil, and generally preferring a loose and sandy ground, wherein it can easily burrow. There is a little village near Oxford, where the Mole Cricket is frequently found, its favourite residence being a wide piece of waste ground, covered with sand, which in some places is blown into hillocks by the wind, and in another is hollowed into pits by the sandman's spade. Grass tries to grow at intervals, and here and there its spreading roots bind together the loose soil, and it is in this curious locality that the Mole Cricket loves to dwell.

To procure the insect is no easy matter, for it always burrows to some considerable depth when the soil is so loose, and a labourer with a spade would find much difficulty in disinterring it. The recognised method of procuring these insects is, to mark their holes by day and to visit them at dusk, just when the insects, which are nocturnal in their habits, are beginning to be lively. A long and pliant grass-blade is then pushed into the hold, the end is grasped in the jaws of the offended inhabitant, and both grass-blade and Mole Cricket are drawn out together. By some persons the Mole Cricket is thought to be a wholly carnivorous insect, injuring the roots merely by its endeavouring to force a passage through them, and not by its desire to eat them. This theory was supported by sundry experiments, whereby the Mole Cricket was proved to be able to subsist on several substances, such as meat and insects, specimens having been feed upon ants alone. Dr. Kidd, however, found that they throve well upon potato, and the best entomologists have decided that vegetable food is their proper diet, though they are able to eat animal food, and on some occasions seem to prefer it. A very decided proof that the Mole Cricket is even in its wild state a carnivorous being, is afforded by the fact that in the stomach have been found the relics of various insects.

Just as the mole constructs a habitation distinct from its ordinary galleries, so does this insect form a chamber for domestic purposes apart from the tunnels which ramify in so many directions. Near the surface of the ground a really large chamber is constructed, measuring about three inches in diameter, and nearly one inch in height. It is made very neatly, and the walls are carefully smoothed. Within this chamber the Mole Cricket deposits its eggs, which are generally from two to three hundred in number, and yellowish in colour. As the chamber lies so near the surface of the ground, the genial sunbeams are able to raise the temperature sufficient for the hatching of the eggs, which in bue course of time produce the tiny young, little white creatures, very like the parent in shape, except that they have no wings. They do not attain the perfect state until the third year. The reader will at once see that this chamber is analogous to the cavities made by the Dusky Ant which has been described on page 149.

It is a rather remarkable fact that one species of this family burrows, not into earth, but into wood. Its form very much resembles that of the wood-burrowing beetles, the body being long, and cylindrical, the legs very short and fitting into cavities at the sides of the body. Its scientific name is *Cylindrodes Campbellii.* This is one of the oddest-looking insects that can be conceived, and really bears no small resemblance to three inches of a blacklead pencil.

The black-bodied FIELD CRICKET (*Acheta campestris*) is also one of the burrowing Orthoptera, working tunnels of considerable depth, and living in them during the day. By night it comes out of its home and sits at the mouth, chirping away foı hours together. The banks at the side of a road or lane are favourite resorts of the Field Cricket, and I have noticed the insect peculiarly plentiful in the roads and lanes between Ramsgate and Margate. Like the mole cricket, it is of a very combative nature, and may be drawn out of its tunnel by the simple process of pushing a grass-stem down the burrow. It is said that in France it is captured in rather a curious manner, an ant being tied to a thread and dropped into the hole. Being partly carnivorous, the cricket seizes the ant for the purpose of eating it, and is immediately dragged out of its house by the thread.

GRYLLUS DEPOSITING EGGS.

IN the accompanying illustration is shown one of the well-known grasshoppers in the act of depositing its eggs. Its popular name is the WART-BITER, and its scientific title *Gryllus*, [or *Decticus*] *verrucivorus*, this name being given to it because its bite is thought to have the property of removing warts.

If the reader will refer to the illustration, he will see that the end of the body is furnished with a long, double-bladed instrument, technically called the ovipositor, or egg-layer. This curious instrument, which is of course only found in the female insect, is of very great comparative length, and is used for the purpose of placing the eggs in a convenient spot. Pressed closely together, the blades form an admirable boring instrument, but when the required hole is made, the blades separate so as to permit an egg to pass between them, and guide it to the exact spot where it is to lie. The insect does not place many eggs in one spot, but after depositing some ten or twelve eggs, she goes off to another locality and repeats the process until her store is exhausted. She thus contrives to spread her offspring rather widely over the ground, and avoids the danger of losing the entire brood by a single accident.

When the young insects are first hatched, they are nearly white, and of very small dimensions, being no larger than ordinary gnats. At the left hand of the engraving may be seen a small cavity, in which the young Grylli have just been hatched, several of them being shown of their natural size. The well-known Great Green Grasshopper, which is sometimes found in our hedges and nut-trees, and so often frightens the ignorant, is closely allied to the insect which has just been described, and has very similar habits.

The terrible Migratory Locust, which passes over the country in such countless hosts, is also a partial burrower, laying its eggs beneath the surface of the ground. It is stated by one naturalist, that the eggs are placed in cells something like the chambers of the mole cricket, the cell itself being about an inch and a half deep, and the entrance to it being a nearly horizontal tube of earth, coated with a kind of glutinous secretion. Sometimes the eggs themselves are enveloped with this glutinous substance, and are stuck together in masses of determinate shape. South America is peculiarly rich in these egg-masses, many varieties of which may be seen in Mr. Waterton's collection. The young do not attain their wings for three years, and during that period are called in Southern Africa by the popular and expressive name of voet-gangers, or foot-goers.

BEFORE leaving the earth-burrowers, it is necessary to mention the larva of the common May-fly, or Ephemera. Sometimes this larva hides itself under stones, but it often burrows under the muddy banks, and there constructs a very curious habitation. If a portion of the mud be carefully removed, it will be seen to be perforated by a series of holes, a few being nearly circular, but the greater part oval, the long diameter being horizontal, in order to suit the peculiar shape of the inhabitant.

These are the habitations of the Ephemera grub; and if the block of mud be laid open, so as to exhibit longitudinal sections of the holes, the spectator will perceive that each hole is double, the two tubes lying parallel to each other, and being in fact only one tube bent upon itself.

Mr. J. Rennie, in his "Insect Architecture," mentions a curious modification of these tubes:—"In the bank of the stream at Lee, in Kent, we had occasion to take up an old willow stump, which, previous to its being driven into the bank, had been perforated in numerous places by the caterpillar of the goat-moth (*Cossus ligniperda*). From having been driven amongst the moist clay, these perforations became filled with it, and the grubs of the Ephemera found them very suitable for their habitations; for the wood supplied a more secure protection than if their galleries had been excavated in the clay. In these holes of the wood we found several empty, and some in which were full-grown grubs."

OUR last example of the earth-burrowing insects is a truly remarkable one. It is scarcely possible to conceive any mode of life more curious than that which is passed by the insect which now comes before our notice—a mode of life so strange and unique, that if it had been related by one observer only, no matter how trustworthy he might have been, his testimony would have been rejected by nearly every man of science. I allude to the celebrated insect known as the ANT-LION (*Myrmeleon formicarius*). In its mature state, it presents nothing worthy of remark, except, perhaps, the elegance of its form, and the delicacy of its wide gauzy wings, which much resemble those of the common Dragon-fly. But in its larval condition it is truly a wonderful being.

Though predaceous, and feeding chiefly on the most active insects, it is itself slow, and totally unable to chase them; and were it not furnished with some quality which serves it in the lieu of speed, it would soon die of hunger. The very look of the larva is enough to make the observer marvel as to its method of obtaining food. Thick, short, soft, and fleshy, the body is supported on six very feeble legs, of which the hinder pair only are employed for locomotion, and these can only drag it slowly backwards. Indeed, the general outline of the body and head bears no small resemblance to that of a fat-bodied garden spider. So feeble are its limbs, that they are practically of very little use in locomotion, and even when they are cut off, the creature can move nearly as well as when they were in their places. From the front of the head project a pair of long, slender, curved mandibles, which give the first intimation that the grub has anything formidable in its nature. These mandibles are curiously made, being deeply grooved throughout their length, and permitting the maxillæ, or inner pair of jaws, to play up and down them.

Inert and helpless as it may seem, this grub is a ruthless destroyer of the more active insects, and, moreover, seldom catches any but the most active. Choosing some sandy spot, where the soil is as far as possible free from stones, it begins to form the celebrated pitfalls by which it is enabled to entrap ants and other insects. Depressing the end of its abdomen, and crawling backwards in a circular direction, it traces a shallow trench, the circle varying from one to three inches in diameter. It then makes another round, starting just within the first circle, and so it proceeds, continually scooping up the sand with its head, and jerking it outside the limits of its trench. By continuing this process, and always tracing smaller and smaller circles, the grub at last completes a conical pit, and then buries itself in the sand, holding the mandibles widely extended.

Should an insect, an ant, for example, happen to pass near the pitfall, it will be sure to go and look into the cavity, partly out of the insatiable curiosity which distinguishes ants, cats, monkeys, and children, and partly out of a desire to obtain food. No sooner has the ant approached the margin of the pitfall, than the treacherous soil gives way, the poor insect goes tumbling and rolling down the yielding sides of the pit, and falls into the extended jaws that are waiting for it at the bottom. A smart bite kills the ant, the juices are extracted, and the empty carcase it jerked out of the pit, and the Ant-lion settles itself in readiness for another victim.

Sometimes, when a more powerful insect, such as a large wood-ant, or beetle, or perhaps a hunting spider, happens to fall into the pit, the Ant-lion does not obtain a meal on such easy terms. The victim has no idea of surrendering at discretion, but tries to scramble up the sides of the pit, and in its furious exertions, it brings down the sand in torrents, filling up the pit, making the slope of the sides shallower, and so rendering its escape easy. Then there is a battle between the Ant-lion and its intended prey, the one bringing the sand into the pit and the other flinging it out again so as to restore the steepness of the sides, and to deepen the pit.

Sometimes a quantity of the sand flung by the Ant-lion happens to fall on the escaping victim, knocks it over, and enables the devourer to grasp it in the terrible jaws, which never open but to reject the dead and withered carcases; sometimes the insect is tired before the Ant-lion, and suffers itself to be captured; and sometimes, though very rarely, it succeeds in making its escape. In either case, the pitfall is quite out of shape, and instead of re-arranging it, the Ant-lion deserts it and makes another. Some writers have said that the Ant-lion flings the sand at its escaping prey with deliberate aim and intention. It does nothing of the kind, but only tosses the sand out as fast as its head can work, without aiming in any direction, or having any idea except to prevent the pit from being filled up.

Its earth-burrowing life does not cease until it assumes the perfect state. When it has passed its full time in the larval condition, and is about to change into a pupa, it spins a silken cocoon of a globular form, and therein remains until it is about to assume its perfect condition. The pupa then bites a hole through the side of the cocoon, and projects its body half out of the aperture. The pupal skin then withers, bursts, and the perfect insect emerges. Scarcely has it taken the first few breaths of air, than its abdomen, which before was short, so as to be included within the cocoon, extends to nearly three times its original length, so as to resemble that of the dragon-fly; the curious antennæ unroll themselves, the wings shake out by degrees their beautiful folds, and in a short time the lovely insect is ready for flight It is scarcely possible to imagine a more complete contrast than that which is exhibited by the larva and the perfect insect, and if the two were placed side by side, no one who was not aware of the circumstances would think that they are but two stages of the same insect.

If the reader will refer to the illustration on page 128, he will see a section of the pitfall, with the Ant-lion at the bottom, and a couple of ants falling into the trap. The Ant-lion belongs to the same order of insects as the dragon-fly, which it so much resembles.

CHAPTER IX.

WOOD-BORING INSECTS.

BEETLES—The usual form of the Wood-borers—The SCOLYTUS and its ravages—Mode of forming the Tunnels—Curious instinct—Theories respecting the Scolytus—Worm-eaten Furniture, its cause, and the best method of checking the Boring Insects—Ginger and Cork-borers—The "Petrified" Man—The MEAL-WORM and its ravages—Weevils—The PALM-WEEVIL of Jamaica—Its development and uses as an edible—Its Cocoons—The WASP-BEETLE, its shape, colours, and tunnelling powers—The MUSK BEETLE—Its beauty and fragrance—Difficulty of detecting the Musk-Beetle—Its Burrows and their inmates—The RHAGIUM and its Cocoon—The HARLEQUIN BEETLE—Wood-boring Bees—WILLOW-BEE, its Tunnel and mode of making the Cells—Food of the Young—The POPPY BEE—The PITH-BORING BEES and their Habits—Structure of the cells and escape of the Young—Economy of labour—Shell-nests of Bees—Wonderful adaptation to circumstances—How the Bee burrows—The HOOP-SHAVER-BEE—Gilbert White's description of its habits—The SIREX and its Burrow—Its ravages among fig-trees—Formidable aspect of the insect—The two British species—CARPENTER BEE—Mode of making its burrow—Methodical labour—Food of the Young—How to make a ceiling—Number of cells in each burrow—The Carpenter Bee of Australia—The PELOPÆUS as a Wood-borer—Its tunnel, and mode of making cells—The SAPERDA—Damage caused to aspen and other trees—A useful parasite—The Goat Moth—Wood Leopard Moth—Clear-wings and Honey-comb Moths.

WE now leave the earth-burrowers, and proceed to those insects which tunnel into wood and other substances. The Hymenoptera are again the best burrowers in wood as they are in the earth, but, as some of the beetles are notable wood-borers, and we shall only mention a few of them, we will take them first in order.

BEETLES generally burrow while in their larval state, though there are some that do so when they have attained their perfect form, and are able to bore their way through wood or into the ground with wonderful ease. All the boring beetles are formed in such a manner that an entomologist can at once detect their habits from their shape. The combination of the cylinder and the sharp-edged screw, is well known to be the best form of boring tool, whether under the name of auger, gimlet, or centre-bit, and it will be found that the harder the substance into which the insect burrows, the more cylindrical is its shape. The dors, clocks, and other earth-boring beetles, depart from that form, but when we come to look at the scolytus, the ptinus, and other wood-borers, we cannot but notice how very cylindrical they are in their shape.

Perhaps there is no wood-boring beetle which is known so well as the little insect which is called *Scolytus destructor.* I am not aware that it has a popular name that will distinguish it from other small beetles which bore into wood.

SCOLYTUS.

The accompanying illustration will probably call to the mind of the reader, the insect which now comes before our notice. If he should have examined the bark of certain trees, particularly that of the elm, he will often have seen that it is perforated with circular holes, very like those which are drilled into worm-eaten furniture, but of rather larger diameter. When I was a very little boy and first saw these holes, I thought that they had been made by shot, and in trying to pick out the shot with my knife. made the discovery that the holes were not due to firearms, but to insects. The pleasure of the discovery nearly compensated for the disappointment concerning the shot, the possession of which seemed to my boyish mind to be a manly trait of character, and calculated to raise me in the eyes of my playfellows.

If the bark be cut through, and then raised with the knife, the curious radiating system of tunnels will be exposed to view, and the observer will notice that, however these tunnels may vary in size and direction, they all agree in these points; firstly, that they radiate nearly at right angles from a single cylindrical tunnel; and secondly, that they are very small at their base, and gradually increase to their termination. The cause of this formation is as follows:—

The mother insect enters the bark in search of food, and burrows deeply into the tree, sometimes boring into the substance of the wood itself, but generally cutting a tunnel between the wood and the bark. She then deposits her eggs regularly along the cylindrical tunnels, and in most cases retreats to the entrance, and there dies, her body forming a natural stopper. In due time the eggs are hatched, producing a number of very minute white grubs, which immediately begin to feed, the substance of the tree being the only diet of this insect in every stage of existence. Urged by a wonderful instinct, each grub arranges its body at a right angle with the burrow in which it was hatched, and so eats its way steadily outwards.

When the grubs have made some progress, the wisdom of this arrangement becomes evident. As they increase in size, the burrows necessarily increase with them, so that if they had all started parallel with each other, the tunnels would coalesce and the grubs be unable to procure their proper amount of food. As however, the tunnels radiate like the spokes of a wheel, they very seldom interfere with each other, their radiation more than keeping pace with their increasing size. It will easily be seen by reference to the illustration, that if a number of these beetles attack a tree, the bark is gradually separated from the woody portion, and that, as in all exogenous trees the nourishment is derived from the bark, the tree must die as soon as the functions of the bark are suspended.

Settlers in any new colony are well aware of this fact, and when they want to kill a tree, they do so by simply removing a rather wide ring of bark from the trunk, and thus cutting off the supply of nourishment. The tree is thus starved to death, and in the following year, a fire applied to the trunk is able to burn it through, and bring down the tree with scarcely any expenditure of labour by the settlers. This mode of killing a tree is technically called "girdling" it. In proportion, therefore, to the amount of bark removed, the tree sickens, from defective nourishment, and if once the bark be separated all round the trunk, the tree will instantly die.

The reader may probably be aware that some of our most skilful naturalists have thought that the Scolytus is not so culpable an insect as is generally supposed, and that it does not attack trees until their race is run, and they have begun to show symptoms of decay. There is great truth in this conjecture, for it is beyond a doubt that if a tree be seriously injured, and begin to droop, the Scolytus is sure to make a lodgment before very long. Girdled trees, for example, are almost always attacked by this beetle as soon as the effects of the injury are apparent. But, though the female may not lay her eggs in healthy trees, there is little doubt but that she and her mate have aided, in no small degree, in bringing the tree to so diseased a condition. For, as has already been mentioned, the food of the adult, as well as of the imperfect insect, consists of the bark and wood, and in boring the tree for the purpose of feeding, the numerous Scolyti can but enfeeble its constitution, and so bring it to that state of ill health which renders it a fit cradle for the immature beetles.

There is hardly a grove or a park in the neighbourhood of London where the ravages of the Scolytus are not painfully apparent, and in Greenwich Park especially, some of the finest trees are riddled with the cylindrical tunnels of this destructive insect. There are several species of Scolytus, each affecting certain trees, so that there is scarcely any tree that can hope to escape from the jaws of some member of this family.

THE well-known worm-eaten appearance of furniture is caused by certain beetles belonging to another family. As may be seen from the dimensions of the tunnels, the insects are very small, and their bodies are nearly cylindrical. The ravages which these beetles cause are fatal to all who happen to possess old furniture, but Mr. Westwood mentions that one common species, *Ptilinus pectinicornis*, completely destroyed a new bedpost, in the short space of three years. There is but one known method of killing the insects which have already taken possession, and of preventing others from following their example, namely, by injecting a solution of corrosive sublimate into the holes, and then treating the whole of the surface with the same poisonous liquid. I need perhaps scarcely mention, that insects which are popularly called Death-watches, belong to this family. Not only do furniture and timber suffer from the attacks of the Ptilinus, but articles of dress and food are also injured by them. Specimens of natural history are often spoiled by the holes which are drilled through them by the beetles; and stationers sometimes suffer from the voracious insects, which bore holes through their wafers, fix them together, and there undergo their transformations within them. One species is obnoxious to wholesale druggists, on account of the damage which it does to the ginger. In some cases, half the ginger is drilled with holes, and rendered quite unsaleable. It is not, however, lost entirely, because it is reserved for the mill, and is then sold as ground ginger, the insects and their grubs being reduced to powder together with the ginger which they have not consumed. Such specimens are of course not exhibited to the general gaze, as the public would be very cautious of purchasing ground ginger if they knew what it contained. In the British Museum, however, may be seen several pieces of ginger completely eaten away by the beetle, and numerous examples of the insect itself are placed in the same tray. The little beetles which eat cork, and are so mischievous in the cellar, belong to the genus *Mycetophagus*. They will eat rotten wood or fungi, but always prefer cork, and in some cases have not only caused much expense by forcing the proprietor to re-cork all his bottles, but have sometimes destroyed the cork so completely, that the wine has escaped.

The reader may remember that a so-called "petrified man" was brought from Australia, and exhibited in London during 1862. Having very great doubts about the petrifaction of a human being, I went to see it, and at a glance perceived that it was no petrifaction at all, but simply a moderately-good example of a desiccated body, such as are common enough in museums, and sometimes occur even in this country. The exhibitor stoutly asserted that it was a petrifaction, but as I noticed the tunnels of sundry Ptilini in various parts of the head, body, and limbs, there could be no doubt but that the body had not been changed into stone.

Cylindrical holes of small size may be often seen in the bark of oak-trees, from which dart certain long-bodied little beetles, with beautifully-fringed antennæ, and shaped much like the common skip-jack beetles. These insects belong to the genus *Melasis.*

The common Meal-Worm may be placed with the wood-borers, for it is able to gnaw its way through almost any bread that can be made, and, as sailors too well know, feeds upon ship-biscuit, and drills it full of holes. Old sailors can never eat a biscuit without mechanically knocking it on the table, a custom which they have learned on long voyages, serving to shake the "maggots" out of the biscuit. The meal-worm is the larva of a beetle, called *Tenebrio molitor,* a long-bodied, small-headed insect, with very long wing-cases, and very slender and rather short antennæ. To bird-fanciers it is invaluable, serving to keep in health the nightingale and several other delicate birds, and those who keep vivaria are also indebted to the meal-worm, as affording food to sundry of the lizard tribe. Even the perfect insect will eat the biscuit, and is nearly as voracious as the larva.

THERE is a genus of weevils called Calandra, which is remarkable for the great diversity of size among its members, some, such as the dreaded grain-weevil of England (*Calandra granaria*), being very small, and scarcely exceeding the eighth of an inch in length; and another, the PALM WEEVIL (*Calandra palmarum*), being a really large beetle, nearly two inches in length. This insect is equally injurious to the sugar-cane and the palm-tree, the larva burrowing into the centre of the plant and eating away its substance. This larva is very large, very fat, and very heavy, and is slightly curved. The natives consider it as one of their greatest delicacies, and have some peculiar fashion of cooking it. They call it by the name of Grugru.

While I was examining the beautiful collection of insect habitations in the British Museum, a gentleman looked on, and presently pointed to a larva, apparently that of some sphinx-moth, and saying that he knew the insect well, and had often eaten it, stating at the same time that it was taken out of a palm-tree. The label attached to the specimen corroborated this assertion in a measure, for the palm-tree was the locality from which it had been taken.

This larva, which is called Tuchutó—I spell the name phonetically, my informant never having seen the word in print—is eaten either cooked or raw, the latter being the usual method among gourmands, who think that, like an oyster, the Tuchutó ought to be eaten without any aid from the fire. The correct mode of eating it is, to hold it neatly by the head, between the finger and thumb, to put the whole of the body into the mouth, and then to bite it off, just as a strawberry is eaten, and its flavour much resembles that of marrow. The grubs are procured in the following manner:—A cabbage-palm is cut down, and allowed to lie for at least a fortnight; at the expiration of that time, the palm is split open, and in the interior are found the Tuchutó grubs.

To return to the Palm Weevil. When the great, unwieldy larva has attained its full growth, it constructs a large cocoon, made of the stringy-fibres which run along the stem of the palm, twisted and intertwined so as to form a strong place of refuge during the time of its helplessness. It is a remarkable fact, that many wood-boring insects are in the habit of enclosing themselves in a strong cocoon before they change into the pupal condition, though the locality in which they live might seem to render them independent of any such protection. It is, however, very possible, that the object of the cocoon may be to save the inhabitant from some other wood-boring insect, which might happen to drive its tunnel through the helpless pupa, and that the cocoon might contain some ingredients which are distasteful to the intruder, and would compel it to turn aside and choose another path. Were it not for some such protection, another insect might get into the burrow made by the weevil grub, follow it up, come upon the pupa while still inert and incapable of resistance, and either eat it, or at all events inflict a serious injury upon it.

THERE is a large group of beetles, which, in consequence of their extremely long antennæ, are called by the name of Longicornes. In some species the antennæ attain a wonderful length, as for example, in *Lamia ædilis,* the antennæ of which are five times as long as the head and body together. We have several examples in our own country, some of them being remarkable for the beauty of their colours, as well as for the elegance of their forms. The common WASP BEETLE (*Clytus arietis*) is a very good example of the longicorn beetles. It may be seen upon the hedges, gently slipping in and out with a curiously fussy movement, that very much resembles the restless gestures of the insect from which it takes its name. Its slender shape and yellow-striped body are indeed so wasp-like, that many persons are afraid to touch one of these beetles lest they should be stung.

The early life of the Wasp Beetle is spent entirely in darkness, the grubs burrowing into wood, and therein undergoing their transformations. They are curious little beings, white, roundish, but flattened; the rings of which the body is made are deeply marked, the segments nearest the head are much larger than those which compose the abdomen, and the head itself is small, but armed with a pair of jaws that remind the observer of wire nippers, so sharp are their edges, and so stout is their make. Old posts and rails are favourite localities with this beetle, and the grubs can almost always be obtained where timber has been left for any length of time in the open air.

ANOTHER well-known boring-beetle, is the large and beautiful insect which is popularly called the MUSK BEETLE (*Cerambyx moschatus*). Nearly an inch in length, with long and gracefully-curved antennæ, and slender and elegant in shape, it would always command attention, even if it were not possessed of two remarkable characteristics, colour and perfume.

To the naked eye, and in an ordinary light, the colour of this beetle is simply green, very much like that of the malachite. But, when the sun shines upon its elytra, some indications of its true beauty present themselves, not to be fully realized without the aid of the microscope and careful illumination. If a part of an elytron be taken from a Musk Beetle, placed under a half-inch object glass, and viewed through a good binocular microscope, by means of concentrated light, the true glories of this magnificent insect become visible. The general colour is green, but few can describe the countless shades of green, gold, and azure, that are brought out by the microscope, and no pencil can hope to give more than a faint and dull idea of the wonderful object. Neither do its beauties end with its colours, for the whole structure of the insect is full of wonders, and from the compound eyes to the brush-soled feet, it affords a series of objects to the microscopist, which will keep him employed for many an hour.

The odour which it exudes is extremely powerful; so strong, indeed, that I have often been attracted by the well-known perfume as I walked along a tree-fringed wood, and, after a little search, discovered the insect. It is no easy matter to find the Musk Beetle, even when it is close at hand, for its slender body lies so neatly along the twigs, and its green colour harmonizes so well with the leaves, that a novice will seldom distinguish the insect. A practised eye, however, looks out for the antennæ, and is at once attracted by their waving grace. By a series of experiments which I have made on this beetle, I have proved that the scent can be retained or emitted at the will of the insect, and it is a rather remarkable fact, that it is often stronger after the death of the beetle than during its life. The Musk Beetle is easily kept alive, provided that it is well supplied with water, and that a little sugar and water be occasionally given to it. The mode of feeding is very curious, as are many of the habits of the insect.

The larva of the Musk Beetle is a mighty borer, making holes into which an ordinary drawing-pencil could be passed. Old and decaying willow-trees are its favourite resort, and in some places the willows are positively riddled with the burrows. If such a tree be sawn open longitudinally, a curious scene is presented to the spectator. In some spots, the interior is hollowed out by nearly parallel burrows, until it looks as if it had been tunnelled by the shipworm, while sections are made of burrows that turn suddenly aside, or gradually diverge towards the yet uneaten parts of the timber. In some of the holes will be found the long white grubs, in others the pupa may be seen lying quiescent, while a perfect beetle or two may possibly be discovered near the entrance of the holes. Nor are the Musk Beetles the only tenants of the tree, for there is generally an assemblage of woodlice, centipedes, and other dark-loving creatures, which have crawled into the deserted holes, and taken up their abode within the tree.

If the reader will refer to the accompanying illustration, he will see that in the upper right-hand corner is represented a beetle within a curiously-woven cell. This beetle belongs to the genus Rhagium. As long as the insect remains in its larval condition, it differs in little from the wood-boring larva. When, however, it is about to change into the pupal state, it makes a

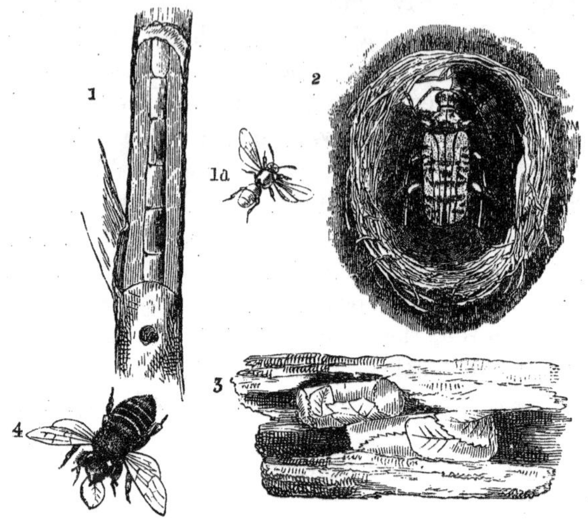

RHAGIUM, ETC.

beautifully-worked cocoon in which it spends the time which intervenes between the change into the pupa and that into the perfect insect. The cocoon is made of woody fibres, which the larva bites and tears away, and the hollow in which the cocoon rests is usually in the bark. The fibres are rather long and narrow, as may be seen by reference to the illustration, which represents the cocoon and insect of the natural size. As the woody fibres are of a pale-straw colour, the cocoon presents an agreeable contrast to the sombre hues of the bark in which it is bedded. When the insect has attained its perfect form, its first care is to escape from the dwelling which has served it so well through its long period of helplessness; and by means of the sharp and powerful jaws with which it is furnished, it gnaws a hole through the side of the cocoon and so escapes into the open air. In the illustration, the beetle is represented in the act of making its way through the cocoon.

THE magnificent insect which is known to entomologists as the HARLEQUIN BEETLE (*Acrocinus longimanus*) also belongs to the wood-burrowers, and in the British Museum may be seen a piece of a tree in which the beetle is still lying, its enormously long limbs packed up into a very small compass. Exotic wood-burrowing beetles are very plentiful, and there is no doubt that several rare or doubtful British species have been introduced as grubs within foreign timber, and that they have made their way out after importation. The Docks are celebrated for the arrival of such distinguished strangers, and a fine collection of exotic beetles has been made by searching the cargoes of timber, after they have been discharged from the ships in which they were brought over the sea.

WE now come to the wood-boring bees, the name of which is legion, and a few examples of which will be now described and figured.

Immediately below the cocoon of the Rhagium may be seen a tunnelled branch, containing two curiously formed cells. These cells are made of rose-leaves, and are the work of the ROSE-CUTTER BEE (*Megachile Willoughbiella*), or WILLOW BEE, as it is often called, because its burrows are so frequently made in decaying willow-trees. This species is very common in most parts of England, and is therefore a good example of the wood-boring bees. The method by which the nests are made is very curious. After the insect has bored a hole of suitable dimensions in some old tree, she sets off in search of materials for the cells, and mostly betakes herself to a rose-bush, or laburnum-tree. She then examines one leaf after another, and having fixed on one to her mind, she settles upon it, clinging to its edge with her feet, and then, using her feet as one leg of a pair of compasses, and her jaws as the other, she quickly cuts out a nearly semicircular piece of leaf. As she supports herself by clinging to the very piece of leaf which she cuts, she would fall to the ground, when the leaf was severed, did she not take the precaution of balancing on her wings for a few moments before making the last cut. As soon as the portion of leaf is severed, she flies away with it to her burrow, and then arranges it after a truly curious fashion.

Bending each leaf into a curved form, she presses them successively into the burrow, in such a manner that they fit into one another, and form a small thimble-shaped cell. At the bottom of the cell she places an egg and some bee-bread, this substance being composed of pollen mixed with honey, and then sets to work upon another cell; and in this manner she proceeds until she has made a series of cells, some two inches in length. When the cells are first made, the natural elasticity of the leaf renders them firm, and as they become dry and stiff in a few days, they are then so strong that they can be removed from the burrow, and handled without breaking.

There is another bee allied to this genus, that employs the petals of the scarlet poppy for this purpose, but unfortunately it is not a native of England. Another species of burrowing bee, Megachile centuncularis, seems rather capricious in its choice of burrows, at one time making its tunnel into an old post or decaying tree, at another into the mortar of old walls, at another into the ground. It is extremely variable in size, sometimes barely exceeding a quarter of an inch in length, and sometimes reaching twice that size. Mr. Smith mentions that this is perhaps the most widely distributed bee in the whole family of Apidæ, extending even as far northwards as Hudson's Bay.

On the left hand of the same illustration may be seen a figure of one of the pith-boring bees, many species of which inhabit this country. This is the insect termed Hylæus dilatatus, and in order to assist the reader in identifying it, a figure of the insect itself is given on the right hand of the nest. Usually, the Hylæus is obliged to bore its way through the pith by sheer labour, but it will always avoid such a task if possible, and make its cells in a hollow already existing. Such has been the case with the nest from which the illustration was drawn, which was made in the hollow stalk of a hemlock. Here let me mention that all hollow stalks and twigs are likely localities for the nests of insects, and that towards the autumn a rich collection may often be made by the simple process of examining all such objects, and splitting them carefully with a knife. Even the reeds and rushes of the river are apt to contain nests, some of them being exceedingly rare.

The particular species of Hylæus which is here figured is by no means common, appearing to be a very local insect, confined to certain spots, and seldom seen except in those favoured localities. Very many species of pith-boring insects are known, most of them inhabiting the dry twigs of the bramble and garden rose. If at the cut end of a branch a round hole be found in the pith, the observer may be sure that a nest of some kind is within. Generally, on carefully laying the branch open, there appears a whole series of cells, one above the other, and in such a case, the cells which are farthest from the aperture always contain the larvæ of female insects, those nearest the entrance being the males.

Sometimes the nests which are found in the bramble contain the larvæ of Osmia leucomelana, a pretty little bee, scarcely more than a quarter of an inch in length, black in colour, with a very glossy abdomen, and a white, downy look about the legs. Five or six cells are made in each branch, and the perfect insect appears about the month of June.

Other bees of this genus are extremely clever in saving themselves labour. Although they can dig industriously when obliged to do so, they will never exert this power without compulsion. The smaller species are very fond of making their cells inside straws, and a thatched roof often contains thousands of nests, which are unsuspected by man, and only discovered by the tomtits and other birds, whose sharp eyes soon detect the hidden insect, and whose ready bills pull the straws out of the thatch, and pick the larvæ from their cells. Nail-holes in garden walls are often filled with cells, and so are the auger-holes in old rails and posts, from which the wooden pins have fallen.

Several species select localities even more remarkable, and make their nests in the empty shells of snails. The common banded snail is a favourite with these bees, and in the British Museum may be seen a whole series of such nests. The number of cells necessarily varies with the size of the snail shell and the number of its whorls, but on the average four or five cells are found in each snail shell. The process of forming the cells is very simple. First, the bee deposits a quantity of pollen and honey, then she places an egg upon the pollen, and then she makes a partition with vegetable fibres torn by her teeth and kneaded firmly together. Lastly, the whole opening of the cell is closed by a wall formed of clay, tiny bits of stick, and small stones, and then the bee goes off in search of another shell. These shells may often be found under hedges, in moss, hidden by grass, and on examination the nests of bees will frequently be seen in them.

In the Museum there is a beautiful example of ingenuity on the part of the builder. Instead of choosing the cell of the banded snail, she has taken that of the great garden snail, and filled it with her cells. Thus it is evident that the shell is too large for the formation of a single cell, and the little architect has, with the greatest ingenuity, evaded the difficulty by placing two cells side by side. When, however, the smaller whorls of the shell were filled, and the bee approached the opening of the shell, the space was too great even for the two cells placed side by side. This difficulty was, however, overcome as readily as the former, and instead of placing the cells perpendicularly, the bee laid them horizontally, and thus filled up the space.

When the Osmia burrows into wood, she sets to work in a very deliberate manner. "A bee," writes Mr. F. Smith, "is observed to alight on an upright post, or other wood suitable for its purposes. She commences the formation of her tunnel, not by excavating downwards, as she would be incommoded with the dust and rubbish which she removes; no, she works *upwards*, and so avoids such an inconvenience. When she has proceeded to the length required, she proceeds in a horizontal direction to the outside of the post, and then her operations are continued *downwards*. She excavates a cell near the bottom of the tube, a second and a third, and so on to the required number. The larvæ when full fed have their heads turned upwards. The bees which arrive at their perfect condition, or rather those which are first anxious to escape into day, are two or three in the upper cells—these are males; the females are usually ten or twelve days later. This is the history of every wood-boring bee which I have bred, and I have reared broods of nearly every species indigenous to this country."

ONE of the wood-boring bees is especially worthy of notice, because some of its habits were remarked a century ago by Gilbert White, who did not know its name, but chronicled its method of obtaining padding for the nest. We will call it the HOOP-SHAVER (*Anthidium manicatum*). It is one of the summer insects, seldom appearing before the beginning of July, and is a rather stout-bodied insect, greyish black, with yellow lines along the sides of the abdomen. The last segment of the male is notable for its termination in five teeth. Its length is rather under half an inch, and it is a very remarkable fact that, contrary to general usage among insects, the male is larger than the female.

This bee seldom takes the trouble of making its own burrow, but takes advantage of the deserted tunnel of some other insect, such as the musk-beetle or the goat moth. When she has selected a fitting home, she enlarges it slightly at the end, and then goes in search of soft vegetable fibre wherewith to line it. The mode of procuring the fibre is thus mentioned by White. "There is a sort of wild bee frequenting the garden campion for the sake of its tomentum, which probably it turns to some purpose in the business of nidification. It is very pleasant to see with what address it strips off the pubes, running from the top to the bottom of a branch, and shaving it bare with the dexterity of a hoop-shaver. When it has got a vast bundle, almost as large as itself, it flies away, holding it secure between its chin and its fore-legs."

After performing this part of her duty, she makes a number of cells, using the same material, together with some glutinous substance, places an egg in each cell, and then leaves them. When the larvæ have obtained their full dimensions, they spin separate cocoons within the cells, and in the following summer the perfect insects make their appearance.

If the reader will visit any fir wood, and look out for the dying and dead trees which are sure to be found in such places, he will probably see that many of them are pierced with round holes, large enough to admit an ordinary quill. These are the burrows of a splendid insect called Sirex gigas by entomologists. Whether it has any popular name I do not know, but I have never been able to discover one, although I have shown specimens of the insect in many parts of England.

This is the more extraordinary, because it is a really splendid creature, nearly as large as a hornet, having wide wings, a bright yellow and black body, and a long firm ovipositor, so that from the head to the end of the ovipositor it measures an inch and three quarters in length. So unobservant, however, is the general public, that nine-tenths of those to whom I showed it declared that it was a wasp, and the remainder thought it to be a hornet. A very reduced figure of the insect is shown in the illustration, and will give a good idea of its general form. In size it is exceedingly variable, some specimens being twice as large as others.

SIREX.

The Sirex is a terrible destroyer of fir-wood, in some cases riddling a tree so completely with its tunnels that the timber is rendered useless. In a little fir-plantation about two miles from my house, there are a number of dead and dying trees, and almost every tree shows the ravages of this destructive insect. The absence of external holes is no proof that the Sirex has not attacked the tree, for they are only the doors through which the insect has escaped from the tree into the world.

The mode in which the Sirex carries on its operations is simple enough.

With the long and powerful ovipositor the mother insect introduces her eggs into the tree, and there leaves them to be hatched. As soon as it has burst from the eggs the young grub begins to burrow into the tree, and to traverse it in all directions, feeding upon the substance of the wood, and drilling holes

XYLOCOPA CAPENSIS. PELOPÆUS SPIRIFER. SAPERDA POPULNEA.

of a tolerably regular form. Towards the end of its larval existence it works its way to the exterior of the trunk, and there awaits its final change, so that when it assumes its perfect form it has only to push itself out of the hole, and so finds itself in the wide world. The insects may often be seen on the trunks of the trees, clinging to the bark close to the hole out of which they have emerged. Sawyers frequently destroy the grubs while they are cutting trees into planks, but as many of the larvæ escape the saw, they remain in the boards, and afterwards emerge in houses, to the great consternation of the inmates. Should such a circumstance occur in the house of any reader of this book, he may at once infer that the wood from which the insect escaped was not properly seasoned. I have known the Sirex to be so numerous that it has fairly driven the inhabitants out of their rooms, for it has a very ferocious appearance, and, on account of its long ovipositor, is always thought to be armed with a venomous sting of peculiar potency.

Two species of Sirex inhabit this country. In general appearance they are very similar to each other, but the species just mentioned is nearly twice the size of its relative. They inhabit similar localities, and I have now before me a piece of a fir tree in which are the holes of both species. These insects are not only interesting in their habits, but they furnish many beautiful objects for the microscopist.

IN the accompanying illustration, we have three excellent examples of wood-boring insects. In the centre of the drawing is seen a portion of a tunnel, which is completely hollowed out, and divided into cells. This is the dwelling which is constructed by the splendid South African CARPENTER BEE (*Xylocopa Capensis*), a wood-borer of great power. She sets about her work in a curiously systematic manner, each action being exactly calculated, nothing left to chance, and all useless labour saved.

When the insect has fixed upon a piece of wood which suits her purpose, usually the trunk or branch of a dead tree, an old post, or a piece of wooden railing, she bores a circular hole about an inch and a half in length, and large enough to permit her to pass. Suddenly, she turns at an angle, and drives her tunnel parallel to the grain of the wood, and makes a burrow of several inches in length. None of the chips and fragments are wasted, but are carried aside and carefully stored up in some secure place, sheltered from the action of the wind.

The tunnel having now been completed, the industrious insect seeks rest in change of employment, and sets off in search of honey and pollen. With these materials she makes a little heap at the bottom of the tunnel, and deposits an egg upon the food which she has so carefully stored.

Having now shown her powers as a burrower and a purveyor, she exhibits her skill as a builder, and proceeds to construct, above the inclosed egg, a ceiling, which shall be also the floor of another cell. For this purpose, she goes off to her store of chips, and fixes them in a ring above the heap of pollen, cementing them together with a glutinous substance, which is probably secreted by herself. A second ring is then placed inside the first, and in this manner the insect proceeds until she has made a nearly flat ceiling of concentric rings. This ceiling bears some resemblance to the operculum of the common water snail. The reader will probably remember, that the ceilings constructed by the ant are made on similar principles. The thickness of each ceiling is about equal to that of a penny.

The number of cells is extremely variable, but on the average each tunnel contains seven or eight, and the insect certainly makes more than one tunnel. As each tunnel generally exceeds a foot in length, and the diameter is large enough to admit the passage of the wide-bodied insect who makes it, the amount of labour performed by the bee is truly wonderful. The jaws are the only boring instruments used, and though they are strong and sharp, they scarcely seem to be adequate to the work for which they are destined.

In the illustration, the upper part of one of these tunnels is shown, and in the two uppermost cells the egg has not been hatched. In the lower cells the young larva is given, in order to show the attitude in which it passes its early life. When all is completed the entrance is closed, with a barrier formed of the same substance and in the same manner as the ceilings.

As far as is yet known, no member of the genus Xylocopa is indigenous to this country.

Among the insect nests in the British Museum is a fine example of the burrow made by a Brazilian member of this genus, *Xylocopa grisescens.* It is a very large insect, and the hole is so wide that it will easily admit a man's thumb. There is in the same case a nest of a curious Australian bee, *Lestis bombylans.* The insect is very like a humble bee, but remarkable for the bright steel-blue of its body, and the absence of hair. The burrow runs through a branch, but is not in the centre, having at least three-fourths of the thickness of the wood on one side, and only one-fourth on the other. All the cells are filled with bee-bread.

IN the upper right-hand corner of the illustration may be seen a curious-looking insect, having its abdomen at the end of a very long footstalk. This is the *Pelopæus spirifer,* an insect belonging to the family Sphegidæ. It is rather a pretty insect, though formidable in aspect, the body and abdomen being black, and the limbs and long footstalk bright yellow. The genus Pelopæus is widely spread over the hot parts of the world, and its members are rather diverse in their habits. Many of them are notable builders, using mud as the material with which they make their nest, and will therefore be described under the head of Builders. The present species, however, takes rank as a wood-burrower, and, as may be seen from the illustration, makes a long tunnel inside a branch, and divides it into cells. True to the instinct which seems common to its kind, the Pelopæus does not make the divisions of wood-chips, as is the case with the Xylocopa, but uses mud for the purpose. Instead of storing her cells with bee bread, she captures spiders, and with their bodies supplies her future young with food, just as is the case with so many earth-boring hymenoptera.

THE last figure in the illustration represents one of the wood-boring beetles, and is given in order to show the curious effect which is produced by its tunnels. The beetle mostly prefers the twigs of the aspen as a home, and its presence can always be detected by the swollen aspect of the injured twig, looking as if the tree were affected with gout. This is a British insect, *Saperda populnea* by name, and in some places it is extremely injurious to the aspen and poplar, always choosing the second or third year's wood, and effectually spoiling its further growth. The larva has a small, flat head, a suddenly enlarged thorax, and a body tapering regularly to the tail.

Of this genus there are many species, each taking its own tree. *Saperda cylindrica*, for example, prefers the pear, plum, and other stone fruit, though it is sometimes found in the nut. It always completes its larval existence in the centre of the stem. Another species prefers the oak. It is fortunate for the owners of woods that the Saperda is greatly persecuted by a parasitic fly belonging to the genus Tachina, and that its numbers are much thinned by the dipterous usurper.

THE Lepidoptera number among their ranks some of the most destructive wood-boring insects that inhabit this country.

There is, perhaps, no insect which makes so large or so ramified a burrow as the common GOAT MOTH (*Cossus ligniperda*). This insect is far more plentiful than is generally supposed, but as in its larval and pupal state it is deeply buried in some tree trunk, and in its perfect condition seldom ventures to fly by day, not one in a thousand is ever seen by the eye of man. This moth breeds in several trees, such as the willow, the oak, and the poplar, the first-mentioned tree seeming to be its chief favourite. Kent is one of the counties wherein this moth is found in greatest profusion, and in the fields round my house there is scarcely a willow of any size which has escaped the ravages of the Goat Moth caterpillar.

The larva of the Goat Moth derives its name from the very powerful and rank odour which it exhales, and which is thought to resemble that of the he-goat. This odour is not only strong but enduring, and for several years after the insect has vacated its burrow the disagreeable scent is plainly perceptible. I have now before me some specimens of the burrow of this creature, and although a very long time has evidently elapsed since the larvæ inhabited them, their odour is quite strong, and can be perceived at a distance of several feet. The pocket in which I placed them, after removing them from the tree, has never lost a rank reminiscence of its contents. As is the case with the musk-beetle, the Goat Moth can be often discovered by means of its odour, and any one who is acquainted with the scent will be attracted by the well-known emanation, and point at once to the tree whence it issues.

The larva is by no means a prepossessing creature, either to the eye or the nostrils, and though some persons believe that it was the famous Cossus, or tree-grub of the Romans, which was thought so great a delicacy by the ancients, I cannot believe that any palate could have attained so very artificial a condition as to endure this repulsive creature, much less to consider it as a dainty.

It grows with wonderful rapidity, being when it has reached its full size seventy-two thousand times heavier than when it was hatched; its segments are deeply marked, and in colour it is of a mahogany-red above, and yellowish below. The whole surface is smooth and polished, and, as may be presumed, considering the life which it leads, its muscular strength is enormous. Not only are the large and trenchant jaws extremely thick and strong, but the development of muscle is singularly great; and the head is of a wedge-like shape, so that the creature can force itself even through hard wood. It feeds entirely upon the substance of the tree in which it takes up its residence, and leaves in its tunnels a considerable amount of *débris.* As the creature increases in size, its tunnel increases in diameter; and it is an amusing task to cut up an old and soft-wooded tree, and follow the caterpillar through its manifold windings.

It lives for some three years in the larval condition, and during the winter it lies dormant in an ingeniously made cocoon, constructed from wood-chips and silken thread, a large store of which can be produced by this caterpillar. Several cocoons are now before me, which I took from a willow tree in Erith marshes. Out of a great number of specimens I have selected four, in order to show the different dimensions of the cocoons. The largest is two inches and a quarter in length, and rather more than an inch in width. In shape it is nearly cylindrical, except at the ends, which are rounded. One of them is intact, but the other has a round hole through which the larva has emerged. It is composed of wood-chips of various sizes, looking like ordinary sawdust, which are loosely, though thickly, fastened upon a silken framework. Near one end of the cocoon the chips are very heavily massed, for what purpose seems doubtful. Rough, however, as is the exterior of the cocoon, the inside is quite smooth and soft, not unlike the interior of the tube made by the trapdoor spider.

The smallest cocoon is barely an inch in length, and is made of much smaller chips, fastened together so strongly that the cocoon retains its cylindrical form when handled, whereas the larger specimen is so loosely made that it collapses under the least pressure. The other two are intermediate in point of size, but precisely similar in point of construction. Besides them there is a specimen of the cocoon in which the creature undergoes its last change. This is of far stronger texture than either of the others, being quite hard, like papier-maché, and dark and polished within.

Generally, just before the moth emerges, the chrysalis works itself along, so that it partially projects from the hole, thus enabling the insect to escape at once into the outer world. In some instances, however, this is not the case, and in the present specimen the empty chrysalis shell may be seen, its shattered sides showing the manner in which the inclosed moth made its exit. The hole through which the moth emerged from the cocoon is of a wonderfully small size, considering the dimensions of the perfect insect, and its sides are very ragged and irregular. Like the other cocoons, it is strongly imbued with the characteristic odour, which has attached itself so strongly to my fingers that careful ablution will be needed before I shall venture to produce my hands in society.

The course taken by the larva is most erratic. Sometimes it travels just below the bark, and then turns suddenly, and dives into the very heart of the tree. It is much given to making these sudden changes, and frequently returns nearly on its former track, a mere shell of wood dividing the two passages. The winter cocoons may be found in various parts of the tree, some deeply buried in the wood, others near the exterior. The pupal cocoon is, as far as I know from personal experience, very near the exterior of the tree, so that when the perfect insect is freed from its enwrapments, it is obliged to traverse only a very short distance before finding itself in the open air.

To keep this larva until it passes through its different stages is by no means a difficult process, and demands less trouble than is given by the generality of caterpillars. The larva is easily found; for whenever an old willow-tree has a powerful scent, and exhibits certain round holes in its bark, the Goat Moth caterpillar is sure to be an inmate. Cut down the tree, and with a saw and chisel break it up, until the caterpillars are found. The chisel is used as much for ripping as for cutting; and the use of the mallet should be avoided as much as possible, lest the jarring strokes should injure the caterpillar. To the ears of the searcher, a few blows with a mallet on the chisel do not appear very formidable; but to an animal inclosed in the wood the shock of every stroke must be terrible, on account of the sound-conducting power of the substance in which it lies.

When one or two of the caterpillars have been obtained, they should be placed in a metal box, as their sharp jaws would soon force a passage through a wooden prison, and a quantity of the wood should be inclosed with them. No further trouble is required; for as they eat the dry wood, and do not need to be supplied daily with fresh food, as is the case with the leaf-eating caterpillars, they are quite content, and feed and grow nearly as well as if they had been left in undisturbed possession of the tree in which they were hatched.

Should any of them die, or should a sufficient number be captured to spare a specimen or two, dissection should be employed, in order to show a very peculiar structure, which is probably of use in enabling the caterpillar to make its way through the wood. When laid on its back, and opened, the general structure of the interior resembles that of the silkworm, save that the muscles are far more strongly defined, and the silk-producing organs are comparatively smaller. Towards the head, however, a pair of organs are seen, which at once arrest the attention. A couple of membraneous sacs—something like the honey-bag of the bee, only very much larger, and long in proportion to their width—lie on either side of the neck, if we may so term the first few segments of the larva. They are generally rather flaccid, so that their sides are wrinkled longitudinally. From the upper part of each of these proceeds a delicate thread, which the microscope shows to be hollow; and which is, in fact, a duct leading into the mouth. These sacs contain a very fœtid fluid, which is supposed to moisten the wood, and partly to soften it, so that it can be shredded with comparative ease.

It is hardly necessary to refer the inquiring reader to Lyonnett's magnificent work on this subject, as the book is a model of skill and perseverance. Still, books are chiefly useful as guides, and not as substitutes for practical experience; and though the beautiful work just mentioned may be perfectly familiar, it ought only to act as a guide to the investigator, and not to supply the place of actual dissection.

ANOTHER moth belonging to the same family is also a notable borer into wood. This is the pretty insect called the WOOD LEOPARD MOTH (*Zeuzera æsculi*), a rarer moth than that which has just been described, though not by any means a scarce insect when the entomologist knows when and where to look for it.

The larva of the Wood Leopard Moth is rather pretty in colour, being whitish-yellow, spotted regularly with black, and having a reddish-brown patch at either extremity. It lives in the interior of many trees, avoiding, however, those which have a very hard-grained wood. Most of the forest trees are subject to its attacks, and the ordinary fruit-trees of our gardens, such as the apple, pear, chestnut, and walnut, are often seriously damaged by this pretty but destructive insect. Like the goat moth, it makes a strong cocoon, in which it can lie safely throughout its pupal condition, and, as with that insect, the walls of the cocoon are rather rough outside and smooth within. When the cocoon is quite dry it is very brittle, and is apt to snap if carelessly handled. This cocoon may often be found when the trees are cut up for firewood; and as it generally lies very near the exterior, a strong pocket-knife will sometimes disclose it.

The perfect insect is remarkably pretty, considering the simplicity of its colouring, which consists of black upon white. The former colour, however, is so disposed, that the wings look as if they were made of the minutest miniver, and the feathery antennæ add considerably to its beauty.

SOME of the most elegant and curious British Lepidoptera are also among the most destructive.

The various species belonging to the remarkable family Ægeriadæ, properly called Clear-wing Moths, are terrible enemies to the gardener, as well as to the landowner, their larvæ feeding upon the pith, and generally preferring the young wood to that of a more advanced growth. In some cases they live in the roots, and are quite as destructive as their relations who prefer the branches. All the Clear-wings are distinguished by the fact that the greater part of their wings is simply membraneous and transparent, without the beautiful feathery scales that are worn by the Lepidoptera as an order. Some of them resemble hornets, others are often mistaken for wasps, while several species are wonderfully like gnats, and as they fly about in the sunshine may readily be mistaken for these insects.

Of one of these insects, *Ægeria asiliformis,* known to collectors as the Breeze-fly Clear-wing, Mr. J. Rennie writes as follows: "We observed above a dozen of them, during this summer, in the trunk of a poplar, one side of which had been stripped of its bark. It was this portion of the trunk which all the caterpillars selected for their final retreat, not one having been observed where the tree was covered with bark. The ingenuity of the little architect consisted in scraping the cell almost to the very surface of the wood, leaving only an exterior covering of unbroken wood, as thin as writing-paper. Previous, therefore, to the chrysalis making its way through this feeble barrier, it could not have been suspected that an insect was lodged under the smooth wood. We observed more than one of these insects in the act of breaking through this covering, within which there is besides a round moveable lid, of a sort of brown wax."

The last-mentioned peculiarity is worthy of special notice, because it is not a general feature in the history of the Clear-wings. Just when they are about to change into the pupal form, they usually nibble a hole through the exterior of the branch, and then make a partial cocoon out of the *débris,* taking care to place themselves so that the head is towards the orifice. The abdominal segments of the chrysalis are furnished with points directed backwards, so that by alternately extending and contracting the abdomen, the creature is pushed onwards. When it is going to break out of its chrysalis case it uses these little points, and forces itself partially through the hole, thus allowing the perfect moth to issue at once into the world.

All gardeners should beware of one very pretty little species, the Gnat Clear-wing (*Ægeria tipuliformis*), which is often to be found upon currant bushes, sitting itself upon the leaves, enjoying the warm sunbeams, and ever and anon opening and closing its fan-like tail. The larva of this insect lives in the young shoots of the currant, and in some seasons damages the crops considerably.

With two more species of lepidopteran burrowers, we must close our list, one of them boring into wood and the other into wax.

The first of these insects, *Tinea granella,* is sometimes called the Wolf Moth. It is a very small insect, and is closely allied to the common clothes moth, so deservedly hated by fur-dealers, careful housewives, and keepers of museums. The larva of this insect feeds upon the corn, covering it at the same time with a tissue of silken threads. The most curious portion of the life of this insect is, that after the larva has finished eating the corn, it proceeds to the sides of the granary, and there burrows into the wood, making its holes so closely together that, if the timber had been taken out of the sea, the Gribble would have had the credit of the tunnels. Nothing seems to stop this little creature, and it bores through deal planks with perfect ease, making its way even through the knots without being checked either by the hardness of the wood, or the abundance of turpentine with which the knots in deal are saturated. This is the more astonishing, because turpentine is mostly fatal to insects, and a little spirit of turpentine in a box will effectually keep off all moths and beetles.

In these burrows the larvæ change into the pupal state, and there remain until the following summer, when they emerge in hosts, ready to deposit their eggs upon the corn, and raise up fresh armies of devourers. Another singular fact is, that after these caterpillars have lived for so long upon corn, their tastes should change so suddenly as to induce them to take to wood, and wood moreover which is never free from turpentine, however well it may be seasoned.

THE last of our burrowers is the HONEY-COMB MOTH, belonging to the genus Galleria. Two species of this genus are known in England, both of which are plentiful in this country.

These moths live in the comb of the hive bee, and when once they have succeeded in depositing their eggs, the combs are generally doomed. The envenomed stings of the bees are useless against these little pests, for though their bodies are soft they take care to conceal themselves in a stout silken tube, and their heads are hard, horny, and penetrable by no sting borne by bee. I once had a very complete case of honey-comb utterly destroyed by the Galleria moths, which draw their silken tubes through and through the combs, ate up even my beautiful royal cells, devoured all the bee-bread, and converted the carefully chosen specimens into an undistinguishable mass of dirty silk, *débris* and moths, both dead and living.

Not long ago, one of my friends, who was about to deliver a lecture on the structure of the bee's cell, and who had got together a collection of combs for the purpose of illustration, came to me in dire distress, and showed me the combs, all covered with the tunnels of the Galleria moth. The damage which they had done was very great, but their presence was discovered in time to prevent them from utterly destroying the combs. After all the caterpillars that could be captured had been destroyed, a wide-mouthed bottle containing spirit of turpentine was placed in the box, and speedily killed the survivors, while a bath in a solution of corrosive sublimate protected the remaining combs against a future brood of Galleria moth.

Although there are still in my list many names of burrowing insects which have not yet been described, it is necessary that we should take our leave of the burrowers, and proceed to the next chapter.

CHAPTER X.

PENSILE MAMMALIA.

THE HARVEST MOUSE—Its appearance—Reason for its name—Mouse nests—Home of the Harvest Mouse—A curious problem—Food of the Harvest Mouse, and its agility—The SQUIRREL—Its summer and winter "cage"—Boldness of the Squirrel—Materials for the nest, and their arrangement.

THERE are not many mammalia which make pensile nests, and we are, therefore, the more pleased to find that one of the most interesting inhabits this country. This is the well-known HARVEST MOUSE (*Micromys minutus*), the smallest example of the mammalia in England, and nearly in the world.

This elegant little creature is so tiny that, when full-grown, it weighs scarcely more than the sixth of an ounce, whereas the ordinary mouse weighs almost an entire ounce. Its colour is a very warm brown above, almost amounting to chestnut, and below it is pure white, the line of demarcation being strongly defined. The colour is slightly variable in different lights, because each hair is red at the tip and brown at the base, and every movement of the animal naturally causes the two tints to be alternately visible and concealed.

It is called the Harvest Mouse, because it is usually found at harvest time, and in some parts of the country it is captured by hundreds, in barns and ricks. To the ricks it would never gain admission, provided they are built on proper staddles, were it not that it gets into the sheaves as they stand in the field, and is carried within them by the labourers. Other mice, however, are sometimes called by this name, although they have no fair title to it; but the genuine Harvest Mouse can always be distinguished by its very small size, and the bright

HARVEST MOUSE.

ruddy hue of the back and the white of the abdomen. Moreover, the ears of the Harvest Mouse are shorter in proportion than those of the ordinary mouse, the head is larger and more slender, and the eyes are not so projecting, so that a very brief inspection will suffice to tell the observer whether he is looking at an adult Harvest Mouse, or a young specimen of any other species.

Mice always make very comfortable nests for their young, gathering together great quantities of wool, rags, paper, hair, moss, feathers, and similar substances, and rolling them into a ball-like mass, in the middle of which the young are placed. I have seen many of these nests, and only once have known an exception to the rule, when the mouse had made its nest of empty and broken nutshells. The Harvest Mouse, however, surpasses all its congeners in the beauty and elegance of its home, which is not only constructed with remarkable neatness, but is suspended above the ground in such a manner as to entitle it to the name of a true pensile nest. Generally, it is hung to several stout grass-stems; sometimes it is fastened to wheat-straws; and in one case, mentioned by Gilbert White, it was suspended from the head of a thistle.

It is a very beautiful structure, being made of very narrow grasses, and woven so carefully as to form a hollow globe, rather larger than a cricket ball, and very nearly as round. How the little creature contrives to form so complicated an object as a hollow sphere with thin walls is still a problem. It is another problem how the young are placed in it, and another how they are fed. The walls are so thin that an object inside the nest can be easily seen from any part of the exterior; there is no opening whatever, and when the young are in the nest they are packed so tightly that their bodies press against the wall in every direction. As there is no defined opening, and as the walls are so loosely woven, it is probable that the mother is able to push her way between the meshes, and so to arrange or feed her young.

The position of the nest, which is always at some little height, presupposes a climbing power in the architect. All mice and rats are good climbers, being able to scramble up perpendicular walls, provided that their surfaces be rough, and even to lower themselves head downwards by clinging with the curved claws of their hind feet. It is also a noticeable fact, that the joint of the hind foot is so loosely articulated that it can be turned nearly half round, and so permits great freedom of movement. The Harvest Mouse is even better constructed for climbing than the ordinary mouse, inasmuch as its long and flexible toes can grasp the grass-stem as firmly as a monkey's paw holds a bough, and the long, slender tail is also partially prehensile, aiding the animal greatly in sustaining itself, though it is not gifted with the sensitive mobility of the same organ in the spider, monkey, or kinkajou.

As the food of the Harvest Mouse consists greatly of insects, flies being especial favourites, it is evident that great agility is needed. In order to show the active character of the quadruped, one of the Harvest Mice is represented in the act of climbing towards a fly, on which it is about to pounce. Under such circumstances, its leap is remarkably swift, and its aim is as accurate as that of the swallow. Even in captivity, it has been known to take flies from the hand of its owner, and to leap along the wires of its cage as smartly as if it were trying to capture an insect that could escape.

The Harvest Mouse is tolerably prolific, and in the airy cradle may sometimes be seen as many as eight young mice, all packed together like herrings in a barrel.

There is another well-known British mammal which, at all events at one season of the year, may be classed among those creatures who build pensile nests. This is the common Squirrel, so plentiful in well-wooded districts, and so scarce where trees are few.

The Squirrel is an admirable nest-builder, though it cannot lay claim to the exquisite neatness which distinguishes the harvest mouse. As is well known, the Squirrel constructs two kinds of nests, or "cages," as they are popularly called, one being its winter home, wherein it can remain in a state of hybernation, and the other its summer residence. These two nests are as different as a town mansion and a shooting-box, the former being strong, thick-walled, sheltered, and warm, and the other light and airy. The winter cage is almost invariably placed in the fork of some tree, generally where two branches start from the trunk. It is well concealed by the boughs on which it rests, and which serve also as a shelter from the wind. The summer cage, on the contrary, is comparatively frail, and is placed nearly at the extremity of slender boughs, which bend with its weight, and cause the airy cradle to rock and dance with every gust of wind.

As if conscious of the impregnable situation which it has chosen, the Squirrel takes no pains to conceal the summer cage, but builds it so openly, that it can be seen from a considerable distance; whereas the winter home requires a practised eye to detect it. So confident is the animal in the strength of its position, that it can scarcely be induced to leave the nest, and will sit there in spite of shouts and stones, provided that the missiles do not actually strike the nest. A well-aimed stone will generally alarm the cunning little animal, and cause it to make one of its rapid rushes to the top of the tree. The materials of the Squirrel's cage are very similar to those of an ordinary bird's nest, consisting of twigs, leaves, moss, and other vegetable substances. Its structure is tolerably compact, though it will not endure rough handling without being injured.

In this aerial nest the young Squirrels are born, making their appearance in the middle of summer, and remaining with their mother until the following spring. There are generally three or four young; and though the nest appears to be so slight, it is capable of sustaining the united weight of young and parents. The Squirrel does not seem to make more nests than can be avoided, and, like many nest-builders, inhabits the same domicile year after year, until it is quite unfit for occupation. Should the nest be assailed while the young are still helpless, the mother takes them in her mouth one by one, leaps away with them, and deposits them in some place of safety. The materials of which the nest are made are grass, moss, and leaves, together with a few twigs, and the shape is nearly spherical. The winter cage, however, is most irregular in form, being accommodated to the space between the boughs in which it is built, and is very thick and warm.

The amount of materials collected for this purpose is surprising. All of them are large and thick-walled, but in some, which are probably old nests, with the accumulation of years upon them, the mass of dried vegetable substances is almost incredible. I have looked into many a winter cage, and on one occasion, when the nest was so hidden that those below could hardly see it, I pulled out whole armfuls of moss, leaves, and grass, and threw them to the ground, where they made a heap like a haycock. The spectators said it looked like the conjuror's trick of producing shawls, flowers, and goblets out of an empty hat. The nest had been deserted for some time, and all the materials were matted together by repeated rains.

CHAPTER XI.

PENSILE BIRDS.

WEAVER BIRDS and their general habits—RED-BILLED WEAVER BIRD—Its bovine friends—Its use to the buffalo—Other parasitic birds—The SPOTTED-BACKED WEAVER BIRD—Its nest, and variable method of construction—The MAHALI WEAVER BIRD—Shape of the nest—Singular defence—Theories respecting the structure—Habits of the bird—Remarkable nests of Weavers—Account of Weavers engaged in nest-building—Very curious contrivance—The GOLD-CAPPED WEAVER—Structure and situation of the nest—The TAHA WEAVER BIRD—Locality selected for its nest—Destructiveness to crops—The PALM SWIFT—Its general habits—The nest and its variable structure—Silk-cotton—The TAILOR BIRD—Antiquity of handicrafts—Structure of the nest—The FAN-TAILED WARBLER—Singular method of fixing its nest—The PENDULINE TITMOUSE—Its habits and food—Remarkable nest and its form.

ALTHOUGH the majority of nest-making birds may be called Weavers, there is one family to which the name is *par excellence* and with justice applied. These are the remarkable birds which are grouped together under the name of Ploceidæ, all being inhabitants of the hot portions of the old world, such as Asia and Africa. The last-mentioned continent is peculiarly rich in Weaver Birds, as may be seen from a glance at the plate which accompanies this description, on which are shown a number of species, together with their nests.

For the most part, the Weaver Birds suspend their nests to the ends of twigs, small branches, drooping parasites, palm-leaves, or reeds, and many species always hang their nests over water, and at no very great height above its surface. The object of this curious locality is evidently that the eggs and young should be saved from the innumerable monkeys that swarm in the forests, and whose filching paws would rob many a poor bird of its young brood. As, however, the branches are very slender, the weight of the monkey, however small the animal may be, is more than sufficient to immerse the would-be thief in the water, and so to put a stop to his marauding propensities. It is well known that the monkey race are very fond of a little bird, mouse, or egg, and that they have such a predilection for blood, that they will snatch the feathers out of parrots' tails, in order to suck the raw and bleeding quills.

Snakes, too, also inveterate nest-robbers, some of them living almost exclusively on young birds and eggs, are effectually debarred from entering the nests, so that the parent birds need not trouble themselves about either foe. Although they may repose in perfect safety, undismayed by the approach of either snake or monkey, they never can see one of their enemies without scolding at it, screaming hoarsely, shooting close to its body, and, if possible, indulging in a passing peck. Such a scene is depicted in the illustration, where Weaver Birds of several species have united in their attacks upon a monkey that is endeavouring to rob a nest, and has met with a suitable fate.

We will now proceed to examine the several species, together with their nests and general habits.

OUR first example of the African Weavers is the RED-BILLED WEAVER BIRD, one of the most plentiful of its kind. Its scientific name is *Textor erythrorhynchus*, and it is remarkable for attending the buffaloes wherever they go. Should the buffalo be driven from any locality, as is often the case when civilization begins to make its mark on a country, the Red-billed Weaver Bird also disappears, and is only to be found in those parts of the land where its huge associate can live in security.

The reason for this peculiarity is, that the bird finds the greater part of its food upon the buffalo, catching and devouring the various parasites and insects which always accompany these animals. Wherever the buffalo exists, there the Weaver Bird may be seen, flitting about the animal as unconcernedly as if it were carved out of wood, perching on its head and pecking among the hair, settling on the massive horns and leaping at passing flies, while ever and anon it makes a dash along the back, digs away at the thick hide, and presently sits quietly on the buffalo, eating something which it has just secured.

The buffalo has very good reason to encourage the presence of its feathered allies, for not only do they free it from the troublesome insects, but they are always vigilant, and serve to detect

AFRICAN WEAVERS.

danger. As soon as the bird perceives, or fancies that it perceives, anything that is suspicious, it ceases from feeding, and looks anxiously about. Should its suspicions prove correct, the bird flies in the air with the peculiar whirring sound that is indicative of danger, and which is known to the buffalo as well as to itself. As soon as the signal of danger is thus given, the buffalo dashes away into the thickest underwood, accompanied by its faithful friends. The reader must not suppose that every buffalo has its bird, or that even a herd of buffaloes must necessarily be accompanied by the Weavers. Sometimes a large party of buffaloes may be without a single bird; only where buffaloes do not exist, these Weavers are not to be found.

There are several other birds which are in the habit of attaching themselves to animals, such as the well-known zic-zac, which befriends the crocodile; the beef-eater, which perches upon the rhinoceros; and a congener, which is found upon the buffalo. This last bird, however, pays more attention to the "wurbles," or larvæ of the bot flies, which burrow into the skin, and make such ugly holes in the hide. All these birds feed upon the parasites and other creatures that are found upon the animal which they affect; and in every case they become watchful guardians, their own alarm being communicated to the larger animal.

TOWARDS the upper part of the illustration may be seen a number of roundish nests, hung on branches in several rows. These are built by the SPOTTED-BACKED WEAVER BIRD, and are slightly variable in the method of their construction, some having the entrance nearly at the bottom, and others more towards the sides. They are all, however, constructed of similar materials, and the different position of the mouth is evidently intended merely as an accommodation to circumstances. Their eggs are not numerous, seldom exceeding four, and their colour is delicate green, something like those of our common starling. The bird is not very plentiful, and seems to be rather limited in its range, not appearing westward of the district called Kaffir-land. Its scientific name is *Ploceus spilonotus*, and it is sometimes called the Yellow-crowned Weaver Bird.

All the pensile birds are remarkable for the eccentricity of shape and design which marks their nests; although they agree in one point, namely, that they dangle at the end of twigs, and dance about merrily at every breeze. Some of them are very long, others are very short; some have their entrance at the side, others from below, and others, again, from near the top. Some are hung, hammock-like, from one twig to another; others are suspended to the extremity of the twig itself; while others, that build in the palms, which have no true branches, and no twigs at all, fasten their nests to the extremities of the leaves. Some are made of various fibres, and others of the coarsest grass-straws: some are so loose in their texture, that the eggs can be plainly seen through them; while others are so strong and thick, that they almost look as if they were made by a professional thatcher.

A good example of the last-mentioned description of nest is the Mahali Weaver-Bird of South Africa (*Pliopasser Mahali*). Although the architect is a small bird, measuring only six inches in total length, the nest which it makes is of considerable size, and is formed of substances so stout, that, when the edifice and the builder are compared together, the strength of the bird seems quite inadequate to the management of such materials.

The general shape of the nest is not unlike that of a Florence oil-flask, supposing the neck to be shortened and widened, the body to be lengthened, and the whole flask to be enlarged to treble its dimensions. Instead, however, of being smooth on the exterior, like the flask, it is intentionally made as rough as possible. The ends of all the grass-stalks, which are of very great thickness, project outwards, and point towards the mouth of the nest, which hangs downwards; so that they serve as eaves whereby the rain is thrown off the nest, and possibly serve also as a protection against foes, though the latter theory has not yet been corroborated by observation.

It is true that the grass stems protrude from the nest like "quills upon the fretful porcupine;" but that they really afford any obstacle to the attacks of a snake or a monkey I cannot believe. If the snake were able to get at the nest at all, it could glide into the aperture, with an upward curve of the flexible body, without troubling itself about the spikes; and if a monkey were to reach the nest, it seems to me that the projecting grass-stems would rather assist than deter it from taking the eggs, as one hand could steady the nest by holding the spikes, while the other was thrust into the aperture. Other nests, moreover, though exposed to the same enemies, and even when placed upon the same trees, do not possess this remarkable armature; and it is hardly to be supposed that if this abattis-like exterior were absolutely needful in the defence of the inmates, it would not be given to all the birds which build under similar conditions. The same may be said of the nests of the Pyrgitæ. There are many structures among animal habitations, the use of which seems to be problematical; and until the case in point be decided by observation, it must remain an open question.

Dr. Smith remarks that the nests of certain Pyrgitæ—*i.e.* little birds which are popularly called sparrows in South Africa—are armed in a similar manner, but with sticks and twigs, instead of grass.

The Mahali is a very sociable bird, being seldom seen alone, and usually assembling in flocks, which sometimes congregate on the ground, and at others assemble in the branches. It is equally sociable in the disposition of its nests, twenty or thirty of these curious structures being often found gathered closely together on the branches of a single tree. Although its colours are not brilliant, it is a pretty bird, the back being of that peculiar brown which is called "liver" by dog-fanciers, and the under parts white, a long patch of snowy white also passing over each cheek. It is about as large as our common starling, the total length being rather more than six inches.

Perhaps the most singular-looking nest made by these birds is that of a rather small, yellow-coloured species (*Ploceus ocularius*), a figure of which may be seen in the left-hand lower corner of the illustration. This nest looks very like a chemist's retort, with the bulb upwards—or, to speak more familiarly, like a very large horse-pistol suspended by the butt. The substance of which it is made is a very narrow, stiff and elastic grass, scarcely larger than the ordinary twine used for tying up small parcels, and interwoven with a skill that seems far beyond the capabilities of a mere bird.

The following account of Weaver Birds engaged in nest-making has been forwarded to me by Captain Drayson, R.A. who has frequently watched the whole proceedings:—

"The bird that builds these nests is colonially termed the Yellow Oriole. The ingenious little creature is nearly as large as a thrush, and is of a bright yellow colour, except the ends of the wings, which are of a brownish hue. It is gregarious; and when a good locality has been found, several hundred nests will be suspended from some dozen trees, within a few yards of each other. The most pliant branches are invariably selected, from which the nest is suspended; and in all cases the end of the nest overhangs the stream, so that any additional weight would bring the nest into the water.

"The birds make a great disturbance when building, there being usually a regular fight in order to secure the best places. In building, the birds first commence by working some stout flags or reeds from the branch, so as to hang downwards. They then attach the upper part of the nest to the branch, so as to form the dome-like roof. By degrees they complete the globular bulb, still working downwards, and, lastly, the neck is attached to the body of the nest. Great skill is required to keep the neck even and open, and yet no machine could accomplish the work better than do these ingenious little architects. The upper part of the nest is very thick and firmly built, more than twice as thick as the neck, and the material of which it is made is far stronger. In some instances I have seen one nest attached to another; and when this is the case, the second builder strengthens the first nest, and then attaches his own work thereto.

"Should by chance a hawk or monkey venture into the vicinity of a colony of birds, it is chased and chirped at by hundreds of these little creatures, who make common cause against the intruder, and quickly drive him off. During the building of the nests, the river side is a most interesting place, as the intelligence and diligence of the birds are most remarkable."

If the hand be carefully introduced up the neck of one of these nests, its admirable fitness for the nurture of the young birds is at once perceived. When merely viewed from the outside, the nest looks as if it would be a very unsafe cradle, and would permit the young birds to fall through the neck into the water. A section of the nest, however, shows that no habitation can be safer, and even the hand can detect the wonderfully ingenious manner in which the interior is constructed. Just where the neck is united to the bulb, a kind of wall or partition is made, about two inches in height, which runs completely across the bulb, and effectually prevents the young birds from falling into the neck.

Although the nests are seen in considerable numbers, the feathered architect does not seem to be a particularly sociable bird, seldom being seen in flocks, and, as a general fact, the male and female associate together and keep all other birds at a distance. The eggs are generally three in number, very pale blue with a few brownish spots, the spots being chiefly gathered towards the larger end. The parent birds are very assiduous in their household cares, and each sits alternately until the eggs are hatched. So absorbed are they in their task that they can be captured alive, merely by grasping the lower end of the neck with one hand, and then cautiously introducing the other hand into the nest. Perhaps this want of caution may arise from the nature of the nest, and the birds being free from all ordinary danger; and if the nest had been open, like that of most birds, the inhabitants would probably be as timid as is usually the case with birds when disturbed in their nests.

BELOW the first-mentioned nest, and nearly in the lower centre of the illustration, may be seen the beautiful nest of the GOLD-CAPPED WEAVER BIRD, *Ploceus icterocephalus*, the figure being drawn from a specimen in my own collection. The nest of this bird is notable for the extreme neatness and compactness of its structure, for it can endure a vast amount of careless handling, and still retain its beautiful contour. The nest was taken from the banks of a river near Natal, and was suspended from two reeds, so as to hang over the water, and at no great distance from the surface.

The whole structure is apparently composed of the same plant, namely, a kind of small reed, but the materials are taken from a different portion of the plant, according to the part of the nest for which they are required. The whole exterior, as well as the walls, are made of the reed-stems, woven very closely together, and being of no trifling thickness. There is a considerable amount of elasticity in the structure, and the whole nest is so strong that it might be kicked down stairs, or be thrown from the top of the Monument, without much apparent deterioration. The interior, however, is constructed after a very different fashion. Instead of the rough, strong workmanship of the exterior, with its reed-stems interlacing among each other, as if woven by human art, and its pale yellow hue, the inside exhibits a lining of flat leaves, laid artistically over each other so as to form a soft, smooth resting-place, but not interlacing at all, being held in their place by their own elasticity. Their colour is of a pale bluish grey, and the contrast which they present to the exterior is very strongly marked. In size the nest is about as large as an ordinary cocoa-nut—not quite so long, though broader.

Mr. Swainson mentions that in one specimen of the nest made by this bird there was a peculiarity about the opening. "The aperture is lateral, but not upon the top, so that it serves the purpose of a window to the inmates, who are sheltered overhead by the convex top of the nest. There is something very ingenious in the construction of this opening, which is not, as it at first appears, round, but semicircular, the arch being bound round with a stronger band than usual, and the plane or base much stronger, and composed of straight pieces of the stalks of the grass, evidently for the purpose of giving to that part on which the birds perched greater strength and substance."

In the right-hand lower corner of the illustration is a nest of another species of Weaver Bird, the pretty TAHA WEAVER, *Euplectes Taha.*

This species, though plentiful, is rather limited in range, and, according to Dr. Smith, is not seen southwards of lat. 26°. Northwards of that line, however, it was found in numbers, associating in large flocks, and generally haunting the neighbourhood of rivers. In some places, the trees that grew near the rivers were filled with crowds of the Taha Weaver Bird. In some localities, where the ground is cultivated, the Taha Weaver is more plentiful than is liked by the natives, for it is very destructive among the gardens; and, in places where it is very numerous, a continual watch must be kept lest the crop should be utterly destroyed. An allied bird, *Euplectes oryx,* is equally destructive in the summer months.

Although the Taha Weavers are mostly found among trees, at the commencement of the breeding season they leave the branches, and retire to the reeds that fringe the river sides, and upon these reeds they build their pensile nest. The plumage of the Taha Weaver varies greatly according to the season of the year, the yellow of the summer coat being freely interspersed with brown dashes in the winter. Even the beak is said to change its colour, and to be lighter in the summer than in the winter.

As in the illustration of African Weaver Birds so many nests and their architects are introduced, I will give a brief summary of its contents.

In the right-hand upper corner are seen the curious nests of the Mahali Weaver, accompanied by the birds themselves. Just below the Mahali are several rows of nests pendent from boughs. These are the homes of the Spotted Weaver, and are represented as attacked by monkeys, which are being ducked for their pains, and will not succeed in reaching the nests. The monkey is the vervet, *Cercopithecus pygerythrus*, commonly called the green monkey, and is of the species that so frequently accompany organ-grinders in this country. Assailing and bickering at the monkeys are several other species of Weavers. Some of the Spotted Weavers are defending their homes, and are aided by three other species, *Ploceus subaureus*, known by its lighter hue; *Hyphantornis textor*, distinguished by its dark bill; and *Textor erythrorhynchus*, known by its light bill.

Rather in the background, and in the centre, are some nests of Ploceus Capensis, woven with a palm-leaf. In the left-hand lower corner is the long, retort-shaped nest of the pretty Yellow Weaver; in the corresponding right corner is the Taha Weaver; and hanging over the water at the bottom of the illustration is the habitation of the Yellow-capped Weaver.

In Jamaica there is a bird, which would not allow any illustration of its size and beauty, but is nevertheless a most interesting species. This is the Palm Swift (*Tachornis phœnicolia*), easily known by the broad white belt on its black body, something like the white band on the common house martin.

As is implied by its generic name, the Palm Swift is celebrated for its very great speed, which it exhibits by its darting flight over the grass savannahs. As, moreover, it resides in Jamaica throughout the whole year, it gives every opportunity for observing its habits.

The nest of this bird is very curious, and always pensile; and, though it can never be mistaken for that of any other bird, it is built after a very diverse fashion. Usually, it is fastened to the spathe of the common cocoa-nut palm, being cemented to the leaf so firmly, that if it be pulled away by force, the outside integument of the leaf comes away also. The nest is ingeniously hidden in the leaf, so that it would not be noticed by an ordinary observer, were it not that in some cases the bird is so very liberal of its materials that their superabundance betrays its presence. The nest is made of cotton and feathers, the cotton forming the exterior and the feathers the lining. The walls of the nest are very strong, though flexible, and something like felt, being firmly compacted, and containing an enormous mass of downy feathers, in the middle of which the eggs are laid.

The cotton is of a very short staple, and is not the substance used in commerce, but the produce of the silk-cotton trees belonging to the genus Bombax. These trees are of very great size, the trunks running nearly a hundred feet in height, without a branch, and being even more than proportionately thick. The cotton is of very trifling use in commerce, the staple being not much more than an inch in length, and is chiefly employed for stuffing mattresses and pillows. The native tribes of Guiana use it for the tiny arrows which they project through their long and slender air-guns, fastening it upon the head of the arrows so as to make them fit the tube. A quantity of this cotton is now before me, and it is evident that the very qualities which render it useless for commercial or mechanical purposes are precisely those which are best adapted for the structure of the nest. It is remarkably fine in texture, being almost silky to the touch; and, instead of becoming inextricably entangled, as is the case with ordinary cotton-wool, it cannot be handled without leaving a number of short fibres on the fingers. Its usual colour is yellowish, but occasionally it is nearly white.

Several nests are often found in each spathe; and it is a curious fact that, in such cases, they are agglutinated together with the same substance that fastens them so firmly to the leaf, and are connected by a kind of gallery, which runs along the side, and communicates with each nest. It is thought that the bird occupies the same nest repeatedly, after the manner of swallows and martins, and that it does not desert the tenement until the spathe becomes detached and falls to the ground, after the custom of its kind. ·Fallen spathes are plentiful under the palms, and in them the nests of the Palm Swift are frequently seen.

Sometimes the Palm Swift choses another tree, and builds its nest in the palmetto, a palm belonging to the genus *Chamœrops*. In such cases, the nest is of a different shape to that which is found in the cocoa-palm, something resembling in form the india-rubber tobacco-pouches which are now so common. The exterior of the nest is loose and woolly, instead of being firm and compact; and in some instances it is so very loose, that it looks just like a doll's wig. The eggs of the Palm Swift are white.

THE man who first invented sewing in all probability thought that he had discovered, or rather created, an art which was entirely new, and that to him alone was due the credit of perceiving the virtues of a fibre thrust through holes.

The capabilities of his invention he could not be expected to foresee, inasmuch as he would in all probability limit its powers to the decoration rather than the clothing of his own person. In process of time he might comprehend that, by means of the needle and thread, a number of small leaves or skins might be made to serve the same purpose as a single large one, and as his instruments improved, so would his work. There are, it is true, certain nations who have been acquainted with the art of sewing from time immemorial, and never seem to have made the least progress in it. The native Australian, for example, displays wonderful ingenuity in making thread from the sinews of the kangaroo's tail, and needles from the emu's bones; but there his invention seems to have stopped, and up to the present time, the junction of a couple of kangaroo skins, or the sewing together of a few "opossum" furs, seem to be the limits of his powers. Still, in other countries, the needle and thread have, as a rule, exhibited a regular improvement, until they have culminated in the sewing-machine of the present day. Had, however, some good genius enabled the original founder of the art to foresee its effect upon the world, he might well have been proud of his discovery, the earliest of human arts.

The respectable guild of tailors, indeed, were wont to attribute to their mystery an antiquity surpassing that of any other handicraft, and, on the strength of a certain passage in Genesis, claimed Adam as the first tailor. As to the smiths and musicians, the tailors looked down upon them as of comparatively recent origin, and considered even the mysterious order of Freemasons as modern upstarts. Had they been moderately skilled in ornithology, they might have claimed a still older origin, on the grounds that, long before man came on the earth, the needle and the thread were used for sewing two objects together.

THE TAILOR BIRDS.

The wonderful little bird, whose portrait is accurately given in the accompanying illustration, is popularly known by the appropriate title of TAILOR BIRD, its scientific name being *Orthotomus longicaudus.* The manner in which it constructs its pensile nest is very singular. Choosing a convenient leaf, generally one which hangs from the end of a slender twig, it pierces a row of holes along each edge, using its beak in the same manner that a shoemaker uses his awl, the two instruments being very similar to each other in shape, though not in material. These holes are not at all regular, and in some cases there are so many of them, that the bird seems to have found some special gratification in making them, just as a boy who has a new knife makes havoc on every piece of wood which he can obtain.

When the holes are completed, the bird next procures its thread, which is a long fibre of some plant, generally much longer than is needed for the task which it performs. Having found its thread, the feathered tailor begins to pass it through the holes, drawing the sides of the leaf towards each other, so as to form a kind of hollow cone, the point downwards. Generally a single leaf is used for this purpose, but whenever the bird cannot find one that is sufficiently large, it sews two together, or even fetches another leaf and fastens it with the fibre. Within the hollow thus formed the bird next deposits a quantity of soft white down, like short cotton wool, and thus constructs a warm, light, and elegant nest, which is scarcely visible among the leafage of the tree, and which is safe from almost every foe except man.

There are several nests of the Tailor Bird in the British Museum, one composed of several leaves, and the other in which one leaf is used. It is a pity that in all instances the leaf has been plucked from the twig on which it grew; and it is to be wished that when other specimens are brought to England the twig will be cut off, and that if the leaf should fall off, it may be replaced on the spot whereon it grew. Beautiful as is the detached nest, it does not give nearly so vivid an idea of its object as if it were still suspended to its branch.

The Tailor Bird is a native of India, and is tolerably familiar, haunting the habitations of man, and being often seen in the gardens and compounds, feeding away in conscious security. It seems to care little about lofty situations, and mostly prefers the ground, or lower branches of the trees, and flies to and fro with a peculiar undulating flight. Many species of the same genus are known to ornithologists.

THE tailor bird is not the only member of the feathered tribe which sews leaves together in order to form a locality for its nest. A rather pretty bird, the FAN-TAILED WARBLER (*Salicaria cisticola*) has a similar method of action, though the nest cannot be ranked among the pensiles.

This bird builds among reeds, sewing together a number of their flat blades in order to make a hollow wherein its nest may be hidden; but the method which it employs is not precisely the same as that which is used by the tailor bird. Instead of passing its thread continuously through the holes, and thus sewing the leaves together, it has a great number of threads, and makes a knot at the end of each, in order to prevent it from being pulled through the hole. A description and beautiful figure of this bird may be seen in Gould's "Birds of Europe," Vol. ii.

The odd little titmice can be admitted among the Pensile Birds, as one of them constructs a habitation as purely pensile as any which has yet been mentioned, and which yields in beauty to none. This is the Penduline Titmouse (*Ægithalus pendulinus*), a native of Southern and Eastern Europe. As is the case with all its family, it is a little bird, scarcely exceeding four inches in length, and being marked with pleasing though not very brilliant colours. In general habits it resembles the bearded titmouse of England, haunting the sides of streams, and feeding upon the seeds of aquatic plants, as well as upon the various insects, larvæ, and small molluscs that are found so plentifully in the water.

The chief point of interest in this bird is, however, concentrated in its nest, which is made in a flask-like shape, and is mostly suspended to the extremity of some twig that overhangs the water. Willows, and other trees that are fond of the water, are favoured residences of this curious little bird. The larger end of the nest hangs downwards, so that at a little distance it looks like a huge pear with a rather long stem. The material of the nest is the cottony down of the willow and poplar, and the opening is always at the side. The position chosen is not invariably at the end of a twig, as the nest is sometimes found among the reeds, hidden by their thick stems from observation.

CHAPTER XII.

PENSILE BIRDS (CONTINUED).

Australian Pensiles—The Yellow-throated Sericornis—Its habits—Singular position for its nest—Conscious security—The Rock Warbler—Shape and locality of its nest—The Yellow-tailed Acanthiza—Its colour and song—Supplementary nests—The Pinc-pinc and its home—Supposed use of the supplementary nest—The Singing Honey-Eater and its nest—The myall or weeping acacia—Various materials—The Lunulated Honey-Eater—A new material—The Painted Honey-Eater, its habits and nest—The art of preservation—Nests and their branches—The colour of eggs—The White-throated Honey-Eater and its habits—Its curious nest—Locality of the nest—The Golden-crested Wren, and the resemblance of its nest to those of the Honey-Eater—The Swallow Dicæum—Its song and beauty of its plumage—The nest, its materials, form, and position—The Malurus and its nest—The Hammock Bird—Singular method of suspending the nest—The White-shafted Fantail—Strange form of the nest—The appendage or tail of the nest.

Some very remarkable instances of pensile birds' nests are found in Australia, and for many of them we are indebted to the patient and careful research of Mr. J. Gould, from whose skilful works on ornithology several illustrations háve been, by permission, copied.

A very curious instance is found in the nest of the Yellow-throated Sericornis (*Sericornis citreogularis*), a rather pretty, but not a striking bird. The general colour is simple brown, and, as its name imparts, the throat is of a citron-yellow. The only remarkable point in the colour, beside the yellow throat, is a rather large patch of black, which envelops the eye and passes down each side of the neck, nearly as far as the shoulders. It is the largest of its genus, and, although not rare, is seldom seen except by those who know where to look for it, as it is scarcely ever observed on the wing, but remains among the thick underwood, flitting occasionally between the branches, but mostly remaining on the ground, where it pecks about in search of the insects on which it feeds.

The reason for its mention in this work is the singular

structure of its nest, which is described by Mr. Gould in the following words :—

"One of the most interesting points connected with the history of this species is the situation chosen for its nest.

PTILOTUS SONORUS. ENTOMOPHILA PICTA. ENTOMOPHILA ALBOGULARIS.
SERICORNIS CITREOGULARIS. ORIGMA RUBRICATA.

"All those who have rambled in the Australian forests must have observed that, in their more dense and humid parts, an atmosphere peculiarly adapted for the rapid and abundant growth of mosses of various kinds is generated, and that these mosses not only grow upon the trunks of decayed trees, but are often accumulated in large masses at the extremities of the drooping branches. These masses often become of sufficient size to admit of the bird constructing a nest in the centre of them, with so much art that it is impossible to distinguish it from any of the other pendulous masses in the vicinity. These bunches are frequently a yard in length, and in some places hang so near the ground as to strike the head of the explorer during his rambles; in others, they are placed high up on the trees, but only in such parts of the forest where there is an open space entirely shaded by overhanging foliage. As will be readily conceived, in whatever situations they are met with, they at all times form a remarkable and conspicuous feature in the landscape.

"Although the nest is constantly disturbed by the wind, and liable to be shaken when the tree is disturbed, so secure does the inmate consider itself from danger or intrusion of any kind, that I have frequently captured the female while sitting on her eggs, a feat that may always be accomplished by carefully placing the hand over the entrance—that is, if it can be detected, to effect which, no slight degree of close prying and examination is necessary.

"The nest is formed of the inner bark of trees, intermingled with green moss, which soon vegetates; sometimes dried grasses and fibrous roots form part of the materials of which it is composed, and it is warmly lined with feathers. The eggs, which are three in number, and much elongated in form, vary considerably in colour, the most constant tint being a clove-brown, freckled over the end with dark umber-brown, frequently assuming the form of a complete band or zone; their medium length is one inch, and their breadth eight lines."

If the reader will bear in mind the remarkable shape of this and a few other nests, he will see, in a future page, how wonderful is the resemblance between the pensile nests of birds and insects.

PENSILE birds do not always suspend their nests to the branches of trees, but in some instances choose exactly the localities which appear to be the most unsuited for the purpose. Still keeping to Australia, we may find a most wonderful example of a pensile nest near mountain courses. The bird which makes it is called, indifferently, the ROCK WARBLER, or the CATARACT BIRD (*Origma rubricata*), because it is always found where water-courses rush through rocky ground. So attached is the bird to these localities, that it is never seen in the forest, nor ever has been observed to perch upon a branch. The generic name, *Origma,* is derived from a Greek word, signifying a rock or a precipice, and is more appropriate than are many scientific titles.

It is a small bird, no larger than our sparrow, and is soberly coloured, the general hue being brown, relieved by a dull red on the breast, something like that of the female robin. It has a melodious though not very powerful note; but its chief claims to admiration are founded upon the extraordinary nest which it builds. In general shape this nest somewhat resembles a claret jug without a handle, having a long, slender neck and a globular and suddenly-rounded bulb.

It is suspended from the rocks in sheltered places, and wherever an overhanging ledge of rock affords protection from the elements, there the strange nests may be found. Just as the martins take a fancy to some favoured spot, and build whole rows of nests on one side of some particular house, utterly disdaining neighbouring houses, which, to all appearance, afford exactly the same advantages, so do the Rock Warblers affect some particular rock, and hang their nests by dozens in close proximity to each other. The material of the nest is the long moss which is plentiful in the country; and, as may be seen from the illustration, the entrance is near the centre of the rounded bulb. In consequence of the material of which the nest is constructed, it is very rough on the exterior, though smooth and comfortable enough within.

Australia certainly produces some of the most singular objects in the world. Among the many varieties of birds' nests which are found in this region, there is one of a very curious form, resembling very greatly a common cottage loaf, and being in fact a double nest, one being placed upon the other.

The bird which makes this nest is termed the Yellow-tailed Acanthiza (*Acanthiza chrysorrhœa*), and is not uncommon in different parts of Australasia. It is rather a neat-looking bird, the colours being beautifully blended together. The back and upper parts are greenish, like the hue of our common wood-wren, and below it is pale yellow, while there is a patch of bright golden yellow at the base of the tail. As if to contradict the popular idea that the birds of Australia have no song, the Yellow-tailed Acanthiza sings a bright, cheerful note, very like that of the goldfinch, so that it is in all points a pleasing little bird.

It is seldom seen on the wing, or, at all events, seldom flies to any distance, as it prefers to remain on the ground, or in the bush, and when disturbed will fly for a few yards and then settle again. It is generally found in small flocks, consisting of six or ten in number, and as it is by no means timid, will allow itself to be approached closely before it takes alarm.

The nest is a very remarkable structure. In most cases it is formed as has just been mentioned, a little nest being stuck on the large one. The materials of which it is made are grass, wood, and leaves, and the structure is rather loose and careless. Generally it is suspended from the delicate mimosa branches, especially in Van Diemen's Land; but when it builds in gardens, as is often the case, it mostly prefers a low shrub for that purpose. Unfortunately for the bird, the bronze cuckoo has a predilection for its nest, and lays its eggs therein. Whenever this is the case, the parasitic bird takes entire possession of the nest, and no other young are found in it.

The supplementary nest is not invariably present, and both the size and shape are extremely variable. The reader may perhaps remember that the PINC-PINC of Africa (*Drymoica textrix*) has a similar custom, constructing a supplementary roosting-place upon the nest. The home of the Pinc-pinc is of much firmer structure than that of the Yellow-tailed Acanthiza, being made of vegetable fibres, interwoven so strongly and elaborately that a thick, felt-like substance is produced. The entrance to the nest is formed in a tubular shape, and projects for an inch or two, so as to look like a spout, and near the entrance is constructed a rounded projection on which a bird can repose.

Some persons think that the male bird uses this perch, and that he posts himself by the entrance in order to act as a sentry and to keep guard over the inmates. It is more probable however, that the projection is used, not so much as a resting-place for the male, although he may possibly take a fancy for sitting in the fresh air rather than in the nest, as a perch on which the bird can settle before it passes into the tubular entrance. This supposition is borne out by the fact that there are mostly several of these perches on each nest, so that the whole structure assumes a rather awkward and irregular aspect. The nest is of very large dimensions when compared with its architect, being on the average four inches in diameter.

There is another species of Acanthiza (*Acanthiza reguloides*) which lives in Australia, and builds a nest very similar in its materials and the general principle of its structure to that of the bird already described, except that the supplementary nest is not present.

A MOST beautiful pensile nest is made by the SINGING HONEY-EATER (*Ptilotus sonorus*), a species which is spread over a large portion of Australasia.

Here we have another example of an Australian singing bird, for the melody of this creature is so loud, so full, and so rich in tone, that Mr. Gould compares it to that of the missel thrush. It is a soberly-coloured bird, though easily identified, the back being pale brown, the top of the head yellow, and a deep black patch passing over the eye and turning downwards along the side of the neck. It is a lively bird, as are all those feathered creatures which feed chiefly on insects, and even in mid-winter its melodious song may be heard in full vigour.

There is a very common tree in Australia, popularly called the myall, known to scientific botanists as *Acacia pendula.* The twigs of this tree are long and very slender, and the leaves are so narrow and delicate that at a little distance they look more like grass-blades than the leaf of a tree. The reader may remember that this is a characteristic of all drooping or "weeping" trees, the leaf and the twig being slender in proportion to each other. The weeping birch and the weeping willow of our own country are good examples of this peculiarity.

Thus, as both the leaves and the twigs of the myall are extraordinarily long and slender, the tree is chosen by many birds which build pensile nests, as will be seen in the course of this volume. It seems a tree that was made for the express purpose, because the long and slender twigs serve the double purpose of affording a firm attachment for the nest and suspending it where no ordinary foe can reach it, while the delicate leaves give their aid in fastening the nest to the twigs, and at the same time serve to conceal the structure from prying eyes.

Although the general structure of the nest is the same in all parts of the country, the materials necessarily differ. In New South Wales, the external shell of the nest is formed of very fine dry stalks, not thicker than twine, while the lining is composed of fibrous roots, matted together with spiders' webs. It is fastened by the rim to the twigs, and as a few of the slender twigs occasionally are interwoven into the nest, it hangs quite securely. In Western Australia, the nest is made of grasses, which, although green when first woven, become white and dry in a short time. The grass is mingled with the hair of the Kangaroo and the fur of some phalangist, vulgarly called opossum, which serve to mat the grass together and to make it impervious to the wind and rain; and the interior is neatly lined with grasses and vegetable down.

THERE are many Honey-Eaters in Australia, all of which are easily known by the hairy tuft at the end of the long tongue, which is used for licking the sweet juices out of flowers. The entomological reader may perhaps remember that the tongue of the hive bee is constructed on precisely the same principle, being long, slender, mobile, and fringed with hair at its top.

Many of them construct nests which may fairly be reckoned among the pensile, and one of the prettiest among the number is that which is built by the LUNULATED HONEY-EATER (*Melithreptes lunulatus*). The bird is easily recognised by the white crescentic mark which runs round the back of the neck, the horns pointing upwards towards the opening of the mouth, and forming a striking contrast with the black hue of the head and neck.

The nest of this bird is very like that of the Singing Honey-Eater, but is mostly suspended to the thinnest twigs which grow at the summit of the enormous Eucalypti-trees. Owing to the great height at which it is placed, and the leaves which surround it, none but an experienced eye can detect it. The walls of the nest are ingeniously made of the inner rind or "liber" of the stringy-bark and other gum-trees, a material which is not unlike the "bass" with which all gardeners are so familiar. The hair of various animals is mixed with the bark, and since sheep have been introduced into Australia, the bird has always availed itself copiously of their wool, finding that it can be worked well into the nest, and serves to bind the materials firmly together. As the nest is always hung by the rim to the twigs, strength of substance is an absolute necessity, and the toughness of fibre and the felting property of the wool make it a most valuable addition to the building materials used by the bird.

For the lining of the nest, the Lunulated Honey-Eater retains the materials to which it has always been accustomed, and uses the fur of the phalangists, which has the advantage of being very soft, very warm, of retaining its elasticity, and of not adhering to the claws of the inmates, as would be the case with wool.

THERE is another of these pretty birds, called the PAINTED HONEY-EATER, on account of the variety of its colouring. Its scientific name is *Entomophila picta.* The general colour of this handsome bird is rich brown above, with the exception of a yellow patch on the base of the tail, and white, slightly spotted, below. A characteristic mark of the species is a little patch of pure white just by the ears.

This handsome species inhabits the interior of New South Wales, and does not confine itself merely to a diet of sweet juices, but feeds much on small insects. The generic title, Entomophila, is composed of two Greek words, which signify insect-lover, and is given to this bird, and several other Honey-Eaters, on account of their insect-eating habits. The birds are extremely active, and devote much of their time to the pursuit of insects on the wing, in which occupation they have a great resemblance to our well-known fly-catcher. They sit on a branch, keeping a careful watch, and whenever an insect passes near, they dart into the air, catch it, and return to their post. They are generally seen in pairs, and are very playful, chasing each other merrily, and spreading their tails so as to show the white colour. When on the wing, they are so like the common goldfinch that they might easily be mistaken for that bird, the patchy distribution of the colour, and the white spot on the face, adding greatly to the resemblance. The nest of this bird is a beautiful example of the pensiles, and on looking at a specimen it is impossible to restrain a feeling of regret that the art of preservation as it now stands will not permit us to retain the branch and its delicate leaves in all their lovely greenery, their long, spear-like blades affording so beautiful a contrast to the dry and withered substance of which the nest is made.

I may perhaps throw out a hint to collectors of birds' nests, that they would always increase the value of the nest by retaining as much as possible of the branch on which it was placed, as the interest greatly depends upon the precise relation which the nest bears to its locality. None, for example, can properly appreciate the extreme beauty of the nest built by the chaffinch, until they see the exquisite manner in which the exterior is covered with mosses and lichens, which exactly resemble in colour the bark of the branches amid which it rests. The pretty cup-like nest of the goldfinch, the domed structure of the long-tailed titmouse, the basin-shaped home of the thrush, and the clumsy structure of the rook, are all so beautifully adapted to the situation which they occupy, that to remove them from their surroundings is to deprive them of half their value.

Although the leaves cannot be induced to retain their form or colour, and always become crisp, and dry, and shrivelled, and brown, the branches still keep their form, and, if properly managed, may be made to retain their position. The best plan for restoring the nest to its original appearance is to substitute for the dried foliage a new set of artificial leaves, which are now made so true to nature that they can scarcely be distinguished from their living models. Only, it is to be hoped that the arsenical green will not be used, not only on account of its poisonous qualities, but also of its peculiar hue, which is quite unlike that of living leaves. The life-like appearance of the bark can easily be restored by the judicious use of colour, moistened wafers, and varnish.

The eggs, too, should always be made to appear in their natural hues, which in many instances are lost when the contents are removed. This is invariably the case with all of the smaller eggs where the shell is not deeply coloured; and in some instances, such as the egg of the kingfisher, the swift, the dipper, and the sand-martin, the colour of the egg is changed from delicate pink to chalk-white. I always renew the colour of these eggs by injecting a mixture of carmine and gamboge—a single drop is sufficient for a small egg; and in order to prevent it from drying in streaks and blots, I hold it over a spirit-lamp, or before a fire, and turn it continually until it is quite dry. An unblown egg should be kept as a model whereby the colour can be precisely determined; and when it is properly done, the effect is very beautiful. A pure white egg, like that of the kingfisher, is much improved by heating the shell, after the colour is dry, and then injecting a little boiling wax, so as to back up the colour, and restore the beautiful translucence of the unblown egg. A tiny scrap of silver-paper should then be fastened over the orifice, in order to prevent dust from entering.

To return to our Honey-Eaters. The material of which the nest of the Painted Honey-Eater is composed is fine fibrous roots, interwoven very artfully, but loosely, and being of so frail a structure, that much care is required to remove it without damage. It is fastened by the rim to the delicate twigs of the beautiful weeping acacia (*Acacia pendula*), whose long lanceolate leaves droop over and nearly cover it. It is a very small nest in proportion to the size of the bird.

STILL keeping to the same interesting family of birds, we find among the pensile builders another species of Honey-Eater.

The WHITE-THROATED HONEY-EATER (*Entomophila albogularis*) is rather like the Painted Honey-Eater, being brown above, white below, and having a yellow patch on the base of the tail. It is, however, easily distinguished from its congener by the peculiarity from which it derives its name—viz. a large patch of pure white in the front of the throat, extending as far as the eyes. The top of the head is greyish blue, and the breast is buff.

It is a lively, active little creature, ever on the move, and delighting to flit from branch to branch, but not caring to make long flights. As it flies from one bough to another, it utters a musical little song, much like that of the goldfinch, and continues to sing for a considerable time. It detests wind, and is mostly seen in the thick bush, and loves to frequent the masses of mangroves which edge bays and creeks, because the air is comparatively still. In these places may be found its curious nest, which is about as large as a breakfast-cup, and very much of the same shape. It is made of the delicate paper-like bark of the Melaleucæ, and various vegetable fibres, with which it is ingeniously hung to the branches. The broad, thin bark causes it to be very smooth on the exterior. For the lining, the bird is not indebted to any animal or bird, but uses grass-blades, which are neatly laid, and form a soft resting-place for the eggs.

The nest is placed very low, being often found scarcely two feet from the water, in that point resembling the nest of the African weaver birds, which have already been described. It is always hung near the extremity of a branch, and invariably is so placed as to be under the protection of a spray of leaves, which act as a roof whereby the rain is thrown off.

In order that these singular Australian nests which have been described may be compared with each other at a glance, five of the most remarkable examples have been placed in the same illustration, and by comparing the description with the figures, a better idea will be obtained than if each had formed the subject of a separate illustration.

BEFORE proceeding to describe another remarkable pensile builder of Australia, I must draw the attention of the reader to a bird of our own country, which oftens builds a pensile nest, in some respects resembling that of the White-throated Honey-Eater. Want of space forbids me to introduce an illustration of this exquisite little creature, including a figure of its nest, which equals in beauty the home of many foreign birds. As, however, this nest is tolerably familiar, and examples can always be obtained, I have preferred to insert figures of the nests made by exotic and less known birds.

The GOLDEN-CRESTED WREN (*Regulus cristatus*), whose form and colours are so well known as to require no description, builds a beautifully neat little nest, thickly lined with feathers, in which the minute little nestlings can lie securely. The nest is always placed under the protection of a natural roof, a spray of leaves being a favourite spot. Almost invariably the nest is fairly suspended, and in several instances I have noticed that three branches were used for the purpose.

WE will now return to our Australian birds.

There is a genus of very small birds, called Dicæum, which is spread over many parts of the world, and finds several representatives in Australia. All are interesting birds; but as the present work only treats of birds as the architects of their nests, it is necessary to select one which builds a pensile habitation. This is the SWALLOW DICÆUM (*Dicæum hirundinaceum*), a bird scarcely as large as our common wren, and glowing with brilliant colours, the whole of the upper part being deep, glossy blue-black; the throat, breast, and under tail-coverts of a fiery scarlet; and the abdomen pure white. It has a very sweet though low and inward note, so faint as scarcely to be audible from the tops of the trees, but continued for a long time together.

Artificial aids to vision are required in order to watch the

SWALLOW DICÆUM.

habits of the Dicæum, for it loves the tops of the tallest trees, where its minute body can scarcely be seen without the assistance of glasses. The Casuarinæ are favourite trees with this bird, which is fond of flitting about the branches of a parasitic plant called loranthus, which bears viscid berries. It is not precisely known whether the bird haunts the loranthus for the sake of the berries or of the insects, but as the Dicæum is one of the insect eaters, the latter supposition is probably correct.

It is very seldom if ever seen on the ground, and its flight among the upper branches is quick, sharp, and darting.

The nest of the Swallow Dicæum is as pretty as its architect, and its ordinary shape can be seen in the accompanying illustration, though the plain black and white of a wood engraving can give but little idea of its full beauty. In colour it is nearly pure white, being made of the cotton-like down which accompanies and defends the seeds of many plants, and this material is so artfully woven that the nest almost looks as if it were made from a piece of very white cloth. It is always purse-like in form, though its shape is slightly variable, and is suspended by the upper portion to the twigs at the very summit of the tree. Generally it hangs its nest upon the parasitic plant which has already been mentioned, but it often selects the Casuarinæ, or the delicate twigs of the myall or weeping acacia, for that purpose. The average number of eggs is five, and their colour is greyish white thickly powdered with small brown specks. Their length is about three quarters of an inch, and their breadth rather less than half an inch.

There is a nest of another Australian bird which has some resemblance to that of the Swallow Dicæum, namely, the home of the species called *Malurus cyaneus,* one of a rather large group of birds, which are peculiar to Australia. Like the habitation which has already been described, the nest of the Malurus is placed very high on the tree, and is purse-like in shape, having an aperture in the side through which the bird can pass. The Malurus belongs to the same group of birds as the remarkable emeu-wren, so well known for its long, hair-like tail-feathers, and its odd custom of holding that appendage erect as it trips over the grass. The Malurus has the same habit, though its tail is comparatively short, and does not attract much attention.

In a previous page it has been mentioned that a bird was undoubtedly the first tailor, and used needle and thread ages before man had invented such implements. We now come to a bird which may be accepted as the first hammock-maker, its nest being made of a hammock-like shape, and slung just as a seaman slings his oscillating couch. Scarcely any more comfortable bed could be invented, provided that it be properly suspended, and the bird certainly deserves our gratitude, if it be only for the fact that it *might* have given the first hint on the subject.

It is one of the Honey-Eaters, and is called the LANCEOLATE HONEY-EATER (*Plectorhynchus lanceolatus*), on account of the

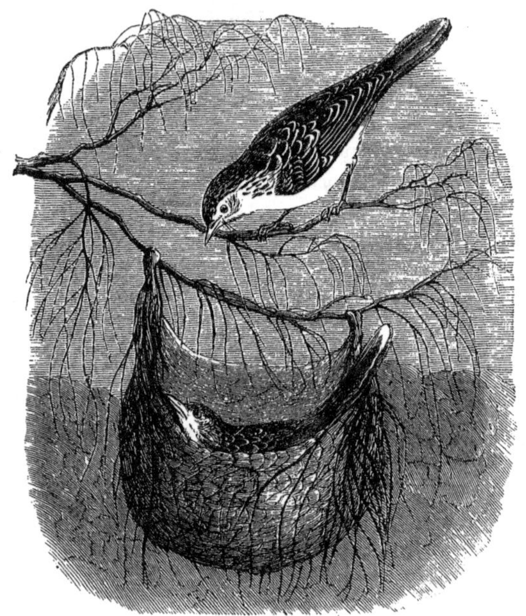

LANCEOLATE HONEY-EATER. (*Plectorhynchus lanceolatus.*)

shape of its feathers. It is not a brilliantly coloured bird, its hues being only brown and white, diversified by a black line down the middle of each feather. It does not seem to be a very lively bird, being accustomed to sit on the very top of some lofty tree, such as an acacia or eucalyptus, and to remain almost motionless in one spot. So still and quiet is it that it would hardly be seen, were not its presence betrayed by an occasional powerful and shrilly-sounding whistle. Its food consists partly of insects, and partly of the pollen and sweet juices of flowers.

The wonderful nest of this bird was found by Mr. Gould on the Liverpool Plains, overhanging a stream, and being a beautiful example of the pensiles. The materials of which it is made are grass and wool, intermingled with the pure white cotton of certain flowers. As the reader may see, by reference to the illustration, it is hung from a very slender twig, and only suspended at opposite extremities of the rim, the tree selected being the myall, or weeping acacia. The nest is rather small in proportion to the bird, and is very deep, so that when the mother is sitting on her eggs, or brooding over her young, she is obliged to pack herself away very carefully, her tail projecting at one side of the nest and her head at the other.

OUR last example of the Australian pensile nests is one which is made by the WHITE-SHAFTED FANTAIL (*Rhipidura albica*), a native of Van Diemen's Land and the southern and western portions of Australia. It is rather a pretty bird, being boldly marked with black and white, and is remarkable for the fact that the shafts and tips of the tail-feathers are pure white, the central feathers only excepted. It derives its popular name of Fantail from its habit of spreading its tail like a fan while descending, and as the tail is very broad, the action has a really remarkable effect.

The nest of this bird is of a figure not very easy to describe, but an idea of it may be formed from a common wine-strainer, with a very long and straight spout. The nest is attached to a branch rather below the middle of the cup, so that the long spout hangs down like a tail, quite independent of the bough. What can be the object of this appendage no one knows, and there is no purpose that it can even be imagined to fulfil, except perhaps that it may serve as a conductor. Like many other pensile nests, it is placed at a low elevation, and hung over water. Sometimes, however, it is found in a forest where no stream runs, but even in such a case it is suspended not many feet from the ground, though high enough to guard it against the attacks of any ordinary foe.

The materials of which the nest is made are the delicate inner bark of the gum-tree, together with mosses, and the soft down obtained from the tree-fern. These substances are interwoven with tough spiders'-web, which has the effect of binding them firmly together. This remarkable nest is mentioned in the present place because its peculiar shape bears some resemblance to certain pensile nests formed by the humming birds, and which will presently be described.

The bird itself is a lively and amusing little being, not only active on the wing, but singularly bold and confiding in character, betraying little fear of man, and even entering houses when engaged in chasing insects. These attributes, however, entirely disappear during the breeding season, when the little bird becomes as shy, as suspicious, and as timid as it was formerly bold and confiding. It cannot endure that a human being should even approach its nest, and in order to draw off his attention, acts after the manner of the lapwing, and by feigning lameness endeavours to decoy the intruder in another direction. The White-shafted Fantail rears at least two broods in a season, and has occasionally been known to produce a third. There are only two young in each brood, so that the parents are not subject to very hard work when rearing their offspring.

These birds are generally seen in pairs, but are not gregarious, and, as far as is known, they are permanent residents in Australia, merely shifting their quarters at the different seasons.

CHAPTER XIII.

PENSILE BIRDS (CONTINUED).

American Pensile Birds—Humming Birds, and the general structure of their nests—The LITTLE HERMIT, its colour, habits, and nest—The GREY-THROATED HERMIT and its hardihood—The PIGMY HERMIT and its seed-nest—The LONG-TAILED HUMMING BIRD—Mode of building its nest—The WHITE-SIDED HILL STAR—Curious method of suspending its nest—The SAPPHO COMET—The CHIMBORAZIAN HILL STAR—Curious locality—Its habits, food, and nest—The SAWBILL and its singular nest—Habits of the Sawbill—The BRAZILIAN WOOD NYMPH—Use made of its plumage and its nest—The RUBY AND TOPAZ HUMMING BIRD—Stuffed Skins—The AZURE CŒREBA, its colour, nest, and habits—The BALTIMORE ORIOLE—Reason for its name—Its beautiful nest, and curious choice of materials—Familiarity of the Baltimore Oriole—The ORCHARD ORIOLE, or BOB-O'-LINK—Various forms of nest—Why called Orchard Oriole—The CRESTED CASSIQUE, its size, form, and colours—Its remarkable nest—Difficulty of obtaining nests—The GREAT CRESTED FLYCATCHER, and its use of serpent-sloughs—The RED-EYED FLYCATCHER, or WHIP-TOM-KELLY—Low elevation of its nest—The WHITE-EYED FLYCATCHER, its nest, and fondness for the prickly vine—The PRAIRIE WARBLER, its habits and nest—The PINE-CREEPING WARBLER—The Asiatic pensiles—The BAYA SPARROW—Its colour and social habits—Singular form of the nest.

HAVING now taken a cursory glance at the pensile nests constructed by the feathered inhabitants of Africa and Australia, we again cross the sea and come to America. There are many pensile builders among American birds, and chief among them are the exquisite little creatures called the HUMMING BIRDS, which are peculiar to America and her islands.

Among the multitudinous species of this wonderful group of birds are very many examples of pensile nests, that mode of structure being, indeed, the rule, and any other the exception. As is the case with the nests of the Australian birds, some are suspended from twigs, others from rocks, and others again from leaves, the last-mentioned plan being the most common. It is evident that, in order to enable a nest to be fastened to a leaf, some very tenacious substance must be employed; and this is found in the webs of various spiders, some of which are of wonderful strength and elasticity—as strong, indeed, as the silken lines of our well-known brown-tailed moth, which, though tightly stretched, can be pulled without breaking, and spring back to their former position like a harp-string. There is also a great variety in spiders' webs, so that the birds can procure at will the long elastic threads with which the materials of the nest can be tied together, or the soft felt-like substances with which the moss, bark, and fibres can be interwoven, so as to form a firm and wet-resisting mass.

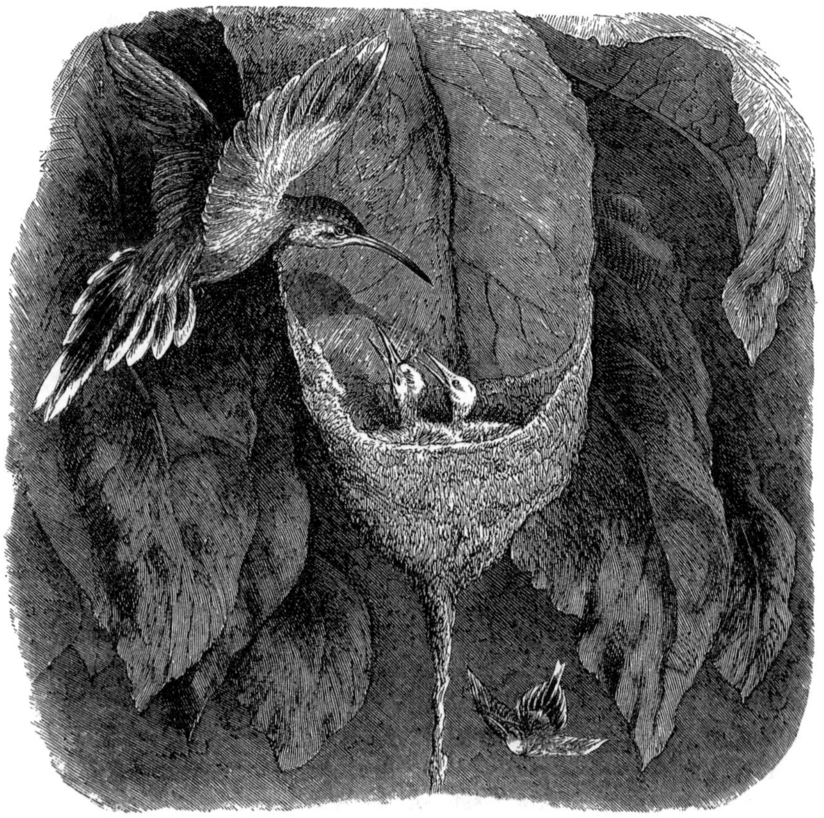

LITTLE HERMIT. (*Phaëthornis eremita.*)

Our first example of the pensile Humming Birds is the beautiful species called the LITTLE HERMIT (*Phaëthornis eremita*), a bird which is known by the warm ruddy colour of the under parts, and the black crescent on the breast. There are many species of Hermit Humming Birds, inhabiting Venezuela and the Caraccas, and choosing those districts where the flora is most abundant. They are all remarkable for two peculiarities, the first being the form of the tail, which is regularly graduated, the two central feathers being the longest, and the others diminishing on either side. The second peculiarity is, that the two sexes are nearly alike in their colouring, contrary to the usual custom among humming birds, the male of which is generally brilliantly clad, and the female quite plain and sombre. All those Hermits whose habitation is known build a curiously formed nest, funnel-shaped, and attached to the end of some drooping leaf.

The example which has been chosen for illustration affords a good idea of the form which is generally followed, and as may be at once seen, closely resembles that of the fantailed warbler, which has already been described.

The nest which is here figured was attached to the very extremity of the leaf, so that the long tail hung down freely. The materials of which it was composed were the silky fibres of plants, the cotton-like down of seed vessels, and some other substance, which is supposed to be fungus, and is of a woolly texture. All these materials were interwoven with spider's-web, by means of which the nest was attached to the leaf at the end of which it swings. The bird almost invariably chooses some dicotyledonous leaf for its pendant home.

Other nests made by birds of the same genus are worthy of a passing mention.

First, there is the pretty nest of the Grey-throated Hermit (*Phaëthornis griseogularis*), a very tiny bird, of comparatively sober plumage, reddish brown being the predominant hue. This species is found in Ecuador, and is seen at an elevation of six thousand feet above the level of the sea. Indeed, the depth of cold which these fragile little beings can endure is really surprising, many species being found only on the highest mountains, and one bird, the Chimborazian Hill Star, inhabiting a zone that is never less than twelve thousand feet, and seldom more than sixteen thousand; above the level of the sea. Immediately above the last-mentioned elevation the line of perpetual snow begins, and though the bird can exist just below it, the absence of vegetation prevents it overpassing that line.

The nest of the Grey-throated Hermit is made of moss fibres and the same silken threads that have already been mentioned, and is fastened to a leaf. It does not, however, hang from the extremity, but is fastened against the side of the leaf, and its tail, if we may so call the lengthened appendage, is not free, but attached to the leaf in the same manner as the nest.

Another species, *Phaëthornis Eurynome*, makes its nest of the tendrils of certain creepers, together with delicate root-fibres, and attaches it to the leaf of some palm by means of cobwebs.

Our last example of this group is the tiny species called the PIGMY HERMIT (*Phaëthornis pygmæus*), a pretty little creature, though scarcely a brilliant one, and decorated with green-bronze above and warm red below. The nest of this species is fastened to a leaf, like that of the grey-throated hermit, and is also deep and cup-shaped, with an appendage so long as to give the whole nest a shape resembling that of a funnel. It is remarkable for the great use of which this little architect makes of seeds, the exterior being covered with downy seeds, and the interior lined with similar down, and the delicate fibres of flowering plants.

THE reader will remember that on page 228, it was mentioned that the general form of the nest made by the white-shafted fantail was similar to that constructed by one of the humming birds. The species in question is the *Hylocharis cyaneus*, and the nest resembles that of the fantailed warbler in two points; first, the manner in which it is suspended, and next, the formation of the appendage. Instead of being fastened to the side of a leaf, as is the case with the habitations of the hermit humming bird, this nest is placed upon a twig, so that it is supported by the cup, and the appendage hangs freely below. In other respects the nest is similar to those which have been described.

As the reader may like to know how the tiny architect achieves her graceful task, I introduce a passage wherein Mr. Gosse describes, with the vivacity of an eye-witness, the manner in which the female LONG-TAILED HUMMING BIRD (*Trochilus Polytmus*), builds its pensile home. The nest is made of fine moss, cotton fibres, spider's-web, and studded externally with lichens, and is one of the true pensiles, sometimes being found over water, and in one instance overhanging the sea waves, suspended to a twig of wild vine. The account is as follows :—

" Suddenly I heard the whirr of a humming bird, and on looking up, I saw a female Polytmus hovering opposite the nest with a mass of silk-cotton in her beak. Deterred by the sight of me, she presently retired to a twig a few paces distant, on which she sat. I immediately sank down among the rocks as gently as possible, and remained perfectly still. In a few seconds she came again, and after hovering a moment, disappeared behind one of the projections, whence in a few seconds she emerged again and flew off. I then examined the place and found to my delight a new nest.

" I again sat down on the stones in front, where I could see the nest, not concealing myself, but remaining motionless, waiting for the bird's reappearance. I had not to wait long ; a loud whirr, and there she was, suspended in the air before her nest. She soon espied me, and came within a foot of my eyes, hovering just in front of my face. I remained still, however, when I heard the whirring of another just above me, perhaps the mate, but I durst not look towards him, lest the turning of my head should frighten the female. In a minute or two the other was gone, and she alighted again on the twig, where she sat some little time preening her feathers, and apparently clearing her mouth from the cotton fibres, for she swiftly projected the tongue an inch and a half from the beak, continuing the same curve as that of the beak.

" When she arose, it was to perform a very interesting action, for she flew to the face of the rock, which was thickly clothed with soft downy moss, and, hovering on the wing as if before a flower, began to pluck the moss until she had a large bunch of it in her beak. Then I saw her fly to the nest, and having seated herself in it proceeded to place the new materials, pressing and arranging and interweaving the whole with her beak, while she fashioned the cup-like form of the interior by the pressure of her white breast, moving round and round as she sat. My presence appeared to be no hindrance to her proceedings, although only a few feet distant ; at length she left the place and I left also."

The bird whose proceedings are thus vividly described is a lovely species, remarkable for the very great length of the two central tail-feathers, which are very narrow, and twice as long as the bird from the point of the beak to the root of the tail. They cross each other as the bird is at rest, and their colour is deep purple-black. The rest of the plumage is most lovely, the upper parts being green, with a golden gloss, and the throat and lower parts emerald-green. The top of the head is deep velvet-like black, and surmounted with a small plume. These are the colours of the adult male, the female being without the two long feathers in the tail, the top of the head brown, and the throat and breast white slightly speckled with green.

In the accompanying illustration may be seen figures of the nests made by three different species of humming birds, each of which is remarkable for some peculiarity of structure, though they are all pensile.

The first of these nests is that which is made by the White-sided Hill Star (*Oreotrochilus leucopleurus*); a native of the Andes of Acoucagua, inhabiting a zone of very great elevation, seldom being seen less than ten thousand feet above the level of the sea. With the exception of a bright emerald-green gorget, it is rather a dull-coloured bird, the prevailing hue being brown. The nest is shaped something like a hammock, not unlike that of the lanceolated honey-eater, described and figured on page 226, and is fastened, not to a twig or a leaf or a branch, but to the side of a rock, being suspended by one side so as to leave the remainder free.

As is the case with the generality of humming birds' nests, cobwebs are employed for the purpose of fastening the structure to the object to which it hangs. The materials of which the nest is made, are chiefly moss, down, and feathers, the feathers being profusely stuck on the outside.

This is not the only humming bird which hangs its nest from rocks, for the lovely Sappho Comet (*Cometes sparganurus*), sometimes called the Bar-tailed Humming Bird, on account of the dark bars which cross its tail, has a similar custom.

This splendid bird inhabits Bolivia, and is a very familiar and bold little creature. The nest is made chiefly of vegetable fibres and moss, and furnished with a long appendage similar to those which are made by so many humming birds, for no conceivable reason. The nest is lined with hair, probably that of the viscacha, one of the llamas, and is hung against the side of a rock or a wall, sometimes being attached to the wall itself, but generally suspended from some twig or hanging root. The bird always selects some spot where the nest can be sheltered by an overhanging ledge of rock, probably because the very loose structure of the nest requires that some such precaution should be taken. The portion of the nest that rests against the wall is always looser than the remainder of the structure. There are two eggs, of a rather large size for humming birds to lay, being about half an inch in length.

THE CHIMBORAZIAN HILL STAR (*Oreotrochilus Chimborazo*) also hangs its nest against perpendicular rocks.

This singular bird, like its congener, the White-throated Hill Star, prefers elevated situations, and is found at still higher altitudes. Specimens are never found less than twelve thousand feet above the level of the sea, and are frequently seen at the astonishing elevation of sixteen thousand feet, where it thrives in spite of the extreme cold which continually reigns and checks the magnificent flowering plants which are so plentiful towards the foot of the mountain. As may be seen from the name of the bird, it lives upon Chimborazo, and feeds upon the juices of the yellow *Chuquiraqua insignis*, an alpine plant with large blossoms.

It is not a brilliantly coloured bird, the general colours being pale dusky green, relieved by a splendid emerald patch upon the chest. The nest of the Chimborazian Hill Star is made chiefly of lichens, and is hung against the sides of some perpendicular rock, where it is sheltered by a shelf overhead. There is another Hill Star which much resembles this species, but does not possess the green patch on the throat. This species also inhabits a volcanic mountain, being confined within a narrow zone of some two hundred yards in width. It is a remarkable fact, that though this species, which takes its name of PICHINCHIAN HILL STAR from the mountain on which it resides, is placed within thirty miles of the Chimborazian Hill Star, neither species is ever found upon the mountain which is appropriated to its congener. Pichinchia is in the republic of Ecuador.

THERE is a very remarkable nest made by one of these birds, called the SAWBILL HUMMING BIRD (*Grypus nævius*), because the slender bill is notched in a saw-like fashion on the edges of both mandibles. These serrations do not reach along the whole bill but only to a short distance from the tip. In appearance it is not so strikingly beautiful as many of the humming birds, and is chiefly remarkable for its reddish throat dotted with white, and having a black mark down its centre. There are several humming birds which have the serrated edges to the mandibles, and are in consequence called Sawbills, but the present species is the most worthy of notice with respect to its nest. It is only found in the south of Brazil.

SAWBILL HUMMING BIRD. BRAZILIAN WOOD NYMPH. WHITE-SIDED HILL STAR.

The nest of the Sawbill is made of fine vegetable fibres, woven together so as to look like an open network purse, the outer walls being so loosely made as to permit the eggs and lining to be visible. Leaves, mosses, and lichens are also woven into the nest, and are packed rather tightly under the eggs. The edge, however, is always left loose. The nest is suspended at the end of some leaf, usually that of the palm.

Mr. Gould mentions that the bird is found in the depths of virgin forests, and is most plentiful about thirty miles from Nova Fribergo, in the months of July, August, September, and part of October. It is generally seen darting round the orchidaceous plants which flower so richly in that fertile climate, and is a rather noisy bird, uttering loud and piercing cries, and making a great whirring sound with its wings as it dashes through the air. It is very strong and energetic on the wing, and is seldom seen to alight. That the Sawbill feeds on insects has been satisfactorily proved, by the presence of small beetles in the throat of newly killed birds; and to judge by its actions, the hovering flight and frequent stoop like that of the falcon, the bird feeds also on flies and other winged insects.

ALTHOUGH it is necessarily impossible to describe or even enumerate one tithe of the interesting nests made by humming birds, I must cursorily mention one or two more of the most curious examples. One of these birds is the BRAZILIAN WOOD NYMPH (*Thalurania glaucopis*), a species which is perhaps more persecuted than any other, its singular beauty causing its plumage to be sought after.

The feather on the crown of the head and front of the throat are of the most lovely azure, and are largely used by the inmates of several convents at Rio Janeiro for the purpose of being made into the beautiful feather flowers which the nuns manufacture so skilfully. Thousands of these birds are slaughtered merely for the crest and gorget, but so prolific are they, and so ingeniously do they hide their nests, that the persecution of many years has scarcely diminished their numbers. Moreover, fortunately for the preservation of the species, the colours of the female are so dull and sober, that her feathers are of no value, and she is allowed to escape the fate that befalls the more brightly coloured male. It is a lively little bird, and when alarmed utters a hurried cry, sounding like the word, " Pip, pip, pip," very sharply pronounced.

The nest of the Brazilian Wood Nymph is exceedingly pretty, and is hung to the tip of some delicate twig, generally that of one of the creeping plants which trail their long stems so luxuriantly over the branches of the great forest trees. The walls of the nest are made of vegetable fibres, generally taken from the fruit of some palm, and upon the outside are fastened many patches of flat lichen, so that the whole nest, which is very long in proportion to its width, may easily escape detection.

THE second species is to be found in every collection of humming birds, and even the glass cases of these creatures which are sold in the shops, are seldom without a specimen of the RUBY AND TOPAZ HUMMING BIRD (*Chrysolampis moschitus*). It derives its name from the rich ruby red which decorates the crown of the head, and the fiery topaz which blazes on the gorget. This species has a very wide range of residence, being found throughout Bahia, all the Guianas, Trinidad, and the Caraccas, and is killed by thousands for the sake of its plumage. I was about to say for the sake of its skin, but as that expression would imply that the humming birds seen in cases are all skinned and stuffed it cannot be rightly used.

A stuffed humming bird is very seldom seen, though thousands are annually sold under that name. In fact, the birds are so tiny, and the amount of flesh is so small, that very few persons care to take the trouble and run the risk of skinning such minute creatures, and content themselves with removing the inside, supplying its place with cotton, inserting wires, as is customary in birds stuffed according to the present fashion, fixing the birds in appropriate attitudes, and then drying them, trusting to the feathers to cover deficiencies. Of course the soft and rounded contours are lost by so rough a process, but as the general public that buys stuffed birds is too uncritical to perceive such defects, and too indifferent to trouble themselves about them, even when pointed out, the professional taxidermists have no inducement to waste their time upon tedious and unremunerative work.

WE now leave the Humming Birds, and pass to other inhabitants of America.

Still keeping to Brazil, we come upon another pensile bird, called the AZURE CŒREBA (*Cœreba cyanea*). This beautiful little creature scarcely yields to any of the gorgeous humming birds in the glory of its plumage, and far exceeds many of them in the fiery brilliance of its hues. Blue is the chief colour in this Cœreba, and, strange to say, different qualities of blue are found in the same bird, without jarring with each other, so wonderfully are they dispersed and so artistically are the various shades separated by velvet-black stripes and patches. The greater part of the body is rich azure, with the exception of a velvet-black stripe that runs round the crown of the head, and widens into a patch on the back of the neck. The quill-feathers of the wing are also black, and a black streak is drawn from the corner of the mouth to the neck, enveloping the eye in its course.

Separated from the azure blue of the body by the black streak just mentioned, a large patch of feathers on the top of the head glows and flashes with metallic splendour, and is of a vivid verditer blue.

The nest of the Azure Cœreba is pear-shaped in form, the hollow for the eggs and young being in the large rounded portion, and the slender part of the pear representing the "tail" of the nest, which is long and slender, like that of many birds which have already been mentioned, except that instead of being solid and pointed, it is hollow and has the opening to the nest in the extremity. In order, therefore, to reach the nest proper, the bird is obliged to enter from below and climb up the hollow shaft, as is the case with some of the African weaver birds. The substances of which the nest is made are long vegetable fibres and slender grasses, and the manner in which these simple materials are woven into so beautiful a nest is remarkably ingenious, and may challenge comparison with the architecture of any other bird.

The Azure Cœreba is a small bird, about the size of our sparrow, but with a long, slender, and slightly-curved beak, as is mostly the case with the large and important family to which it belongs. It feeds chiefly on insects, and may be seen busily engaged among the flowers of its native land, flitting from one blossom to another, and daintily extracting the minute insects that endeavour to conceal themselves within the recesses of the petals.

Still keeping to America, we may see more examples of pensile nests. Two differently-shaped specimens are given in the accompanying illustration, in order that they may be compared with each other.

The first in order is that of the Baltimore Oriole (*Yphantes*

Baltimore), a pretty bird, coloured with orange and black in bold contrast to each other. Its name is derived, not from any particular locality, but from the orange and black of its plumage, those being the heraldic colours of Lord Baltimore, formerly proprietor of Baltimore. It does not receive the full colouring until its third year, the orange hues being simply yellow at the

CRESTED CASSIQUE. BALTIMORE ORIOLE.

end of the second year, and having no red in them until the last moult is completed. So far, indeed, is it from belonging to any particular locality, that it is spread over a very wide range of country, inhabiting the whole of America from Canada to Brazil. The Baltimore Oriole goes by many names; some, such as Golden Robin and Fire Bird being in allusion to its plumage, and others, such as Hang-nest and Hanging Bird, from the beautiful pensile nest which it makes.

The general shape of these nests is much the same in every specimen, and a good idea of it may be formed from the illustration, which was taken from a nest in my own possession. It is always pensile, and is hung by the rim to the under side of some slender bough, usually at a considerable elevation from the ground. It is almost entirely made of vegetable fibres, and is so strongly constructed, that, although it had been knocked about for some years in the neglected spot whence I rescued it, and was once crushed into a shapeless mass at the bottom of a wine hamper by a careless servant, and covered with soot and dust, it has retained its form, and shows perfectly well how the fastening to the branches was managed.

The materials of the nest are, however, extremely variable, the the bird having a natural genius for nidification, and being always ready to take advantage of any new discovery in architecture. One of these nests, described by Wilson, was deeper in proportion than the specimen which has been figured, being five inches in its widest diameter and seven in depth, the opening being contracted to two and a half inches. Various materials, such as flax, tow, hair, and wool, were woven into the walls, which were strengthened by horsehairs, some two feet in length, sewn through and through the fabric. Cow's hair was also employed for the bottom of the nest, and, like the walls, was sewn together with long horsehairs.

The same writer remarks, that "so solicitous is the Baltimore to procure proper materials for his nest, that in the season of building, the women in the country are under the necessity of narrowly watching their threads that may chance to be out bleaching, and the farmer to secure his young grafts; as the Baltimore, finding the former, and the strings which tie the latter, so well adapted for his purpose, frequently carries off both. Or, should the one be over heavy, and the other too firmly tied, he will try at them for a considerable time before he gives up the attempt. Skeins of silk and hanks of thread have often been found, after the leaves were fallen, hanging round the Baltimore's nest, but so woven up and entangled as to be entirely irreclaimable.

"Before the introduction of Europeans, no such materials

could have been obtained here; but, with the sagacity of a good architect, he has improved this circumstance to his advantage, and the strongest and best materials are uniformly found in those parts by which the whole is supported."

This bird is very fearless, and, like some other species, is fond of the society of mankind, building in gardens and orchards, and piping its mellow notes within the very streets, in calm defiance of the roar and rattle of town life. This fearlessness of disposition enables observers to watch its proceedings very closely, and in general the bird is found to begin its nest by working the strongest threads or strings round a forked branch, so as to mark out the entrance, and then by weaving the remainder of the nest upon the strings. The neatness and strength of construction are, however, very variable; and it is suggested by Wilson that the inferior nests are probably made by young and inexperienced birds, their architectural powers increasing with practice.

A CLOSELY allied species, the ORCHARD ORIOLE, or BOB-O'-LINK (*Xanthornis varius*), is equally notable for its skill in nest-building—if such a word may be used of a structure which is begun at the top and carried downwards, after the fashion employed in Laputa.

It is a pretty bird, but not so pretty as the Baltimore Oriole, and the tints are very differently disposed, scarcely any two individuals having the colours in exactly the same places. Like the Baltimore Oriole, it is extremely variable in different stages of its existence, the young male bearing great resemblance to the mature female, and not attaining its full beauty until its third year. When adult, the whole of the head, neck, upper part of the back, breast, wings, and tail, are deep black, and a rich ruddy chestnut hue occupies the remainder of the breast, the under parts of the body, and part of the wing-coverts, some of which are tipped with white. The young male and the adult female are yellowish olive above, instead of black, with brown wings, and yellow on the breast and abdomen; while the male of the second year has much the same colours, but is known by a patch of black over the head and on the throat, together with a few chestnut feathers on the flanks and abdomen. It is smaller than the Baltimore Oriole, and more slenderly made.

The nest of this bird is almost as variable in structure as is its architect in colour, its form being accommodated to the situation in which it is placed. When fastened to a tolerably stout branch, its depth is less than its diameter, and it is firmly tied in several directions to prevent the wind from upsetting it. But when it is slung to a long and slender branch, over which the wind has great power, and which is swung to a distance of fourteen or fifteen feet in a smart breeze, the nest is made of much greater depth, and is of a lighter construction. The weeping willow is a favourite tree with this bird, as the drooping leaves conceal the nest effectually, and the delicate twigs can be gathered together so as to support the entire circumference of the entrance.

Wilson remarks, in allusion to these nests, that they "exhibit not only art in the construction, but judgment in adapting their fabrications so judiciously to their particular situations. If the actions of birds proceeded, as some would have us believe, from the mere impulses of that thing called *instinct*, individuals of the same species would uniformly build their nests in the same manner, wherever they might happen to fix it; but it is evident, from those just mentioned, and from a thousand such circumstances, that they reason, *à priori*, from cause to consequence, persistently managing with a constant eye to future necessity and convenience."

The popular name of Orchard Oriole is given to this species because it is a familiar and bold bird, not in the least fearing the vicinity of man, but rather seeming to find a protection therein, and loving to build its pensile nests in orchards. As is the case with many British birds, it long had an evil reputation which it did not deserve, and was thought to devour the ripe fruit of the trees in which the nest was placed. Cultivators now know better, and are aware that, so far from being a foe, it is one of their best friends, eating vast numbers of the noxious insects which infest fruit trees, and saving many a crop by its exertions to procure food for itself and young family.

Indeed, one of the nests has been observed to be completely overshadowed by a large bunch of apples, which had grown over the entrance, and had absorbed more than half the space through which the bird was accustomed to enter its home. Yet, although the destruction of the fruit would have been a positive convenience to the Oriole, not a single apple was touched, and the bird slid in and out of its nest as cautiously as if it were aware of the value set on the fruit, and determined not to injure it.

On the left hand of the Baltimore Oriole's nest is represented a very curious structure swaying in the wind, long, purse-like, and having the entrance near the top. This is the nest of the Crested Cassique, or Crested Oriole (*Cacicus cristatus*), and the bird itself is seen clinging to the lower part of the nest.

There are several species of Cassiques, all of which are natives of tropical America, and build nests of a similar structure. The Crested Cassique is the largest of the genus, equalling the common jackdaw in size, and its nest is larger and more striking than that of any other species. It loves the tallest trees, and may be seen actively traversing the branches in search of food, pecking here and there in haste as it trips along, or passing from one tree to another with a rapid darting flight, snapping at insects as it dashes through the air. Like the preceding species, it is fond of human society, and builds its pensile nest close to the habitation of man, so that its customs can be easily watched.

The bird is a handsome creature, the greater part of the body being rich chocolate, the wings dark green, and the outer tail-feathers bright yellow, this colour being displayed conspicuously as the bird flies, particularly when it makes a sharp turn in the air and is obliged to spread its tail-feathers rapidly. The beak of this species is very remarkable, being of a green colour, and extending far up the forehead. The head is adorned with a long pointed crest, from which its popular name of Crested Oriole is derived. In some favoured spots these birds are quite plentiful, producing a beautiful effect, as the variegated plumage gleams among the foliage, while the bird is engaged in its active quest after food.

The nest of the Crested Cassique is of great length and, as may be seen by the illustration, has the entrance like that of a pocket. The opening is rather small when compared with the size of the nest itself, and the bird always dives head foremost into its home, its yellow tail flashing a last golden gleam before it disappears. The nest is strongly built, and the materials are rather coarse, not in the least resembling the delicate and neatly rounded fibres of which many of the weaver nests are made. These nests often exceed a yard in length, and owing to their great size, are very conspicuous, as the wind sways them backwards and forwards from the bough.

The same may be said respecting the nests of other Cassiques, and the stay-at-home reader is often apt to wonder why the traveller does not ascend every tree on which he sees a nest, and bring it down. There are two reasons why such nests are not so common in European museums as their number would seem to promise. One reason is, that the trees are not easily climbed. Some of them run to a height of eighty or a hundred feet without a bough; others have stems of great girth and wondrous smoothness, so that to ascend them is as difficult as to climb a greased pole at a fair; others again, which do not appear to present any difficulties, have their stems beset with thorny spikes, from an inch to two inches in length, as strong as nails and as sharp as needles.

Supposing, however, that the traveller is a practised climber, and always carries with him a rope and climbing spurs, and that by dint of the pointed spurs sticking into the tree, and the strong leather gaiters repelling the thorns, and the rope enabling him to pull himself upwards, he has arrived at the branches, he still finds many an obstacle to overcome. In the first place, distances are mightily deceptive when viewed from below, and a nest which appears from the ground to be close to a certain branch, is found really to be some yards on one side, and as many above.

Most birds, especially the tropical birds, have a custom of placing their nests at the very ends of boughs, where the twigs could not sustain the weight of a monkey, much less that of a man; so that the adventurous climber finds himself scarcely nearer his object than when he stood upon the ground. Such nests can only be obtained by skilfully throwing a rope around the branch to which they are hung, drawing it up, severing it as near the nests as possible, and then lowering the whole to the earth.

Supposing, however, that he has successfully overcome his difficulties, and has been able to reach the nest, he still finds himself in a very awkward position on account of the multitudinous insects which swarm upon tropical trees, and the majority of which can either sting or bite savagely. There are many kinds of wasps, larger, fiercer, and more irritable than the little yellow and black insect which terrifies us so much in this country, and these creatures have a habit of fixing their nests among the branches, where they are concealed by the leaves, and cannot be seen by the climber until he nearly strikes them with his hands.

But the very worst of all his foes are the ants and termites, which infest the trees to a wonderful degree. The ants are of various kinds. There are arboreal ants, which make their nests among the branches, and there are terrestrial ants, which make their home under the earth, but ascend the trees in search of insects or to procure materials for their subterranean abode.

The termites, again, are found on many trees, and in some instances actually hollow out the branches, so that when the climber grasps a bough, for the purpose of hauling himself up by it, the treacherous branch breaks in his hands, and pours out a flood of angry insects, all provided with means of offence, and anxious to wreak their vengeance on the enemy. Even the natives, accustomed as they are to these pests of their woods, and versed in every method of foiling them, confess themselves worsted by the ants, and are often forced to yield the point to their tiny foes.

In some cases, they attack so fiercely, that the unlucky climber is perforce obliged to descend the tree with all speed, and envelop himself in smoke in order to rid himself of his adversaries; or, whenever a river flows beneath the branches, the tortured native is fain to fling himself into it, and to drown off the myriad insects who are burying their jaws, or stings, or both, in his flesh. A naturalist's labours in a tropical forest are very pleasant reading at home, but they are not quite so pleasant to perform, even setting aside the chances of fever, and snake bites, and the certainty of being sucked by thousands of mosquitos, sand flies, and other winged plagues.

Before leaving the American pensile birds, we must briefly notice one or two other species. The Flycatchers of all countries are generally notable for the beauty or eccentricity of their nests, one of the oddest being that of the Great Crested Flycatcher of America, which always uses the cast slough of snakes when building its nest. The reason no one seems to know, though several opinions have been offered; one person thinking the snake-slough is peculiarly grateful to the young birds which are intended to lie upon it; and another, that the presence of the cast slough acts as a scarecrow, and frightens away obnoxious birds. One conjecture is as good as another, and both are absurdly bad.

The species which we have now to notice is the RED-EYED FLYCATCHER (*Muscicapa olivacea*) popularly known as "Whip-Tom-Kelly," from its peculiar articulate cry, which is said to bear a strangely exact resemblance to the words "Tom Kelly, Whip-tom-kel-ly," and is uttered so loudly and briskly, that it can be heard at a considerable distance. It inhabits a tolerably wide range of country, being found from Georgia to the St. Lawrence, and in many parts is plentiful.

The nest of the Red-Eyed Flycatcher is small and very neatly made, and, contrary to the usual custom of pensile nests, is placed near the ground, seldom at a height of more than five feet. Bushes and dwarf trees, such as dogwood or saplings, are usually chosen by the bird when it looks about for a branch wherefrom to hang its nest. A wonderful array of materials is employed by the feathered architect, which makes use of bits of hornets' nests, dried leaves, flax-fibres, strips of vine bark, fragments of paper and hair, and binds all these articles firmly together with the silk produced by some caterpillars. The lining is made of fine grasses, hair, and the delicate bark of the vine.

The nest is wonderfully strong, so compact indeed, that after it has served the purpose of its architect, it is usurped by other birds in the following year, and saves them the trouble of building entire nests of their own. Even the mammalia receive some benefit from the nest, for the field-mouse often takes possession of it, and rears its young in the pensile cradle.

An allied species, the WHITE-EYED FLYCATCHER (*Muscicapa cantrix*), builds a very pretty pensile nest, and uses so much old newspaper in the construction of its home, that it has gone by the name of the POLITICIAN. The other materials used in the structure of the nest are bits of old rotten wood, vegetable fibres, and other light substances, woven together with wild silk, and the lining is mostly of dried grasses and hair.

The form of the nest is nearly that of an inverted cone, and it is suspended by part of the rim to the bend of a species of smilax, that is popularly called the prickly vine, and which grows in low thickets. The bird is very fond of this smilax and rarely chooses any other tree for the reception of its nest, so that the home of the White-Eyed Flycatcher is not very difficult to find; moreover, the bird is so jealous and so bold when engaged in rearing its young, that it betrays the position of the nest by scolding angrily as soon as a human being approaches the thicket, and by dashing violently at the intruder with impotent rage.

ANOTHER pensile species is the PRAIRIE WARBLER (*Sylvia minuta*), a bird which, as its specific name denotes, is of very small size, not reaching five inches in total length.

It is a lively little bird, but withal deliberately cool in its movements, flitting about among the foliage and grass with a quick, though jerking, regular movement and yet inspecting every leaf and blade with perfect composure; chirping feebly all the while, and allowing itself to be watched without betraying any alarm. The nest of this little bird is unusually small, even when the size of the feathered architect is taken into consideration, and when dry weighs scarcely a quarter of an ounce. The materials of which it is made are moss, mixed with rotten and very dry wood, fastened together with caterpillar-silk, and the lining is made of very fine and delicate fibres of grape-vine bark.

OUR last example of American pensile birds is the PINE-CREEPING WARBLER (*Sylvia pinus*), a pretty little species, which has many of the actions that characterize the titmice, flitting among the branches like these birds, and hanging head downwards from the twigs while looking for insects. Sometimes it runs along the ground, and is equally active there; and when disturbed, it flies upwards, and clings to the trunk of the nearest tree, the whole movement being so peculiar that the bird can be distinguished at a long distance.

The Pine-Creeping Warbler is found in the pine-woods of the Southern States, where it assembles in little flocks of twenty or thirty in number. Its nest is suspended from the horizontal fork of some small branch, and is made of strips of grape-vine bark and rotten wood, tied firmly together with caterpillar-silk. Sometimes the bird finds a hornet's nest, and rightly considering that the substance of which it is made is the driest and lightest rotten wood that can be obtained, robs the insect, and builds its own nest with the spoils. The interior of the nest is lined with the fine roots of plants and dry pine-leaves, which latter materials afford a softer bed than their shape seems to indicate.

As we are near the end of our list of pensile birds, we must turn to Asia for a specimen as remarkable as any which has yet been mentioned. This is the nest of the Baya Sparrow, sometimes called the Toddy Bird, a native of several parts of India, and found in Ceylon.

BAYA SPARROW

As may be seen by the illustration, the nests are variable in shape, and hang close to each other; indeed, the birds are very sociable in all their manners, and fly about in great numbers, flocks of thousands flitting among the branches and displaying their pretty plumage to the sun. They have no song, and can only chirp in a monotonous manner; but the want of song finds its compensation in the brilliancy of the plumage, which is mostly bright yellow, the wings, back, and tail being brown. They are particularly fond of the acacias and date-trees, and choose the branches of those trees for the suspension of their nests.

Sometimes the nest is only made for incubation, sometimes it is intended merely as an arbour in which the male sits while the female incubates her eggs, and sometimes it consists of the nest and arbour united, producing a most curious effect. This "arbour," in fact, serves precisely the same purpose as the supplementary nest of the pinc-pinc and other birds which have already been described.

CHAPTER XIV.

PENSILE INSECTS

The Hymenoptera—Australian Insects—The CREMATOGASTER and NEGRO-HEAD—The GREEN ANT, its Habits and Nest—An African species—Pensile Ants of America—The ABISPA, and its remarkable Nest—Ingenious entrance—The TATUA, or DUTCHMAN'S PIPE—Structure and Shape of its Nest—Firmness of the Walls—Average number of Cells in each Tier—The Common WASP as a Pensile Insect—Gigantic Nest—Union of three Colonies—Character of the Wasp—The NORWEGIAN WASP—Structure and Locality of its Nest—Classification of the Wasps—The CAMPANULAR WASP and the NORTHERN WASP—The CHARTERGUS or PASTEBOARD WASP—Mode by which the Nest is suspended—Method of Structure—Meaning of the Name—Enormous Nest from Ceylon—Various Wasp Nests—The POLISTES as a Pensile Insect—Singular Nest in the British Museum—The GIBBOUS ANT—Honey Wasps, the general characteristics of their Nests—The MYRAPETRA—Its singular Nest—Structure of the Walls and use of the Projections—The NECTARINIA—Why so called—Locality of the Nest—Size of the Insect—The TRIGONA and its Nest—Ichneumon Flies—Different species of MICROGASTER, and their Habitations—The PERILITUS—Weevils—Beautiful Cocoon of Cionus—The EMPEROR MOTH and its Home—The ATLAS MOTH and other Silk Producers—The HOUSEBUILDER MOTH and its movable Dwelling—The TIGER MOTH and its Hammock—The CYPRESS-SPURGE MOTH—Various Leaf-rollers—Suspended Cocoon—LEAF-BURROWERS and their Homes—The SPIDER.

WE now leave the birds, and proceed to the insects which make pensile nests. Some of them, such as those which will be first described, do not become pensile architects until they have attained their perfect state; while many others form their nests, either as a place of refuge during their larval life, or as an asylum in which they can rest while in the transition state of pupa.

Just as the Hymenoptera are the best burrowers, so are they the best insect artizans when the nests are suspended, and we shall therefore take them first in order. The reader will probably recall to mind during the perusal of the following pages, that several admirable examples of pensile nest-makers are not mentioned. The reason for their temporary omission is, that some of them make their nests of mud, and will therefore be described under the head of Builders; while others make their joint homes on so large a scale that they will be considered under the head of Social Nest-makers.

UPON the large illustration will be seen several examples of pensile nests; and, as many Australian insects are remarkable for the beauty and singularity of the pensile nests which they build, I have selected three of the most remarkable instances for illustration. Adhering to the principle which has been followed throughout the work, the scene of the drawing has been laid in Australia, and the general contour of the country, the peculiar foliage, the animals which enliven the scene, and the singular manner in which a wooded district is often dotted with trees, have been carefully represented.

In the upper corner of the drawing is seen the large nest of a remarkable ant, called *Crematogaster læviceps.* I do not know whether this species has any particular name, but in the Brazils an allied species goes by the name of Negro-head Ant, because the nest is round, like the bullet-shaped head of a Negro, and is covered on the exterior with little projections that are supposed to resemble the close woolly hair.

When the ant runs about, it has a curious habit of holding its abdomen so high in the air that it curves over the back and overhangs the thorax, a peculiarity which has earned for the genus the name of Crematogaster, or "hanging-belly." At first sight the nest bears a close resemblance to the pensile habitation of certain wasps, but when subjected to a nearer examination it proves to be even more complicated, being composed of multitudinous curved and intricate ramifications, all leading to the interior galleries and cells.

There are other ants which have the habit of carrying the abdomen erect, such as *Myrmica Kirbii,* and *Formica elata.* The former of these insects makes its nest in the branches of trees, and composes it of cowdung, having the art of spreading that singular material into thin flaky masses, which overlap each other like the tiles of a house. There is a separate roof to the nest which is partly domelike, and projects on all sides beyond the circumference of the nest. The latter insect fixes its nest in the thicker branches, and forms it of mixed mud and leaves.

AT the foot of the illustration is seen another rounded nest,

ŒCOPHYLLA VIRESCENS. ABISPA EPHIPPIUM. CREMATOGASTER LÆVICEPS.

also made by an ant, called *Œcophylla virescens.* Travellers know it by the name of the GREEN ANT; a title which is very insufficient, as it embraces several other species. The name of Œcophylla is compounded of two Greek words, the former signifying a house, and the second a leaf, and is given to this insect because it makes its home of dead leaves.

This ant is sometimes very troublesome to travellers, who may unconsciously disturb one of the nests that hang among the branches, nearly concealed by the leaves. The ants come pattering down like hail-drops, and in a moment he will be covered with a whole swarm of them, seeking for unprotected parts which they can wound, and having a special faculty for getting down the neck.

The nest is about eight inches in diameter, and is made in a very singular manner. The general mass of its substance is composed of leaves which have been cut by the ants and masticated until they form a coarse pulp, something like that which is made by the wasp and hornet, except that the material is green leaves instead of wood fibres. With this substance the nest is formed, and is hung among the thickest foliage, being sustained not only by the branches, but by the leaves, which are worked into the nest and in many parts project from its outer wall. The outside of the nest is easily to be distinguished from that of the Crematogaster by the smoothness and regularity of its walls. A species of this genus inhabits Africa, and was discovered by Mr. Foxcroft, who noticed that whenever the ants were disturbed, they ran about the outside of their nest so fast and in such numbers, that their pattering steps on the papery covering of the nest deluded him into the idea that rain was falling on the leaves above.

BEFORE describing the third nest in the illustration, which is the workmanship of a wasp, I will briefly mention one or two remarkable instances of pensile nests made by ants. One species, *Formica bispinosa,* which inhabits Central America, makes use of the silk-cotton which is produced by the seed-vessels of the cotton-tree (*Bombax ceiba*), and makes it into a sponge-like mass, which much resembles amadou, and, like that substance, is extremely valuable for stopping violent discharges of blood.

Another ant, *Formica merdicola,* rivals the Myrmica Kirbii in the singularity of the material which it uses in the construction of its nest, employing horse-dung for that purpose, and fixing its home either on the stems of reeds, at some distance from the ground, or on the spiny trunks of certain palms. There are also ants which form their nests from vegetable hairs ; such as the *Formica molestans,* which employs extremely minute hairs, and makes with them a nearly globular nest, which is placed in the petioles and vesicles of different plants.

We now proceed to the third figure in the illustration, placed upon the tree near the centre. This represents the remarkable nest of *Abispa Ephippium,* an Australian insect, belonging to the wasp tribe.

The nest is not very large, being about three or four inches in diameter, and rather more in height, exclusive of the entrance-tunnel. The material is clay, kneaded and masticated by the insect until perfectly plastic, and then moulded into a very remarkable form.

The exterior view of the nest presents a curious outline, showing the pipe through which the insect enters, and which reminds the observer of the tube constructed by several pensile birds. Strange as is the external appearance of the nest, a longitudinal section shows a still more extraordinary construction of the interior. The tube does not merely act as an entrance, but is carried about an inch into the interior of the nest, possibly in order to prevent the young insects from falling out before they are fit to cope with the world. The bottom of the nest through which it passes is nearly flat, and the whole shape of the edifice is not unlike a large clay thimble, with the opening closed by a circular flat cake of hard mud.

Attached to the ceiling of the nest is a single layer of cells, arranged without any particular order or regularity; and it is a curious fact, that only a single wasp has been observed in the act of building the nests, or making the interior arrangement.

This is not the only insect that makes entrance tubes to its nests; for the *Trypoxylon aurifrons,* a native of the Amazons, has been noticed to build similar entrances, though much shorter. This insect will be again mentioned, under the head of Builders.

In the accompanying illustration may be seen two specimens

of a remarkable pensile nest that is made by a wasp called *Tatua morio,* an insect which is notable for having the basal segment of the abdomen narrowed into long and slender foot-stalks, not unlike that of the Eumenes, and others.

The nest of this species is made of the papery substance used by many wasps, except that the material is so hard and

TATUA MORIO.

smooth as to resemble white cardboard. The general form of the nest is shown in the engraving, being somewhat like a sugar-loaf, *i.e.* a round-topped cone with a flat bottom. It is found in several parts of Central America; and in Guiana the nest goes by the popular name of "the Dutchman's pipe," being supposed to bear, in shape and dimensions, some resemblance to the pipe-bowl celebrated by Washington Irving. The exterior walls are so hard, firm, and smooth, that they can withstand any vicissitudes of weather, neither the fierce storms that blow in those regions, nor the torrents of rain that occasionally fall, having any power over an edifice so well protected.

The tiers of cells are variable in number; a rather remarkable fact, as the floors are made before the cells are built. In a good specimen of this nest in the British Museum there are only four tiers of cells. How many tiers are completed before the insects begin to affix cells to them, or whether the cells are made as soon as the floors are finished, are two points in the history of this wasp which have not yet been decided. These floors extend completely to the walls, to which they are fastened on all sides, and the insects gain admission to the different floors by means of a central opening which runs through them all.

In Mr. Waterton's museum, at Walton Hall, are several specimens of these nests, one of which is cut open so as to show the interior, as well as the central aperture, the whole of the bottom being cut away and raised like the lid of a box. The substance of this nest resembles thin brownish pasteboard, and, as is the custom with most of the wasp tribe, the cells are placed with their mouths downward, the nurses being enabled to attend to their charges by remaining on the floor of the next tier of cells. Taking one row of cells as an average, I counted twenty-four from the central aperture to the circumference, thus giving a tolerable notion of the number of cells in each tier. The aperture is not precisely in the middle, so that some rows of cells are necessarily larger than others, but I purposely selected a row which seemed to afford a fair average.

THE COMMON WASP (*Vespa vulgaris*) figures in several capacities. It has already been mentioned as a Burrower, deserves notice as a Social Insect, and must now be briefly described as a builder of pensile nests.

In the splendid museum at Oxford, there is an object which never fails to attract the notice of visitors, whether entomologists or not. It is a square glass case, some four feet in height by two in width, and the interior of this large case is almost entirely filled by a single wasp's nest. This enormous nest resembles a turnip in shape, but with the addition of a large knob at the top, by means of which it is suspended.

Its origin is sufficiently remarkable. On the 18th of July, 1857, this nest was found at Cokethorpe Park, Oxfordshire, being then of moderate dimensions, and measuring about five inches in diameter. It was taken from the ground, and hung near the window of a dwelling-house upon the ground floor, so as to give the inmates facility for procuring food. There was no danger in the experiment, for, as has been mentioned on page 146, the wasp is really a good-natured insect, unless irritated, and can be watched as safely as the hive bee.

In order to induce the labourers to work with more assiduity, the wasps were supplied with food in the shape of sugar and beer, of which mixture they consumed a large amount, their daily allowance being a pound of sugar to a pint of beer, and the aggregate weight being two pounds. Under such favourable auspices they built their nest at a wonderful rate, when they were suddenly reinforced after a singular manner. It so happened that on the first floor of the house two other wasps' nests had been placed. The workers of these nests were not fed like their kinsmen below, and in consequence, about the end of August they deserted their own house, and united with the more favoured wasps on the ground floor. The three colonies having thus joined their forces, the nest grew with marvellous rapidity, and at last attained the gigantic size which has already been mentioned.

In shape it is very irregular, as though the turnip to which it was compared had been made of a soft yielding substance, and had been thrown down and roughly handled. The entrance is close to the bottom of the nest, and a little on one side, and just by the opening the nest is flattened, and seems as if it had been pinched by some giant finger and thumb. For this singular structure we are indebted to Mr. S. Stone of Brighthampton.

THERE are also certain British wasps which always make pensile nests, though none of them are so complicated or so finely constructed as those of the pasteboard wasps of hotter climates.

These are popularly called TREE WASPS, and the best known among these pensile wasps is the insect which is sometimes known as *Vespa Britannica*, but which is now named *Vespa Norwegica*, and may therefore be called the NORWEGIAN WASP.

I may here mention that, until a very late period, the history of the wasp—whether British or foreign—was in dire chaos, the species, sexes, and varieties being so confounded together, that even the best entomologists could make nothing of them.

In Mr. Westwood's admirable "Classification of Insects,' published in 1840, the following passage occurs, showing how keenly an accomplished entomologist could feel the want of sound information on a difficult subject. In Vol. II. of that work, page 248, Mr. Westwood remarks as follows: "The specific differences of the British species of wasps require a more minute investigation than has yet been given to them. This can only be done by studying the habits of the different species, in conjunction with individuals of the different sexes from the nest of each. Thirty years ago the necessity for such an inquiry was pointed out by Latreille, who added 'Utinam exergat alius Kirby, qui hanc familiam elucubret' (*i.e.* 'Would that another Kirby would arise, who would elucidate this family'). But the wasps still remain in as great or greater confusion than they were at that period."

Since that time, the "other Kirby" has arisen in the person of Mr. F. Smith, who has disentangled the knotty confusion in which the wasps were enveloped, and has recorded his observations in the Catalogue of Hymenoptera in the British Museum, published by order of the trustees in 1858, some forty-eight years after Latreille had invoked assistance.

Of the species in question Mr. Smith remarks that it is rare in the South and West of England, but is not uncommon in Yorkshire and plentiful in Scotland. It seems to be a nocturnal insect, for a collector of lepidoptera found that when "sugaring" trees at night, for the purpose of attracting moths, numbers of these wasps settled on the sweet bait, and not only were more numerous than the lepidoptera, but actually resented any attempts at dislodgment.

The nest of this insect is always pensile, and is hung from the branches of a tree or shrub, the fir and gooseberry being the favourites. A pretty specimen in our own collection was taken from a gooseberry-tree in a garden, and another similar nest was found at no great distance. One of these nests I presented to the British Museum, and the other is now before me. It is very small, only having one "terrace," in which are thirteen cells, arranged in five rows, four being in the central row, and the rest graduating regularly. It is almost as large as a well-sized turnip radish, and something of the same shape, supposing the radish to be suspended by the root, and to be cut off just below the leaves. The outer envelope is composed of three layers overlapping each other, which are very fragile, considering the work they have to perform.

The wasp itself is prettily marked, and although it is variable in colouring, can be recognised by the black anchor-shaped mark on the clypeus, and the squared black spot on the segments of the abdomen.

ANOTHER species of British Tree Wasp is the CAMPANULAR WASP (*Vespa sylvestris*), a species which has received a multitude of scientific names, but which is not variable in colour as that which has just been mentioned. Though it has a wider distribution than the Norwegian wasp, it is scarcely so plentiful an insect, and is remarkable for an occasional habit of making a subterranean nest like that of the common wasp. The NORTHERN WASP (*Vespa borealis* or *arborea*), is another of the pensile wasps, and is mostly found in the North of England and Scotland. Its nest is built in fir-trees. I may perhaps mention that the tree wasps may always be distinguished from their subterranean brethren by the colour of the antennæ, workers and females having the scape black in the ground wasps, and those which build in trees having it yellow in both sexes.

THE nest of the tatua, which has recently been described, must not be confounded with that of the PASTEBOARD WASP (*Chartergus nidulans*), although both insects inhabit the same country, and the nest of the latter bears a great external resemblance to the pendulous nest of the tatua. But when examined closely, this nest is seen to have a remarkable addition to its structure, the hole through which the branch is passed being very large, so as to permit the nest to swing freely in the wind. In most specimens of these nests the hole is simply made through the thick upper end of the structure but in a few examples the pasteboard-like substance is so moulded that it looks as if a ring had been added to the top of the nest.

The dimensions of the Chartergus' nest are extremely variable, each structure appearing to be capable of unlimited enlargement. The mode by which the wasps increase the size of their pensile home is equally simple and efficacious. When the number of the inhabitants becomes so large that a fresh series of cells is required, the insects enlarge their home with perfect ease, and at the same time without destroying its symmetry, a point which is

CHARTERGUS NIDULANS

often forgotten when human architects undertake the enlargement of some fine old edifice. Taking the bottom of the nest as the starting-point, they build upon it a series of cells, taking care to add another row or two to the circumference, so as to increase the diameter in proportion to the length. They then add fresh material to the outer wall, which is lengthened so as to include the new tier of cells, and then the bottom is closed with a new floor, which in its turn will become the ceiling of the next tier of cells.

These nests are therefore permanent; unlike the habitations of the common British wasps, which are only used for a single season and then deserted, the few surviving females seeking their winter quarters elsewhere, and always choosing some fresh spot for the nucleus of a fresh colony. On the average, a well-sized nest of the Chartergus is about one foot in length and of proportionate width, a few being found of larger dimensions and many of smaller. Now and then a positive giant of a nest is discovered where the colony has not only been undisturbed, but surrounding circumstances have been favourable to its continued increase. The name Chartergus is derived from two Greek words, signifying paper-maker.

One of the largest, if not the very largest of these pasteboard nests that has yet been discovered, was found in Ceylon, attached to the inside of a huge palm-leaf, and was of the astonishing length of six feet. Now, to form an idea of a nest six feet in length is not very easy. It is as easy to write the words six feet as six inches, but the idea which is to be conveyed is another matter, the cubical measurement being absolutely enormous.

The gigantic wasp's nest which has lately been described is so conspicuous an object that, although it is only a little more than three feet in length, no one can enter the room without noticing it. But a nest six feet in length is so huge as scarcely to be credited except from actual sight. Such a nest could hardly be taken through an ordinary doorway, and there are few houses of the modern build which could receive it into any room except through the window after both sashes have been removed. We all know how conspicuous among ordinary men is one who measures six feet in height, and we shall form a better idea of the nest in question, if we reckon it to be equal in length to a "six-foot" man, and of course to occupy much more space, on account of its bell-like shape.

Mr. Westwood mentions the nest of an allied species of wasp, which is about eight inches in diameter, and is so hard and smooth on the exterior, that it almost seems to be made of pottery instead of vegetable fibre. This nest is in the museum of the Jardin des Plantes in Paris.

I HAVE already mentioned that there are many genera of nest-making insects, whose habitations are in some degree similar, and yet present such salient points of difference that they must be classed under different heads. Such, for example, is the strange genus Polistes, which is spread over a large portion of the globe, and which makes so singular a variety of nests. However different they may be, there is always one point of union among them; that the cells are exposed to the air without any covering at all, and in consequence, are made of stouter material than those of ordinary wasps, which protect the cells from the weather by a covering.

Many of this species make a nest of a nearly circular shape, and attach it sideways to branches, walls, trunks of trees or other supports; but there is a very curious nest in the British Museum which is made on a totally different principle, the combs looking as if they were soft, flexible, and hung carelessly over a twig. There are three of these remarkable combs, having the cells very like those of the common hive bee, both in shape and size, but all being of a dark brown hue. The cells are laid on their sides, like those of the bee, and the combs are long and narrow, looking like one large comb cut into three strips. This curious nest came from Siam.

In the accompanying illustration are represented two nests, both from tropical America, and both found in similar localities. These are the habitations of two species of wasp, which are remarkable for their honey-making powers.

In the year 1780, a Spanish officer named Don Feliz de Azara was raised from the rank of captain to that of lieutenant-colonel, and sent to Paraguay, in order to decide a dispute concerning the limits of the possessions respectively held by Spain and Portugal.

He was then thirty-four years of age, and being a man of great energy, set to work out the construction of a map of Paraguay. This was a Herculean task, occupying thirteen years in its completion, and forcing De Azara to explore regions before unknown, and to trust himself to the native tribes who had never before seen the face of a white man. While engaged in this occupation, he made a vast collection of notes upon the native tribes of Paraguay, as well as upon the beasts, birds, insects, and vegetation, together with an account of the method by which the Jesuit missionaries established themselves and ruled the country for many years.

After his return to Europe, in 1801, he published the account of his travels, and met with the usual fate of those who first penetrate into unknown countries. His statements were not believed, and among those which raised the greatest discredit was an account of certain wasps which made honey. Some persons said that the whole statement was a fabrication, and others remarked that the honey-making insects were simply bees which De Azara had erroneously considered to be wasps. Time, however, had its usual effect, and De Azara has been proved to be perfectly trustworthy in his remarks. The two specimens which are represented in the illustration are now in the British Museum, and afford tangible proofs that De Azara was right and his detractors wrong.

NECTARINIA. MYRAPETRA.

The right-hand figure represents the nest of a curious insect, named by Mr. Adam White *Myrapetra scutellaris.* The generic title is not very appropriate, being simply a fanciful name, composed of the names of two ancient cities, one called Myra, in Lycia, and the other Petra, the capital town of Arabia Petræa.

It is much to be regretted that this plan of inventing fanciful names and giving them to newly-discovered species should have been so common a practice among systematic zoologists. I hold that both the generic and specific name of every animal and plant should be intelligible, and refer either to some peculiarity of form, habit, colour, or locality. There is no great difficulty in doing so. Greek is a language that affords an inexhaustible supply of compound words, and even if the nomenclator be no scholar, any one who is moderately versed in the classics would compose the desired names if he were only furnished with the necessary information.

The vagaries in which some nomenclators indulge are so absurd as scarcely to be believed. Firstly, they will invent some word that exists in no language whatever, merely because the sound pleases their ears, and they like to amuse themselves with the conjectures of future zoologists. Then, they will divide the word into its syllables, and make new words with them. Then they will break it up into its component letters, and make as many anagrams as can be pronounced.

It is quite bad enough to name a new species after some particular friend, or after your favourite dog, horse, or cat, or after the name of your house; or, as in the present instance, to name an American insect after two defunct cities of Asia and Africa. The former cases show that you have a friend, or a dog, or a cat, or a horse, as the case may be; and the latter affords conjecture that you have a Lempriere's Dictionary. But the allocation of meaningless syllables in order to form a word which was never intended to have any meaning at all, is so utterly senseless and so completely without excuse that no words of reprobation are too strong for it. The very essence of scientific nomenclature is to convey ideas, whereas the names invented by the delinquents in question are chosen just because they convey no ideas at all.

Such persons shelter themselves behind the great name of Linnæus, saying that his fanciful separation of the butterflies into Greeks and Trojans, knights and commoners, was quite as indefensible as their own system, and that the name of an ancient warrior conveys no idea of a butterfly. But in the days of Linnæus, the father of scientific nomenclature, the art was in its infancy, and necessarily crude and imperfect; and there is no doubt that if Linnæus had foreseen the enormous discoveries of later times, he would have carried out fully the plan which he generally followed, and have made all his names descriptive.

Scientific nomenclature is of necessity quite complicated and crabbed enough without the infusion of a meaningless element, and those authors who introduce such terms are doing their best to deter future students of zoology, and to render it a repulsive rather than a fascinating science.

When we look at the remarkable nest which is made by the Myrapetra, one cannot but see a vast number of peculiarities which would have furnished an appropriate name, a name which would have stamped upon the mind something of the character of the insect architect.

This beautiful nest was presented to the Museum in the year 1841 by Walter Hawkins, Esq., and a very elaborate memoir by Mr. Adam White is to be found in the "Annals of Natural History," Vol. VII. page 315.

On looking at the exterior of the nest, our attention is at once excited by the material of which it is made, and the vast number of sharp tubercular projections which stud its surface. In colour it is dark, dull, blackish brown, and its texture somewhat resembles very rough papier mâché. On examining it with a pocket magnifier a matted structure is plainly visible, as if it were made of short vegetable fibres. This appearance accords with the accounts of the natives, who say that it is made from the dung of the capincha, one of the aquatic cavies of tropical America.

The whole of the exterior is thickly studded with projections, varying in size and shape, but being all of some sharpness at the tip. These projections are comparatively few at the top of the nest, becoming gradually more numerous as they approach the bottom, until at last they are set so thickly that the finger can scarcely be laid between them.

The object of these projections is not ascertained. The nest always hangs very low, seldom being more than three or four feet from the ground, and some writers say that the office of the sharp projections is to guard the nest from the attacks of the felidæ and other honey and grub-loving mammalia. Such may indeed be the true explanation, and indeed it is so obvious that no one could avoid seeing it. But I very much doubt whether a far better explanation is not in store, and I cannot see why the Myrapetra should stand in need of such protection, when the nest of the Nectarinia, which is placed in precisely the same conditions, is perfectly smooth and defenceless.

One use of the projections is evidently for the double purpose of concealing and protecting the entrance. On looking at the nest from above, no entrance is visible, and it is not until after a close examination that the openings are found. They are concealed under a row of the projections, which overhang them like the eaves of a house, and effectually keep off the rains which fall in such heavy torrents during tropical storms. The material of which these projections are made is the same as that of which the walls of the nest are built, except that it is very much thicker and harder, the various layers being hardly distinguishable, even with a good magnifier.

The interior of the nest is as remarkable as its exterior.

When cut open longitudinally, an operation which was carefully performed by Mr. White, a very curious sight presents itself. The nest is filled with combs, all very much curved, and these curves accommodating themselves beautifully to the general form of the nest. At the top is a nearly globular mass of brown paper-like substance, which is apparently the nucleus of the nest. The first comb closely surrounds this globular mass, leaving only a small interval between them, so that it forms part of a hollow sphere, and a section of it would present a form like that of the capital letter C laid on its back.

The rest of the combs follow in regular order, the curve of each becoming shallower, until the last is but slightly depressed in the centre. They are carried to the sides of the nest and thereto attached, except in a few places, where an open space is left between the edge of the comb and the side of the nest, so as to allow the wasps to have access to the different tiers of cells. As is the case with most of the wasp tribe, the tiers are single, and the mouths of the combs are all downwards.

The depth of the cells, and consequently the thickness of the combs, varies according to their position in the nest, the upper cells being the largest, and those below the smallest. The longest cells are from five to seven lines in length, and the shortest, about two lines. The material of which they are made is the same as that of which the exterior is formed, and is of quite as dark a colour. In texture, however, it is much slighter, being very thin and paper-like. These cells extend to the very edges of the combs, of which there are fourteen in the present specimen. The length of the nest is sixteen inches, and its diameter in the widest part is one foot.

In the upper combs was discovered a quantity of honey, which, when it was found, was hard and dry, of a deep brownish red, and without either taste or scent. De Azara mentions that himself and some of his men ate the honey of the Myrapetra, and that it was of a deleterious character. Another species of honey-making wasp, *Polistes Licheguana,* a native of Brazil, was discovered by M. St. Hilaire, who mentions that it lays up in the nest a large provision of honey, which is very injurious to mankind, on account of the poisonous plants from which it is taken. *Polistes gallica* also fills its cells with honey, which, however, does not seem to be poisonous.

Within the nest were found also the remains of insects. There was the body of a black fly, which belongs or is allied to the genus *Bibio,* and the remains of a neuropterous insect, which apparently belongs to the genus *Hemerobius.*

The Myrapetra itself is of variable size, the largest being about four lines in length, and rather more than half an inch in expanse of wing. It is of a dusky brown colour, and is remarkable for having the first joint of the abdomen very much lengthened and narrowed, so that it somewhat resembles the same organ in the Pelopæus.

At the left hand of the same illustration may be seen a rather large globular nest, suspended from the boughs. This nest is shown in the position which it usually occupies, namely, hidden in the dark recesses of the Brazilian forest, amid the varied vegetation which grows so profusely in the hot and wet parts of the country which the insect frequents.

The name of the species which makes this nest is *Nectarinia analis,* a title which is significant and appropriate enough, but which is rather unfortunate, inasmuch as it has already been applied to a genus of birds, the well-known honey-suckers of Africa and India, which are so frequently mistaken for humming birds, on account of their small size, their brilliant plumage, their slender beaks, and their fondness for flowers.

This is not nearly so beautiful a nest as that which has just been described, the combs being devoid of regularity, and piled upon each other, as if the insect had no settled plan on which to work, and put each comb in any place where there happened to be room for it. Irregular, however, as the structure may seem, it is not without a kind of order, for though the combs look as if they had been placed in a heap, and then rolled together, so as to assume a partially spherical shape, they are at all events made with the intention of forming that shape, so that they may be included under a single covering. In the specimen in the British Museum, the outer wall of the nest has been broken away in several places, so as to permit the combs to be seen.

The entrance for the insects is very small, and when the respective dimensions of the wasp and the nest are taken into consideration, it seems really wonderful that when the inhabitants enter their house, they do not lose themselves in the intricate windings through which they pass from one comb to another. The wasp which makes this nest is bee-like in form, and very small, not a quarter of an inch in length, and bearing some resemblance to those tiny solitary bees that are seen so plentifully upon dandelions and various umbelliferous flowers.

The nest is always hung near the ground, quite as low as that of the Myrapetra, and is suspended from the slender twigs and long, delicate leaves which are woven into its substance, and in many places pierce completely through the nest, and project through the outer covering. It is, however, destitute of the sharp projections which guard the home of the latter insect, and as the outer wall is both thin and fragile, it would fall an easy prey to any insect-eating animal that might take a fancy to it. I cannot but think that this utterly defenceless state of the Nectarinia's nest affords a proof that the spikes upon the habitation of the Myrapetra are not for the purpose of defending the nest against the attacks of enemies.

As is the case with the Myrapetra, the cells are made with walls much firmer than those of our English wasps or hornet, which are only intended to hold successive generations of young, and in consequence are made of a comparatively flimsy material, only strengthened very slightly at the entrance. Were honey to be placed in the cell of any known British wasp it would immediately soak into the walls of the cell, and thence escape by slow degrees, but as the young grub, which is the only tenant of the cell, is without feet and is not in the least formed for locomotion, a very slight partition is sufficient to control its movements.

The grub does nothing but hold to the end of the cell with its piercers, open its mouth for food, and occasionally protrude or withdraw itself in a very slight degree ; and its utter immobility in the larval and pupal states affords a strange contrast to the restless and fussy activity which actuates it after it has attained its perfect form.

As is generally known, the nests of wild honey-bees are placed in the hollows of trees. Mr. Cotton, the well-known apiarian, remarked, when discussing the comparative merits of straw and wooden hives, that in a state of nature the bee never builds in a truss of straw, but in a hollow tree. Now, although I quite concur with that author in his partiality for the wooden hive, I cannot see that the illustration which he employs has anything to do with the subject, or that it affords the least proof on either side of the argument. Wild bees are not very likely to find trusses of straw in the woods, and those trusses conveniently hollowed to receive them. But I do think that if a few common straw hives were set in the woods, the bees would be as likely to take up their habitation in them as in the hollows of trees.

Still, among honey-bees, of which there are several species, the custom of nesting in hollow trees is almost universal. Bee-hunters, whether biped or quadruped, whether man, bird, ratel, or bear, search for their sweet spoil in the trees, and know by experience when a tree is likely to contain honey-combs. But in certain parts of tropical America the bees change their habits.

There is a genus of wild honey-bees, named *Trigona*, the members of which are notable for their bold departure from ordinary bee customs. They make their nests at the tops of the branches, it is true, but they do not place their combs inside the hollow trees, of which there is great store in the woods. The Trigonas make nests of a pear shape, and of tolerable size, and hang them at the very summit of trees and at the end of the slenderest twigs, so that even the agile monkeys of that land, aided with their long, prehensile tails, are unable to reach the nest.

It is somewhat remarkable that the habit of this insect should be so different from the usual custom, and the more so that a closely allied species, inhabiting the same country, and which possibly belongs to the same genus, makes its nest in trees according to the ordinary type, and places its combs within the hollows of decaying trees. The honey of this bee is described as being very sweet and richly flavoured, so richly in fact, that very little of it can be taken.

A CREATURE is upon our list of pensile insects, which may also be reckoned among the social or parasitic insects, but which makes its habitation in such a manner that its proper place is among the pensiles. This is the pretty little ichneumon which is known to entomologists as *Microgaster alvearius.* The name Microgaster is of Greek origin and signifies 'little belly,' this being a very appropriate name for this insect, whose abdomen is of very small dimensions, and indeed appears to be just a little supplementary growth which might be removed without causing any inconvenience to the insect. It belongs to the same genus as a very common insect called *Microgaster glomeratus,* which will be duly described when the parasitic animals are under consideration.

With regard to this insect, I have been rather fortunate, having found many specimens of the nests, and bred from them several hundred insects.

Although plentiful enough in certain places, the BURNET ICHNEUMON, as I shall venture to call this species, is very local, and while abounding in one place may never be seen in another spot at the distance of a very few hundred yards. I give it the popular name of Burnet Ichneumon, for the same reason—comparing great things with small—that Caius Martius bore the title of Coriolanus and Publius Cornelius Scipio was termed Africanus—namely, that it destroys so many Burnet Moths.

In its perfect state the Ichneumon looks like a rather small gnat, and would probably be mistaken for that insect by a non-entomological observer. When examined through an ordinary magnifying glass, it is seen to possess a wondrous beauty which no one could ever suspect when looking at it with the unaided eye. The body and head are of a pale yellow colour, except the prominent compound eyes, which are dark blackish brown. The head is round and rather small, but the thorax is of enormous size, quite as proportionately large as the chest of a man would be did it project some eighteen inches in front and reach to his heels.

In singular contrast to the huge thorax is the very tiny abdomen, which is of a retort shape, curved, and fixed in the upper surface of the thorax by its smaller end. Indeed, the abdomen bears the same relation to the thorax, that the "tick" in the capital letter Q does to the whole of the letter. The limbs are long, and, when the size of the insect is considered, are singularly powerful, especially the last pair of legs. We think the legs of the kangaroo are enormously large in proportion to the size of its body, but they must be doubled in length as well as in thickness to equal those of the Burnet Ichneumon. The fore-limbs are not so very large, but they are long and possessed of great clasping power, aided by the hooked feet.

What then is the use of such powerful limbs? The habits of the insect supply the answer.

As is the case with many ichneumon flies, this insect—which, by the way, is not a fly but a near relation to the bee and ant—deposits its eggs upon caterpillars, boring holes in their skin with its pointed ovipositor, which is the analogue of the bee's sting, and inserting its eggs in the perforations. As may naturally be imagined, the caterpillar has a very strong objection to this proceeding, and when the ichneumon settles upon it, and begins to use her weapon, twists and wriggles about like a captured eel.

Now the strong limbs of the ichneumon come into play. Minute as is the insect when compared to the caterpillar, bearing about the same relationship that a rabbit bears to an elephant, the legs are so long that they can include a considerable portion of the skin in their embrace, and so strong that they can retain their hold in spite of the contortions with which the caterpillar tries to rid itself of its persecutor. Retaining her place, therefore, the ichneumon deposits a great number of eggs in the poor caterpillar, and then goes to find another victim.

I am not sure whether or not the ichneumon makes a separate wound for every egg. If so, the feelings of the caterpillar are not to be envied, for I have found nearly a hundred and fifty ichneumon larvæ in the body of a single caterpillar. No wonder that the persecuted being endeavours to fling off the creature that is inflicting so many wounds. The numerous short and bristle-like hairs with which the legs are thickly clad, are doubtless useful in retaining the hold of the insect.

The long and slender many-jointed antennæ are also covered with a thick down which has an iridescent effect when the light plays on those organs, and during the life of the insect has a most beautiful effect, owing to the restless, quivering movements which characterise all the antennæ ichneumons, and which at once serve to distinguish those insects at a glance.

The chief beauty, however, of the insect lies in the wings To the naked eye they are simply colourless, transparent appendages, with a little black spot on the outer edge of the upper pair. But when placed under a magnifier, and with the light properly directed upon them, they blaze out in iridescent glory that almost fatigues the eye with its resplendence. One of these insects is now under the microscope before me, a low power of only thirty-six diameters being used, so that each wing appears to be about three inches in length, and in order to give an idea of the extraordinary colouring of these apparently transparent organs, I will describe as far as I can the appearance of the right-hand upper wing.

The material of which it is made is a translucent membrane, appearing single with this low power, but shown by a higher power to be double. The wing is traversed by numerous nervures to support it, as the tracery of a Gothic window supports the glass, and which divide it into numerous compartments, technically called cells, each of which is known by name to entomologists. The whole of the membrane is covered with very minute hairs, dotted at regular intervals, like the holes in perforated zinc, and as each of these hairs is in fact a minute prism, they break up the light into the well-known prismatic colours.

Upon the outer edge of the wing is a triangular black spot, which is not transparent, and serves as a foil to show off the lovely colours by which it is surrounded. The whole upper part of the wing is pale yellow, passing, by the gentlest imaginable transition, through delicate rays into lively pink, of the character termed "rose-carmine." Towards the lower edge of the pink, a slight infusion of blue steals in, being first purple and then changing to azure. Here the colours are abruptly cut by a nervure and one of the large cells next comes into view. This cell is wonderfully beautiful, for the colours are no longer subdued, as is the case with the upper part of the wing, but are startling from their extreme brilliancy. The circumference of the cell is emerald, enclosing three distinct centres of colouring which seem to divide the cell into three parts. The upper division consists of a large emerald patch, changing in the centre by imperceptible gradations to golden green. Immediately below the green comes a patch of fiery ruby, edged on one side with azure and on the other with golden yellow. The third, or lowermost division, is chiefly blue, edged on one side with ruby and on the other with golden yellow. Thus, we have in this one cell three centres of colour, each centre being one of the three primary colours, and changing by degrees to the secondaries and tertiaries. The next cell is coloured in a similar manner, except that the colours which form the centre of the divisions in the last-mentioned cell form their circumferences in this case; and the base of the wings fades off into delicate shades of pink and rays of golden yellow like the tip.

Now it must be borne in mind that the microscope has nothing to do with the *production* of these colours, but is limited to their exhibition. These wondrous colours already exist, although they are on so small a scale that the unassisted eye fails to separate them, and so they are blended together and appear to be colourless. I mention these apparent platitudes because, while exhibiting the microscope, I have found many persons falling into the error of supposing that the wondrous beauties which they see are due either to the excellence of the instrument or the skill of the operator.

After the Burnet Ichneumon has laid the eggs she leaves them to be hatched in the animal, which is generally, but not always, the caterpillar of the Burnet moth (*Anthrocera filipendulæ*), itself a pensile insect. This is not always the case, as one of my group of Burnet Ichneumons proceeded from the body of a caterpillar belonging to the geometridæ. It was too much shrivelled for identification, but it was about as large as the larva of the swallow-tailed moth.

In the body of the caterpillar they live until the larval stage is nearly completed, and then they burst on all sides through the skin of their victim, proceed to a small twig and there weave a number of cocoons. These cocoons are about the eighth of an inch in length, cylindrical in shape, set closely side by side and fastened firmly together, so as to form a flattish mass extremely variable in shape and size, the latter depending on the number of cocoons. One of these masses now before me consists of one hundred and seventeen cocoons, and its shape is that of a segment of a circle, fixed to the twig by the flat side.

The ends of the cocoons are both closed, but when the young Ichneumon is hatched it makes its exit by cutting a circular flap from one end of the cocoon, pushing the flap outwards and then creeping into the air. The insects are quite indifferent as to the end of the cocoon through which they escape, and in the example before me nearly two-thirds of the creatures have escaped out of one end and the remaining third out of the other.

The texture of these cocoons is very firm and stiff, and the silken material is so closely fitted together as to be completely waterproof. The microscope shows that the exterior of the cocoon is composed of white silken fibres matted tightly together, and rather rough, while the inside of the circular flap shows that the interior of each cocoon is smooth, hard, and of a pale yellow hue.

The longest and largest cells occupy the centre of the mass, while those at either end are shorter, smaller, and fewer, being about one-fifth of the entire number. Knowing the customs of most hymenopterous insects, we may conclude that the females occupy the centre and the males the extremities.

THERE is a very remarkable pensile cocoon constructed by the larva of another hymenopterous insect belonging to the same family as the Burnet ichneumon, and placed in the genus *Cryptus*.

The insects of this genus seem to construct a strange variety of cocoons, some being white, some yellow, and some banded and mottled with black. The most remarkable forms, however, are those in which the cocoon is attached to a thread some inches in length, the other end of which is fastened to a bough or a leaf. Réaumur, who discovered these curious objects, found that when the cocoon was detached from the branch and laid on the table it sprang to a distance of several inches, probably because the enclosed insect was able to bend itself and then suddenly straighten the body.

Réaumur believes that the Ichneumons which make these pensile cocoons are parasitic on the processionary caterpillars, because he found them plentiful near the nests of these insects.

How the cocoons are made and suspended is quite a mystery. Mr. Westwood offers a suggestion that, before changing into the pupal state, the insect spins its thread to the required length, and, while still suspended at the end of its rope, spins the cocoon, which thus becomes fastened to the thread. In a future page will be described a cocoon woven on a similar plan, but made by the caterpillar of one of the moths.

Mr. Westwood mentions that when examining the cocoon of the Cryptus, he found that it was composed of three distinct layers, that on the exterior being composed of loose silk, which could be wound off like that of the common silkworm, but that the two interior layers were very shining, smooth, and of a gummy membranous texture, thus agreeing with the cocoons of the Burnet ichneumon.

Our last example of the pensile nests formed by the hymenoptera is a truly remarkable one. For some time I could scarcely decide upon its place in the present work, whether it was to be ranked as an example of the pensiles, social insects, or builders. On account, however, of the locality which is chosen for it, and the peculiar method by which it is attached to the branch, I have decided upon placing it among the pensile nests.

It has already been mentioned that the members of the genus Polistes are in the habit of building their cells in the open air, and leaving them without covering to defend them.

The shape, material, and arrangement of the comb is extremely variable; some, as that which has already been mentioned, hanging their cell-masses to the branches, just as if a number of bee-combs were simply hitched on the twig by the simple process of boring a hole in the upper part of the comb, and pushing the twig through it; others, again, make their cells of mud, in a nearly globular shape, and fasten them on the branches like so many berries. The species, however, which make the cells represented in the illustration, is one of the most remarkable, and so elegant is the form of the combs, and so singular the method of their attachment, that I have had them drawn nearly of the natural size.

Generally, the shape of the comb is nearly round, as is seen in the upper figure of the illustration. The cells are remarkable for their radiating form, the bases being a trifle smaller than the mouths, a peculiarity which would hardly be noticed in a single cell, but which produces the spreading outline when a number of them are massed together.

NESTS OF POLISTES

Some of the cells, those in the middle for example, are much longer than the others, and in the specimens in the British Museum many of them are closed at the mouth, showing that the insect is within, and has not yet attained its perfect state. Those on the circumference, however, are much shorter, and are entirely empty, not having been yet occupied. It is very possible that these cells would have been lengthened had the insects been left to themselves.

Although the circular shape is mostly the rule with these combs, so that they look something like withered dahlias or chrysanthemums, it is not the invariable form. If the reader will look at the lower figure in the illustration, he will see that it is much wider than long, and is apparently composed of two of the circular combs fixed together.

Now comes the curious part of the structure. The combs are not fastened directly to the branches, but are attached to footstalks which spring from their centre, and are firmly cemented upon the branch or twig. How wonderfully the insect must manage the comb so that it shall be balanced on this slender footstalk! To preserve the equilibrium of even an empty comb would be difficult enough, but when the cells are filled with fat, heavy grubs, the difficulty must be multiplied with every one.

The footstalks are made of the same *papier-mâché* like substance as the cells, only the layers are so tightly compressed together that they form a hard, solid mass, very much like the little pillars which support the different stories of an ordinary wasp's nest, but of much greater size. The position of the combs is extremely variable, some being nearly horizontal, and others perpendicular, as shown in the illustration. These nests came from Bareilly in the East Indies.

HAVING now completed our notice of the pensile hymenoptera, we turn to another order of insects.

We can hardly expect to find that any of the beetles can be ranked among the pensile insects, their appearance and general habits being opposed to such an idea. The variety of nests made by the hymenoptera lead us at once to conjecture that some of them may be pensile, for it is at least likely that the little architects which can construct the marvellous system of the honeycomb, or the complicated galleries of the ant's nest, or contrive the wonderful homes of the leaf-cutter bees, would be also able to make nests which could be suspended from leaves or branches. But there is nothing in the general history of beetles which could lead us to place them among the pensile insects, a rank, however, which can be taken by a very few species, most of which belong to a single group.

This group is that of the Curculionidæ, or Weevils, and there are a few species of these long-snouted beetles which make for themselves certain pensile habitations of a most elegant form. Two genera of Weevils are remarkable for the beauty of their cocoons, namely, the Hypera and Cionus.

If the reader should desire to possess specimens of these cocoons, he cannot do better than procure some seeds of the common species of Verbascum, say the Great and White Mulleins (*Verbascum Thapsus* and *Lychnitis*), and sow them in sandy or gravelly soil. The beetles of the genus Cionus feed on the mulleins, and when they are about to change into the pupal state, do not trouble themselves to leave the plant upon which they have been feeding. So fond are these beetles of the Verbascum, that Mr. Stephens found on a solitary plant, which was growing in a garden at Ripley, all the five species of the genus.

During the month of August the larva may be found in the flowers and seeds, and one species burrows into the leaves themselves, getting between the two membranes and feeding on the soft green parenchyma. When the larva are about to enter the pupal state, they cease from feeding, and spin for themselves cocoons of a most remarkable shape. The cocoons are very small, being on the average about as large as sweet peas, and nearly as globular. They are constructed of a rather stiff and glutinous thread, which is so wonderfully twined as to form large open meshes of a nearly circular form.

The cocoon is very firm and elastic, feeling and looking very much like those hollow spheres and cylinders that artists in hair are so fond of making. The open meshes are so large that the enclosed pupa can be seen through them, so that there is but little protection from the elements. A very good idea of the general appearance of the cocoon may be obtained from the toys which are made from nuts by neat-handed schoolboys, by the simple process of boring them full of holes until the shell is reduced to a kind of wooden network with circular meshes.

All the beetles of the genus Cionus are pretty little creatures, very hard shelled, nearly as globular in form as the cocoon, and marked with dark patches and streaks.

The cocoons of the genus Hypera are also made with open meshes, and of a similarly stiff thread, but the form is oval instead of round. The larva of the Hypera is long and narrow, having its rings or segments very deeply cut, covered with bristle-like hairs, and having some light lines along the back and sides. The "Charançon de la Patience" of De Geer, is a beetle of this genus. In both cases the cocoons are affixed to the under side of the leaves, whether they are attached to the mullein or the heath, so that they are not readily seen, except by careful observers, who know where to look for them. In the insect room at the British Museum there is a beautiful series of these delicately formed cocoons, still adhering to the dry and shrivelled leaves of the plant on which the beetle had fed.

We now come to the pensile lepidoptera, of which a number of specimens will be mentioned. They all belong to the moths, the pensile butterflies being content with suspending themselves by a couple of threads, without taking the trouble to build or spin a nest.

One of the most beautiful of these nests is the cocoon of the common Emperor Moth (*Saturnia pavonia-minor*). The moth itself is very beautiful, with its broad, soft-plumaged, pink-eyed wings, but it is even equalled by the larva in beauty of colour, a phenomena not very usual among the lepidoptera. The rings or segments of the caterpillar are rounded and deeply cut, and are remarkable for the tufts of golden-coloured bristles with which they are covered, each tuft springing from a raised and rounded tubercle. The body itself is of a beautiful leaf green.

The cocoon which is made by this remarkable insect is extremely beautiful, though its beauty does not appear to a careless observer. Some twenty years ago, when I first began to study practical entomology, and had no access to the books that were then published on the subject, I took to breeding every caterpillar that could be found, not having the least idea what kind of being would issue from it. Among them was a caterpillar which struck my fancy so much, by its green body and golden tufts, that I made a coloured drawing of it, and constructed for its benefit a separate cage, wherein it lived for some little time, and then spun a silken cocoon of a flask-like shape, very rough and loose on the exterior.

Some time afterwards, upon looking into the box, I saw a beautiful moth clinging to the side. How the creature had gained admission I could not conceive, for the cocoon seemed to be perfectly intact, and to exhibit no signs that an insect had broken through the walls. Concluding, however, that the moth might have crept into the box without my knowledge, or might have been placed there by some kind friend, I set it, and watched the cocoon as usual. After a whole year had passed, I thought that there must be something wrong, and so took out the cocoon and carefully cut it open.

The mystery was at once explained. Within were the cast shell of the chrysalis, and the dried shrivelled skin of the caterpillar, crushed up into a very small space, but recognizable by the hairy tufts. The manner in which the moth had escaped was also evident. Taking as our model a common Florence oil flask, from which three-fourths of the neck have been removed, we shall obtain a clear notion of the method by which the cocoon is made, so as to allow the egress of the moth, and at the same time to show no aperture through which the creature had emerged.

Let us suppose the material to be stiff, bristle-like hair, and that the body of the flask is made stiff and firm by cementing the hairs together, while they project loosely at the neck. Now, let us further suppose that these projecting hairs are all bent inwards, so as to cross each other slightly, and we shall have a tolerably correct idea of the manner in which the cocoon of the Emperor Moth is made. It will be seen, that if a creature try to push its way out from the inside, the hairs will yield and allow it to pass, but that if any insect tries to push its way in from the outside, the converging hairs are pressed tighter together, and effectually debar it from gaining admission.

This beautiful structure is not visible until the observer strips away a thick, loose coating of yellow-white silk which covers the cocoon, and probably acts as a non-conductor of heat as well as a protection from the weather. This cocoon may be found upon the plant on which the insect feeds, but the best method of procuring perfect specimens is by searching for the caterpillars and feeding them until they change.

ON the right hand of the accompanying illustration may be seen a large moth flying downwards, and just above it are a couple of oval objects attached to a slender bough. This moth is that magnificent insect the ATLAS MOTH (*Saturnia Atlas*), and the oval objects are the cocoons which are spun by its larva.

The Atlas Moth belongs to the same genus as the emperor moth, which has just been described, and is a truly splendid insect, though without the beautiful colours which decorate the emperor. Creamy white, soft yellow, and pale brown are the chief tints of the Atlas Moth, but they are so beautifully blended, the plumage is of so downy a softness, and the expanse of wing is so great, that the Atlas holds its own even amid the more vividly coloured lepidoptera of its own country.

OIKETICUS AND ATLAS MOTH.

There are many members of this genus scattered over the different parts of the earth, the finest and largest specimens being found between the tropics. In all the species the antennæ of the males are remarkable for their beauty, being deeply feathered, and shaped something like a spear-head with a triangular blade, and in many examples there is a loose membranous talc-like spot in the middle of the wing.

The cocoons of the Atlas Moth are made of silken thread, much like that of the common silkworm, the cocoon being large in proportion to the size of the moth, and the quantity of silk is necessarily very great. Although the thread is not so fine or glossy as that of the ordinary silkworm, it is strong, smooth, and serviceable, and capable of being woven into fabrics of much utility.

The well-known Eria silk of India is produced by an insect closely allied to the Atlas moth, *Attacus ricini.* This silk is very loose in texture, and, being without gloss, has a rather flimsy look. In reality, however, it is possessed of peculiar strength.

One large species of silk-producing moth, also allied to the Atlas, is the AILANTHUS SILKWORM. The Acclimatisation Society is endeavouring to introduce this useful insect into this country, and so to make England a silk-producing country. We have not sufficient mulberry-trees to feed silkworms in such numbers as would make their employment profitable, and thus the ordinary silkworm is rather beyond our reach. But the insect in question feeds on the *ailanthus glandulosus,* a tree which has been imported from China, and thrives wonderfully in the open air. In March, 1864, I saw a young sapling about three feet in height that had sprung from a seed sown in March of the previous year.

This insect is very hardy, and after it has been hatched and fed for a little time like the ordinary silkworm, it is laid on the growing leaves and left to shift for itself. The caterpillar is nothing of a wanderer, and does not attempt to straggle from the tree, being content to stay and make its cocoon among the branches. The moth is coloured like the Atlas, being mostly of a greyish yellow, with some markings of dull violet, and some spots of black and white. The caterpillar is green marked with black.

There is a North American species of moth—also one of the Atlas moth's numerous allies—which displays a wonderful piece of ingenuity in suspending its cocoon. This is the insect called *Saturnia Promethea,* which lives in the sassafras-tree. The cocoon is placed within the leaf of the tree and secured by a strong web; but as the leaf would fall before the moth could escape, a strange instinct is implanted in the insect, which fastens the stem of the leaf to the branch by sundry silken threads, so that although it may wither and part from the branch it cannot fall to the ground.

WE now pass to the second insect represented in the illustration. This is the HOUSE-BUILDER MOTH (*Oiketicus Sandersii*), an insect which is common in many parts of the West Indies, in several places being so plentiful that the sight of its long pendent domiciles is anything but pleasant to the proprietor of a garden.

Out of five species of insects belonging to this singular genus, the present has been selected, because on the whole its habitation is more remarkable than that of any other species. Some of them make their nest in a much stiffer form than is depicted in the engraving, taking pieces of slender twigs and forming them into hollow cylinders, the twigs being laid parallel to each other, very much like the rods in the old Roman fasces, which were borne by the lictors before the consuls. So close indeed is the resemblance, that by some writers the insects have been called Lictor Moths.

The reader will observe that in the illustration the nest is shown as depending from the caterpillar, part of which protrudes from its mouth and the other part is hidden. This attitude is given because it is that in which the insect is generally seen. While young the caterpillar is so strong, and the house is so light, that it can carry the tail nearly upright.

Scraps of wood mixed with fragments of leaves are the materials which are used, and they are bound together very firmly by the silken threads with which so many caterpillars are endowed, whether they belong to the butterflies or moths. There is a tolerable degree of elasticity about it, especially at the mouth, which is slightly expanded so as to assume an irregular funnel-like shape, and can be drawn together at will by means of the silken threads attached to its circumference. The caterpillar has thus two means of guarding itself from attacks If it is still clinging to a branch, it can retreat into the house and press the mouth so firmly against the branch that it is closed effectively, just as a limpet shelters its soft body by pressing the top of the shell against the rock. Or, if detached, it can pull the lips together and thus shut itself up in its strange house as completely as a box tortoise in its shell.

Not only does the creature reside in this nest during its larval condition, but also passes the pupal stage in it, and sometimes the whole of its life. As soon as it ceases from feeding, and is about to become a pupa, it retires far into its cell, shuts up the mouth, throws off its last caterpillar skin, and there remains until the larva has become a perfect insect. Should the moth be of the male sex, it creeps out of the domicile and speedily takes to wing, employing itself in the great object of its life, that of seeking a mate.

In ordinary cases, to find a mate seems to be no difficult task, but the House-builder Moth has no ordinary obstacles to overcome. The female never leaves her cell, for she would be more helpless as a moth than as a caterpillar. Among the British moths we have several species in which the females are wingless, but at all events they do look like moths which have been deprived of wings, and are able to move about with tolerable freedom. Of these wingless females, the common Vapourer moth (*Orgyia antiqua*), is a familiar example, its fat, rounded abdomen and little truncated rudiments of wings being known to all collectors.

But the female House-builder Moth is as utterly helpless a being as can well be conceived. She has not the least vestige of wings, and but the smallest indications of legs or antennæ. None but an entomologist would take her for a lepidopterous insect, or even for an insect at all, for she looks like a fat, down-covered grub, with very feeble limbs, which can scarcely support the body, and with antennæ that merely consist of a few rounded joints, entirely unlike the beautiful feathered forms which decorate the male.

So utterly unlike a moth is this creature, that our most skilful entomologists are much perplexed as to the position which the insect ought to occupy. Mr. Westwood states that they are "the most imperfect of all lepidopterous insects, and even less favoured than their larva, which they considerably resemble;" while

Mr. Newman expresses still stronger opinions, and asserts that the Oiketici ought to be removed from the lepidoptera altogether, and placed with the Phryganeidæ, or caddis flies, whose dwellings are wonderfully similar to those of the Oiketici.

The Oriental idea that feminine delicacy is only to be maintained by concealing the face, seems to have been borrowed from the House-builder Moth, which is a perfect model of female excellence, according to Oriental notions, always staying at home, always hiding her face, and always producing enormous families. Perhaps the male may be attracted to the female by some peculiar instinct, for the eyes can have little to do with the discovery, she being so closely shut up in her house, and never leaving it till the day of her death. Many British insects, such as the well-known oak-egger moth, have this curious power, and the male has even been known to enter a pocket in which was a female shut up in a box.

There is an allied genus, named Psyche, found in England, the males of which have their wings partly transparent, rather long and sharply pointed, and the females are without wings at all.

The larva of this insect also makes a hollow case, and behaves in a very curious manner before it assumes the pupal condition. First, it fastens the mouth of the case firmly to the leaves or branches of the plant on which it has been feeding, and then withdraws itself into the case. Should it be a male larva, it turns completely round, so that its head coincides with the opening at the lower end of the case, through which it makes its escape when fully developed. The female moth, however, behaves like that of the Housebuilder, and although she also fastens the mouth of her case to the tree, she never leaves her home, and therefore does not need to alter her attitude.

The name of Psyche certainly seems to be misapplied in this instance. In our minds the name of Psyche conveys an idea of the utmost grace and delicacy—two attributes which sculptors and painters have in vain endeavoured to embody. If, therefore, we hear that a certain insect is named Psyche, we certainly expect to see a bright and elegant creature, delicate in form and pleasing in colour. Whereas, when the domicile is opened and the real Psyche comes to view, nothing can be more disappointing than the fat, awkward, shapeless grub which has been glorified with such a name.

ONE of our commonest moths makes a really beautiful pensile nest, though it is hardly appreciated as it should be. I allude to the well-known TIGER MOTH (*Arctia caja*), whose scarlet, white, and brown robes are so familiar to every one who cares for insects, or who happens to possess or take an interest in a garden.

In two of its stages the insect is very common. In the larval condition it is popularly known as the Woolly Bear, in consequence of the coating of long bristle-like hairs with which its body is profusely covered, and which project like the quills of a porcupine, or the spines of a hedgehog, whenever the creature rolls itself up, a movement which it always makes when alarmed. So elastic are the hairs, that the caterpillar may be thrown from a considerable height without suffering any injury, and in all probability their formidable appearance serves to deter foes from meddling with it.

Certain enemies, however, care nothing for this hairy defence, but swallow the caterpillar without hesitation. Chief among these foes is the cuckoo, which feeds largely on the caterpillar of the Tiger Moth, and in consequence is subject to a very remarkable phenomenon. The interior of the gizzard had long been known to be lined with hair, which was thought to be a natural and ordinary growth peculiar to the species. It was however discovered—I believe by John Hunter—that these hairs are those of the Tiger Moth, the points of which have worked themselves into the coats of the organ in which they were found. Hunter employs this fact as an illustration of the power and peculiar movement of the gizzard.

Doubts have been thrown upon the accuracy of Hunter's statement; but the question has been set at rest by two facts. In the first place, cuckoos that have been held in confinement do not possess the hairy lining; and in the second place, the microscope proves that the hairs are those of a caterpillar, allied at least to the Tiger Moth, if not belonging to the insect itself.

When the caterpillar has ceased feeding, and is about to become a pupa, it ascends some convenient object, and then spins a beautiful cocoon, shaped very much like the grass hammocks made by the natives of tropical America, and bearing a considerable resemblance to them in general form, as well as in the loose and open meshes. So long, indeed, are the meshes made, that the inclosed insect can be seen through the network, from the time that the old wrinkled skin is cast off and pushed away in a heap by the white and shining chrysalis, to the time when the chrysalis shell is in its turn shattered, and the perfect moth creeps slowly into the air, all dull, and sodden, and bewildered, with its undeveloped wings looking like four mottled split peas rather than the beautiful members which they soon become, when the air has passed into their vessels, and their multitudinous folds have been shaken out.

I hope that none of my readers will kill a Tiger Moth in either of its stages. It does no harm to the gardener, and has quite enough foes of its own; the ichneumon flies piercing it in spite of its long bristles, and the cuckoo, together with other birds, revelling in so large and juicy a morsel. It is a special favourite of mine, this great moth, for I have kept so many hundreds of them, and have admired the wondrous details of their anatomy so often, that I am always glad to say a kind word for a creature which has afforded me so much amusement and instruction.

AMONG the pensile insects may be reckoned the beautiful BURNET MOTH (*Anthrocera filipendulæ*), an insect which has already been mentioned, while treating of the pensile hymenoptera.

This insect, which is well known for its splendid colours of deep velvet green, and blazing scarlet, is also notable for the shape of its antennæ, which are so swollen towards the tips as to induce many persons to reckon the insect as a butterfly rather than a moth.

The shape of the cocoon of the Burnet Moth is not unlike that of the tiger moth, but its material and position are very different. The cocoon of the tiger moth is slung horizontally, in hammock fashion, while that of the Burnet is set perpendicularly, and fastened to the upper part of a grass stem, one side being firmly pressed against it. The substance of the cocoon is quite opaque, greyish, rather stout, very tough, and having the silken threads, of which it is chiefly made, so conspicuous, that many persons take the cocoon to be the work of a spider.

Sometimes in a field, or even in a limited portion of a field, these cocoons are so numerous that at a little distance they look almost as if they were the seeds of the plant rather than the cocoons of an insect. In such cases the moths themselves may generally be found near the cocoons, sometimes being on the ground and sometimes on the wing. These moths are peculiarly liable to the attacks of the ichneumon flies, for not only does the Burnet ichneumon make them its special prey, but I have seen a large percentage of the cocoons bored full of holes, which show that one of the parasitic hymenopteras has laid its eggs in the caterpillar, that the young have been developed, and made their escape to continue the work of destruction, and that the caterpillar which nurtured them is lying dead within its useless cocoon.

THERE are others of our finest and yet commonest moths which make to themselves pensile habitations in which they pass the long time of helplessness when they are in the pupal state. Anything more utterly helpless than the pupa of certain moths cannot well be imagined, their only protection consisting either in their hiding-place or the sheltering armour in which the creature is enveloped.

The fur-clad DRINKER MOTH, for example (*Odonestis potatoria*), spins a cocoon which bears some resemblance in its texture to that of the Burnet moth, though it is rather looser in structure and is of much larger dimensions. The general colour of the cocoon is grey, with a few brownish mottlings here and there, and in form it is spindle-shaped, being widest in the centre, and diminishing to a point at either extremity. Conspicuous as this cocoon appears to be when exhibited in a glass case, it is anything but conspicuous in the position wherein it is placed by the insect. I have bred at least two hundred moths from the caterpillar, and though the space was necessarily limited, many of the cocoons escaped observation until after the moth had been developed and made its escape.

Like the Burnet moth, the Drinker is very liable to the attacks of ichneumons. There is now before me a cocoon which was made in 1846, and is preserved as one of the first instances of an entomologist's disappointment. As it now lies on its slab of white cardboard, it looks as if a charge of dust-shot had been fired through it, no less than seventeen minute holes being perceptible on one side alone, each hole representing at least one ichneumon fly which had made its escape after fulfilling its destructive mission.

THE handsome OAK EGGER MOTH (*Gastropacha quercus*) affords another example of the pensile cocoon. Of these insects also I have had great numbers; and some specimens of the moth, chrysalis, and cocoon are now before me, the cocoon unchanged by the eighteen years which have elapsed since it was made, but the moth sadly faded, after the manner of its kind when exposed to the action of light. This insect, by the way, is one of those which suffer the most from the fumes of sulphur, a lesson which I long ago learned from experience. Having been told that the best method of killing moths was to expose them to the fumes of burning sulphur, I invented an apparatus which would cause the insects to be enveloped in dense fumes, while the heat of the burning sulphur was carried off in another direction.

Of its efficacy as a means of destruction no complaint can be made, inasmuch as it destroyed the insect in a very few moments; but as it likewise discharged the colours, its use was soon given up. All the beautiful scarlets lost their tone, and became pale orange, and in the case of the Oak Egger and similar moths, the warm dun of the wings changed to dirty yellow. Moreover, the sublimated sulphur was sure to rest upon the wings, and to destroy their delicacy.

Camphor, which is so largely and so wrongly used in cabinets, is liable to the same objection. Its volatility is extreme, a large lump vanishing in a wonderfully short time when exposed to the air. The pieces of camphor used in cabinets continually need renewal, and the question frequently arises, Where has the camphor gone? The answer may be found in the dimmed glass, on which a deposit has been left, and which is so difficult to be cleaned, as well as on the inclosed insects, the lustre of whose bodies is sadly marred by the same substance.

Large as is the caterpillar of the Oak Egger moth, it is contracted into a comparatively small chrysalis when it assumes the pupal state, and makes a cocoon which only allows enough space for the pupa and the cast larval skin. The form of the cocoon is egg-shaped, whence the name of Oak Egger, and its substance is rather peculiar, being thin, hard, and rather brittle when quite dry. Externally it is surrounded by a loose layer of silken threads, by means of which it is attached to the plant on which it hangs; but the cocoon itself is smooth, very much the colour of half-charred paper, and in spite of its brittleness is possessed of some elasticity.

The manner in which the insect packs itself in so narrow a cell is most ingenious, and a cocoon may well be sacrificed in order to show the method by which this feat is achieved. If a cocoon be opened longitudinally, the chrysalis will be seen to fill the whole of the interior. On examining it more closely, the cast skin is seen to envelop the whole abdomen of the pupa, being pushed down in folds so as to fit closely round the pointed abdomen, and to occupy as little space as possible.

When the moth escapes from the cocoon, it breaks away quite a large hole at the end next the head, and slips out of the chrysalis shell with great ease, by lifting up a large flap which covers the legs and the head, and which gives way at the line of demarcation which separates it from the wings. In consequence of this arrangement, the pupa shell and the cast caterpillar skin remain in exactly the same position, and by means of a little ingenuity the raised flap can be replaced and fastened so as to give no indications that the insect has ever broken it. These cocoons are far more conspicuous than those of the Drinker moth, and are attached rather lightly to the stems of various plants.

There is a smaller insect, popularly called the LITTLE EGGER MOTH (*Eriogaster lanestris*), which spins a cocoon of a similar structure, except that the walls are of even harder and more uniform texture, scarcely larger than a wren's egg, and of a substance which looks almost as if it were made of the same material as the egg. When broken, it is found to be even more brittle than that of the larger insect. Owing, in all probability, to the exceeding closeness of the structure, which would exclude air from the inhabitant, it is perforated with one or two very tiny and very circular holes, which look just as if some one had been trying to kill the insect by piercing the cocoon with a fine needle or pin.

Even from the outside these perforations are visible, but they are much more evident when the cocoon is opened. The object of these holes is, however, conjectural, and it would be a useful experiment to stop them with wax, in order to see whether the inclosed insect could be developed when the air was thus excluded. I believe that there are none of these holes in the cocoon of the large Oak Egger Moth, and if there be any such perforations, they are so minute as to escape notice.

If the reader will refer to page 274, he will see an account of certain cocoons which are made by hymenopterous insects, and suspended by a single thread from the branches. In Mr. H. W. Bates's work on the natural history of the Amazon River, there is a most interesting account of a pensile cocoon also suspended by a single thread, but which is the work of a lepidopterous insect. It will be seen that Mr. Bates was able to see the insects spin the cocoon, and his account exactly tallies with Mr. Westwood's conjecture as to the method by which the creature manages to produce a hollow cocoon at the end of a single thread. Mr. Bates's account is as follows :—

"The first that may be mentioned is one of the most beautiful examples of insect workmanship I ever saw. It is a cocoon, about the size of a sparrow's egg, woven by a caterpillar in broad meshes, of either buff or rose-coloured silk, and is frequently seen in the narrow alleys of the forest, suspended from the extreme tip of an outstanding leaf by a strong silken thread, five or six inches in length. It forms a very conspicuous object, hanging thus in mid-air. The glossy threads with which it is knitted are stout, and the structure is therefore not liable to be torn by the beaks of insectivorous birds, while its pendulous position makes it doubly secure against their attacks, the apparatus giving way when they peck at it. There is a small orifice at each end of the egg-shaped bag, to admit of the escape of the moth, when it changes from the little chrysalis which sleeps tranquilly in its airy cage. The moth is of a dull slaty colour, and belongs to the Lithosiidæ group of the silkworm family (*Bombycidæ*).

When the caterpillar begins its work, it lets itself down from the tip of the leaf which it has chosen, by spinning a thread of silk, the thickness of which it slowly increases as it descends. Having given the proper length to the cord, it proceeds to weave its elegant bag, placing itself in the centre, and spinning rings of silk at regular intervals, connecting them at the same time by means of the loose thread; so that the whole, when finished, forms a loose web, with quadrangular meshes of nearly equal size throughout. The task occupies about four days: when finished, the inclosed caterpillar becomes sluggish, its skin shrivels and cracks, and there then remains a motionless chrysalis of narrow shape, leaning against the sides of its silken cage."

Some other lepidopterous insects suspend themselves by single threads, but most of them make their habitations of leaves, so that, when suspended, they do not attract much attention, looking like chance leaves that have fallen from the branches and caught in a stray piece of spider's web. Sometimes these nests are made from single leaves, the edges of which are drawn together by the silken threads spun by the caterpillar that takes refuge within, and sometimes they are made from several leaves, which are fastened to each other by similar threads. Some of these pensile nests are inhabited by a number of caterpillars, which live together in perfect harmony. Such nests are not uncommon in tropical countries, and one or two of them will be described in the chapter on Social Insects. One traveller describes some of these nests by comparing them to the white paper bags in which grapes are tied, when ripe, in order to preserve them from wasps and other marauders.

He also mentions that the interior contained a quantity of green leaves, which afforded food to the inhabitants, but does not tell us whether the leaves were actually growing on the tree and surrounded by the nest, or whether they had been cut from the boughs outside, and carried into the interior by the inhabitants. The latter supposition is implied, but it can hardly be a correct one, as it is directly contrary to our present knowledge of the habits of caterpillars. I believe that no lepidopterous larva is known to fetch food from a distance, and to store it for future consumption. As far as we know at present, the caterpillar has not the least thought for the morrow, but simply devours the leaves where they grow.

There are many species, such as the larva of the common Brown-tail Moth (*Porthesia auriflua*), or of the Small Ermine Moth (*Yponomeuta padella*), which travel by day to considerable distances in their search after food, and return at night to their common habitation, guided by the threads which they continually spin as they crawl along. But no caterpillar is known which is gifted with the instinct of cutting off leaves and bringing them home for food, and we may therefore infer that the leaves in question were growing on the branches, and that the nests had been purposely spun round them.

THERE are, however, one or two species of British insects belonging to the lepidoptera, which do cut off leaves and use them for the construction of the cocoon, though they do not employ them for food. These insects are moths, belonging to the genus Acronycta, and popularly called SPURGE MOTHS, on account of the plant on which they reside. One of these species makes a really curious pensile cocoon from the leaves of the cypress spurge (*Euphorbia cyparissias*), rather a scarce perennial plant about a foot in height, growing about woods and the borders of the fields. The leaves of the stem are lance-shaped, and those of the branches almost linear, like grass blades, and it is of these latter that the insect makes its habitation.

About October, the caterpillar begins to make its house, and does so in a very curious manner. Detaching a leaf from a branch, it fastens one end to the stem, and then bends the leaf so as to form a loop, and fastens the other end in a similar manner. A number of the leaves are placed nearly parallel to each other, so that when they are firmly woven together they form a bag-like cocoon, fixed to the stem of the plant by one side, and being upright like that of the burnet moth. Its texture is, however, very unlike that of the burnet, being loose, almost wholly composed of vegetable matter, and comparatively flimsy.

It has well been remarked that the strength, or at all events, the weather-resisting power, of a cocoon depends upon the length of time which is occupied by the insect in undergoing its transformation, those creatures which only spend a few weeks in the pupal state being content with a mere web or hammock of silk, while those which pass the winter in the pupal condition make habitations which are comparatively substantial.

This rule, however, is not without its exceptions, as we find the pupæ of several butterflies, the common cabbage butterfly, for example, merely hung against walls, &c., without any protection around them. Instinct leads them to choose such spots as can best afford them shelter, as every one knows who has a tool-house or a summer-house in the garden, but there are many cases in which no such protection can be found, and the insects are forced to content themselves with the southern side of a tree trunk, or the least windy side of a paling.

The caterpillar of the Spurge Moth is rather prettily marked, being striped longitudinally with white, red, and brown, relieved with black, and furnished with some scanty tufts of hair on each segment.

One species of insect suspends the cocoon by a thread at each end, so that the resemblance to a hammock is exact. This is the ***Argyromiges*** *autumnella,* one of the minute moths called micro-lepidoptera. The larva of this species is naked. It is a native of England.

We now pass to the enormous variety of caterpillars which are popularly called Leaf-rollers, because they make their homes in leaves which they curl up in various methods.

Some use a single leaf, and others employ two or more in the construction of their nests. Even the single-leaf insects display a wonderful variety in their modes of performing an apparently simple task. Some bend the leaf longitudinally, and merely fasten the two edges together, while others bend it transversely, fixing the point to the middle nervure. Some roll it longitudinally, so as to make a hollow cylinder corresponding with the entire length of the leaf, while others roll it transversely so that the cylinder is only as long as the leaf is wide, and a few species cut a slit in the leaf and roll up only a small portion of it.

The leaf-roller caterpillars belong to numerous species, and are plentiful enough, too plentiful indeed to please the gardener, who finds the leaves of his favourite trees curled up and permanently disfigured by these little marauders. All of them are of small size, and some so minute that the mere fact of their ability to roll up a leaf is something wonderful.

They flourish best during mild and rather rainy seasons, because the leaves are charged with moisture, and are so soft that they can easily be rolled, and moreover, contain a plentiful supply of food. During the present year, 1864, the Leaf-rollers have suffered greatly, the continual drought having dried up the leaves, rendering them both stiff and innutritious. The lilacs in my garden, which are usually covered with these cylindrical nests of Leaf-roller caterpillars, are comparatively free from them, and the few which exist are very poor specimens, several having been abandoned in a half-made state. In the lilacs of a friend, however, where the soil is about one hundred and twenty feet lower than my own garden, there are plenty of Leaf-roller nests, the ground being much moister than in more elevated situations, and being, moreover, on a different soil.

The mechanics of the Leaf-roller nest are very curious, and will be presently mentioned.

One of the most common among the Leaf-rollers is the pretty Oak Moth (*Tortrix viridana*), which must not be confounded with the oak egger moth already mentioned. It is a little creature with four rather wide delicate wings, the upper pair of a soft leaf green, and the under pair of a greyish hue. In some seasons, the moths, or rather their larvæ, are so plentiful that great damage is done to the oak forests, tree after tree being so covered with them that scarcely a leaf escapes destruction, and the growth of the tree is consequently checked.

Like all Leaf-rollers, they feed on the green substance, or *parenchyma* of the leaf, and being ensconced within their tubular home can eat without fear of molestation. They are not very much afraid even of the small birds, for as soon as a bill is pushed into one end of the leafy cylinder, the caterpillar hastily "bundles" out of the other—there is no other word which so fully expresses the peculiar action of the larva—and lowers itself towards the ground by a silken thread which proceeds from its mouth. In fact, it acts like a spider in similar circumstances.

Where these insects are plentiful, an absurd effect can be produced by tapping the branches of oak trees with a stick. As the stroke reverberates through the branch, the leaves, which appear to the casual passenger to be in their ordinary condition, give forth their inhabitants, and hundreds of tiny caterpillars descend in hot haste, each lowering itself by a thread and dropping in little jerks of an inch or two each. Some of them are more timid than the others, and descend nearly to the ground, but the general mass of them remains at about the same height. Another tap will cause them all to drop a foot or two lower, the stroke being felt even at the end of the suspending thread, and by administering a succession of such taps they will all be induced to come to the ground. There they will wait a considerable time, but presently one of them will begin to re-ascend, working its way upwards along the slender and scarcely visible line as easily as if it were crawling upon level ground. The least alarm will cause them to drop again, for they are then very timid, but if allowed to remain in peace, they speedily reach their cells and enter them with a haste that very much resembles the quick jerk with which a soldier-crab enters the shell from which he has been ejected.

If a tolerably smart breeze be blowing, the sight is still more curious, for the caterpillars are swung about through very large arcs, and, if the wind be steady, are all blown in one direction, so that their line forms quite a large angle with the level of the leaf to which the upper end is attached. The caterpillars, however, seem to be quite indifferent in the matter, and ascend steadily, whether the line be simply perpendicular, or whether it be violently blown about by the wind.

At the proper season of year, the moths are as plentiful as the larvæ, and a shake with the hand will cause a whole cloud of the green creatures to issue forth, producing a strangely confused effect to the eye as they flutter about with an uncertain and devious flight. A sweep with an ordinary entomological net will capture plenty of them, but in a few minutes they all disappear, some of them returning to the branches whence they had come, and others dropping to the ground. During the summer of 1864 they were very plentiful in Darenth Wood, the heavy growth of oaks giving them every encouragement.

THE insect which commits such devastation on the lilacs is generally the little chocolate-coloured moth called the LILAC MOTH (*Lazotœnia ribeana*), though there are other allied species which infest the same plant. Any one may see the damaged leaves for himself, and therefore I shall not particularly describe them, but pass at once to the mechanical powers which are involved in the task of curling the elastic leaf into cylindrical form.

Compare the size of the lilac leaf and of the newly hatched caterpillar, the latter being about as large as the capital letter I.

That so minute a creature should roll up the leaf by main strength is of course an impossibility, and the method by which that consummation is attained is so remarkable an instance of practical mechanics that I must describe the operation at length.

If the reader will procure one of the rolled leaves, he will see that the cylindrical portion is retained in its place by a row of silken threads, which are individually weak, but collectively strong, holding the elastic leaf as firmly as Gulliver was held by the multitudinous cords with which he was fastened to the ground. That they should hold the cylinder in shape is to be expected, but the manner in which the cylinder is made is not so clear. The following is the process :—

First, the caterpillar attaches a number of threads to the point and upper edges of the leaf, and fastens the other ends to the middle of the leaf itself. It now proceeds to perform an operation which is precisely similar to the nautical method of "bowsing" up a rope. In order to "bowse" a rope taut, two men are employed, one of them pulling the nearly tightened rope at right angles so as to bend it, while the other continually belays it to the cleats. Now, the caterpillar performs precisely this operation, but without requiring the aid of an assistant, the "bowsing" being performed by its feet, and the belaying by its spinneret. By thus hauling at, and tightening each line in succession, the caterpillar bends the leaf over slightly, and then attaches a fresh series of threads to keep it in its place. By repeating this process, and by continually adding fresh lines, the creature fairly bends the leaf into a hollow cylinder, and then crawls inside to enjoy its well-earned home.

I may here point out that the whole process of rolling the leaf affords an admirable example of mechanics as exhibited in nature, and that it is achieved by the well-known principle of exchanging space and time for power. Although the caterpillar cannot by any exertion of strength roll up the leaf in one minute, it is enabled to do so by dividing the work into a multitude of parts, and taking much longer time about it, just as a man who cannot lift a single weight of a thousand pounds may do so with ease by dividing it into ten parts, and in consequence, by taking up a considerable time in lifting the separate parts.

Again, in the silken bands which hold the rolled and elastic leaf in its place, we have an excellent example of accumulated power; neither of the threads being alone capable of enduring the tension, but their united strength being more than sufficient for the task.

As soon as the caterpillar has entered its new home, it begins to feed, eating the green substance of the leaf, and generally leaving the nervures untouched. Sometimes the caterpillar lives for so short a time that a single leaf is sufficient for its subsistence; but there are some species which are obliged to repeat the task more than once.

THERE are other insects which also make their habitations in leaves; but, instead of rolling up the leaf and living inside the cylinder, they make their way between the two membranes, and there remain until they have undergone their transformation.

The reader must often have seen the leaves of garden plants and trees, especially those of the rose, traversed by pale winding marks, that look something like the rivers upon a map, and having mostly a narrow dark line running exactly along the middle. These curious marks are the tracks which are made by the various leaf-mining insects, while eating their way through the leaf in which they pass their larval state. In most cases, when the insect has completed its term of larval existence, one end of the track is found to be greatly widened, and to contain either the pupa itself or its empty case.

The track differs considerably in shape, according to the insect which makes it. Sometimes it winds about in the middle of the leaf, crossing itself more than once in its progress. Sometimes it proceeds in a nearly straight line across the leaf, and very frequently, especially in deeply-cut leaves, it follows the outline, keeping to the edge, and not trenching at all on the central portions.

Insects belonging to three orders are known to make these curious habitations; namely, the Lepidoptera, the Coleoptera, and the Diptera. Of these, the Lepidoptera are by far the most numerous, and belong to that group which is called, on account of their very minute dimensions, the Micro-Lepidoptera. These are all little moths, so small that on the wing they can scarcely be recognised as moths, and look more like little flies. They are all very beautiful, and many of the species are truly magnificent when seen through a microscope, their plumage glittering as if made of burnished gold and silver. Indeed, one genus in which these leaf-miners are comprised, is named Argyromiges, a title based on a Greek word signifying silver.

The species which is most common in the leaves of the rose-tree is the RED-HEADED PIGMY (*Microsetia ruficapitella*). The larva of this insect seems not to possess even the rudiments of legs, and forces itself through the leaf by means of certain projections of the skin, which are sharp and angular, and serve as instruments of progression, like the abdominal scales of the serpent and the bristles of the earthworm. A species which is found in the leaf of the oak is known to collectors by the name of CRAMER'S PIGMY (*Argyromiges Cramerella*). The caterpillars of the DAGGER MOTH (*Diurnea*), also live between the membranes of leaves, and are remarkable for the last pair of feet, which are shaped like a couple of very minute battledores. These feet are spread out greatly in the act of walking, and the creature is further aided in its progress by the hair-covered warts upon the body.

As for the beetle leaf-miners, they are to be found among the weevils; and it is a remarkable fact that one of these insects belongs to the genus Cionus, which has already been mentioned on page 278, as the weaver of certain beautiful pensile cocoons.

Of the Diptera, the CELERY FLY (*Tephritis onopordinis*) is a good example. The larva of this really pretty fly, with its green eyes and black and white spotted wings, feeds not only on the celery but on the parsnip, and does great harm to both plants. Gardeners often employ little boys to examine the celery plants, and whenever they find a "blister," as they technically call it, to crush the inclosed maggot between the fingers. The colour of this larva is pale green, so that it is not readily seen even when the blister is opened. If allowed to have its own way, the larva remains in the leaf until it has finished its eating, and then descends into the ground, where it changes into the pupal state, and remains until the following spring. In such a case, the leaves are often much damaged, the blisters being yellowish white, and the leaf itself drooping and half withered.

Our last examples of pensile nests are taken from the Arachnida, being formed by several species of spiders.

It may perhaps be necessary to remark that the threads with which the spiders make their webs are in some respects similar to those which are produced by various caterpillars, and in other respects are exceedingly dissimilar. In both cases, the threads are formed from a semi-liquid secretion, which is produced in the internal organs, is forced through minute apertures at the will of the animal, and hardens into a thread as soon as it comes in contact with the air.

Here, however, the resemblance ceases. The threads of the caterpillar are double, or rather are composed of two lines fused together throughout their length, the two half-lines proceeding from a large silk-secreting tube at either side of the body, and uniting at the mouth, where they become fused together by passing through a short tube common to both. The threads of the spider are much more complex, each being formed of a vast number of smaller lines, which are produced from a peculiar organ termed the "spinneret," which is placed at the extremity of the body. In consequence of its position, the spider always hangs with its head downwards while lowering itself by means of its line.

The spinnerets are externally like little rounded projections, arranged in pairs, and four, six, or eight in number. They are variable in shape, mostly being rounded, but sometimes being so long that they have been mistaken for feelers. The spinnerets are covered with a multitude of very minute hair-like appendages, which are, in fact, the tubes through which the liquid secretion is forced into the air. All the threads which proceed from these tubes are joined into a single line; and it will be at once seen that very great strength is obtained by making the line compound instead of single.

The best known of these creatures is the common GARDEN SPIDER (*Epeira diadema*), sometimes called the GEOMETRIC SPIDER, whose beautifully radiated net is so familiar that its general shape requires no description. Suffice it to say, that the spider exhibits wonderful skill in placing its web, making a framework of very strong threads or ropes, and then spinning the net itself between them. Very great elasticity is thus obtained, for the threads are exceedingly elastic; so that, although stretched tolerably tightly, they will yield to pressure, and immediately recover themselves. This property is very needful, in order to enable them to resist the wind, to which they are so fully exposed.

These spiders have, moreover, a most singular plan of strengthening their web, when the wind is more than ordinarily violent. If they find that the wind stretches their nets to a dangerous extent, they hang pieces of wood, or stone, or other substances to the web, so as to obtain the needful steadiness. I have seen a piece of wood which had been thus used by a Garden Spider, and which was some two inches in length and thicker than an ordinary drawing-pencil. The spider hauled it to a height of nearly five feet; and when by some accident the suspending thread was broken, the little creature immediately lowered itself to the ground, attached a fresh thread, ascended again to the web, and hauled the piece of wood after it.

It found this balance-weight at some distance from the web, and certainly must have dragged it for a distance of five feet along the ground before reaching the spot below the web. There were eight or ten similar webs in the same verandah, but only in the single instance was the net steadied by a weight.

The structure of the beautiful web is very remarkable.

It is nearly circular, and is composed of a number of straight lines, radiating from a common centre, and having a spiral line wound regularly upon them. Now, the structure of the radiating and the spiral lines is quite distinct, as may be seen by applying a microscope of moderate power. The radiating lines are smooth and not very elastic, whereas the spiral line is thickly studded with minute knobs, and is elastic to a wonderful degree, reminding the observer of a thread of India-rubber. So elastic, indeed, is this line, that many observers have thought that the spider has the power of retracting them within the spinnerets, inasmuch as she often will draw a thread out to a considerable length, and then, when she approaches the point to which it will be attached, it seems to re-enter the spinneret until it is shortened to the required length. This, however, is only an optical delusion, and caused by the great elasticity of the thread, which can accommodate itself to the space which it is required to cross.

It is to the little projections that the efficacy of the net is due, for they are composed of a thick, adhesive, and viscid substance, and serve to arrest the wings and legs of the insects that happen to touch the net. In his splendid work on the British Spiders,

Mr. Blackwell has the following remarks upon the structure of the threads:—"As the radii are unadhesive, and possess only a moderate share of elasticity, they must consist of a different material from that of the viscid spiral line, which is elastic in an extraordinary degree. Now, the viscidity of this line may be shown to depend entirely upon the globules with which it is studded, for if they be removed by careful application of the finger, a fine glossy filament remains, which is highly elastic, but perfectly unadhesive. As the globules, therefore, and the line on which they are disposed, differ so essentially from each other and from the radii, it is reasonable to infer that the physical constitution of these several portions of the net must be dissimilar.

An estimate of the number of viscid globules distributed on the elastic spiral line in a net of *Epeira apoclisa* of a medium size, will convey some idea of the elaborate operations performed by the *Epeira* in the construction of their snares. The mean distance between two adjacent radii in a net of this species, is about seven-tenths of an inch; if therefore the number seven be multiplied by twenty, the mean number of viscid globules which occur on one-tenth of an inch of the elastic spiral line, at the ordinary degree of tension, the product will be 140, the mean number of globules deposited on seven-tenths of an inch of the elastic spiral line. This product multiplied by twenty-four, the mean number of circumvolutions described by the elastic spiral line, gives 3,360, the mean number of globules contained between two radii; which, multiplied by twenty-six, the mean number of radii, produces 87,360, the total number of viscid globules in a finished net of average dimensions.

A large net, fourteen or sixteen inches in diameter, will be found by a similar calculation to contain upwards of 120,000 viscid globules, and yet *Epeira apoclisa* will complete its snare in about forty minutes if it meet with no interruption."

These calculations will serve to show the elaborate nature of the webs which we see constantly in our gardens, as well as their value to the architect. The secretion of the liquid from which the lines proceed is a work of time, so that if a spider is forced to spin several nets in rapid succession, it loses all its silk and cannot make a web. To wait until a fresh supply should be secreted would be a terrible privation, and moreover, the want of food would stop the secretion, so that the spider has no other resource than to make war on a weaker spider, drive him out of his net and usurp possession thereof. Such being the case, the spiders are all very chary of using their silk, and never trouble themselves to make webs when a storm is impending They are therefore very excellent barometers, and if the spiders all take to mending their nets or spinning new webs, fine weather is always at hand.

One very remarkable point in the construction of these webs, so exactly true in all their proportions, is that they are executed entirely by the sense of touch. The eyes are situated on the front of the body and on the upper surface, whereas the spinnerets are placed at the very extremity of the body and on the under surface, the threads being always guided by one of the hind legs, as may be seen by watching a garden spider in the act of building or repairing her web. In order that the fact should be placed beyond a doubt, spiders have been confined in total darkness, and yet have spun webs which were as true and as perfect as those which are made in daylight.

A PECULIARLY beautiful pensile cocoon is constructed by a common British spider, scientifically termed *Agelena brunnea,* but which has no popular name. It is really remarkable that, considering the great number of species which inhabit England, so very few should have been sufficiently distinguished to receive popular names. Owing, in all probability, to the foolish dislike towards spiders entertained by most persons. a dislike which has been instilled into their minds at a very early age, these wonderful and interesting creatures are seldom watched, and there are very few persons who really know one spider from another, or who have any idea of their exceeding usefulness when in the places which they were intended to inhabit. Spiders are certainly out of their place in a room, and the housemaid is perfectly justified in exterminating them, but in the garden or the field they should never be injured, but rather encouraged as much as possible.

The species whose beautiful nest will now be described is generally to be found upon commons, especially where gorse is abundant, as it generally hangs its nest to the prickly leaves of that shrub. The cocoon is shaped rather like a wine glass, and is always hung with the mouth downwards, being fastened by the stalk to a leaf or twig of the gorse. It is very small, only measuring a quarter of an inch in diameter, and when it is first made, is of the purest white, so as to be plainly visible among the leaves.

This purity, however, it retains but a very short time, for after the spider has deposited her eggs, which are quite spherical, and about forty or fifty in number, she closes the mouth of the cocoon and proceeds to daub it all over with mud. The moistened earth clings tightly to the silken cocoon, and disguises it so effectually that no one who had not seen it before that operation, could conceive how beautiful it had once been. The muddy cover certainly makes the cocoon less visible, and may probably have another effect, that of protecting the inclosed eggs and young from the attacks of insects that feed upon spiders. Several other species have the habit of daubing their beautiful cocoons with mud.

This species is plentiful in Bostal Common and Bexley Heath in Kent, the profuse growth of gorse being very suitable to its mode of life, and I have several specimens of their nests taken from Shooter's Hill. June is the best month for them, as they may be found both before and after the mud has been applied.

An allied species, *Agelena labyrinthica,* is equally plentiful in similar localities, where its curious webs may be seen stretched in horizontal sheets over the gorse, and having attached to each web a cylindrical tube, at the end of which sits the spider itself. Heath and common grass are also frequented by this spider.

Besides the net or web in which it lives, and by means of which it catches prey, it makes a beautiful cocoon in which the eggs are placed. Externally the cocoon looks like a simple silken bag, perfectly white in colour, and, except in size, somewhat resembling that of the preceding species. It is only when quite freshly made, that the white hue of the cocoon is visible; for after its completion, it is covered with scraps of dry leaves, bark, earth, and other substances. If, however, this cocoon be opened, it is found to contain at least another cocoon within, and often comprises two, of a saucer-like shape, and made also of white silk. These inner cocoons are nearly half an inch in diameter, and contain a very variable quantity of pale yellow, spherical eggs, sometimes fifty in number, but often exceeding a hundred. The inner cocoons are firmly tied by strong lines to the interior of the large sac in which they are inclosed.

CHAPTER XV.

BUILDERS.

Building Mammalia—Definition of the title—Inferiority of the mammalia as architects—The BRUSH-TAILED BETTONG—its structure and colour—The Nest of the Bettong, and its adaptation to the locality—Singular method of conveying materials—Its nocturnal habits—The RABBIT-EARED BANDICOOT, and its habitat—The generic title—Curious form of the ears and feet—Difficulty in discovering its nest—The MUSQUASH or ONDATRA—Its general habits—Its burrowing powers, and extent of its tunnels—The Musquash as a builder—Form and size of its house—Mode of killing the animal by spear, gun, and trap—Its flesh and fur.

WE now take our leave of the Pensiles, and pass to those animals which build, rather than burrow or weave. The materials used by the Builders are variable. In the most perfect examples, earth is the material that is employed, but in many instances other substances such as wood, earth, and sticks are used by the architect.

As a general rule, the mammalia are by no means notable for their skill in the construction of their homes. In making burrows they far excel all the other vertebrates both in the length of the tunnels and in the elaborate arrangement of the subterranean domicile. The mole, for example, is pre-eminent as a burrower and as a subterranean architect, and there are many of the rodents which drive a whole labyrinth of tunnels through the soil. But they are very indifferent builders, and with a few exceptions are unable to raise an edifice of any kind, or to weave a nest that deserves the name.

Our list of Building Mammalia will therefore be a short one, comprising only three species, two inhabiting Australia and one a native of America.

THE first example of the Building Mammalia is the PENCILLED BETTONG (*Bettongia pencillata*), sometimes called the BRUSH-TAILED BETTONG, and often known by the name of JERBOA

KANGAROO. The word Bettong is a native name for a group of small kangaroos that are easily recognised by the shape of their heads, which are peculiarly short, thick, and round, and very unlike the long deer-like head of the larger kangaroos.

The Brush-tailed Bettong is about as large as a hare, and its tail is not quite a foot in length, though it appears longer in consequence of a brush-like tuft of long hair which decorates the end. It is a pretty creature, elegant in shape, extremely active, and the white pencillings on the brown back, the grey-white belly, and the jetty tuft on the tail are in beautiful contrast to each other.

The home of this animal is a kind of compromise between a burrow and a house, being partly sunk below the surface of the ground and partly built above it. The localities wherein the Bettong is found are large grassy hills whereon there is hardly any cover, and where the presence of a nest large enough to contain the animal, and yet small enough to escape observation, appears to be almost impossible. The Bettong, however, sets about its task by examining the ground until it finds a moderately deep depression, if possible near a high tuft of grass.

Using this depression as the foundation of the nest, it builds a roof over it with leaves, grass, and similar materials, not high enough to overtop the neighbouring herbage, and being very similar to it in external appearance. Grass of a suitable length cannot always be obtained close to the nest, and the Bettong is therefore obliged to convey it from a distance. This task it performs in a manner so curious, that were it not related by so accurate and trustworthy an observer as Mr. Gould, it could hardly be credited. After the animal has procured a moderately large bunch of grass, it rolls its tail round it so as to form it into a sheaf, and then jumps away to its nest, carrying the bunch of grass in its tail. In Mr. Gould's work on the Macropidæ of Australia, there is an illustration which represents the Bettong leaping over the ground with its grass sheaf behind it. After the nest has been completed, the mother Bettong is always careful to close the entrance whenever she leaves her home, and pulls a loose tuft of grass over the aperture.

To an ordinary European eye, the homes of the Bettong are quite undistinguishable from the surrounding grass. The natives, however, seldom pass a nest without seeing it, and destroying the inmate. Being a nocturnal animal, the Bettong is sure to be at home and asleep during the daytime, so that when a native passes a nest he always dashes his tomahawk into its midst, thus killing or stunning the sleeping inmates.

THE second Building Mammal on our list is also a native of Australia, and is known by the name of RABBIT-EARED BANDICOOT (*Perameles,* [or *Chœropus*] *castanotis*). Of the two generic names the latter is certainly preferable, as it alludes to the remarkable structure of the limbs. The fore feet are small and delicate, and only two toes are developed. Instead of being furnished with long claws at their extremity the feet are terminated by two short and pointed claws of equal length, and looking exactly like the hoofs of a pig. It is in allusion to this peculiarity that the generic name "Chœropus," or swine-footed has been given to the animal.

It is a rather odd-looking little creature, about as large as an ordinary rabbit, and having ears so long and large that the resemblance to the rabbit is really striking. Owing to the great length of the hind legs, the gait of the animal is rather peculiar, being a kind of mixture between walking and hopping, and when the creature is alarmed, it jumps away with wonderful speed. Specimens of this Bandicoot have lived in England.

The nest which it makes is not unlike that of the Bettong, which has been already described. The animal inhabits the same kind of locality—namely, grass-covered hills, and "scrubs," and builds its nest of grass and leaves, sheltering it if possible beneath a grass tuft or some thick bush. The Rabbit-eared Bandicoot inhabits New South Wales, and the nests are chiefly to be found near the banks of the Murray River. They are, however, so cleverly hidden, and the materials of which they are built are so similar to surrounding objects, that an inexperienced person might almost walk over them without discovering their presence.

WE now come to our last example of the Building Mammalia; namely, the MUSQUASH, or ONDATRA of North America (*Fiber Zibethicus*), sometimes called the Musk Rat.

This animal might have been placed among the burrowers, for it is quite as good an excavator as many which have been described under that title, but as it builds as well as burrows, it has been reserved for its present position in the work.

Essentially a bank-haunting animal, it is never to be seen at any great distance from water, and like the beaver, to which it is closely allied, it is usually to be found either in the river itself or on its edge, where its brown, wet fur harmonizes so well with the brown, wet mud, that the creature can scarcely be distinguished from the surrounding soil. It is seen to the best advantage in the water, where it swims and dives with consummate ease, aided greatly by the webs which connect the hinder toes.

The Musquash drives a large series of tunnels into the bank, excavated in various directions, and having several entrances, all of which open under the surface of the water. The tunnels are of considerable length, some being as much as fifty or sixty feet in length, and they all slope slightly upwards, uniting in a single chamber in which is the couch of the inhabitants. If the animal happens to live upon a marshy and uniformly wet soil, it becomes a builder, and erects houses so large that they look like small haycocks. Sometimes these houses are from three to four feet in height.

The natives take advantage of the habits of the animal, and kill it while it lies on its couch, much after the same manner as is used by the natives of Australia when they pass the house of the Bettong. Taking in his hand a large four-barbed spear, shaped something like the well-known "grains" with which sailors kill dolphins and porpoises, the native steals up to the house, and driving his formidable weapon through the walls, is sure to transfix the inhabitants. Holding the spear firmly with one hand, with the other he takes his tomahawk from his belt, dashes the house to pieces, and secures the unfortunate animals.

As the fur of the Musquash is valuable, and the flesh is considered as good as that of the duck, it is greatly persecuted by hunters, who generally employ one of four methods, two of which require a knowledge of the home. One plan has already been described, and another consists in finding out the different entrances, blocking them up, and then intercepting the animals as they try to escape. Sometimes the gun is used, but not very frequently, as the Musquash is so wary, that it dives at the least alarm, darts into one of its holes, and will not show itself again until assured of safety. The trap, however, is the ordinary means of destruction. This is made of iron, and is set in such a manner that as soon as the animal is caught its struggles cause the trap to fall into the water, dragging after it the Musquash, which is soon drowned.

In its subterranean home the Musquash lays up large stores of provisions, and in the habitation have been found turnips, parsnips, carrots, and even maize. All the roots had been dug out of the soil, and the maize had been bitten off close to the ground. The Musquash is not a large animal, the length of its head and body being only fourteen inches.

I have in my collection a curious bag or pouch made from the skin of the Musquash by a very simple process. The animal has been laid on its back, and the skin divided transversely across the lower part of the abdomen. The body has then been gradually turned out of the skin, all the limbs removed except the paws, and the skull also taken away. The inside of the skin is then dried, and prepared in some ingenious manner so that it serves as a convenient pouch, the slit across the abdomen forming the entrance, the tail acting as a handle for suspension, and the feet dangling as ornaments. For this curious specimen I am indebted to Lieutenant Pusey, R.N.

CHAPTER XVI.

BUILDING BIRDS.

THE OVEN BIRD and its place in ornithology—Its general habits—Nest of the Oven Bird—Curious materials and historical parallel—The specimens in the British Museum—The internal architecture of the nest—Division into chambers—The PIED GRALLINA—The specimens at the Zoological Gardens—Materials and form of the nest—Boldness of the bird—The SONG THRUSH and its nest—The BLACKBIRD and its clay-lined nests—Supposed reasons for the lining—The FAIRY MARTIN—Locality, shape, and materials of the nest—Social habits of the bird—How the nest is built—The RUFOUS-NECKED SWALLOW—Locality and abundance of its nests—Curious habit of the bird—Audubon's account—The RUFOUS-BELLIED SWALLOW—Supplementary nest—How the bird builds—Popular superstition and its uses—The HOUSE MARTIN—Material of its nest—Favourite localities—Ingenuity of the Martin—Adaptation to circumstances—Parasitic intruders, their number, dimensions, and tenacity of life—The SWALLOW—Distinction between its nest and that of the Martin—Why called the Chimney Swallow—TALLEGALLA, or BRUSH TURKEY—The illustration explained—Various names of the Bird—Its singular and enormous nest—How the eggs are laid and hatched—Egress of the young—Remarkable instinct. AUSTRALIAN JUNGLE FOWL—Shape, size, and position of its nests—How the eggs are discovered—LEIPOA or NATIVE PHEASANT—Its mound-nest, and general habits.

AMONG the building birds, there is one species which is pre-eminently superior. Not only is there no equal, but there is no second. This is the OVEN BIRD (*Furnarius fuliginosus*), which derives its popular name from the shape and material of its nest.

The Oven Bird belongs to the family of the Certhidæ, and is therefore allied to the well-known Creeper of our own country. It is about as large as a lark, and is a bold looking bird, rather slenderly built, and standing very upright. Its colour is warm brown. It is very active, running and walking very fast, and is much on the wing, though its flights are not of long duration, consisting chiefly of short flittings from bush to bush in search of insects. It generally haunts the banks of South American rivers, and is a fearless little bird, not being alarmed even at the presence of man. The male has a hard shrill note, and the female has a cry of somewhat similar sound, but much weaker.

The chief interest of this bird centres in its nest, which is a truly remarkable example of bird architecture. The material of which it is made is principally mud or clay obtained from the river banks, but it is strengthened and stiffened by the admixture of grass, vegetable fibres, and stems of various plants. The

OVEN BIRD

heat of the sun is sufficient to harden it, and when it has been thoroughly dried, it is so strong that it seems more like the handiwork of some novice at pottery than a veritable nest constructed by a bird, the fierce heat of the tropical sun baking the clay nearly as hard as brick.

The ordinary shape of the nest may be seen by reference to the illustration, which was drawn from a remarkably fine specimen in the British Museum. It is domed, rounded, and has the entrance in the side. Its walls are fully an inch in thickness, and it looks strong enough to bear rolling about on the ground. This specimen was placed on a branch, but the bird is not very particular as to the locality of its nest, sometimes building it on a branch of a tree, sometimes on a beam in an outhouse, and now and then on the top of palings; generally, however, it is built in the bushes, but without any attempt at concealment. Owing to its dimensions and shape, the nest is extremely conspicuous, and the utter indifference of the bird on this subject is not the least curious part of its history.

Strong as is the nest, it is still further strengthened by a peculiarity in the architecture, which is not visible from the exterior. If one of the nests be carefully divided, the observer will see that the interior is even more singular than the outside. Crossing the nest from side to side is a wall or partition, made of the same materials as the outer shell, and reaching nearly to the top of the dome, thus dividing the nest into two chambers, and having also the effect of strengthening the whole structure. The inner chamber is devoted to the work of incubation, and within it is a soft bed of feathers on which the eggs are placed. The female sits upon them in this dark chamber, and the outer room is probably used by her mate. The reader will remember that several instances of such supplementary nests have already been mentioned. The eggs are generally four in number.

Both sexes work at the construction of the nest, and seem to find the labour rather long and severe, as they are continually employed in fetching clay, grass, and other materials, or in working them together with their bills. While thus employed they are very jealous of the presence of other birds, and drive them away fiercely, screaming shrilly as they attack the intruder.

AUSTRALIA produces the two remarkable birds whose nests are given in the accompanying illustration.

The first of these feathered builders is the PIED GRALLINA (*Grallina Australis*), a bird which has become familiar to the public since its introduction to the Zoological Gardens. A pair of these birds have lived for some time in the Aquarium House, and have always attracted much attention as they fly to and fro in the large inclosure which is dedicated to them. to the dabchicks, kingfishers, wagtails, and other water-loving birds. Owing to the bold contrasts of black and white in their colouring they are very conspicuous, and their restless movements always attract the eye.

Although in its shape the nest of the Pied Grallina does not resemble that of the Oven bird, the materials with which it is constructed are almost identical, consisting of mud and clay, in

FAIRY MARTIN. PIED GRALLINA.

which are interwoven certain sticks, grasses, feathers, and stems of plants, which serve to bind the clay together, just as cow's hair binds together the plaster on our walls. When looking at these nests, the observer is irresistibly reminded of the old Babylonish bricks, in which the grass and straw still remain, and serve to strengthen the ill-burned clay, which in many cases was only dried in the sun. Possibly, if the bird were deprived of such materials, and only furnished with mud and clay, it would be as much at a loss as were the captive Israelites when they were compelled to make bricks without being supplied with straw.

Like the Oven bird, the Pied Grallina makes no attempt to conceal its nest, but places it quite conspicuously on a branch, as is shown in the illustration. It is almost invariably built on a bough which overhangs the water, and in spite of its weight and size, is fixed so firmly to the branch that there is no fear lest it should overbalance itself. The walls of the nest are very thick and solid, and the whole edifice looks very like an exceedingly rude and ill-baked earthenware vessel, just such an one, indeed, as Robinson Crusoe manufactured on his island. The bird is widely spread over Australia, so that its nest may be found in many parts of the country.

I MAY here mention that two of our best known song-birds form a basin-like nest of somewhat similar materials. Every one who has taken the nest of a SONG THRUSH (*Turdus musicus*), will remember that its interior is lined with a cup of a substance that resembles clay, but which is in fact composed chiefly of cowdung and decayed wood. This cup is exceedingly thin, but it is very hard and tough, and is so compact in its structure that it will hold water for some time. Like the mud wall of the Pied Grallina, it is strengthened by sticks and grass, with this difference, that whereas the latter bird incorporates the sticks and straws with the mud, the Thrush works the cup upon the sticks and straws.

The BLACKBIRD (*Turdus merula*), too, has a similar habit, only it employs veritable mud for the purpose, and spreads it in a much thicker layer than the Thrush. The eggs, however, are not placed on the dried mud, but on a layer of very fine grass. The object of this curious lining seems to be still undiscovered. Both the birds build in similar localities, and both make their nests close to the ground. It is possible that the stout walls may prevent the weasel or stoat from tearing the nest away from below, and so catching the young birds, but this is mere conjecture. Even the muddy lining does not repel all such attacks, for I once knew a dog that was in the habit of searching for nests of both these birds, and of eating the eggs and the young. He always obtained his prey by getting under the nest, biting out the bottom, and receiving the contents in his mouth.

THE curious flask-shaped nests which are seen in the illustration are built wholly of clay and mud, and are made by a beautiful little Australian bird, named the FAIRY MARTIN (*Hirundo Ariel*), closely allied, as its generic name signifies, to the swallows and martins of our own country. The bird is spread over the whole of Southern Australia, where it arrives in August, and whither it departs in September.

These remarkable nests are generally to be found upon rocks, and are always close to rivers, but have never been seen within many miles of the sea. Sometimes, however, the bird chooses another locality, and, instead of fixing its nests to the side of a rock, attaches them to the interior of one of the huge hollow trees which are so common in Australia. Now and then it behaves like the martin of England, and builds its nest under the protection of human habitations.

The shape of the nests always resembles that of a flask or retort, and their size is extremely variable, the length of the spouts, or necks, being from seven to ten inches, and the diameter of the bulb varying from four to seven inches. Mr. Gould mentions, in his work on the Birds of Australia, that each nest is the joint work of several birds, six or seven being sometimes employed upon one nest, one sitting in the interior, as chief architect, arranging and smoothing the material, while the others go off in search of mud and clay, which they knead well in their mouths before applying it to the nest.

As is generally the case with clay which is thus kneaded, it becomes very hard when baked in the sun, but, at the same time, is rather slow in drying. When the weather is dry, the bird can only work in the mornings and evenings, because the heat of the sunbeams soon renders the clay too stiff to be worked by the delicate beaks of the birds; and, therefore, in the middle of the day, the Fairy Martins cease from their architectural labours, and do nothing but chase flies. During wet weather, however, when no flies are abroad, and the air is full of moisture, the birds work continually at their nests, and soon complete their labours.

The exterior of the nest is quite as rough as that of the common English martin; but in the interior it is beautifully smooth. The birds do not seem to have any particular care about the point of the compass towards which the entrance looks, but arrange it indifferently in any direction.

The Fairy Martin is a prolific little bird, laying four or five eggs, and rearing two broods in a year.

There is an American Swallow which builds a nest very similar in form to that of the Fairy Martin. This is the RUFOUS-NECKED SWALLOW (*Hirundo fulva*), whose nests are made of mud, and flask-shaped, but have a wider and shorter neck than is the case with the nest of the Fairy Martin. On account of its gregarious propensities, it is sometimes called the REPUBLICAN SWALLOW. Wherever a favourable spot is found, such as a perpendicular rock with an overhanging shelf, the nests are built in profusion, being placed so close to each other that the rock is almost covered with them.

The birds are also gregarious on the wing as well as in nesting, as will be seen by Audubon's remarks upon their habits:—"About sunset they begin to flock together, calling to each other for that purpose; and in a short time presented the appearance of clouds moving towards the lakes on the mouth of the Mississippi, as the weather and wind suited. Their aërial evolutions before they alight are truly beautiful. They appear at first as if reconnoitering the place, when, suddenly throwing themselves into a vortex of apparent confusion, they descend spirally with astonishing quickness, and very much resemble a *trombe* or water-spout. When within a few feet of the *ciriers*, they disperse in all directions, and settle in a few moments. Their twitterings and the motion of their wings are, however, heard during the whole night.

"As soon as the day begins to dawn, they rise, flying low over the lakes, almost touching the water for some time, and then rising, gradually move off in search of food, separating in different directions. The hunters who resort to these places destroy great numbers of them, by knocking them down with light paddles, used in propelling their canoes." The *cirier* which is here mentioned is the French popular name for the *Myrica cerifera*, a shrub belonging to the same genus as the well-known British shrub called Sweet Gale, or Dutch Myrtle.

Another American bird, the RUFOUS-BELLIED SWALLOW (*Hirundo erythrogaster*), is notable for the nest which it makes. This species follows the example of the Oven Bird in its selection of materials, strengthening the mud walls of its nest with fine hay. The nest is furthermore remarkable for having a supplementary perch, or small nest attached to the larger one, serving as a seat for the male, while his mate is engaged in the business of incubation. On such occasions he is in the habit of pouring forth a lively, though not varied song, being, in fact, a sustained twitter. The shell of the nest is about an inch in thickness, and the mixed mud and hay are arranged in regular layers. Owing to the thickness, and the complicated structure of the nest, a full week is required for its completion. The form of the nest is nearly that of an inverted cone, being flattened on the side which is set against the wall or rock. The bird is of gregarious habits, and twenty or thirty nests are often seen so close together that a finger could scarcely be placed between them.

Fortunately for itself, this bird is protected by popular superstition, which attributes all kinds of ill-luck to the person who kills one of them. Wilson remarks that, in consequence of long immunity, they feel so secure among human habitations, that although the woods may be destitute of them, every farm-house is sure to attract them. There is scarcely a barn in which they will not build; and right glad is the farmer when they take possession of a house, for, according to popular belief, such a building will never be injured by lightning. He further mentions, that a farmer said, that if he were to permit his swallows to be shot, all his cows would give bloody milk; to which remark Wilson merely nodded assent, being unwilling to disturb any feeling, however superstitious, which had for its object the protection of useful birds.

WE have several builders among our British birds, the best known of which is the common HOUSE MARTIN (*Chelidon urbica*), whose nests are so plentiful upon the walls of our houses.

The material of which the nests are built is a kind of mud, which becomes tolerably hard when dry, and is strong enough to exist for a series of years, and to serve for the bringing up of many successive broods. The bird is exceedingly capricious as to the spot which it selects for its residence, some houses being crowded with the mud-built nests, while others are free from them. The points of the compass are always noted by the Martin, for there are some points which it clearly detests, while it is equally fond of others. A wall with a north-eastern aspect is a favourite locality, while a southern wall is seldom chosen, probably because the heat of the meridian sun might dry the mud too quickly, or might cause inconvenience to the young birds.

My own house, however, forms an exception to this general rule, for the Martins have chosen to build on the south wall only, probably because the eaves project so far that after nine A.M. the nests are in shadow. Moreover, there is a narrow ledge, barely an inch in width, which runs under the eaves, and forms a support for the nests. While the Martins were engaged in bringing up their young, I ascended to the nests, and inspected them carefully, much to the indignation of the parent birds, who flew about wildly, darting occasionally out of their nests, and then stopping short and dashing away over the house. The opening of the nest being close against the eaves, the interior could not be inspected; but the touch of the finger showed that the walls were tolerably smooth, forming a great contrast with the rough exterior. The young birds were quite as much alarmed as their parents, and shrank to the very bottom of the nest, where they were quite invisible.

As to the nests themselves, they are exceedingly irregular on the outside, and look as if they had been made of that preternaturally ugly substance called "rough-cast," with which the walls of houses are sometimes disfigured. The material of which the Martin makes its nest is said to be the earth that is ejected by worms; but that this substance does not form the whole of the material is evident from the fact that stones, grass, and feathers are mixed with the mud, together with small twigs and a few fine roots of an inch or two in length.

The Martin is a rather ingenious bird, and is always ready to take advantage of any circumstance which may aid it in building its nest. The inch-wide ledge, for example, which I have just mentioned, has been quite appropriated by Martins, and there is scarcely a part of it which does not bear marks of their labours. At least a dozen nests have been begun and abandoned after a few beakfuls of mud have been put together, probably because the position is so exceedingly advantageous that the birds can scarcely begin in one place without regretting that they have not chosen a neighbouring spot.

There is an interesting account in the "Zoologist," of the unexpected skill displayed by these birds:—"Under the eaves of a house, not so high as to be beyond the reach of any urchin who could procure a rod or fling a stone, a Martin had built its nest, which had more than once been destroyed. There is no doubt that, under ordinary circumstances, these birds would have gone on building their habitation in the same place and manner, if left to themselves and their own resources, although even in such cases some important variation in the structure has been known to have occurred. But, in the present instance, the inhabitants of the cottage were not satisfied to see the labours of their favourite perpetually rendered void, and they set their wits to work, in what manner to secure them from harm.

"The method adopted was, to place a small round basket under the eaves, at the place where the nest had been, as a protection from injury below; but it was attended with the inconvenience that the handle prevented it from being pressed into contact with the stone, while the breadth of the basket was so great as to cause the wet dripping from the eaves to fall within the cavity. It was to obviate this last annoyance that a flat piece of board was laid as a cover to the basket, with the precaution of leaving an opening, not in front, but at the side, for the birds to enter, if they should choose to adopt this new contrivance for their advantage; and they did justice to the kind intentions of their friends by adopting it, and that, too, in a way of their own contrivance. They began by placing a rim of their usual mortar round the basket, at the border where the covering board rested on it; but in thus rendering it safe and close on every side, they observed the precaution of leaving a small hole at the side, by which to enter. In this convenient piece of wicker-work they formed a cradle, in which they were able successfully to rear their brood.

"But this was not all. Another pair of birds had seen the good fortune of their fellows, and they resolved to be sharers in the advantage they were enjoying. The space above the board, and within the arched handle of the basket, was only inferior to the basket itself as a situation for a nest, and there, accordingly, they proceeded to place it. It was formed of clay, in the usual manner, and here, immediately above their neighbours, they successfully hatched their young. . . . The laying hold of a novel, but obvious convenience, to secure an important object, is not the least of the operations of the reasoning powers."

The writer of this notice is quite correct in attributing the performances of these birds to reason, and not to instinct. Instinct would have taught them to make their nests under ordinary conditions, and to raise their clay-built houses against a wall. But the mental process which led them to accommodate themselves to such a change of circumstances as the substitution of a basket for a wall does, undoubtedly, belong to the province of reason, rather than of instinct.

To examine minutely the economy of a Martin's nest is a pleasant task enough, but has its drawbacks, which are very numerous, and may be summed up in one word—vermin.

All birds are liable to the attacks of parasitic insects, but the Martins contrive to harbour such quantities of them that the spectator cannot but wonder how they contrive to live through the constant attacks. The nest itself swarms with them, and so numerous are their hosts that I have found an isolated lump of clay filled with these repulsive insects, though at the distance of eighteen inches from the nest. They are not visible at first, and but for their cast skins would probably attract no notice. But when one of these innocent-looking pieces of mud is removed, and put under a glass in which a few drops of spirits of turpentine have been placed, the vermin come trooping out of every crevice, many in numbers, large in dimensions, and obese in outline.

In one lump of clay about as large as a walnut, I have seen so many parasites that they seemed capable of devouring all the little birds; and when it is remembered that every portion of the nest is equally tenanted, how the inmates can survive for a single night is indeed matter of surprise. Their size is absolutely portentous; for when compared with the birds on which they feed, they are as large as full-grown frogs compared with men. I mention this circumstance in order that my readers may be chary of bringing a Martin's nest into a room, for to introduce

NEST OF TALLEGALLA.

such pests into the house is far more easy than to extirpate them. Most insects are killed at once by inhaling the vapour of turpentine, but I have kept a number of them shut up in a tin box in which some spirits of turpentine had been poured, and after six and thirty hours found them still alive. They certainly dislike the vapour, and it has the effect of stupefying them. But, as soon as they are removed from its influence, the fresh air seems to restore them, and they begin to crawl about again.

THE common SWALLOW (*Hirundo rustica*) also makes a clay-built nest, similar in many respects to that of the martin, but differing in its shape. The nest of the martin is always covered, and entered by an aperture on one side. Mostly it is built immediately under a projecting ledge, which answers the purpose of a roof, but if no such accommodation can be obtained, it covers in the nest with a dome-like roof. The nest of the Swallow, on the contrary, is open at the top, probably because the long forked tail would be crushed if pressed into so small a compass, while the shorter and simpler tail of the martin does not require so much space.

Wherever it can find an old chimney, the Swallow will always build its nest therein, a habit which has gained for the bird the popular title of Chimney Swallow. It will, however, build in many other situations, such as precipitous rocks and quarries, barns, outhouses, and steeples. There are usually five eggs, and the nest is lined with a soft bed of feathers, like that of the martin.

I MUST now refer the reader to the large illustration, wherein is depicted a group of natives engaged in digging eggs out of an earth-heap. This engraving represents a scene of very common occurrence in Australia, and serves to illustrate the habits of the natives as well as of the bird which will presently be described.

In the foreground is a group of natives resting themselves after a successful hunt, the evidences of which are scattered around them. There is the emu with its head in the woman's lap, the kangaroo, the echidna, and the duckbill. The weapons by which they were killed are thrown carelessly on the ground, and comprise the waddy or club, the boomerang, the spear, and the wummerah or throwing stick, by which it is hurled with terrific force. The large wooden shield indicates also that the natives in question consider themselves in danger of hostile tribes. On the upper branches of a tree are seen a pair of those wonderful kingfishers, popularly termed from their cry, Laughing Jackasses, and in the centre of the illustration is seen an old man crouched upon his knees, busily engaged in digging from a large mound some eggs which are arranged nearly in a circle, and are set perpendicularly with their larger end upwards, as if they had been placed there by the Opposition party in Lilliput.

This mound is the work of an Australian bird popularly called the BRUSH TURKEY or TALLEGALLA (*Tallegalla Lathami*), one of a small series of birds which scrape together great heaps of vegetable substances, and lay their eggs in them so as to be hatched by the heat given out during the process of fermentation. A very brief account of these birds will be given, but we will at present confine ourselves to the Tallegalla.

This bird belongs to the order Gallinæ and the family Megapodidæ, or large-footed birds, the name being given to them on account of the very great comparative size of the feet. It is a native of New South Wales, and is generally found in the densest bushes, through which it can make its way with such rapidity that it can scarcely be captured. As the bird is called by many names, I will mention one or two of them, so that the reader may be better able to identify it while reading the accounts of observant but unscientific travellers. The natives sometimes call it Tallegalla, and sometimes Weelah; and it is occasionally named the New Holland Vulture, because the bare head and neck give it a somewhat vulturine aspect.

We will now proceed to the nest itself.

This curious edifice is often of very great size, several cartloads of materials being used, and its dimensions enlarged from year to year. In order to show the general appearance of the nest, an example is shown in the background, with the bird running over it. The mound is conical in shape, and, as may be imagined from its enormous size, is the result of joint labour, several hens uniting in its formation. The method by which it is made is very curious, and Mr. Gould's account of the bird has been fully corroborated by the habits of the birds in the Zoological Gardens.

Tracing a circle of considerable radius, the birds begin to travel round it, continually grasping with their large feet the leaves, and grasses, and dead twigs which are lying about, and flinging them inwards towards the centre. Each time that they complete their rounds they narrow their circle, so that in a short time they clear away a large circular belt, having in its centre a low, irregular heap. By repeating the same process, however, they decrease the diameter of the mound as they increase its height, and at last a large and rudely conical mound is formed.

The next process is to scrape away the middle of the heap until a cavity of nearly two feet is formed, in which the eggs are carefully placed, being set in the peculiar manner which has also been mentioned. They are then covered up, and are hatched by the joint effects of fermentation and hot sunbeams. By adopting this process the bird does not escape any of the cares of maternity, for the male is very watchful over the eggs, being gifted with a wonderful instinct which tells him of the temperature which is proper for them. Sometimes he covers them with a thick layer of leaves, and sometimes he lays them nearly bare, these operations being repeated several times in a single day.

At last the eggs are hatched, but when the young bird escapes from the shell, it does not emerge from the mound, remaining therein for at least twelve hours. Even after it has enjoyed the open air it retires to the mound towards evening, and is covered up like the eggs, only not to so great a depth. It is a remarkable fact that in all cases a nearly cylindrical hole is preserved in the middle of the mound, being evidently intended as a chimney by which the heat may be moderated, and through which gases produced by fermentation may escape. The reader will probably call to mind that in a well-made haystack a central aperture is preserved for exactly the same purpose, the modern farmer having therefore been anticipated by a bird.

A very great number of eggs are placed in the nest, a bushel of eggs being sometimes taken out of a single mound. These eggs are peculiarly well flavoured, and are equally sought by natives and colonists. The Tallegalla has a habit of scratching large holes in the ground while dusting itself after the manner of gallinaceous birds, and these holes often serve to direct the experienced hunter towards the nest itself.

ANOTHER species of mound-making bird is tolerably common about Port Essington. This is the AUSTRALIAN JUNGLE FOWL (*Megapodius tumulus*), which makes earth-mounds of prodigious size, one of them which was measured being no less than fifteen feet in perpendicular height, and twenty feet in diameter. If the reader will measure off twenty feet along the floor of a room, and fifteen feet upon the walls, he will form a conception of the enormous size of these tumuli. These heaps are always placed under shelter, and are sometimes so enveloped in foliage that, in spite of their great size, they can scarcely be discovered. The materials of which they are composed are rather variable, according to the locality, but the general mass consists of leaves, grass, and other vegetable matter.

Vast numbers of eggs are laid in these nests, and are placed at a considerable depth, some of them being as much as six or seven feet from the top of the heap. They are deposited in a curious manner, the bird scratching its way into the heap, laying an egg, and then filling up the hole as she makes her way out again. The natives always use their hands in digging out these eggs, because their fingers can follow the track of the bird, the softer and looser material acting as a guide. A twig is generally used as a probe by which the presence of a hole is detected, but the hands are the only tools which are used in following up the tortuous track, which sometimes proceeds in a straight line, and then turns suddenly at an angle, the bird having come on a stone or some such obstacle which prevents her from continuing in the same line.

It is a remarkable fact that these mounds are always found near the sea, and in one instance a heap was seen on the very shore, only just above highwater mark.

THE curious bird called by the natives LEIPOA, and by colonists the NATIVE PHEASANT (*Leipoa ocellata*), is another of the mound-makers. In order to avoid confusing the mind of the reader, I may here mention that there are three Australian birds which are popularly called pheasants, the one being the Leipoa, and the others the two species of lyre-bird (*Menura*). The Leipoa certainly has a very pheasant-like appearance, both in the general outline of the head and body, together with the pencilled plumage, the long tail being only wanted in order to complete the resemblance.

It is usually found towards the north-west portions of Australia, preferring sandy plains to any other localities.

The mound which is made by the Leipoa is comparatively small, being seldom more than eight or nine feet in diameter, and a yard or so in height. It is made up of mixed sand, soil, leaves and grass, and is sometimes so hard at its lowest portions, that the hands become useless in digging out the eggs, and strong tools are required. In each nest there are usually about a dozen eggs, which are deposited singly in the mound. One nest, however, will afford a large supply of eggs, just as is the case with our domestic hens, for if her nest be repeatedly robbed, the bird continues to lay for a very long time. The eggs are whitish, slightly speckled with dull red. It is a curious fact that a number of ants are always to be found about the nest of the Leipoa, and their presence, together with the hard, strong substance of the lower part of the nest, would lead many persons to suppose that the mound was nothing but a large ant-hill.

CHAPTER XVII.

BURROWING BIRDS—(CONTINUED.)

Nesting of the Hornbills—Dr. Livingstone's account of the KORWÉ, or RED-BREASTED HORNBILL—The LONG-TAILED TITMOUSE—Its general habits—Its use to the gardener—Number of the young—Form and materials of the nest—Localities chosen by the bird—How to prepare the fragile eggs—The MAGPIE—Its domed and fortified nest—The common WREN and its nest—Pseudo-nests and their probable origin—The HOUSE WREN of America—Its habits and mode of nesting—Wilson's account of the bird—Its usefulness and quarrelsome nature—The LYRE BIRD—Origin of its name—Its domed nest—The ALBERT'S LYRE BIRD and its habits—The BOWER BIRD—Why so called—Civilization and social amusement—The remarkable bower—Its materials and mode of construction—Use to which it is put—The Bower Birds in the Zoological Gardens, and their habits—Love of ornament—Meaning of the scientific name—The SPOTTED BOWER BIRD of New South Wales—Its bower—Description of the birds and their place in the present system.

THE reader may remember that in the account of the toucan and its semi-burrowing mode of nesting, it was mentioned that the bird was sometimes in the habit of closing the aperture of its nest with mud. It is a very remarkable fact that both groups of large-billed birds should possess the same habit, and that the HORNBILL of Africa should close its nest with mud like the toucan of tropical America. These groups of birds are somewhat similar in external appearance, the huge beak giving them a kind of family likeness. They are, however, widely distinct in zoological systems, the toucans belonging to the scansorial, or climbing birds, and the hornbills ranking with the touracos, plantain-eaters, and colies.

Like the toucan, the Hornbill makes its nest in the hole of some decaying tree, and one of the species, at all events, seems invariably to reduce the size of the entrance by plastering it up with mud, and leaving only a very little aperture. The following interesting account of the Hornbill and its nest is quoted from Dr. Livingstone's well-known work.

"We passed through large tracts of Mopane country, and my men caught a great many of the birds called KORWÉ (*Tockus erythorhynchus*) in their hiding-places, which were in holes in the mopane-tree. On the 19th (February) we passed the nest of a Korwé, just ready for the female to enter; the orifice was plastered on both sides, but a space was left of a heart shape, and exactly the size of the bird's body. The hole in the tree was in every case found to be prolonged some distance upwards above the opening, and thither the Korwé always fled to escape being caught. In another nest we found that one white egg, much like that of the pigeon, was laid, and the bird dropped another when captured. She had four besides in the ovarium.

"The first time that I saw this bird was at Kolobeng, where I had gone to the forest for some timber. Standing by a tree, a native looked behind me, and exclaimed, 'There is the nest of a Korwé.' I saw a slit, only about half an inch wide and three or four inches long, in a slight hollow of the tree. Thinking the word 'Korwé' denoted some small animal, I waited with interest to see what he would extract; he broke the clay which surrounded the slit, put his arm into the hole, and brought out a *Tockus*, or Red-breasted Hornbill, which he killed.

"He informed me that when the female enters her nest, she submits to a real confinement. The male plasters up the entrance, leaving only a narrow slit by which to feed his mate, and which exactly suits the form of his beak. The female makes a nest of her own feathers, lays her eggs, hatches them, and remains with the young till they are fully fledged. During all this time, which is stated to be two or three months, the male continues to feed her and the young family. The prisoner generally becomes fat, and is esteemed a very dainty morsel by the natives, while the poor slave of a husband gets so lean that, on the sudden lowering of the temperature, which sometimes happens after a fall of rain, he is benumbed, falls down, and dies. I never had an opportunity of ascertaining the exact length of the confinement, but on passing the same tree at Kolobeng about eight days afterwards, the hole was plastered up again, as if in the short time that had elapsed the disconsolate husband had secured another wife. We did not disturb her, and my duties prevented me from returning to the spot.

"This (February) is the month in which the female enters the nest. We had seen one of these, as before mentioned, with the plastering not quite finished; we saw many completed, and we received here the very same account that we did at Kolobeng, that the bird comes forth when the young are fully fledged, at the period when the corn is ripe; indeed, her appearance abroad with her young, is one of the signs they have for knowing when it ought to be so. As that is about the end of April, the time is between two and three months. She is said sometimes to hatch two eggs, and when the young of these are full-fledged, other two are just out of the egg-shells: she then leaves the nest with the two elder, the orifice is again plastered up, and both male and female attend to the wants of the young which are left."

In this curious history of bird architecture, two points are peculiarly interesting, one being the reservation of a higher point whereto the bird may fly in case of invasion, and the other the fact that two broods of young can be in the nest at one time.

Passing from the birds which build with mud, we now come to those which use vegetable substances in their habitations. As examples of such architecture, we shall select the nests of those birds which are able to construct domed habitations, as well as the remarkable structures which are built by the Beaver birds of Australia.

The Long-tailed Titmouse (*Parus caudatus*) constructs a nest which is quite as wonderful in its way as the pensile home of the harvest mouse.

This pretty little bird is very plentiful in England, and owing to its habit of associating in little flocks of ten or twelve in number, and the exceeding restlessness of its character, is very familiar to all observers of nature. These flocks generally consist of the parent and their offspring, for the little creature is exceedingly prolific, laying a vast quantity of tiny eggs in its warm nest, and rearing most of the young to maturity. This is a bird which ought to be cherished by all possessors of fields or gardens, for there is scarcely a more determined enemy to the many noxious insects which destroy the fruits, vegetables, and flowers. Fortunately for ourselves, the Long-tailed Titmouse is very fond of the various sawflies, that work such mischief among our fruit trees, and often lay waste whole acres of gooseberries, and it is no exaggeration to say that to a possessor of an orchard, or a fruit garden of any kind, every Long-tailed Titmouse is well worth its little weight in gold.

Were it only for the beauty and elegance of its form, no one who had an eye for living art could kill the pretty little bird, and reduce the bright, active, happy creature to a mere pinch of ruffled feathers. Were it only for the wonderful structure of its nest, it would be worthy of preservation. But when we come to consider the inestimable and inappreciated services which this tiny bird renders to mankind, we should not only be devoid of

THE LONG-TAILED TITMOUSE.

all gratitude, but likewise of all common sense—which however comes to much the same point—were we willingly to destroy our feathered benefactor.

Although almost every one who lives in the country or who possesses a tolerably large garden in a town is perfectly familiar with this bird, comparatively few are in a position to narrate from personal observation the benefits which it confers upon us. The reason is simple; they do not rise early enough. A Long-tailed Titmouse in early morning, and the identical bird at noon, scarcely seem to be the same creature, so different are its ways. It is a specially early bird, earlier than the sparrow, which is apt to be rather a sluggard as regards leaving its nest, though it sets up its garrulous chirp soon after daybreak. At that hour of the morning the Long-tailed Titmouse seems to cast off fear and diffidence, and allows itself to be watched without displaying much alarm. Indeed, with the aid of a good opera-glass, it may be observed almost as well as if it were in a cage.

As the sun ascends above the horizon, and men and boys begin to go about to their daily work, the Titmouse loses its easy confidence, and will not suffer itself to be approached so calmly as in the early morning. Generally, somewhere about five or six A.M. it leaves the garden and flies afield, and must then be sought far from human habitation. If, however, the garden should happen to be surrounded by walls, and the owner should happen to understand humanity as well as self-interest, the little bird will know that it will not be disturbed, and will remain in its sanctuary throughout the greater part of the day.

The quick, lively movements of the little creature are quite indescribable, so incessant and so varied are its changes of attitude. As it runs about the branches, it seems almost independent of gravity, and is equally at its ease whether its head, back, or breast be upward. It ever and anon utters an odd chirping note, which seems to issue from the bird as if it proceeded from some internal machinery, and were independent of the will of the creature which utters it. The observer should be careful to notice its quick, frequent pecks, and may be sure that every such movement denotes the slaughter of some insect, whether in the stage of egg, larva, pupa, or imago. The little beak is by no means so feeble as it seems, and is able to pick up an insect so small as would escape the observation of human eyes, or to pounce upon and destroy one which many a human being would not care to handle.

All the little flock, which are seen flitting about the trees, darting from branch to branch and tree to tree as if they were little arrows projected from bows, have at one time been inmates of the same nest, the beautiful domed structure which is shown in the illustration. How they are accommodated in so small a space seems quite a mystery, for not only is the hollow of the nest of no great size, but the interior is so filled with feathers and down that the space is still further limited.

The nest of the Long-tailed Titmouse is rather variable in shape, but its usual form is shown in the illustration. Generally, it is rather oval, and has an aperture at one side and near the top, through which the birds can pass. I believe that all domed nests, whether of bird or beast, are constructed by at least two architects, one of which remains within, while the other works from without. This is certainly the case with many creatures, and is probably so with all. The materials of which the nest is made are mosses of various kinds, wool, hair, and similar substances, woven by them with great firmness. It is remarkable that in the construction of this nest, which requires peculiar solidity, the Long-tailed Titmouse uses materials like those which are employed by the humming birds, and binds its nest together with the webs of spiders, and the silken hammocks of various caterpillars. The exterior of the nest is covered with lichens, so that the whole edifice looks very much like a natural excrescence upon the tree or bush in which it is placed, as is the case with the well-known nest of the chaffinch.

Sometimes the form of the nest is rather different from that which has been mentioned, and the structure is flask-shaped, the entrance corresponding to the neck of the flask. Now and then a nest is found in which there are two openings, one near the top in the usual position, and the other on the opposite side and near the bottom. The presence of one or two apertures is probably influenced by the position of the nest and the climate of the locality. If the finger be introduced into the aperture, a charmingly soft and warm bed of downy feathers is felt, *in* which, rather than *on* which, the numerous eggs repose.

The bird will build its nest in various trees, but always chooses a spot where the branches are very close and the foliage dense. The gorse bush is a favourite residence of the Long-tailed Titmouse, and so deeply is the nest buried in the prickly branches, that it cannot be removed without the aid of thick leather gloves, and a sharp, strong knife. Some skill and artistic taste are required in order to secure a good specimen, and it is difficult to hit the happy medium between cutting away too many branches, and retaining so many that the shape of the nest cannot be seen for their luxuriance. I may mention here that such nests are fertile homes of insects and various vermin, and that they ought to be placed in a box with spirits of turpentine for some weeks, and then exposed to strong heat, before the possessor can be sure that all existing insects are dead, and their eggs addled.

The number of eggs is rather variable, but is always great, and on an average, some ten or twelve eggs can be found in a nest. They are so small and so fragile that the novice finds great difficulty in emptying them without breaking their delicate shells. This task may, however, be accomplished with perfect ease and safety if managed in the right way. Each egg should be enveloped in repeated wrappers of silver paper, soaked in a solution of gum arabic, one layer being allowed to dry before the next is added. When they are dry, a little hole is easily drilled on one side by means of a needle, the contents of the egg are then broken up with the same needle, and are then washed out by injecting water through a very delicate glass tube. Any one can make these slender tubes by merely taking a piece of ordinary glass tubing, heating it in a spirit lamp, and drawing the ends apart. It may then be broken off to form a tube of any degree of fineness, and by alternate injection of water and sucking the diluted contents into the tube, the egg will soon be emptied.

We have another well-known bird, which makes a nest as well domed as that of the long-tailed titmouse, though not nearly so pretty nor so elegant. This is the common Magpie (*Pica caudata*) which is one of the very handsomest birds that are indigenous to England. Popularly, the Magpie is thought to be only black and white; in reality there is scarcely a black feather about the bird, its plumage being adorned with steel blue, green, and purple of such intensity that, in certain lights they appear to be jetty black. I may here mention that the wings and tail of the Magpie can be made into beautiful fire-screens, which are light and elegant as well as brilliantly coloured.

The nest of the Magpie is of very large size when compared with the dimensions of the architect, probably on account of the long tail of the mother bird, which cannot be protruded over the edge of the nest, as is the case with many long-tailed birds. It is not merely made of moss and similar soft substances, but the framework is very strongly constructed of sticks, among which are generally interwoven a number of sharp thorns, so that the nest is nearly as unpleasant to the bare hand as a thistle. Moreover, the bird has a way of gathering the thorns round the entrance, so that the hand cannot be inserted into the nest without danger of many wounds. Indeed, the nest is so large, and the eggs lie so far from the entrance, that to extract them is generally a task that cannot be accomplished without the aid of a knife.

Besides the thorny defence, the nest is mostly strengthened by its very position, being generally fixed in the furcation of several stout boughs, so that it can only be approached in certain parts. Moreover, the great height at which the Magpie loves to build the nest renders the operation of robbing it so dangerous, that many a nest escapes because no one has nerve enough to risk the ascent.

The position of the nest, too, conceals its true form so well, that a very practised eye is needed to distinguish it from an ordinary swelling of the bough, or from the heaps of dislodged twigs which are so often found in the forked branches of trees. Deserted nests are very common, and during my bird's-nesting days, I have frequently been disappointed to find that after all the trouble of ascending a lofty tree, the nest was empty, and had clearly been deserted for a year or two. Sometimes the nest is occupied by other creatures, and in some parts of the country, the pine marten has been known to take possession of a deserted Magpie's nest, and to lie therein quite unsuspected until driven out by some accident. Although a lofty tree is mostly chosen by the Magpie, such is not invariably the case, for now and then a low tree, or even a bush, is selected. In any case, however, the branches are sure to be thickly set, so that the nest may be firmly held among the boughs.

ANOTHER of our feathered dome-builders is the common WREN (*Troglodytes vulgaris*). The form and colouring of this bird are too well known to need description, and we shall therefore pass at once to its mode of nesting.

The Wren is rather peculiar in its method of constructing the nest, for though it can build a dome when there is need for it, and generally does so, it does not always choose to take so much trouble, but contents itself with an open nest arched over by a natural dome. Wherever it can find a convenient cavity, it will make its nest therein, building either no dome at all, or one of very flimsy construction, and such nests can generally be found in the holes of ivy-covered walls, under eaves, or among the thickly growing branches of fir-trees.

During the time when the Wren is building its nest, its loud, cheerful voice is heard in full perfection, and so full and powerful are its tones that the tiny bird seems hardly able to produce them. It is but a short song, and is little varied, the bird repeating nearly the same melody time after time within a few minutes. The long-drawn song of the nightingale, or the mellow notes of the thrush, are beyond the power of the Wren, but there are few birds whose song is more enlivening, or which add so much to the pleasure of a country walk. Besides the more formal song, the Wren has a pretty little monosyllabic chirp, which it utters as it pops about the hedges with its peculiar movements, dropping and ascending again with restless activity. The bird is so bold, too, that it will perch on a branch or a paling within a yard or two of the observer, and pour forth its bright song without displaying the least alarm.

As to the materials of the nest, the bird is in no way fastidious, and generally seems to regard quantity rather than quality. Grasses of various kinds usually form the bulk of the nest, together with mosses, lichens, and similar substances. Withered leaves are generally worked into the nest, and I have more than once found specimens which were almost wholly composed of leaves. The size of the nest is wonderfully large, when the dimensions of the tiny architect are taken into consideration, and however large may be the hole in which the Wren makes its nest, it is nearly filled with the mass of grass, leaves, and wool which the Wren has conveyed into it. The interior of the nest is always warmly lined, sometimes with feathers, and sometimes with hair, and in the lining are generally some six or eight little eggs, nearly white, and covered with very minute red specks.

Probably, the very large mass of material is employed in order to defend so small a bird from the inclemency of the season, for the Wren stays with us throughout the year, and in the winter time resides in the same nest which was used as a breeding place during the summer. If an old ivy-covered wall, or a haystack, or an old house, be examined at night, there will often be found certain pseudo-nests in which the Wren hides itself. These curious edifices are raised by the Wren, though they are never used for the legitimate object of a nest, and the reason of their construction is not very evident. In all probability they are the work of young and inexperienced nest-builders, who begin to make their home, and when they have proceeded with their work, find that the locality is unsuitable, and that they must find another spot. The juvenile bird-nester is often woefully disappointed by finding these nests, especially if he finds three or four in a single wall or stack, as is not unfrequently the case.

As is the case with the redbreast and one or two of our more familiar birds, the Wren will sometimes enter houses and build its nest in curtains, on shelves, and similar localities, while the interior of a disused greenhouse or stable loft is nearly sure to be tenanted by a Wren and its little brood.

An allied bird, the HOUSE WREN of Northern America (*Troglodytes œdon*), has very much the same habits, and will generally take possession of any box that is nailed on a wall, or a post where a cat cannot reach it. On account of, probably, the bird-eating snakes, which are plentiful in that country, the materials of the nest are much stronger than in England, and consist generally of twigs and sticks on the exterior and feathers within. Wilson mentions that on a hot June day, a mower happened to hang up his coat in a shed, and left it there for two or three days. When he removed it from the nail on which it had hung, and was putting it on, he found one of the sleeves quite choked up with sticks, grass, and feathers, being the completed nest of a House Wren. The unfortunate little architects were very angry with the man for disturbing their home, and followed him out of the shed, scolding him for the damage which he had unwittingly done to their newly-finished home.

Happily for the little bird, the popular feeling is in favour of its preservation, and in many a garden there is a box for the House Wren, carefully mounted on a pole like one of our barrel pigeon-cotes, and each box having only a little hole by way of entrance, so that no larger and more powerful bird should be able to usurp the comfortable little house. In default of a box, however, the House Wren will put up with very poor accommodation, and make its nest in an old hat nailed under the eaves of a house, or in a flower pot, or in a hollow cocoa-nut or gourd.

There is wisdom as well as kindness in providing a home for the House Wren, for it is one of the insect-eating birds, and when it is thus suited with a house, it remains near the spot, to the manifest advantage of the herbs and fruit.

Of this little bird Wilson gives an interesting anecdote. "A box fixed up in the window of the room where I slept was taken possession of by a pair of Wrens. Already the nest was built and two eggs laid, when one day, the window being open, as well as the room door, the female Wren, venturing too far into the room to reconnoitre, was sprung upon by grimalkin, who had planted herself there for the purpose, and before relief could be given, was destroyed. Curious to see how the survivor would demean himself, I watched him carefully for several days.

"At first he sang with great vivacity for an hour or so, but becoming weary, went off for half an hour; on his return he chanted again as before, went to the top of the house, stable, and weeping-willow, that she might hear him. But, seeing no appearance of her, he returned once more, visited the nest, ventured cautiously into the window, and gazed about with suspicious looks, his voice sinking to a low, melancholy note as he stretched his little neck about in every direction. Returning to the box, he seemed for some minutes at a loss what to do, and soon after went off as I thought altogether, for I saw him no more that day.

"Towards the afternoon of the second day, he again made his appearance, accompanied by a new female, who seemed exceedingly timorous and shy, and who, after great hesitation, entered the box; at this moment the little widower or bridegroom seemed as if he would warble out his very life with ecstasy of joy. After remaining about half a minute in, they both flew off, but returned in a few minutes, and instantly began to carry out the eggs, feathers, and some of the sticks, supplying the place of the latter with materials of the same sort; and ultimately they succeeded in raising a brood of seven young, all of which escaped in safety."

In this little narrative there are two curious points to be noticed, the one that the eggs already laid were turned out, and the other that the new mistress of the house, with a natural jealousy of her predecessor, re-arranged the interior, so as to suit her own ideas of good taste.

As the bird is so useful, the proprietors of gardens would be glad to have a number of families in their domains. This plan, however admirable in theory, is found to be impracticable in fact, the quarrelsome nature of the bird enduring no rival. During the building season, the House Wren sings, fights, and builds with equal energy, and drives away birds of three times his size. The woodpecker is obliged to quit so disturbed a spot, the fussy and active titmice yield to the Wren, and even the blue bird itself, which is also a favoured inmate of the garden, and is furnished with breeding boxes, is obliged to retire from the field, and to allow its tiny antagonist the choice of houses.

AUSTRALIA is proverbially a strange land, and it is only in Australia, or perhaps in Madagascar, that we should look for a wren measuring some seventeen inches in height. Such a bird is, however, to be found in Australia, and is known to the natives by the name of BULLEN-BULLEN, and to the Europeans as the LYRE BIRD (*Menura superba*). It is remarkable by the way that the genius of the Australian language causes many words to be doubled, so that the natives speak of a well known Australian marsupial as the devil-devil, and of a domestic servant as Jacky-Jacky.

New South Wales is the chosen country of the Lyre Bird, which is rather local, and affects certain defined boundaries. Its native name is derived from its peculiar cry, and the popular European name is given to the bird on account of the shape of its tail feathers. The two exterior feathers are curved in such a manner, that when the whole tail is spread they exactly resemble the horns of an ancient lyre, the place of the strings being taken by a number of slender decomposed feathers which rise from the centre of the tail. When the bird is quietly at rest, the tail-feathers cross each other at the curves, and present a very elegant appearance, though not in the least resembling a lyre. In general shape the bird bears some resemblance to a small turkey, except that the legs are longer and more slender, and that the feet do not resemble those of a gallinaceous bird. It is rather remarkable that the egg presents as curious a mixture of the insessorial and gallinaceous aspects as the bird itself.

The nest of this bird is not at all unlike that of the wren, being very much of the same shape, and domed after a similar fashion. The nest is, however, a very rough piece of architecture, composed almost wholly of twigs, roots, and various sticks, which are interwoven in a very loose, but very ingenious manner, so as to form a structure of tolerable firmness, which can be lifted and even subjected to rough treatment without being broken. At first sight it looks like those heaps of dead twigs which are so common in the birch-tree, but a closer inspection shows that there is a certain regularity in the disposition of the sticks, and that the bird is not without method, though that method be not at first apparent.

So rude a structure as this nest would be unsuitable for the tender young, and therefore the whole of the interior is stuffed full of soft feathers. The nest of an allied species, ALBERT LYRE BIRD (*Menura Alberti*) is made in a similar manner, except that the materials are almost wholly small and rather long sticks. Specimens of these nests may be seen in the British Museum. Both the birds are very shy, and cannot be approached without the greatest caution. Like the gallinaceous birds, to which they bear a strong resemblance, they are fond of scratching large holes in sandy soil, sometimes making them nearly a yard in width and eighteen or twenty inches in depth.

In the "corroboring" places, as the natives call them, the Lyre Bird is mostly to be found, and the experienced hunter always watches for the disappearance of the bird into the hole to make his advance. Every now and then it jumps out of the hole, and struts about, mocking with wonderful facility the notes of various other birds, and even imitating precisely those of the laughing jackass. It has, however, a very sweet and powerful note of its own. Each bird makes three or four of these corroboring places, sometimes at a distance of three or four hundred yards from each other.

Dr. Stephenson thinks that the corroboring places are not merely made for amusement, but that they are used as traps in which are caught sundry beetles and other insects, which fall into the pits and cannot get out again. In fact, should this theory be true, the Lyre Bird and the ant-lion have a similar method of trapping their prey in sandy pitfalls, though the former is a bird and the latter an immature insect.

OUR last example of the Building Birds will be the well-known BOWER BIRD of Australia (*Ptilonorhynchus holosericeus*).

Perhaps the whole range of ornithology does not produce a more singular phenomenon than the fact of a bird building a house merely for amusement, and decorating it with brilliant objects as if to mark its destination. Such a proceeding marks a great progress in civilization, even among human races. The savage, pure and simple, has no notion of undergoing more

THE BOWER BIRD.

labour than can be avoided, and thinks that setting his wives to build a hut is quite as much labour as he chooses to endure.

The native Australians have no places of amusement. They will certainly dance their corrobory in one part of the forest in preference to another, but merely because the spot happens to be suitable without the expenditure of manual labour. The Bushman has no place of resort, neither has the much farther advanced Zulu Kafir. Even the New Zealander, who is the most favourable example of a savage, does not erect a building merely for the purpose of amusement, and would perhaps fail to comprehend that such an edifice could be needed. Such a task is left to the civilized races, and it is somewhat startling to find that in erecting a ball-room, or an assembly room, or any similar building, we have been long anticipated by a bird which was unknown until within the last few years. Truly, nothing *is* new under the sun.

The ball-room, or "bower," which this bird builds is a very remarkable erection. Its general form can be seen by reference to the illustration, but the method by which it is constructed can only be learned by watching the feathered architect at work. Fortunately there are several specimens of this bird at the Zoological Gardens, and I have often been much interested in seeing the bird engaged in its labours.

Whether it works smartly or not in its native land I cannot say, but it certainly does not hurry itself in this country. It begins by weaving a tolerably firm platform of small twigs, which looks as if the bird had been trying to make a door mat and had nearly succeeded. It then looks for some long and rather slender twigs, and pushes their bases into the platform, working them tightly into its substance, and giving them such an inward inclination that, when they are fixed at opposite sides of the platform, their tips cross each other, and form a simple arch. As these twigs are set along the platform on both sides the bird gradually makes an arched alley, extending variably both in length and height.

When the bower is completed, the reader may well ask the use to which it can be put. It is not a nest, and I believe that the real nest of this bird has not yet been discovered. It serves as an assembly-room, in which a number of birds take their amusement. Not only do the architects use it, but many birds of both sexes resort to it, and continually run through and round it, chasing one another in a very sportive fashion.

While they are thus amusing themselves, they utter a curious, deep, and rather resonant note. Indeed, my attention was first attracted to the living Bower Bird by this note. One day as I was passing the great aviary in the Zoological Gardens, I was startled by a note with which I was quite unacquainted, and which I thought must have issued from the mouth of a parrot. Presently, however, I saw a very glossy bird, of a deep purple hue, running about, and occasionally uttering the sound which had attracted me. Soon, it was evident that this was a Bower Bird engaged in building the assembly-room, and after a little while he became reconciled to my presence, and went on with his work. He went about it in a leisurely and reflective manner, taking plenty of time over his work, and disdaining to hurry himself.

First he would go off to the further end of the compartment, and there inspect a quantity of twigs which had been put there for his use. After contemplating them for some time, he would take up a twig and then drop it as if it were too hot to hold. Perhaps he would repeat this process six or seven times with the same twig, and then suddenly pounce on another, weigh it once or twice in his beak, and carry it off. When he reached the bower he still kept up his leisurely character, for he would perambulate the floor for some minutes, with the twig still in his beak and then perhaps would lay it down, turn in another direction, and look as if he had forgotten about it. Sooner or later, however, the twig was fixed, and then he would run through the bower several times, utter his loud cry, and start off for another twig.

Why these birds should trouble themselves to make this bower is a problem as yet unsolved. Had the structure served in any way as a protection from the weather, there would have been a self-evident reason for its existence, but the arching twigs are put together so loosely that they cannot protect the birds from wind or rain. Whatever may be the object of the bower, the birds are so fond of it that they resort to it during many hours of the day, and a good bower is seldom left without a temporary occupant.

Ornament is also employed by the Bower Bird, both entrances of the bower being decorated with bright and shining objects. The bird is not in the least fastidious about the articles with which it decorates its bower, provided only that they shine and are conspicuous. Scraps of coloured ribbon, shells, bits of paper, teeth, bones, broken glass and china, feathers, and similar articles, are in great request, and such objects as a lady's thimble, a tobacco-pipe, and a tomahawk have been found near one of their bowers. Indeed, whenever the natives lose any small and tolerably portable object, they always search the bowers of the neighbourhood, and frequently find that the missing article is doing duty as decoration to the edifice.

This species is more plentiful than another Bower Bird which will presently be described. As is the case with many birds, the adult male is very different from the young male and the female in his colouring. His plumage is a rich, deep purple, so deep indeed as to appear black when the bird is standing in the shade. It is of a close texture, and glossy as if made of satin, presenting a lovely appearance when the bird runs about in the sunbeams. The specific name, *holosericeus,* is composed of two Greek words signifying all silken, and is very appropriate to the species. The female is not in the least like the male, her plumage being almost uniform olive green, and the young male is coloured in a similar manner.

ANOTHER species of Bower Bird inhabits New South Wales, and on account of its variegated plumage is called the SPOTTED BOWER BIRD (*Chlamydera maculata*).

The bower which is built by this bird is of very great comparative size, being sometimes a full yard in length, and the arches higher than those of the previous species. Long grass is plentifully interwoven among the twigs, and the decorations of stones, shells, and feathers extend to a considerable distance from either end of the bower. Mr. Gould mentions that the bird places the heaviest stones so as to keep the twigs in their places, and that it will even bring the skulls and bones of the small mammalia to aid in the decoration of its bower.

These birds are allied to the common starling, and belong to a small group of that family which have gained the name of Glossy Starlings on account of their satin-like plumage.

The colour of the Spotted Bower Bird is warm brown, profusely spotted with buff, and upon the back of the neck there is a kind of falling ruff or collar of long feathers which shine like spun glass, and are of a lovely rose pink colour. The generic name "Chlamydera" literally signifies "cloak-necked," and is given to the bird on account of this peculiarity. The classical reader will remember that the chlamys was a short cloak or scarf, that could be thrown round the neck or over the shoulder at the convenience of the wearer.

NESTS OF TERMITE, AFRICA.

CHAPTER XVIII.

BUILDING INSECTS.

The TERMITE, or WHITE ANT—General habits of the insect—African Termites and their homes—Termites as articles of food—Indian Termites—Account of their proceedings—American Termites—Mr. Bates' account of their habits—European Termites—Their ravages in France and Spain—M. de Quatrefages and his history of the Termites of Rochefort and La Rochelle—The EUMENES and its mud-built nest—The TRYPOXYLON of South America—The PELOPŒUS and its curious nest—The MUD-DAUBER WASP—Mr. Goss's account of its habits——The MELIPONA of America—Mr. Stone's Wasp nests and their history—Difference of material—The FORAGING ANTS of South America and their various species—Nests and habits of the Foraging Ants—The AGRICULTURAL ANT of Texas—Dr. Lincecum's accounts of its habits.

WE now pass to the many insects which may be classed among the Builders. The reader will probably notice that several of the true builders are omitted in this department, but will find them under the head of Social Insects.

OF the Building insects the TERMITE, or WHITE ANT, as it is popularly and wrongly called, is the acknowledged head and chief. There are certain other insects which erect habitations which are truly wonderful, but there is not one that approaches the Termite in the size of its building or the stone-like solidity of the structure.

If the reader will refer to the large illustration, he will see that the Termite of Southern Africa can erect nests of very great size. Three of these structures are shown, and a human being has been introduced by one of them in order to show their average height.

The history of the Termites is so complicated, and so full of incident, that I might occupy several hundred pages of this work in describing them and their nests, and yet not have exhausted the subject. I shall, therefore, give a general sketch of the Termites and their habits, and then relate a few details concerning the species which are found in Africa, Asia, America, and Europe.

In the first place, the reader must understand that the Termite is not an ant at all, but belongs to a totally different order of insect, and is allied to the dragon-flies, the ant-lions, the May-flies, and the beautiful Lace-wing flies.

The Termites are social, and, like other social insects, are divided into several grades, such as workers, males, and females, the two latter of which are winged when they reach maturity. The body is oblong and flat, the antennæ short, and the mandibles flattened and toothed, and in most cases extremely long and formidable. Each colony is founded by a single pair, popularly called the king and queen, and the rest of the population consists of developed males and females, which are intended to perpetuate the species and found fresh colonies, and of undeveloped individuals, or neuters, of both sexes. The neuter males are termed soldiers, and are armed with powerful jaws proceeding from enormous heads, and the neuter females are termed workers, and are very small.

There are now before me some specimens of African Termites, the soldiers of which are five oı six times as large as the workers. They are formidable creatures, but they can do little harm beyond inflicting a severe bite, as they are not furnished with stings nor even with poison glands. They can bite through the clothes of an European, and when they swarm upon the bare limbs of the negro, they inflict almost unbearable tortures. The chief duty of the soldier seems to be the defence of the nest; for whenever the walls are broken down the soldiers come trooping out to attack the invader, and being quite unconscious of fear, they will seize on the first strange object that happens to come in their way. There are comparatively few soldiers, their proportion to the workers being only one per cent.

When a pair of developed Termites have settled themselves to form a colony, they share the fate of certain Oriental potentates, and never move out of their royal cell. When the queen is fairly settled, she increases in size so rapidly, that, even if she were set at liberty, she could not crawl an inch. While the head, thorax, and legs retain their original dimensions, the abdomen swells until it is more than two inches long and about three quarters of an inch in width. Thus developed, she produces eggs by the thousand, which are immediately carried off by the workers, who have reserved certain apertures in the royal apartment through which they can easily pass. When the eggs are hatched, the young are carefully watched and tended until they are at last developed into males, females, or neuters, and themselves are able to take part in the manual work.

A full-sized nest of the African Termite is a wonderful structure. Although made merely of clay, the walls are nearly as hard as stone, and quite as hard as the brick of which villa residences are usually built. The form of the nest is essentially conical, a large cone occupying the centre, and smaller cones being grouped round it, like pinnacles round a Gothic spire.

In Anderson's valuable work, "Lake Ngami," there are many detached accounts of the African Termite. He states that he has seen nests which were full twenty feet in height, and had a circumference of hundred feet, and that when the insects were developed and obtained their wings, they issued forth in such hosts that the air seemed as if it were filled with dense and white snow flakes. So strong is the instinct for rushing into the air, that they can scarcely be retained within the nest, and will even pass through fire in order to gain their end.

The nests are always interesting objects, even from the exterior. The walls are so hard that hunters are accustomed to mount upon them for the purpose of looking out for game, and the wild buffalo has a similar habit, the structure being strong enough even to support the weight of so large an animal. The daily labours of the architects can easily be traced, on account of the dampness of the recent clay, so that an approximation can be formed as to the length of time which is occupied in erecting one of the nests. The traveller is always glad to see a large Termite nest, because he is nearly sure to find the surface studded with mushrooms, which are larger and better flavoured than those which our fields produce.

The natives have another motive for looking after the Termite nests, because they eat the inmates, considering them to be a peculiar luxury. The same author whom I have already mentioned, describes a curious interview that he had with Palani, a Bayeiye chief. Wishing to show the chief the superiority of European cookery, Mr. Anderson spread some apricot jam on bread, and offered it to him. The chief took it, and expressed himself much pleased with it, but asserted that Termites were much superior in flavour. In order to catch the Termites in sufficient numbers, the native makes a hole in the nest, and when the workers are congregated for the purpose of repairing the breach, he sweeps them into a vessel, and repeats the operation until he has obtained as many as he wants.

As is the case with the true ants, the Termites only retain their wings for a limited period, using them for the purpose of escaping from the nest, and snapping them off as soon as they have met with a partner. The manner in which the wings are fixed to the body is the same in both groups of insects, and these singular organs are shed by being bent sharply forwards. If a living Termite be caught, and its wings pressed forward with a pin, they will instantly snap off; but if bent backwards, a piece of the body will be torn away before the wings can be removed.

A correspondent of the *Field* newspaper gives a very interesting account of the proceedings of the Termites living in India. After mentioning the peculiar shedding of the wings, he writes as follows:—

"The career of the winged white ant, as far as I have had an opportunity of judging, is as follows:—Soon after the commencement of the first shower which ushers in the rainy season in India, swarms of winged white ants are to be seen issuing from small holes in the earth, in old mud or sunburnt brick walls, and from places of a similar character, in which the original nests may have been located whence these swarms are thrown off. These legions at once attract the attention of all the insectivorous and omnivorous birds in the neighbourhood, and the minahs, crows, and sparrows are on the alert to feast to satiety on the defenceless ants.

"Judging from the appearance of the wings of these ants as they emerge from their earth-home, I should be disposed to think that they do not develop their wings until the dampness of the atmosphere warns them to prepare for action. There is a new, smooth, and glossy appearance about them, not unlike the wings of a young wasp, or the shine of a new hat. The ants vary much in size at this period of their existence; in good damp seasons, and perhaps in favourable localities, they have a well-fed, plump appearance, whereas under unfavourable circumstances they present a slender and measly complexion. They crawl to the mouth of the hole in the first instance, and at once take wing. The males and females take no particular notice of each other until they have made their preliminary flight, which is but short; they soon alight on the ground, or on the dinner-table, as the case may be, making direct for a light if their flight is after dark.

"As soon as they obtain a footing after their descent from their aërial expedition, both males and females commence to run a most headlong and reckless career. Nothing appears to arrest their progress. The female, who is larger and more full in figure than the male, is also slower in her movements. She stops from time to time and performs slow and singular contortions with the hinder portion of her body. I can't help thinking that she gives out some peculiar odour at these times—at all events the males are sensible of the scent of the females, and if they, in their more rapid quartering of the ground, pass over the track of a female, their excitement and activity is redoubled, and they take up the running with singular pertinacity.

"Up to this point both male and female ant retain their wings, and it is as difficult to deprive them of these members as it is to pull the wings from a house-fly. No sooner, however, does the male ant overtake the female, than he makes a dash at the but-too-willing flirt, and seizes (gently, I presume) the extreme end of her plump figure with his jaws. This is the signal to the female that she no longer requires her wings, and at once, with a jerk, both male and female throw from them these now useless incumbrances. Away they go, madam towing my lord, who never quits his hold, but clings to the skirts of his ladylove in a most gallant manner. They race over stock and stone, over garden walk, verandah, or dinner-table, as the case may be, until they jointly fall victims to the ever-watchful birds, beasts, or fishes, who are all on the *qui vive* for the dainty morsel.

"Should they escape all their numerous enemies, and not succeed in being swept from the table by the ever-watchful kitmutgar, the female soon selects for her home, in which to spend the honeymoon of her existence, some spot which seems adapted for the end which she has in view. If she is a prudent, cautious dame, she picks out some soft nodule of earth moistened by the recent rain, and having done so, she communicates to her lord and follower by some means, which to me are inscrutable, that she thinks the lodgings will do.

"The worthy gentleman at once casts off the tow-line, and he and the lady of his affections buckle to without loss of time, and excavate a home for the comfortable reception of the lady. White ants are put to sad shifts at these times. Any dark nook brings them to a halt, and the lee-side of a plate, tray, or the shady side of a candlestick is often selected by the confiding female. In all her arrangements the male appears to acquiesce without demur.

"Vanity and vexation are the natural results of such ill-considered domestic arrangements, and the expiring couple, exhausted with their endeavours to make an earthen nest out of a bamboo tray and a cotton tablecloth, are scattered to the winds by the first servant who may be pleased to screw up energy sufficient to remove the *débris* of the evening meal, to make way for master's breakfast next day. Of the subsequent career of the happy couple it is out of my power to speak, never having had an opportunity of carefully watching their movements."

As to the Termites of Southern America, much information may be obtained from Mr. Bates' valuable work on the natural history of the Amazons. As many of his remarks simply prove the identity of habits between the Termites of the old world and those of the new, I shall say nothing about them, but merely give a brief abstract of his observations.

As with the species which have already been described, the soldiers are the only individuals that fight. When, therefore, the ant-bear tears down the walls of the nest and begins to lick up the inmates, none but the soldiers are killed, they having come out to fight the enemy, while the workers have all run away and hidden themselves underground. In consequence of this fact, the economy of the nest is but slightly disturbed, and after the ant-bear has gone away, the workers begin to raise their walls afresh.

It must be remembered that the nests of the Termite are not confined to the surface, but extend to a considerable distance in the earth, the subterranean galleries being proportionately large to the superimposed nest. Indeed, the greater part of the material with which the walls and galleries are built is brought from below and carried upwards through the nest itself. There is no visible outlet to a Termite's nest, because the insects construct long galleries through which they can pass without suffering inconvenience from the light of day. Both the workers and soldiers are blind; but, in spite of the absence of external visual organs, they are very sensitive to light, and avoid it in every possible way.

The food of the Termite is of a vegetable character, and consists mostly of wooden fibres. They will, however, eat through almost anything, and the traveller in hot climates finds them among his worst troubles. They will cut to pieces the mat on which a man is lying. They will eat nearly all the wood of his strong box, leaving a mere shell no thicker than the paper on which this account is printed. They will devour all his collection of plants, beasts, birds and insects; and a table or any other article of furniture, if left too long in one position, will be utterly ruined by the Termites, which have a fashion of eating away all the interior, but leaving just a thin shell, which looks as if nothing were the matter.

Extirpating them is a difficult task. It is true that, if the mats, clothes, and other household goods are washed with a solution of corrosive sublimate, the Termites will not touch them; but as the articles which can be thus protected are necessarily few in number, the best method is to extirpate them. This can only be done by going to the fountain-head, and cutting off the supply. It is useless to destroy the workers or soldiers, for they are replaced as fast as killed. But, if the queen be destroyed, the supply of eggs is at once stopped, the subjects lose heart, and the colony dies off.

When the adult Termites leave their homes, they often fly in such clouds that they fill the rooms, and even put out the lamps by their numbers. As soon as they touch ground they shed their wings, and then they begin to find how many enemies they have. Of the myriad hosts that pour into the evening air, not one in twenty thousand survives to found a new colony. They have foes above, below, and on every side. The bats and goat-suckers hold high festival on these evenings when the Termites are abroad, and after the insects have cast their wings they are pursued by ants, toads, spiders, and a host of other enemies.

We will now pass to the European Termites, whose history is elaborately given by M. de Quatrefages. Rochefort, Saintes, and

Tournay-Charente have for some years suffered from the ravages of the Termites, and now La Rochelle is invaded by these terrible destroyers. In all probability they were imported by some ship, taken ashore in the boxes into which they had penetrated, and thence spread into the country around. Efforts are being made towards the extirpation of these terrible insects, but nothing seems as yet to have had any great effect. How serious are the damages which they work may be seen from the following account by M. de Quatrefages, in his "Rambles of a Naturalist," vol. ii. p. 346 :—

"The Prefecture and a few neighbouring houses are the principal scene of the destructive ravages of the Termites, but here they have taken complete possession of the premises. In the garden, not a stake can be put into the ground, and not a plank can be left on the beds, without being attacked within twenty-four hours. The fences put round the young trees are gnawed from the bottom, while the trees themselves are gutted to the very branches.

"Within the building itself, the apartments and offices are alike invaded. I saw upon the roof of a bedroom that had been recently repaired, galleries made by the Termites which looked like stalactites, and which had begun to show themselves the very day after the workmen had left the place. In the cellars I discovered similar galleries, which were within half-way between the ceiling and the floor, or running along the walls and extending no doubt up to the very garrets; for on the principal staircase other galleries were observed between the ground-floor and the second floor, passing under the plaster wherever it was sufficiently thick for the purpose, and only coming to view at different points where the stones were on the surface, for, like other species, the Termites of La Rochelle always work under cover wherever it is possible for them to do so.

"MM. Milne-Edwards and Blanchard have seen galleries which descended without any extraneous support from the ceiling to the floor of a cellar. M. Bobe-Moreau cites several curious instances of this mode of construction. Thus, for instance, he saw isolated galleries or arcades, which were thrown horizontally forward like a tubular bridge, in order to reach a piece of paper that was wrapped round a bottle, the contents of a pot of honey, &c.

"It is generally only by incessant vigilance that we can trace the course of their devastations and prevent their ravages. At the time of M. Audoin's visit a curious proof was accidentally obtained of the mischief which this insect silently accomplishes. One day it was discovered that the archives of the Department were almost totally destroyed, and that without the slightest external trace of any damage. The Termites had reached the boxes in which these documents were preserved by mining the wainscoting; and they had then leisurely set to work to devour these administrative records, carefully respecting the upper sheets and the margin of each leaf, so that a box which was only a mass of rubbish, seemed to contain a pile of papers in perfect order."

In the British Museum are several examples of the ravages worked by Termites, one of which is an ordinary beam that has been so completely hollowed and eaten away, that nothing remains but a mere shell no thicker than the wood of a band-box.

Besides the species which were investigated by M. de Quatrefages, there are others in the south of France, and in Sardinia and Spain. One species, *Termes flavicollis,* chiefly attacks and destroys the olives, while in the Landes and Gironde the oaks and firs are killed by another species, *Termes lucifugus.*

As the limits of the work preclude a very lengthened account of any one creature, our history of the Termites must here be concluded, although much interesting matter remains unwritten.

In the accompanying illustration are shown two nests, the two upper specimens on the right hand having been already described. They are made by the little spider called *Agelena brunnea,* and their history will be found on page 303.

The two lower nests are made by a species of solitary wasp, which has no popular name, but is known to entomologists as *Eumenes coarctata.* It is not a large insect, the female being only half an inch in length, and the male rather smaller. The general colour is black, with a fine velvet-like pile on the abdomen, and relieved by lines and spots of yellow. The abdomen is small, and set on a rather short and pear-shaped footstalk, as may be seen by the illustration, which represents the insect of its natural size.

This is one of the species which are tolerably common in certain localities, but as they are very local, may be reckoned among the varieties. Mr. F. Smith, in his "Catalogue of the British Vespidæ," mentions that it has been taken in several parts of Hampshire, Berkshire, near Weybridge, and has been found plentifully at Sunninghill. Probably, the rarity or frequency of this species, as is the case with many others, depends greatly on the eyes which look after it.

EUMENES AND AGELENA.

This little wasp constructs small globular cells of mud, and fastens them to the stems of various plants, the common heath being the greatest favourite, so that heath-covered commons are likely to afford specimens of the nest and its architect. Each nest contains only a single cell, and is only intended to rear a single occupant. The wasp is a very useful insect, as it provisions its nest with the larvæ of small lepidoptera, each Eumenes grub requiring a tolerably large supply of caterpillars.

As is the case with so many insects, the Eumenes is greatly subject to the attacks of parasites, which contrive to deposit their eggs in the larvæ in spite of the hard mud walls of the cell. Mr. Smith mentions that he has had from the nest of the Eumenes, an ichneumon fly belonging to the genus *Cryptus.*

In the accompanying illustration are figured the nests of two insects, both of them natives of tropical America, and both belonging to the hymenopterous order. The upper insect is known to entomologists by the name of *Trypoxylon aurifrons*, but has at present no popular name.

TRYPOXYLON AND PELOPÆUS.

This insect makes a great number of earthen cells, shaped something like those of the last-mentioned species; the cells being remarkable for the form of the entrance, which is narrowed and rounded as shown in the figure. In some cases the neck is so very narrow in proportion to the size of the cell, and the rim is so neatly turned over, that the observer is irresistibly reminded of the neck of a glass bottle. The insect makes quite a number of these nests, sometimes fastening them to branches, as shown in the illustration, but as frequently fixing them to beams of houses. It has a great fancy for the corners of verandahs, and builds therein whole rows of cells, buzzing loudly the while, and attracting attention by the noise which it makes.

THE lower insect is the pretty *Pelopœus fistularis,* with its yellow and black banded body. Both the insects, as well as their houses, are represented of the natural size.

The cell of the Pelopæus is larger than that of the preceding insect, and occupies much more time in the construction, a week at least being devoted to the task. She sets to work very methodically, taking a long time in kneading the clay, which she rolls into little spherical pellets, and kneads for a minute or two before she leaves the ground. She then flies away with her load, and adds it to the nest, spreading the clay in a series of rings, like the courses of bricks in a circular chimney, so that the edifice soon assumes a rudely cylindrical form.

When she has nearly completed her task, she goes off in search of creatures wherewith to stock the nest, and to serve as food for the young, and selects about the most unpromising specimens that can be conceived. Like many other solitary hymenoptera, this Pelopæus stores her nest with spiders, and any one would suppose that she would choose the softest and the plumpest kinds for her young. It is found, however, that she acts precisely in the opposite manner.

In tropical America there is a large group of spiders allied to the common garden spider, but of the most extraordinary shapes and colours. They all possess a hard, shelly covering, polished and shining like that of many beetles, and glittering with bright and radiant hues—blue, crimson, green, and purple being the colours with which they are ordinarily decorated. Their forms are, however, even more remarkable than their colours. The hard and shelly covering is not uniform and smooth, but shoots out into the most extraordinary projections, giving to the creatures a wild and fantastic grotesqueness of aspect, that surpasses even the weird imaginings of Breughel, Cranagh, Callot, and other masters of diablerie in art.

One genus has the abdomen formed in a drum shape, the sides and extremity being covered with short, sharp, and stout spines. Another has the abdomen modified into a ball-like shape, from which radiate sharp spikes, like those of the well-known "calthrop;" while in another genus certain enormous projections issue from the abdomen, two being so large that in volume they exceed the whole of the abdomen and body. In one species they are thick, solid, and palmated, like the horns of the elk; in another they are slender, and curved like the horns of a bull, and there are other species quite as bizarre in form. It is from these creatures, more especially from the first-mentioned, that the Pelopæus selects her victims, and it is evident that the jaws of the young Pelopæus must be exceedingly strong to be enabled to pierce their hard and well-armed bodies. Like the previously-mentioned insect, the Pelopæus makes a loud and cheerful buzzing while engaged in her work of building.

MR. BATES, who has described these two insects, has likewise mentioned a builder insect of the same order, called *Melipona fasciculata*. The genus to which this insect belongs is a very large one, containing some forty-five species, some of which are very common in woods, and being extremely small, measuring only the twelfth of an inch in length, they are very annoying to the traveller, getting into his nostrils and worrying him in various ways. Fortunately, they do not sting, but their bite is very sharp, and if made on a sensitive surface like the lining membrane of the nostril, can inflict very severe pain.

The form of habitation is various, according to the species, but they all use clay for that purpose, kneading it with their mandibles, and then passing it to the hind legs and pressing it into the hair-fringed depression which is popularly called the basket. Some species are accustomed to employ any casual crevice as a nest, stuffing it up with clay, and leaving only a little orifice through which they can pass. Others again make long tubes of clay, with trumpet-shaped mouths, and it is a remarkable fact that a number of the bees are always at the entrance as sentinels, just as is the case with the hive bee when wasps are abroad.

IN the "Zoologist," for 1864, p. 582, is a very interesting description, by Mr. P. H. Gosse, of the proceedings of insects which he appropriately calls the DAUBER WASPS, and which belong to the same genus as the Pelopæus mentioned above. One insect he identifies as Pelopæus flavipes, and the other is probably *Pelopæus spirifer*. One of these insects is now before me, and a very pretty creature it is. In shape it exactly resembles that which is figured on page 353, but the colours are different. The general hue is deep brown-black, very shining in the abdomen, and softened by thick down upon the thorax.

It is, however, not a sombre insect, as the long footstalk of the abdomen is bright yellow, and the limbs are banded with the same lively hue. I strongly advise my readers to peruse this account, because it is full of detail, and contains much useful information about the method of working adopted by the insect, thus giving a clue to the proceedings of other insects which build habitations of similar materials. The length of the account is the reason why it cannot be transferred to these pages, and I must, therefore, give a short abstract.

Having seen many patches of a yellow mud on the walls and rafters, some as large as the closed fist and others of comparatively small dimensions, he asked some boys what they were, and was told that they were the nests of the Dirt-daubers. Finding that as the weather became warm the insects began to build, he set to work and watched them carefully. First he tried their sagacity by boring holes in their cells, in order to see whether the insects would fill them up, and afterwards by inserting foreign substances, such as a tin-tack and a piece of worsted, into the cell. The insect proved herself equal to the occasion, filled up the holes, and pulled out both the tack and the worsted. The next point was, to watch a nest from its foundation, and to see how it was built. The insect always went off, was absent for about a minute, and then returned, bearing in her jaws a lump of clay larger than her own head. The clay was perfectly plastic, and could be spread at once. The method by which the cell is formed must be given in the author's own words :—

"About this time (August 18) the other species of Pelopæus began to be busy fabricating their artful thimble-shaped nests.

"It is difficult to convey by words an idea of their mode of working. The commencement of a cell was by laying down the load and working it into an oval ridge, one extremity of which was to be the apex of the thimble cell. The next load was laid on the ridge, but so as to be higher at the apex than at any other part, and made slightly concave. When the top was made, the work proceeded regularly by additions to the edges, which were smoothly laid on, and always in the same slanting direction that had been given at first, by raising one end of the incipient oval, so that an unfinished cell in any state of progress appears to be a cylinder cut off by a diagonal section.

"This is not casual but invariable, as the ridges remaining plainly mark the precise limits of every load.

"When a little more in length is finished than suffices for a single cell, the work ceases for awhile, an egg is laid in the bottom, though this end is generally uppermost, and spiders are brought in. This species usually, not always, selects a very beautiful species of Tetragnatha, bright green with white spots; and it is worth remarking that spiders are carried both with the jaws and feet, one of the forelegs of the spider being grasped in the mouth, while the body is held under that of the fly, and sustained by the anterior and middle legs and feet, the posterior pair being extended behind, as usual during flight.

"When the first cell is stocked, it is closed up by a transverse partition of mud, and the thimble goes on increasing as before. When finished, one will contain three or even four cells, and then a new one is commenced, adjoining and parallel with it. In both this and the other species, I believe that the enclosed grub eats only the abdomen of the spiders (which are so stung as to be helpless but not dead) as the cephalothorax and legs of each may generally be found afterwards in the cell."

The same writer noticed a remarkable instance of ingenuity in these insects. An empty ink-bottle about an inch and a half in length lay on the table. The neck of this bottle was one day seen to be stopped up with a substance like white pipe-clay, and when this was broken, the bottle was found to be stored with spiders. The fact was, that a Pelopæus had spied out the bottle, and thought that she had a fine opportunity of providing a home for her young without troubling herself to build a regular nest. A day or two afterwards, the Dauber returned to see after the nest, and finding that it had been disturbed, she entered the bottle, took out all the spiders, replaced them with fresh specimens, and then re-closed the mouth. It is evident from this fact, that the insect does not entirely abandon her young when she has completed and closed the nest.

Another curious discovery was also made while watching the Pelopæus. If the reader will refer to the illustration, he will see that the abdomen of the insect is supported on a very long and slender peduncle, or footstalk. Mr. Gosse was naturally anxious to discover how the insect could draw the abdomen out of the pupal skin when it came to change into its perfect condition.

On examining some specimens, he discovered the curious fact, that the pupal envelope did not sit closely to the body, but that it was as wide in the middle as at either end, so that when the insect came to assume its perfect form, the peduncle was quite loose in the centre of the envelope, and the abdomen could be drawn out without any difficulty.

These observations are peculiarly valuable, because they set at rest a question which was raised by several entomologists, who thought that the nests were made by some species of eumenes, and that the Pelopæus was a mere parasite upon them, like the cryptus, and many other of the ichneumonidæ.

If the reader will refer to the large engraving, entitled "Mr. Stone's Wasp-nests," he will see a representation of four square boxes, each containing an object which would hardly be taken for a wasp's-nest at a little distance. Such, however, is the case; and these boxes are four selected examples out of a series of six which were built in Mr. Stone's house, and presented by him to the British Museum. The story of these nests is very remarkable, and shows how much we have to learn concerning the habits and instincts of insects.

In the month of August, 1862, a nest of the common Wasp (*Vespa germanica*), was taken near Brighthampton, and handed over to Mr. Stone, who has long been in the habit of experimenting upon these insects. One extraordinary nest which was built by wasps under his auspices, has already been mentioned on page 256.

The nest was very much damaged by carriage, and Mr. Stone took it entirely to pieces, placing one or two small combs inside a square wooden box with a glass front, and supporting them by a wire which passed through the combs to the roof of the box. He then fixed the box in a window, so as to allow the insects free ingress and egress through a hole in the back.

About three hundred of the workers were then collected, placed in the box, and well supplied with sugar and beer. They immediately began to work, and their first object was to cover the combs with paper. They worked with great rapidity, and in two days had formed a flask-shaped nest, having covered both the combs and the wire, beside plastering large sheets of paper over the sides of the box. They did not attempt to build upon

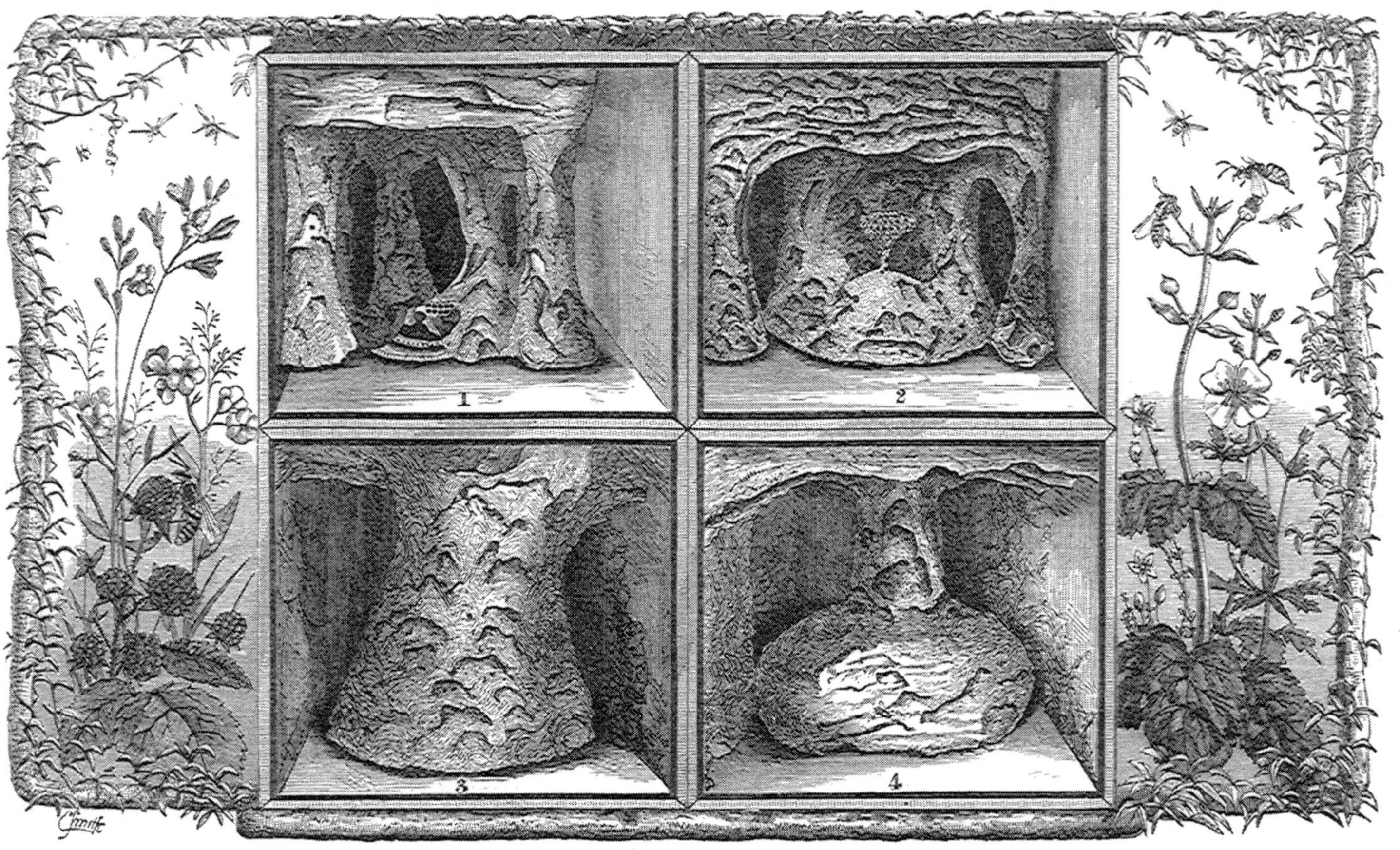

MR. STONE'S WASP-NESTS.

the glass front, because it was frequently moved in order to introduce a supply of sugar. This nest is represented at fig. 4 in the illustration, and one of the wasps is introduced, in order to show the comparative dimensions of the nest and its architects.

As the wasps were building at such a rate, it was evident that they would shortly fill the whole box with a shapeless mass of paper. Another similar box was therefore prepared, and the wasps ejected by tapping the box which was already completed. As soon as they were all out, the second box was substituted for the first, and the wasps crowded eagerly into it and again began their labours. In this box they were allowed to remain for a week, and the result was as is seen in fig. 3. The wasps were now transferred to a third box, in which they laboured for four days, and produced a nest somewhat similar to the others, but not quite so symmetrical.

At this time Mr. Stone fitted up another box with two rows of wire pillars, eight in number, placed with tolerable regularity about two inches apart, and having a piece of comb at the base and summit of each. In this box the wasps remained for fifteen days, and in that time had covered all the wires and most of the combs, and had nearly filled the box with paper.

In order that a more symmetrical structure might be produced, a fifth box was fitted up with wires arranged in a different manner. Four wires were placed across the box, rather in advance of the middle, and two others in front of them. To all these wires a piece of comb was fixed at the base and summit, but between the two central pillars a short wire was placed, having a piece of comb at its summit only. The wasps were transferred to this box, and in the short space of five days, they covered all the combs and wires, and produced the extraordinary structure which is shown in fig. 1, and which looks like a paper imitation of a stalactitic cavern. The insects were ejected from this nest before they had finished their work, and in consequence, a portion of the comb on the small central pillar is still left uncovered.

As this box had been so successful, another was prepared on the same principle, and the wasps were permitted to reside in it for the same number of days, in which time they produced an equally beautiful but rather more massive nest. This specimen is shown at fig. 2 of the illustration. In hopes that the wasps might make a still more splendid nest, a much larger box was fitted up, and the insects transferred to it. As by this time the autumn was closing in, and the weather became cold, the wasps could do but little work, and in a short time they died.

Thus, in the wonderfully short space of thirty-eight days, six elaborate and beautiful nests had been made by a single brood of wasps, and it is probable that if the original nest had been taken at an earlier period of the year, they would have made a still larger number. However, such a feat as they did perform ought to make us look upon the wasp with a more indulgent eye, and although it cannot supply us with honey, as does the bee, it can certainly rival that useful insect in industry.

On looking at this beautiful series of nests, the observer cannot but admire the manner in which the instinct of these creatures is made subservient to human reason. Their instinct teaches them to cover all their combs with a thick mass of paper, the reason being, although they may not know it, that a certain uniformity of temperature is needed for the well-being of the eggs and young. If, therefore, combs are placed conveniently for the insects, they will assuredly cover them according to their instincts, and will as surely take advantage of wires or any other supports to which they can attach the fragile substance of which the nest is made.

Mr. Stone has made other experiments upon wasps, and has kindly sent me the following account of his proceedings:—

"I have a beautiful series of their nests of this season's production (1864), from specimens which are the work of two or three hours, to those which have occupied as many months.

"But my working communities in a semi-domesticated state within the house, have for the last few weeks been going the wrong way. Earlier in the season, I had as many as ten colonies of various species at work in the different windows of the house which I have for some years used for the purpose, all of which went on satisfactorily for some time, but the sugar with which they were fed, at length attracted a vast number of strangers, which crowded into the various boxes, and at first impeded, and ultimately put an end to the work. Before this event happened, one extraordinary nest had become advanced as far as I wished;

and a second, which was still more extraordinary, almost as far as I desired. The facts connected with these nests are as follows:—

"I had a working community of *Vespa germanica* in the left-hand corner of a window on the ground-floor, and another in the right-hand corner. When these nests had increased in size to four or five inches in diameter, I chloroformed the insects, removed the shell or covering of each nest from the combs, putting aside the coverings for specimens. In order to remove the combs, I had to cut out a piece from the outside, and when this was neatly united again, the empty shells had all the appearance of perfect nests, with this advantage, that they contained nothing which required drying in an oven in order to prevent decomposition, which must have been done had the combs, with their complement of grubs, &c., been allowed to remain in the nests. This plan I always adopt when it is practicable. I then returned the combs to the boxes from which they were respectively taken, and introduced the workers, still in a comatose state from the effects of the chloroform. As soon as they recovered from their stupor, they set to work at constructing fresh coverings.

"I now brought home a nest of *Vespa vulgaris*, with its inmates. This was placed for work in a box in the left-hand corner of a room immediately over the one just mentioned. Soon after this, I perceived that the newly-formed covering to the nest of the *V. germanica* in the left-hand corner of the window below, was beginning to assume a variety of curious colouring. On clipping away the covering, when it became sufficiently advanced for another specimen, I found that numbers of workers from the nest of *V. vulgaris*, situated in the window above, had actually joined themselves to this nest, and had been working with is original inmates.

"Not only had they been working in concert with them, but they had been depositing eggs in the cells, as is proved by the fact that numbers of young specimens of *V. vulgaris* were afterwards bred from the combs contained in the nest of *V. germanica.* I do not know whether you are aware that worker wasps have the power of producing fertile eggs without contact with the other sex; yet such I have proved over and over again to be the case.

"Well, having again, as above stated, removed the covering from this nest, I took away the lower comb and reduced the nest somewhat in size, placing them in a box thirteen inches in length, and arranged in such a way that the workers should necessarily produce a vase, or rather a goblet-shaped nest. This they did, and a splendid object it is, being, as before, the joint work of two species of wasps, the one, *V. vulgaris*, using, as it invariably does, decayed wood (such as is commonly called touchwood), and the other, *V. germanica*, using sound wood, or sound vegetable fibre of some kind, in the fabrication of its paper. Thus they gave to the coverings of both these nests an extraordinary beauty, from the variety and charming distribution of the colours with which they were enriched.

As none of the workers from the nest of *V. vulgaris* were ever found to attach themselves to the nest of *V. germanica*, which was situated in a similar corner of the window below, I conclude that they made no mistake as to the corner of the window in which their nest was situated, but miscalculated the height of the window. As they entered the strange nest with food and building material, they were not molested, but allowed to join peaceably in the work of the nest.

"Widely different would have been the case had they entered it for the purpose of pillage; for, though wasps will not interfere with strange individuals of their own species, even when they come with thievish intentions, they instantly seize all individuals of a different species, if their intentions appear suspicious.

"I have since met with another instance of the kind.

"Two nests were situated almost close together, in a drain at Cokethorpe Park, one belonging to *V. vulgaris* and the other having been originally the property of *V. germanica*. It would, however, appear that at an early period in the season, workers from the former nest had attached themselves to the latter, their numbers increasing as the season advanced. Judging from the appearance of the nest, and from the amount of work done by each species, it was easy to see that at the end of August, when I dug it out, the number of individuals of each species was almost equal. There is no possibility of mistaking the work of one species for that of the other, and the distinction is apparent at a glance.

"Apart from the interest attached to nests of this description, no examples of which had been, as far as I am aware, obtained by any naturalist, their beauty of colouring is so remarkable, as to render them objects of general admiration. If, too, as I apprehend must have been the case, the workers belonging to the colony of *V. vulgaris* mistook their neighbours' house for their own, the entrances being so near together, it is rather extraordinary that those belonging to the other species should not have made a similar mistake. They appeared, however, not to have done so, or if they did, the mistake must have been rectified as often as it occurred, for no work of theirs was to be found in the nest of *V. vulgaris*."

Before closing the history of the wasps, I may mention that the two species, *Vespa germanica* and *Vespa vulgaris* are so similar to each other in shape and colour, that an unpractised eye cannot readily discern the distinction between them. Specimens of both these wasps are now before me, and when placed side by side the difference is clearly evident. The yellow colour predominates in the former insect, and the dark bands of the abdomen are much narrower. In the female *Vespa germanica*, there are three black spots on the basal margin of the first segment of the abdomen.

In Mr. Bates's valuable work on the natural history of the Amazons, there is an interesting account of the proceedings of certain ants belonging to the genus *Eciton*, and which are popularly classed together under the name of Foraging Ants. These insects have often been confounded with the Saüba or parasol ant, which has already been described, although they belong to different groups and have different habits. The native name for them is Tauóca. There are many species belonging to this genus, and I shall therefore restrict myself to those which seem to have the most interesting habits, giving at the same time a general sketch of their character. I regret that, as in so many other cases, the lack of popular names forces me to employ the scientific titles by which the insects are known to naturalists.

Although in the Ecitons there are the three classes of males, females, and neuters, these neuters are not divided into two distinct sets as in the termites, but are found in regular gradations of size. The real Foraging Ant is *Eciton drepanophora*, and it is this insect which is so annoying and so useful to house builders. The ants sally forth in vast columns, at least a hundred yards in length, though not of very great width. On the outside of the column are the officers, which are continually running backwards and forwards, as if to see that their own portions of the column are proceeding rightly. The proportion of officers to workers is about five per cent., or one officer to twenty workers, and they are extremely conspicuous on the march, their great white heads nodding up and down as they run along.

ECITON.

One of the large workers is now before me, and a most formidable insect it looks. Its head is round, smooth, and very large, and is armed with a pair of enormous forceps, curved almost as sharply as the horns of the chamois, and very sharp at the points. Their length is so great, that if straightened and placed end to end, they would be longer than the head and body together. They are beset with minute hairs, which, when viewed under the microscope, are seen to be stiff bristles, arranged in regular rings round the mandibles. The thorax and abdomen are but slender, and the limbs are long, giving evidence of great activity. In the dried specimen, the colour of the insect is yellowish-brown, becoming paler on the head, but when the creature is alive, the head is nearly white. The eyes are very minute, looking like little round dots on the side of the head, and being so extremely small, that they can scarcely be perceived without the aid of a magnifying-glass. The half-inch power of the microscope shows that they are oval and convex, but as they are set in little pits or depressions, they do not project beyond the head. The hexagonal compound lenses, which are generally found in insects, are not visible, and the eye bears a great resemblance to that of the spider.

The difference in dimensions of the workers is very remarkable. The specimen which I have just described, measures a little under half an inch in length, exclusive of the limbs, while another specimen is barely half that length, and in general appearance much resembles the familiar ant, or emmet of our gardens.

The presence of these insects may be always known by the numbers of pittas, or ant-thrushes, which feed much upon them, and which are sure to accompany a column of Foraging Ants on the march. The ant-thrushes are odd, short-tailed birds, with stout bodies, and a remarkably long hind claw. Some of this species are decorated with colours of wonderful brilliancy, glittering with blue, green, copper-red, and purple, and having a peculiar silken gloss. Others are soberly clad in simple brown and white, and such are the birds which usually accompany the Foraging Ants on their march.

As soon as the experienced inhabitants of tropical America see the ant-thrushes, they rejoice in the coming deliverance, and welcome the approaching army. The fact is, that in those countries insect life swarms as luxuriously as the vegetation, and there are many insects which, however useful in their own place, are apt to get into houses, and there multiply to such an extent, that they become a real plague, and nearly drive the inhabitants out of their own homes. They are bad enough by day, but at night they issue from the nooks and crevices where they lay concealed, and make their presence too painfully known.

There are insects that bite, and insects that suck, and insects that scratch, and insects that sting, and many are remarkable for giving out a most horrible odour. Some of them are cased in armour as hard as crab-shells, and will endure almost any amount of violence, while some are as round, as plump, as thin-skinned, and as juicy as over-ripe gooseberries, and collapse almost with a touch. There are great flying insects which always make for the light, and unless it is defended by glass, will either put it out, or will singe their wings and spin about on the table in a manner that is by no means agreeable. The smaller insects get into the inkstand and fill it with their tiny carcases, while others run over the paper and smear every letter as it is made. There are great centipedes, which are legitimate cause of dread, being armed with poison fangs scarcely less venomous than those of the viper. There are always plenty of scorpions; while the chief army is composed of cockroaches, of dimensions, appetite, and odour such as we can hardly conceive in this favoured land. As to the lizards, snakes, and other reptiles, they are so common as almost to escape attention.

For a time these usurpers reign supreme. Now and then a few dozen are destroyed in a raid, or a person of sanguine temperament amuses his leisure hours, and improves his marksmanship, by picking off the more prominent intruders with a saloon pistol; but the vacancies are soon filled up, and no permanent benefit is obtained. But when the Foraging Ants make their appearance, the case is altered, for there is nothing that withstands their assault. As soon as the pittas are seen approaching, the inhabitants throw open every box and drawer in the house, so as to allow the ants access into every crevice, and then retire from the premises.

Presently the vanguard of the column approaches, a few scouts precede the general body, and seem to inspect the premises, and ascertain whether they are worth a search. The long column then pours in, and is soon dispersed over the house. The scene that then ensues is described as most singular. The ants penetrate into the corners, peer into each crevice, and speedily haul out any unfortunate creature that is lurking therein. Great cockroaches are dragged unwillingly away, being pulled in front by four or five ants, and pushed from behind by as many more. The rats and mice speedily succumb to the onslaught of their myriad foes, the snakes and lizards fare no better, and even the formidable weapons of the scorpion and centipede are overcome by their pertinacious foes.

In a wonderfully short time, the Foraging Ants have completed their work, the scene of turmoil gradually ceases, the scattered parties again form into line, and the procession moves out of the house, carrying its spoils in triumph. The raid is most complete, and when the inhabitants return to the house, they find every intruder gone, and to their great comfort are enabled to move about without treading on some unpleasant creature, and to put on their shoes without previously knocking them against the floor for the purpose of shaking out the scorpions and similar visitors.

In the illustration a column of Foraging Ants is seen winding its way through a wood. Every one who is accustomed to the country takes particular care not to cross one of these columns. The Foraging Ants are tetchy creatures, and not having the least notion of fear, are terrible enemies even to human beings. If a man should happen to cross a column, the ants immediately dash at him, running up his legs, biting fiercely with their powerful jaws, and injecting poison into the wound. The only plan of action in such a case, is, to run away at top speed until the main body are too far off to renew the attack, and then to destroy the ants that are already in action. This is no easy task, for the fierce little insects drive their hooked mandibles so deeply into the flesh that they are generally removed piecemeal, the head retaining its hold after the body has been pulled away, and the mandibles clasped so tightly that they must be pinched from the head and detached separately.

There seems to be scarcely a creature which these insects will not attack, and they will even go out of their way to fall upon the nests of the large and formidable wasps of that country. For the thousand stings the ants care not a jot, but tear away the substance of their nest with their powerful jaws, penetrate into the interior, break down the cells, and drag out the helpless young. Should they meet an adult wasp, they fall upon it, and cut it to pieces in a moment.

Another species, *Eciton prædator*, does not form long and narrow columns, but marches in a broad and solid phalanx. It is but a little creature, no bigger than the common red ant of England. It is, however, of a brighter red colour, and when a phalanx of these ants ascends a tree, the vast multitudes spread over all the trunk and branches in such numbers, that the tree looks as if a blood-red liquid was being poured over it.

There is another Foraging Ant which forms in broad columns when on the march. This is *Eciton legionis*, a species which is not so common as either of the preceding, and appears only to be seen on the wide sand plains of Santarem.

These insects sometimes attack the nest of one of the large burrowing ants. Mr. Bates mentions that on one occasion, he watched a large army of Foragers begin their attack upon the nest of an ant, some specimens of which he desired to procure. The Foragers set to work with wonderful skill, arranging themselves into two distinct sets of labourers, one set digging into the ground and taking out large pellets of earth, and the other set receiving them from their comrades and carrying them away.

While watching the proceedings of the soldiers when repairing the Thames river-wall after the terrible explosion near Belvedere, I was strongly reminded of the Foraging Ants and their method of working. The parallel was exact in every respect. The officers stood here and there and directed the efforts of their men, while the workers were arranged in regular lines, one set of men digging out the clods of earth, and a second set receiving them and handing them to the spot where they were wanted. I could but fancy that if an observer had been poised at some height above the beach in a balloon, watching the soldiers at work, and had previously seen an army of Ecitons engaged in sinking a shaft, he would have seen the insects and their labours precisely reproduced in the human beings, art having at last discovered a process which was in full operation before man knew how to handle a weapon or a tool.

After Mr. Bates had watched the proceedings of the ants for some time, he took a trowel, and opened the ground with it. The clever insects at once took advantage of this aid, and dashed into the breach by thousands, pouncing on the luckless inhabitants and carrying them off in their jaws. So bold and so quick were they, that Mr. Bates could scarcely manage to secure a single specimen, and even when he had caught an ant, the Foragers would pull it out of his fingers.

The same observer has known them to sink their shafts to a depth of ten inches, invariably succeeding in their raid upon the nest. The materials of which the nest is made they pull to pieces, and carry the fragments home, together with the inmates. When the nest is completely sacked, the invaders move out in small lines, which march to join the main body, and soon unite with it. The discipline of the community is really wonderful. Each insect knows its own place and its own work, and so perfect is the organization, that during the busy season, the long train resolves itself into two distinct columns, one going out to search for food, and the other returning home laden with spoil.

Another species, *Eciton rapax*, is also in the habit of attacking the nest of various ants. In habits it is very similar to the preceding insect, and is chiefly remarkable for being the largest of all the Ecitons, the average length being half an inch.

The last species of Eciton which I shall mention, is that wonderful species which is called the BLIND ANT (*Eciton erratica*).

I have already mentioned that in Eciton drepanophora the eyes are very small, but in the Blind Ant they are absolutely wanting, the horny covering of the head being quite uniform, without the least trace of external eyes. Some naturalists have thought that the Blind Ant may possess organs of vision, and that the horny covering of the head is sufficiently transparent to admit the light. That the insects can distinguish light from darkness is perfectly true, for they display the greatest uneasiness when brought into the light, and therefore it is possible that the optic nerves may be in some degree affected, though there are no external organs of vision. But the covering of the head is certainly too thick and too opaque to permit distinct vision, and that the insect can see an object clearly is manifestly impossible.

These insects are wonderful builders, and bear a great resemblance to the termites in many of their proceedings. The reader will remember that the termites always choose to work under cover, and proceed from one place to another by means of galleries, which they construct with great rapidity. The Blind Ant acts in a precisely similar manner, constructing long galleries through which it travels. These galleries are of small dimensions, though their length is almost unlimited, and they are made in a very flimsy manner, without the use of any cement. If the observer wishes to capture some of the large-headed officers, he can easily do so by breaking down any part of the gallery. As soon as the unwelcome light streams in, the soldiers are seen to come slowly out, moving their great heads from side to side, and opening their powerful jaws with silent menace. If they are not further disturbed, they will retire into the gallery, and the breach is soon mended by the workers.

These galleries cannot be called tunnels, because they are built upon the surface of the earth, and do not penetrate below it, and ought perhaps to be called "covered ways," rather than galleries.

I HAVE intentionally reserved the last place among the builders for an insect which is certainly the most wonderful of them all; not only raising an edifice, but clearing a space around, and preparing it for a garden. This insect is called by Dr. Lincecum, the discoverer of its habits, the AGRICULTURAL ANT, and its scientific name is *Atta malefaciens*. As the reader will perceive, it is allied to the parasol ant, which has been already described.

This remarkable insect is a native of Texas, and until a few years ago, its singular habits were unknown. Dr. Lincecum, however, wrote a long and detailed account to Mr. Darwin, who made an abstract of it, and read the paper before the Linnean Society, April 18th, 1861. This abstract may be found in the Journal of that Society, and is as follows :—

"The species which I have named 'Agricultural' is a large, brownish ant. It dwells in what may be termed paved cities, and like a thrifty, diligent, provident farmer, makes suitable and timely arrangements for the changing seasons. It is, in short, endowed with skill, ingenuity, and untiring patience, sufficient to enable it successfully to contend with the varying exigencies which it may have to encounter in the life-conflict.

"When it has selected a situation for its habitation, if on ordinary dry ground, it bores a hole, around which it raises the surface three and sometimes six inches, forming a low circular mound, having a very gentle inclination from the centre to the outer border, which on an average is three or four feet from the entrance. But if the location is chosen on low, flat, wet land, liable to inundation, though the ground may be perfectly dry at the time the ant sets to work, it nevertheless elevates the mound, in the form of a pretty sharp cone, to the height of fifteen to twenty inches or more, and makes the entrance near the summit. Around the mound, in either case, the ant clears the ground of all obstructions, and levels and smooths the surface to the distance of three or four feet from the gate of the city, giving the space the appearance of a handsome pavement, as it really is.

"Within this paved area, not a blade of any green thing is allowed to grow, except a single species of grain-bearing grass. Having planted this crop in a circle around, and two or three feet from the centre of the mound, the insect tends and cultivates it with constant care, cutting away all other grasses and weeds that may spring up amongst it, and all around outside the farm-circle to the extent of one or two feet more. The cultivated grass grows luxuriantly, and produces a heavy crop of small, white, flinty seeds, which under the microscope very closely resemble ordinary rice. When ripe, it is carefully harvested and carried by the workers, chaff and all, into the granary cells, where it is divested of the chaff and packed away. The chaff is taken out and thrown beyond the limits of the paved area.

"During protracted wet weather, it sometimes happens that the provision-stores become damp, and are liable to sprout and spoil. In this case, on the first fine day, the ants bring out the damp and damaged grain, and expose it to the sun till it is dry, when they carry it back and pack away all the sound seeds, leaving those that had sprouted to waste.

"In a peach orchard not far from my house is a considerable elevation, on which is an extensive bed of rock. In the sand-beds overlying portions of this rock are fine cities of the Agricultural Ants, evidently very ancient. My observations on their manners and customs have been limited to the last twelve years, during which time the inclosure surrounding the orchard has prevented the approach of cattle to the ant-farms. The cities which are outside the inclosure, as well as those protected in it, are at the proper season invariably planted with the ant-rice. The crop may accordingly always be seen springing up within the circle about the 1st of November every year. Of late years, however, since the number of farms and cattle has greatly increased, and the latter are eating off the grass much closer than formerly, thus preventing the ripening of the seeds, I notice that the Agricultural Ant is placing its cities along the turn-rows in the fields, walks in gardens, inside about the gates, &c., where they can cultivate their farms without molestation from the cattle.

"There can be no doubt that the particular species of grain-bearing grass mentioned above is intentionally planted. In farmer-like manner the ground upon which it stands is carefully divested of all other grasses and weeds during the time it is growing. When it is ripe, the grain is taken care of, the dry stubble cut away and carried off, the paved area being left unencumbered until the ensuing autumn, when the same 'ant-rice' reappears within the same circle, and receives the same agricultural attention as was bestowed upon the previous crop—and so on, year after year, as I *know* to be the case, in all situations where the Ants' settlements are protected from graminivorous animals."

After receiving this account, Mr. Darwin wrote to Dr. Lincecum, asking him whether he thought that the Ants planted seed for the next year's crop, and received the following answer: "I have not the slightest doubt of it. And, my conclusions have not been arrived at from hasty or careless observation, nor from seeing the Ants do something that looked a little like it, and then guessing the results. I have at all times watched the same ant-cities during the last twelve years, and I know that what I stated in my former letter is true. I visited the same cities yesterday, and found the crop of ant-rice growing finely, and exhibiting also the signs of high cultivation, and not a blade of any other kind of grass or seed was to be seen within twelve inches of the circular row of ant-rice."

CHAPTER XIX.

SUB-AQUATIC NESTS. VERTEBRATES.

Fishes as architects—The STICKLEBACKS and their general habits—The FRESH-WATER STICKLEBACKS—A jealous proprietor—Punishment of trespassers—Form and materials of the nest—Use of the nest—Cannibalistic propensities—The FIFTEEN-SPINED STICKLEBACK, and its form—Its curious nest—Mr. Couch's description of a nest in a rope's end—Fishes of Guiana—The HASSAR or HARDBACK, and its place in zoology—Nest of the Hassar—Parental watchfulness—Singular position of the nest—Habits of the Hassar.

As a rule, FISHES display but little architectural genius, their anatomical construction debarring them from raising any but the simplest edifice. A fish has but one tool, its mouth, and even this instrument is of very limited capacity. Still, although the nest which a fish can make is necessarily of a slight and rude character, there are some members of that class which construct homes which deserve the name.

The best instances of architecture among the Fishes are those which are produced by the STICKLEBACKS (*Gasterosteus*), those well-known little beings whose spiny bodies, brilliant colours, and dashing courage make them such favourites with all who study nature. There are several species of British Sticklebacks, but as the fresh-water species all make their nests in a very similar manner, there will be no need of describing each species separately.

These fishes make their nests of the delicate vegetation that is found in fresh water, and will carry materials from some little distance in order to complete the home. They do not, however, range to any great extent, because they would intrude upon the preserve of some other fish, and be ruthlessly driven away.

When the male Stickleback has fixed upon a spot for his nest, he seems to consider a certain area around as his own especial property, and will not suffer any other fish to intrude within its limits. His boldness is astonishing, for he will dash at a fish of ten times his size, and, by dint of his fierce onset and his bristling spears, drive the enemy away. Even if a stick be placed within the sacred circle, he will dart at it, repeating the assault as often as the stick may trespass upon his domains. Within this limit, therefore, he must seek materials for his nest, as he can hardly move for six inches beyond it without intruding upon the grounds of another fish. This right of property only seems to extend along the banks and a few inches outwards, the centre of the stream or ditch being common property. Along the bank, however, where vegetation is most luxuriant, there is scarcely a foot of space that is not occupied by some Stickleback, and jealously guarded by him.

Although the nests of the Stickleback are plentiful enough, they are not so familiar to the public as might be expected, principally because they are very inconspicuous, and few of the uninitiated would know what they were, even if they were pointed out. Being of such very delicate materials, and but loosely hung together, they will not retain their form when they are removed from the water, but fall together in an undistinguishable mass, like a coil of tangled thread that had been soaked in water for a few weeks.

The materials of which the nest is made are extremely variable, but they are always constructed so as to harmonize with the surrounding objects, and thus to escape ordinary observation. Sometimes it is made of bits of grass which have been blown into the river, sometimes of straws, and sometimes of growing plants. The object of the nest is evident enough, when the habits of the Stickleback are considered. As is the case with many other fish, there are no more determined destroyers of Stickleback eggs than the Sticklebacks themselves, and the nests are evidently constructed for the purpose of affording a resting-place for the eggs until they are hatched. If a few of these nests be removed from the water in a net, and the eggs thrown into the stream, the Sticklebacks rush at them from all sides, and fight for them like boys scrambling for halfpence. The eggs are very small, barely the size of dust-shot, and are yellow when first placed in the nest, but deepen in colour as they approach maturity.

FIFTEEN-SPINED STICKLEBACKS AND NESTS.

THERE is a well-known marine species of this group, called the FIFTEEN-SPINED STICKLEBACK (*Gasterosteus spinachia*), a long-bodied, long-snouted fish, with a slightly projecting lower jaw, and a row of fifteen short and sharp spines along the back. This creature makes its nest of the smaller algæ, such as the corallines, and the delicate green and purple seaweeds which fringe our coasts.

Sometimes, indeed, it becomes rather eccentric in its architecture, and builds in very curious situations. Mr. Couch, the well-known ichthyologist, mentions a case where a pair of Sticklebacks had made their nest "in the loose end of a rope, from which the separated strands hung out about a yard from the surface, over a depth of four or five fathoms, and to which the materials could only have been brought, of course in the mouth of the fish, from the distance of about thirty feet. They were formed of the usual aggregation of the finer sorts of green and red seaweed, but they were so matted together in the hollow formed by the untwisted strands of the rope, that the mass constituted an oblong ball of nearly the size of the fist, in which had been deposited the scattered assemblage of spawn, and which was bound into shape with a thread of animal substance, which was passed through and through in various directions, while the rope itself formed an outside covering to the whole."

THIS is not the only fish that is known to construct a nest.

In the fresh waters of tropical America there is a genus of fish belonging to the Siluridæ, and named *Callichthys*, from the beauty of the species. The fishes of this genus have four very long barbules hanging from the upper lip and pointing backwards, and are all mailed except part of the belly. Their general colour is green-brown, and they do not reach to any great size, eight inches being their usual length. They are generally very fat, and are much valued by the natives of Guiana, who live so much upon fish. The native name for this fish is HASSAR, and the European residents call it the HARDBACK, in allusion to its coat of shining mail.

To the naturalist, however, the chief point of interest in these fish is the fact that they are in the habit of constructing nests which are quite as well formed as those of the stickleback, and are made of grass-blades, straws, and leaves. These nests are very plentiful in the little muddy streamlets that intersect the sugar marshes, so that the habits of the fish can be easily watched. The parent fish is very jealous of the eggs, and waits near them until they are hatched, and the young family committed to the water. The natives are well aware of this habit, and catch the fish readily by insinuating a net or even a basket under the water near a nest, and then raising it quickly, when the parent fish is mostly found in the net.

Perhaps the most curious part of the economy of this fish is the fact that the nest is not placed *in* the water, but in a muddy hole just above the surface. This habit, however, accords with the qualities of the fish, which is remarkably independent of water, and can travel over land from one pond to another, led by some mysterious instinct, which we of higher powers cannot comprehend. During the dry season the Hassar is in the habit of burrowing into the mud, and there residing until the welcome rain sets it free. Those who know the customs of the fish are therefore able to procure it at almost any period of the year, digging for it in the dry season, and fishing for it in the wet months.

CHAPTER XX.

INVERTEBRATES.

A Pool and its wonders—The WATER SPIDER—Its sub-aquatic nest—Conveyance of air to the nest—The diving-bell anticipated—Character of the air in the nest—Mr. Bell's experiments upon the Spider—Life of the Water Spider—The HYDRACHNA—The CADDIS FLIES and their characteristics—Sub-aquatic homes of the Larva—Singular varieties of form and material—Life of a Caddis—Description of nests in my own collection—Fixed cases, and modification of Larva—Singular materials for nest-building—CORALS and their general history—The Coral of commerce—Development and extension of the Coral—How fresh colonies are founded—Various Corals and their growth—Submarine tube-makers—The SERPULÆ and their general habits—The Operculum of the Serpula—The TEREBELLÆ and their submarine houses—The CADDIS SHRIMP—Remarkable analogy.

WHEN I was a very little boy, I was accustomed to spend much time on the banks of the Cherwell, and used to amuse myself by watching the various inhabitants of the water. Animal life is very abundant in that pleasant little river, and there was one favourite nook where a branch of a weeping-willow projected horizontally, and afforded a seat over the dark deep pool, one side of which was abrupt and the other sloping.

Here the merry gyrini ran their ceaseless rounds, and the water-boatmen rowed themselves in fitful jerks, or lay resting in a contemplative manner on their oars. Now and then an unlucky insect would fall from the tree into the water, and then uprose from the dark depth a pair of dull eyes and a gaping mouth, and then, with a glitter as of polished silver, the dace would disappear with its prey. In the shelving part of the pool the caddis-worms moved slowly along, while the great dyticus beetle would rise at intervals to the surface, jerk the end of his tail into the air, and then dive below to the muddy bottom. This spot was much favoured by the nursemaid, for she had no trouble in watching me, as long as I could sit on the branch and look into the water. True, I might have fallen into the river, but I never did; and even had that accident occurred, it would have wrought no harm, except wet clothes, for I could swim nearly as well as the water-insects themselves.

Close under the bank lived some creatures which always interested me greatly. Spiders they certainly were, but they

WATER SPIDER.

appeared to have the habits of the water-beetle—coming slowly to the surface of the water, giving a kind of flirt in the air, and then disappearing into the depths, looking like balls of shining silver as they sank down. I had been familiar with these creatures for years before I met with them in some book, and learned that they were known under the name of WATER SPIDER (*Argyronetra aquatica*).

This Spider is a most curious and interesting creature, because it affords an example of an animal which breathes atmospheric air constructing a home beneath the water, and filling it with the air needful for respiration.

The sub-aquatic cell of the Water Spider may be found in many rivers and ditches, where the water does not run very swiftly. It is made of silk, as is the case with all spiders' nests, and is generally egg-shaped, having an opening below. This cell is filled with air; and if the Spider be kept in a glass vessel, it may be seen reposing in the cell, with its head downwards, after the manner of its tribe. The precise analogy between this nest and the diving-bell of the present day is too obvious to need a detailed account. How the air is introduced into the cell is a problem that was for some time unsolved. The reader is probably aware that the bubbles of air which are to be seen on sub-aquatic plants are almost entirely composed of oxygen gas, which is exuded from the plant, and which is so important an agent in purifying the water. Some zoologists thought that the air which is found in the cell of the Water Spider was nothing but oxygen that had been exuded from the plant upon which the nest was fixed, and that it had been intercepted in its passage to the surface. In order to set the question at rest, Mr. Bell, the well-known naturalist, instituted a series of experiments upon the Spider, and communicated the results to the Linnean Society. The experiments were made in 1856, and Mr. Bell's remarks are as follows:—

"No. 1. Placed in an upright cylindrical vessel of water, in which was a rootless plant of *Stratiotes*, on the afternoon of November 14. By the morning it had constructed a very perfect oval cell, filled with air, about the size of an acorn. In this it has remained stationary up to the present time.

"No. 2. *Nov.* 15. In another vessel, also furnished with Stratiotes, I placed six Argyronetræ. The one now referred to began to weave its beautiful web about five o'clock in the afternoon. After much preliminary preparation, it ascended to the surface, and obtained a bubble of air, with which it immediately and quickly descended, and the bubble was disengaged from the body, and left in connexion with the web. As the nest was, on one side, in contact with the glass, inclosed in an angle formed by two leaves of the Stratiotes, I could easily observe all its movements. Presently it ascended again and brought down another bubble, which was similarly deposited.

"In this way, no less than fourteen journeys were performed, sometimes two or three very quickly one after another; at other times with a considerable interval between them, during which time the little animal was employed in extending and giving shape to the beautiful transparent bell, getting into it, pushing it out at one place, and amending it at another, and strengthening its attachments to the supports. At length it seemed to be satisfied with its dimensions, when it crept into it and settled itself to rest with the head downwards. The cell was now the size and nearly the form of half an acorn cut transversely, the smaller and rounded part being uppermost.

"No. 3. The only difference between the movements of this and the former was, that it was rather quicker in forming its cell. In neither vessel was there a single bubble of oxygen evolved by the plant.

"The manner in which the animal possesses itself of the bubble of air is very curious, and as far as I know, has never been exactly described. It ascends to the surface slowly, assisted by a thread attached to the leaf or other support below and to the surface of the water. As soon as it comes near the surface, it turns with the extremity of the abdomen upwards, and exposes a portion of the body to the air for an instant, then with a jerk it snatches, as it were, a bubble of air, which is not only attached to the hairs which cover the abdomen, but is held on by the two hinder legs, which are crossed at an acute angle near their extremity, this crossing of the legs taking place at the instant the bubble is seized. The little creature then descends more rapidly and regains its cell, always by the same route, turns the abdomen within it, and disengages the bubble.

"No. 4. Several of them, when I received them, had the hair on the abdomen wetted, and I placed them on some blotting-paper until they were dry. On returning them to the water, two remained underneath a floating piece of cork, and the hair, being now dry, retained the pellicle of air which is ordinarily observed. One of the two came out of the water, attached the cork to the glass, and wove a web against the latter, against which it rested about a quarter of an inch above the surface of the water. After remaining there about two days, it resumed its aquatic habits, and, like all others, formed its winter habitation."

Water Spiders are now familiar to us on account of the widespread fashion for aquaria, but so thoroughly have the ditches and streams been ransacked by professional dealers, that the creature has become quite rare in spots where it was once plentiful.

The Water Spider places her eggs in this cell, spinning a saucer-shaped cocoon, and fixing it against the inner side of the cell and near the top. In this cocoon are about a hundred eggs, of a spherical shape, and very small. The cell is a true home for the spider, which passes its earliest days under the water, and when it is strong enough to construct a sub-aquatic home for itself, brings its prey to the cell before eating it.

The colour of the Water Spider is brown, with a greyish surface caused by the thick growth of hair which covers the body, and with a very slight tinge of red on the cephalothorax. The reader must not confound this creature with another Arachnid that is sometimes called the Water Spider (*Hydrachna cruenta*), and is of a bright scarlet colour, with a peculiar velvety surface.

THERE is an order of insects which is especially dear to anglers; not so much to fly-fishers, as to those who like to sit and look at a float for several consecutive hours. This order is scientifically termed TRICHOPTERA, or Hair-winged insects, and the various species of which it is composed are classed together under the familiar title of CADDIS FLIES.

These insects may always be known by the peculiar leathery aspect of the body, and by the coating of hair with which the wings are covered, the long hairs being spread over the whole surface, and standing boldly out like a fringe round the edge. They all have long and slender antennæ, and in some genera, such as Mystacida, these organs are nearly three times as long as the head and body, reminding the observer of the lovely Japan moths (Adelæ) whose delicate antennæ wave and glitter in the sunbeams like stray threads of spider's web. For the perfect insect the angler cares comparatively little. Imitations in hair, feather, and silk are useful to the fly-fisher, and are known to anglers by the eccentric nomenclature by which such imitations are called. It is the larva in which the angler delights, and it is chiefly of the larva that our present description will treat.

We will now trace the life of the Caddis Fly from the egg to the perfect insect.

In the breeding season, the female may be observed to carry about with her a double bundle of little greenish eggs, probably in order to expose them for a certain time to the warm sunbeams before they are immersed in the water. This curious bundle is a long oval in shape, and is bent sharply in the middle, its extremities being attached to the abdomen of the insect. When her instinct tells her of the proper time, she proceeds to the water, and attaches the eggs to the leaf of some aquatic plant, often crawling down the stem for several inches. The Caddis Fly is quite at home on the water, and, unlike the dragon flies, which are quite helpless when immersed, can run on the surface with considerable speed, and on occasion can swim below the surface with scarcely less rapidity.

They may often be observed in the act of running on the water, and while they are thus employed, they often fall victims to some hungry fish, which is attracted by the circling ripples occasioned by the movement of the limbs. Fly-fishers, who are acquainted with the habits of fishes and insects, take advantage of their knowledge, and by causing their imitation Caddis Fly to ripple over the surface, or even to sink beneath it, like the veritable insect, delude the unsuspecting fish into swallowing a hook instead of a fly.

In process of time the eggs are hatched, and the young larvæ then proceed to construct houses in which they can dwell. These houses are formed of various materials and are of various shapes, and, indeed, not only does each species have its own particular form of house, but there is considerable variety even in the houses of a single species. In the accompanying illustration are shown a number of the nests formed by the Caddis Fly in its larval state, together with the perfect insects. All the figures have been drawn from actual specimens, some of which are in the British Museum, and others in my own collection. The materials of which the nest is made, depend greatly on the locality in which the insect is hatched, and in a rather large series of Caddis nests now before me, there are some very remarkable instances of the manner in which the insect has been obliged to adapt itself to circumstances. The most common style of case is that which is composed of a number of sticks and grass stems laid longitudinally upon each other like the fasces of the Roman consuls. Of these I have specimens of various sizes and shapes, some being barely half an inch long, while others measure four

CADDIS.

times that length, the sticks being sometimes placed so irregularly, that the home of the architect is not easily seen. The creatures are not at all particular about the straightness of the sticks, but take them of any degrees of curvature, as in one of the examples represented in the illustration, where the stick is not only curved, but has a large bud at the end.

Another case is made of the hollow stem of some plant, apparently that of a hemlock, to which are attached a few slips of bark from the plants. Next comes a series of cases in which the Caddis larva has contrived to secure a great number of cylindrical grass stems and arranged them transversely in several sets, making one set cross the other so as to leave a central space, in which the little architect can live. One or two cases are made wholly of bark, apparently the cuticle of the common reed, a plant which is very common in the Cherwell, whence the cases were taken. In all probability these strips of cuticle have been dropped into the river by the water rats while feeding on the reeds.

Several cases are made entirely of leaves, mostly taken from the white-thorn, which grows in great quantities along the banks of the above-mentioned river. Then, there are cases which are equally composed of sticks and leaves, these materials generally occupying opposite ends of the case. There is another series of cases made up of fine grass, apparently the *débris* of hay which had been blown into the water during the summer, and having the materials laid across each other like the needles of a stocking-knitter. Most of these cases are balanced by a stone.

Next come a number of cases which are composed of small shells, those of the Planorbis being the most common, and having among them a few specimens of the Limnæa, or pond-snail, and many separate valves and perfect shells of the fresh-water mussel. The Caddis larva is an incorrigible kidnapper, seizing on any shell that may suit its purpose, without troubling itself about the inhabitant. It is quite a common occurrence to find four or five living specimens of the Planorbis and Limnæa affixed to the case of a Caddis larva, and to see the inhabitants adhering to the plants and endeavouring to proceed in one direction while the Caddis is trying to walk in another, thus recalling the well-known episode of the Tartar and his captor. In these cases the cylindrical body is made of sand and small fragments of shells bound together with a waterproof cement, and the shells are attached by their flat sides to the exterior.

There are also several cases which are made entirely of sand cemented together, some being cylindrical and others tapering to a point, like an elephant's tusk. There are also examples of mixed structures, where the Caddis has combined shells with the leaf and twig cases, and in one of these instances, the little architect has bent back the valves of a small mussel, and fastened them back to back on its house. Beside these, there are one or two very eccentric forms, where the Caddis has chosen some objects which are not often seen in such a position. The seed-vessels of the elm are tolerably common, but I have several specimens where the Caddis has taken the operculum of a dead Pond-snail and fastened it to the case; and there is an example where the chrysalis of some moth, apparently belonging to the genus Porthesia, has been blown into the water from a tree over-hanging the stream, and seized upon by a Caddis as an unique ornament for its house. These latter examples were found in a stream in Wiltshire, and the tusk-like sand-cases were found in a disused stone quarry in the same county.

Various experiments have been tried upon the larva of the Caddis, in order to see its mode of building. A lady, Miss Smee, has been very successful in this pursuit, and has forced the Caddis larvæ to build their nests of the most extraordinary subtances, such as gold-dust, crushed glass, and other substances. They would not, however, use beads, or anything where the surface was smooth and polished.

In this remarkable sub-aquatic home the Caddis larva lives in tolerable security, for the head and front of the body are clothed in horny mail, and the soft, white abdomen is protected by the case. The food of the Caddis is generally of a vegetable nature, though there are one or two species which live partly, if not entirely, on animal food. When the larva has lived for its full period, and is about to change into the pupal condition, it closes the aperture of its case with a very strong net, having rather large meshes, and lies securely therein until it is about to change into the winged state. It then bites its way through the net with a pair of strong mandibles, comes to the surface of the water, breaks from its pupal envelope, and shortly takes to flight. The larger species crawl up the stems of aquatic plants before leaving the pupal skin, but the smaller merely stand on the cast skin, which floats raft-like on the water.

There are one or two species whose cases are not movable, but are fixed to the spot whereon they were made. In order, therefore, to compensate for the immobility of the case, the larva has a much larger range of movement. In the ordinary species, the creature holds itself to the extremity of the case by means of hooks at the end of its body, which can grasp with some force, as any one knows who has pulled a Caddis larva out of its house. But when the case is fixed, the abdominal claspers of the larva are attached to a pair of long foot-stalks, so that the creature can extend its body to some distance from the entrance of the tube.

We now turn to the warmer seas, and shall there find some most magnificent examples of subaquatic homes.

Our first examples of these will be the wonderful creatures which are classed together under the general term of Corals, and which are so familiar to us either in a manufactured state or as ornaments for the drawing-room. How vast are their submarine labours is evident from the enormous "Coral-reefs" which they raise, and which form great islands whereon an army can live, and inlets wherein a fleet can ride securely at anchor.

Before proceeding further in the history of the Coral and its submarine home, we will see how it extends itself with such wonderful rapidity, and what is the process that enables fresh colonies to establish themselves, and existing colonies to spread themselves—both these operations being conducted on different principles.

How the Coral grows is a problem which was unsolved until a comparatively late period. Not only were naturalists ignorant of its development, but they did not even know in what kingdom to place it, whether vegetable or mineral. Opinions were long divided on this point, the men of greatest reputation inclining to the belief that it was mineral, while a very few thought that it must be vegetable, and that the flower-like rays of the polype were veritable submarine blossoms. But when a more careful observer announced that the Coral was really the production of an animal, both parties united in ridiculing his theory, and for a while the animal origin of Coral was put aside by the scientific world. Truth, however, prevailed, as it always will do, sooner or later, and every one now-a-days knows that Coral is the production of animated beings.

Still, although the general fact is known, its details are not so familiar.

In the active season of Coral life, that is, from May to August or September, millions of young Corals are launched into the world. When they first pass into the sea from the mouth of their parent, they are tiny flask-shaped beings, covered with minute cilia, by means of which they can pass through the water with some rapidity. They have a kind of mouth at the small end, which corresponds to the neck of the flask, and it is a curious fact that they always swim with the large end forwards.

After a time they change their shape, elongating until they look like little white worms, but still moving about after the same fashion. They pass some time in this phase of existence, and then settle down upon some fixed object, such as a rock, and adhere to it by the enlarged base. Scarcely have they done so than they again alter their shape, and assume a form so different that no one who was not acquainted with the little creatures could recognise them. Instead of being long and worm-like, they now contract themselves in length, while they proportionately increase in width, and look something like an echinus, or sea-urchin.

It is a curious fact, that up to this time they have passed through four stages, and in each stage they change in length. Before they issue from the parent, they are nearly spherical, but when they pass into the sea they assume the flask-like shape. They next elongate themselves considerably, and then suddenly become even shorter than their width.

In this fourth stage, the mouth is surrounded by eight little cushion-like projections, which soon sprout into the beautiful fringed tentacles which give to the animal so flower-like an appearance.

When it has arrived at this stage of existence, the young Coral begins to develop some new and remarkable powers. From various parts of the body spring little projections, which soon exhibit an orifice like the mouth of the original Coral, and in a short time, this aperture is surrounded with cushions, which are developed into tentacles as before. As soon as it comes to maturity, each of these supplementary Corals puts forth similar buds, so that the increase is wonderfully rapid, even by "gemmanation," as this mode of multiplication is called, not to mention the vast numbers of new settlements that are made by the young which are poured from the mouth.

Although we have now learned the method by which the Coral animals are reproduced, we have not ascertained how the solid, stony substance which we call Coral is formed, nor the precise connexion which exists between the animal and the Coral.

If the reader will take up a branch of the ordinary Coral of commerce, he will see that it is slightly grooved or fluted throughout its extent, and that its surface is studded with little projections having star-like discs. Now, if this piece of Coral could be again clothed with the living creature by which it was deposited, we should see a beautiful and a wonderful sight. Next to the stony core lie a series of longitudinal vessels, each vessel corresponding with a groove, and above them lies a confused mass of irregular vessels communicating with each other. At intervals there arise the lovely flowerets of the Coral, the bodies being bright rose-colour, and their arms pure white. These arms or tentacles are in ceaseless motion, and the aspect of a large and healthy branch of coral is imposingly beautiful.

The animal has the power of depositing certain minute calcareous particles, commonly called spicules, which are always of remarkable forms, and are different in the various species of coral. In the common red Coral, they are nearly cylindrical, and armed with projecting knobs covered with angular spikes. These spicules are then bound together by a red cement, and thus the Coral is formed, the fluted branches being deposited under the longitudinal vessels, and the raised projections under the flowerets of the polype. To see the Coral in full vigour it is necessary to visit the spots where it grows, as it dies almost immediately after being taken out of the water, and even if transferred with great care to a vessel, is sure to die in a very short time. After death, the whole of the bark dries up, and fades away, so that it crumbles into powder at a touch, and can be removed by merely rubbing the Coral between the fingers.

Several of the more curious species of Corals and Madrepores are to be seen upon the large illustration, which represents a portion of sea-bed beset with these beautiful zoophytes.

In the upper left hand corner is the common Red Coral of commerce, which has been already mentioned, and in the upper

CORALS AND MADREPORES.

Corallium rubrum.
Tubipora rubeola.
Astrœa viridis.

Clavellaria viridis.
Pœcillopora cespitosa.
Xenia elongata.

centre is a clustered branch of the WHITE CORAL, which is so much used as a drawing-room ornament when placed under a glass shade. In the upper right-hand corner may be seen a singular group of zoophytes with curved stems and flower-like heads. This is the GREEN CLAVELLARIA (*Clavellaria viridis*), one of the most striking examples of the genus.

The Green Clavellaria is very common on the Isle of Vanikoro, and is found in tolerably large masses, adhering to rocks, madrepores and similar substances. All the Clavellarias are of somewhat similar shape, though variable in size and colour, and may be recognised by several conspicuous characteristics. The tubes are nearly cylindrical, but tapering, as shown in the illustration, and forming a footstalk, which is more or less bent.

The texture is somewhat leathery, and is strengthened by innumerable spicules of a calcareous nature, which are agglomerated together in bundles. These spicules are too small to be distinguished except by the microscope; but under a tolerably high power, they are seen to be long and spindle-shaped, sharp at both ends, and being encircled with little rounded knobs, set in regular rows. These tubes are always placed very closely together, but do not adhere to each other, their grooved surfaces being always distinct.

The animal by which this tube is formed is a very pretty one, cylindrical, with eight radiating tentacles of a violet grey. The colour of the tube is green for the upper half, and then changes to brown, so that the contrast of the two colours is very decided. The average length of the tube is two inches.

There are several other species of Clavellaria, among which may be noticed the VIOLET CLAVELLARIA (*Clavellaria violacea*). This is a much smaller species, but is coloured in a more bold and decided manner. In this creature the tubes are dark violet, and the tentacles are bright yellow. They do not, however, project from the tube as boldly as those of the preceding species, but only just show their tips above the entrance, withdrawing them smartly on the least alarm. As they retreat, they slightly contract the orifice, their bodies being fastened to the inside of the tube.

TOWARDS the centre of the illustration, and on the right-hand side, may be seen a remarkable tree-like object, covered with long, tendril-like appendages, each tipped with a radiating beard. This zoophyte is known by the title of *Xenia elongata,* and on account of its singular form, is a very conspicuous species.

Examples of this genus are spread over many of the hotter parts of the world, some being found in the Red Sea, and all notable for the remarkable form of the animal and its submarine home. The present species has been chosen more for the singularity of its form, than the beauty of its colours, which cannot be expressed in the simple black and white of a woodcut. Some species of this genus have the star-like tentacles coloured with blue of various shades, some with rose, and some with lilac, and as in many cases the expanded tentacles are an inch in diameter, the effect of a large mass of these animals in full health is very fine.

Except, however, in their native state, they never can be seen in full health, their constitutions being so delicate that they cannot endure removal from the spot whereon they were developed. If removed from the water, they immediately shrink to half their size, and do not assume their former dimensions, no matter how carefully they are tended. The present species is found on the shores of the Feejee Islands, and in form is certainly the most singular of its genus, the enormously long and slender body at once distinguishing it from any other species. Its colour is simple brown, and the diameter of the tentacles is rather more than three-quarters of an inch.

In the left-hand lower corner of the illustration is a curious globular object, covered with circular and radiated marks, and having a number of flower-headed projections upon the top. This is the Green Astræa (*Astræa viridis*), one of the finest examples of a singular and beautiful group of zoophytes. In this genus the animals are shaped something like the well-known sea anemone, and rather short, having a great number of very small tentacles, which are gathered round a central mouth. The "cups" which these animals form, and in which they live, are rather deep and conical, and their inner surfaces are corrugated into a number of thin walls with beautifully serrated edges.

The animal is rather oddly formed. The body is not quite cylindrical, but is broader at the base than at the top, and is deeply ridged with circular furrows, so that it looks very like the well-known glass bottles which are used for holding salad dressing, and which make up for lack of contents by profusion of glass. In the very centre of the top is placed the mouth, and round it are grouped a vast number of little tentacles that radiate like the flowerets of a daisy or dandelion. In the present species the animal is about half an inch in length, and the animals are clustered together in masses that are often as large as the fist. They are rather variable in shape, but are always more or less globular.

The colour of this species is simple and pleasing. The body of the animal is pale grey-blue, and the tentacles are bright green, so that when a number of the animals are simultaneously protruding themselves, the general effect is very striking. These zoophytes are able to retract themselves almost wholly within their houses, so that nothing is visible except that round the mouth there is a small green circle, which is formed by the projecting tips of the tentacles. This species is found at Vanikoro.

There are many species of Astræa, all very pretty, and some quite beautiful. Among the most conspicuous are the ABNORMAL ASTRÆA (*Astræa abdita*), in which the mouth is scarlet and the tentacles yellow; the ANANA ASTRÆA (*Astræa ananas*), where the tentacles are yellow, but the mouth white; and the CHOCOLATE ASTRÆA (*Astræa fusco-viridis*), when the mouth is green, surrounded with a broad chocolate border, and the tentacles are white.

IN the left centre of the illustration is seen a group of that most beautiful zoophyte which is known as the RED ORGAN-PIPE CORAL (*Tubipora rubeola* [or *syringa*]).

This handsome zoophyte is found chiefly off Carteret, in New Ireland, and is grouped together in masses that are often many yards in diameter. It is usually found in about two or three feet of water, but is sometimes placed so high that at very low tides it is laid bare by the receding waters.

The animal which forms this wonderful tubing is cylindrical, and the tentacles are pinkish, not possessing the brilliant red of the tubes, and in its native state, the animals envelop so completely the upper part of the general mass, that the bright red head is not perceptible. The coral masses are very fragile, and will not bear the pressure of the human foot, crumbling beneath the tread as if they were made of sugar. The tubes are beautifully cylindrical, and do not adhere to each other, being kept asunder by partitions, which precisely resemble the boards through which the pipes of an organ are passed.

They are very thin, though hard, and a rough pressure of the hand will always damage them. My own specimen is now sadly shorn of its original fair dimensions, at least half of its tubes having been broken away by the rude grasp of servants' hands, just as my best specimens of the paper nautilus and other fragile curiosities were damaged before I learned to put them under lock and key.

The animal is not a large one, its length being scarcely greater than that of the distance from one partition to another. The arrangement of the tubes and the partitions looks very complicated, but is, in fact, simple enough. The animal secretes around itself the calcareous substance which forms the tube, but when it has reached to its full extent, it is obliged to leave the cylindrical home in which it had resided. The partition is then secreted by the edge of the mantle, or membrane by which the creature is attached to its tube, and the zoophyte then begins another tube immediately above that which it has quitted.

Sometimes there is a kind of floor that separates the upper tube from the lower, but it is extremely thin, so that a tolerably stout bristle can be pushed through it. The partition is at least twice as strong as the tubes, which are scarcely thicker than the paper on which this account is printed, and is not solid, but perforated with holes just like the machine-made bricks which have lately come into use. In my own specimen there is a curious proof of the abundance of submarine animal life. The group of Organ-pipe Coral has enveloped a piece of White Coral, and has shown a remarkable instance of the manner in which beings so low in the scale of nature can accommodate themselves to circumstances. As if conscious that the coral formed an obstacle which they could not pass in a direct line, they ceased from tube-building when they arrived within a little distance of the coral, and threw up a partition vertically instead of horizontally, so as to envelop the greater part of the coral with the red calcareous substance. Having done so, they then made a new foundation upon the coral, and built a fresh series of tubes, so that when viewed from above, the series of tubes is quite uninterrupted, and no one would imagine that any extraneous substance had intruded into the mass.

The tubes themselves have formed the basis of other submarine habitations, for a moderate magnifying-glass shows that sundry molluscs and molluscoids have settled down upon their exterior, while the white serpentine tubes which creep among the perpendicular pillars, show that the serpulæ and other tube-making creatures have taken up their residence in so well protected a spot.

TEREBELLA CONCHILEGA. SERPULA CONTORTUPLICATA. SABELLA UNISPÍRA. SABELLA ALVEOLARIA.

At the bottom of the sea there are a vast number of wonderful and interesting animals that are known to naturalists as *Tubicolous Annelides, i.e.* Tube-inhabiting Worms. These creatures are true architects, not inhabiting the tubes which have been constructed by other creatures, but making them gradually, and in some instances, getting together sand, stones, mud, shells, and other objects, which they use as materials for their homes. We may call the caddis grubs, which have recently been described, the tubicolous larvæ of fresh water, and very curious are the habitations which they construct. In salt water, however, there is much more variety than in ponds and rivers, and the tubicolæ are many and diversified.

We will begin with those which are most plentiful, and which are best known, because the hardness of their tubes causes them to be preserved long after the inhabitants have perished. If the reader should happen to have used a dredge, or a trawl, he must have brought to the surface great numbers of white shelly tubes, some of them being nearly triangular, and adhering by their whole length to a stone or a shell, and others nearly cylindrical, and rearing themselves from the stone after the first half-inch or so.

These are the work of a curious worm, called by Linnæus the SERPULA, on account of the serpentine manner in which its tube is formed. There are many species of Serpula, and some of the more conspicuous among them will be mentioned.

The Serpula belongs to the Annelidæ, or Ringed animals, a very large group, which includes the earthworm, the leech, the nereis, and many other well-known creatures, all of which have the body composed of a great number of rings, and are destitute of true feet. They breathe either through the skin, by sacs, or by gills, and in the present genus, the respiration is by means of gills, which are always delicate and elegant in form, and mostly brilliant in colour. The body of the Serpula is comparatively short, the tube being frequently six or ten times as long as the animal which made it, and in consequence the Serpula possesses a deep and safe retreat, into which it can withdraw itself whenever threatened by danger.

The very fact, however, that the worm lives in a tube, causes the observer to ask himself how the creature contrives to breathe, and how the long tubular shell is kept clean. A reference to the illustration will explain the former of these difficulties, and the latter will presently be touched upon.

Projecting from the orifice of the shell may be seen a curious fan-like appendage; were the illustration to be coloured, this fan would have been tinted with the brightest hues of scarlet and white, though no brush could give the wondrous delicacy of the feeling or express its semi-pellucid beauty. This "fan" is composed of the gills of the Serpula, and the animal is enabled to breathe by its power of projecting them from the orifice of its tube.

If the reader can procure a group of living Serpulæ and a little sea water, he is strongly advised to do so, as he will understand their structure far better than can be the case if he merely refer to books. The creatures can be procured at any of the numerous shops in which aquaria are sold, and as they are plentiful on our own coast, they can be procured at a very cheap rate. They are terribly apt to die, unless the purity of the water be carefully preserved; but even in that case they can be dissected, and will afford lovely objects for the microscope. There is always an abundance of Serpulæ in the aquarium house at the Zoological Gardens, and the keeper is quite willing to give any advice as to the management of these interesting annelids.

Supposing, then, that a group of Serpulæ has been procured, it should be placed in the aquarium close to the glass, and if possible, the vessel should be one with flat sides, the ordinary fish-globes giving a distorted image. A magnifying-glass should then be fixed so as to command the orifices of the tubes, and all will be ready. It is necessary to fix the glass, because the Serpulæ are strangely sensitive beings, and retreat into their tubes at the slightest alarm. If even a hand be moved suddenly, though at some little distance, the Serpulæ shoot back into their tubes as if propelled by springs, the feathery tentacles collapsing, and the beautiful operculum closing up the entrance.

This lightning-like rapidity of movement is caused by a wonderful array of hooks on the front portion of the body. These hooks are placed on the foot-warts which edge that part of the body, and are wonderfully adapted for catching the membrane that lines the tube. The botanist who looks at such a group as is now before me in the microscope, would be instantly reminded of the seed-vessels of the "wait-a-bit" thorn, with its double array of hooked prickles. Each hook is rendered still more effective by the six or seven teeth into which its inner edge is cut, so that the most formidable array of hooks that the ingenuity of anglers ever devised, and named with appellatives as strange as their shapes, would appear quite harmless by the side of the Serpula's hooks, could they only be magnified to proportionate size. One bunch of these most formidable hooks would seem to be all-sufficient for the purpose, but when it is remembered that every foot-wart has its hooked armatures, amounting to fourteen or fifteen hundred in number, the power of the creature's hold ceases to be surprising.

Perhaps the reader may ask how the animal ever contrives to push itself out of the case at all, seeing that it is held by such a grasp. A further look through the microscope shows that the hooks are affixed to long tendinous bands, of great delicacy, but at the same time of great strength, which enable the animal to protrude the hooks so as to seize the membrane, and to withdraw them when their purpose has been served. The dark-background illumination shows the formation of the hooks in a very clear and beautiful manner.

If possible, the observer should preserve specimens of the hooks for the microscope. He will not want for examples, as the creatures have a habit of coming out of their tubes and dying on the floor of the aquarium, to the great discomfiture of the owner. I find that Deane's gelatine answers very well for the purpose, and the specimens which have just been mentioned are now in perfect preservation, after having been in the gelatine for some six years.

It is very curious to watch the different methods by which the Serpula protrudes and withdraws itself. When it retires into the tube, it vanishes so quickly that the eye cannot follow its movements, but when it protrudes itself, it does so in a very deliberate manner, seeming to feel its way cautiously towards the light, and to be ready to dart back again with or without reason.

The organs by means of which it protrudes itself are placed close to those which withdraw it. Through the foot-warts project a number of stiff, transparent bristles, which are wonderfully like the many-barbed spears of savage nations. I have an arrow of the Tonga Islands which is almost an exact reproduction of a foot-bristle of the Serpula, saving that wood and bone are substituted for the purer material of the bristle. The shaft runs quite straight, but towards the tip, the bristle is flattened and widened into a head just like that of a spear. The head has a very slight curve, is sharply pointed, and is armed with a double row of barbs on the edge. A great number of these bristles are clustered together, so that their united force is really considerable.

Supposing that the animal is fully expanded, and that the observer has succeeded in placing his eye to the glass without alarming the sensitive creature, he will see a wonderful sight. In its way there is nothing which surpasses in beauty the expanded gill-tuft of a Serpula in perfect health. The long feathery gills radiate from the tube in a curve which combines grace and force in no ordinary degree, while the beautiful colours that glow as the blood courses through their translucent substances are far beyond the power of description.

One use of the gills is evident enough. They serve for respiration. But they also answer another purpose, and aid the animal in procuring food. Being necessarily stationary, it cannot roam about in search of food, and its appetite is so great, that it would soon die were it to depend for subsistence on the nourishment which might be brought within its reach by the waves.

If a tolerably powerful glass can be brought within the necessary focal distance, it will be seen that the exterior of the gill-tufts is covered with wonderfully delicate filaments, or *cilia*, as they are technically named, which are continually waving in regular ripples. Their movement constitutes a sharp current of water, which not only washes against the gills and furnishes the requisite supply of air for the regeneration of the blood, but carries the water downwards into the mouth, which opens at the bottom of the gill-tufts. In the water is always a bountiful supply of minute animal organisms, together with other substances, which, although microscopically small, are in the aggregate sufficient to feed the Serpula. This current is kept continually flowing as long as the tentacles are protruded, and thus the creature is enabled to breathe, is supplied with nourishment, and the tube is kept clean by the water current which is perpetually rushing through it.

We now come to that wonderful portion of the animal which is called by naturalists the "operculum," and which is popularly and rightly known as the "stopper." This is the conical appendage which hangs from the tube, and is used in closing its mouth when the gills have been withdrawn. The operculum is known to be a development of one of the antennæ, the other antenna being small, slender, and without any such appendage. To systematic naturalists, the operculum is one of the most valuable parts of the animal, as it is often found in the tube in good preservation, long after all the softer parts of the animal have disappeared. We can learn but little from the tube, as it is very similar in animals belonging to different genera. The value of the operculum was detected by Dr. Philippi, and has been corroborated by Dr. Baird, who has made great use of it in some valuable papers on specimens of Serpulæ which are now in the British Museum.

In one genus, *Eupomatus*, for instance, the operculum is furnished with a number of moveable spikes. In one species, *Eupomatus Boltoni*, these spikes are about twenty in number, and are hard, flat, and calcareous. Their form is very much like that of a hedger's bill-hook, except that they are furnished in the inner edge with several bold tooth-like projections. These spikes are called by Philippi, horns or cornua, and are always deeply toothed, whatever may be the species. The very appropriate name, Eupomatus, is formed from two Greek words, and signifies "beautiful lid," in allusion to the elegant structure of the operculum and its complement of horns.

In another genus, which has been called *Placostegus*, the operculum is calcareous, flat and rounded, looking so like the operculum of some aquatic mollosc, that it might easily be mistaken for that object. The genus is so named on account of the shape of the operculum, and the word is of Greek derivation, signifying Plate-roofed.

One species of this genus, *Placostegus carinatus*, is remarkable for a peculiarity which has been brought forward by Dr. Baird, in a paper read before the Linnæan Society, in April, 1864. "I wish particularly to bring before the notice of the Society the fact that the animal gives out a beautiful dye or colour. The specimens which were the subjects of my examination had been for a number of years in the British Museum, some having been placed there in 1845, and others in 1847. Notwithstanding their having been so long dry, when softened in water, taken out of the tubes and placed in spirits of wine, they imparted to the liquid a beautiful and delicate red tint."

The specific title of *carinatus*, or keeled, is given to the animal because its tube has a decided keel or ridge upon its upper surface. The colour of the animal is blue, and the fan-like tuft of gills is blue, banded with white. An allied species, *Placostegus latiligulatus,* is remarkable for the shape of its tube, which is defended at the orifice by a kind of pent-house or hood, which projects boldly from the upper edge, just like the peak of a boy's cap.

Another genus of Serpulæ, called *Cymospira,* has the operculum horny, elliptical, and furnished with two or more large toothed horns, which are generally placed near the hinder edge. Some of the species are very large, one, which is in the British Museum, being as thick as a man's finger, and being inhabited by a Serpula three inches in length, and more than a quarter of an inch in diameter. Sometimes the horns are long and boldly projecting, and sometimes the tube has a pointed projection like the hood which has already been mentioned. One of the short-horned species which was procured from Swain's Reefs, on the eastern coast of Australia, was always so embedded in corals and madrepores, that the true shape of its tube cannot be ascertained.

Perhaps the most extraordinary example of the operculum is furnished by the genus Pomatostegus, in which the operculum is made in three stories, each smaller than that below it, so that it bears a distant resemblance to the fusee of a watch. In the British Museum is a fine example of this genus (*Pomatostegus Bowerbankii*), in which the operculum rises like the conventional tower of Babel, all the stories being devoid of horns, but covered with short hairs of a fibrous nature. If the reader should happen to be acquainted with conchology, he can form a very good idea of this remarkable operculum by taking up the similar organ in any species of solarium, or staircase shell, except that in the Serpula the stories of the operculum are distinct, and not formed by successive whorls.

The operculum of the animal which is supposed to be under examination does not present any of these singular appendages, but is more or less conical, grooved above in a radiated form, and horny in substance. It is, however, a very beautiful object, if only for the elegant shape, which remains after the softer parts have perished, and is in form so like a wine-glass with beautifully fluted sides, that Dr. Johnston has remarked that it might serve as a pattern for that article.

There is another marine annelid which constructs calcareous tubes, and which is sufficiently interesting to warrant a short notice. If the reader should happen to have in his possession a piece of stone, or an oyster-shell, that has been for some time immersed in the sea, or if he has a piece of the large tangle-weed that is so popular as a barometer, he will see that upon the surface are certain tiny calcareous tubes, scarcely thicker than hairs, and rolled spirally so as to form flat circular objects, about as large as pins' heads. They are firmly fixed to the objects on which they are placed, and are often thought to be the earlier forms of the serpula.

These, however, are distinct animals, called Spirorbis by naturalists, who have noticed that their spiral tubes bear a great resemblance to the planorbis shell, which is so plentiful in our rivers and ponds. A tolerably powerful magnifying-glass is needed before the real nature of the Spirorbis can be made out; but a short examination will show that not only is the little worm furnished with gills or branchiæ which closely resemble those organs in the serpula, but that, like that animal, it can shut up its tube with an operculum of a conical shape.

Another example of a submarine builder may be found in the well-known Terebella of our coasts, sometimes known by the name of Shell-binder. Sandy shoals are the best spots for the Terebella, and in many places there is scarcely a spare yard of sand without its inhabitants. Like the serpula, the Terebella constructs tubes, but, unlike that animal, it makes the tubes of a soft and flexible texture, although the materials which it employs are far harder than those which are used by the serpula. The Terebella has the art of making its submarine tubes of sand, which it agglutinates together with such wonderful power, that if Michael Scott's impish familiar had only been acquainted with natural history, he might soon have learned the art of making ropes of seasand, and have turned the tables on his master.

Should any of my readers be desirous of finding the habitation of a Terebella, he may easily do so by repairing to the nearest sandy shore, and looking under every large stone or piece of rock. There he will probably find some loose tufts of sandy threads, which are fixed to the mouth of a flexible tube, made of the same materials. This tube is the habitation of the Terebella, and by means of a crowbar and a chisel, the animal may generally be procured, together with its home. There are, however, plenty of deserted tubes, and I have often been sadly disappointed by finding that, after a long and laborious digging, nothing but the empty tube was to be found.

Supposing, however, that a specimen is obtained in an uninjured state, the observer can easily watch its method of house-building, by ejecting it from its tubular home, placing it in a vessel filled with sea water, and supplying it with a handful of sand. As clearness of the water is an essential part of success, shell-sand is the best material that can be supplied, and it will be safer to wash the sand thoroughly before placing it in the vessel. A large rough stone should also be placed in the vessel, as the animal always likes to lurk behind some sheltering object while it is engaged in the task of house-building.

Like many other creatures, the Terebella is a night-worker, and during the hours of daylight will retire behind the stone, and crouch in the darkest corner, as if to repose itself after the violent struggles and gyrations which it enacts when it is first taken out of the tube. Until noon is passed, the only sign of life will be the slight movement of the many tentacles which surround the upper lip; but, as the sun declines, the tentacles begin to move more rapidly, and as if they had some purpose to fulfil. In the evening, the worm is in full work; and as Professor Rymer Jones has given a clear and graphic description of its proceedings, I cannot do better than transfer his account to these pages. After remarking on the general habits of the creature, and describing the tentacles, he proceeds as follows :—

"They," *i.e.* the tentacles, "are now spread out from the orifice of the tube like so many slender cords—each seizes on one or more grains of sand, and drags its burden to the summit of the tube, there to be employed according to the service required. Should any of the tentacula slip, the same organs are again employed to search eagerly for the lost portion of sand, which is again seized and dragged towards its destination.

"Such operations are protracted during several hours, though so gradually as to be apparently of little effect; nevertheless, on resuming inspection next morning, a surprising elongation of the tube will be discovered; or, perhaps, instead of a simple accession to its walls, the orifice will be surrounded by forking threads of sandy particles agglutinated together.

"The architect has now retired to repose; but as evening comes, its activity is renewed, and again at sunrise a further prolongation has augmented the extent of its dwelling.

"At first sight, the numerous tentacula seem only so many long, cylindrical, fleshy threads, of infinite flexibility.

"On examining them, however, more attentively, we see that in exercising their special function, the surface which is applied to the foreign objects becomes flattened into twice or thrice its ordinary diameter; and while conveying the sandy materials to the tube, these are seized and retained in a deep groove, which almost resembles a slit; in fact, the tentaculum becomes a flat, narrow riband, folding longitudinally in different places to hold the particles securely.

"Although these organs, when contracted, are collected into a brush scarcely double the thickness of the animal's body, so enormous is their extensibility, that they can be stretched out to the length of four inches, or half the length of the body, thus sweeping the area of a circle eight inches in diameter.

"A thin internal coating, resembling silk, lines the whole tube, and at the same time serves as a real cement to unite and strengthen its innumerable parts. This silk-like material is derived from a glutinous slime, which exudes from the surface of the body of the Terebella.

"Notwithstanding the unrivalled expertness and expedition with which this Annelidan advances its work, it has never been observed to resume possession of its tube when once forsaken. To obtain the shelter of a new dwelling in place of the old, its labours are invariably recommenced from the foundation."

"In *Terebella nebulosa*," writes Dr. Williams, "the tentacula consist of hollow, flattened tubular filaments, furnished with strong muscular parietes, each tentacle forming a band which may be rolled longitudinally into a cylindrical form, so as to inclose a hollow, cylindrical space, if the two edges of the band meet, or a semi-cylindrical space, if they imperfectly meet. This inimitable mechanism enables each filament to take up and firmly grasp, *at any point of its length*, a molecule of sand, or, if placed in a linear series, *a row* of molecules. But so perfect is the disposition of the muscular fibres at the extreme free end of each filament, that it is gifted with the twofold power of acting on the sucking and the muscular principle. When the tentacle is about to seize an object, the extremity is drawn in, in consequence of the sudden reflux of fluid in its hollow interior; by this movement a cup-shaped cavity is formed, in which the object is securely held by atmospheric pressure; this power is, however, immediately aided by the contraction of the circular muscular fibres. Such, then, are the marvellous instruments by which these peaceful worms construct their habitations, and probably sweep their vicinity for food."

THERE are many species of Terebella, each remarkable for some peculiarity in the structure of its habitation. On the eastern coast of England, the preceding species seems to be more plentiful, and affords a notable instance of the manner in which stubborn materials can be rendered useful. Some of our coasts, however, are plentifully stocked with the SHELL-BINDER TEREBELLA (*Terebella conchilega*), a creature which has earned its well-deserved name from its curious habit of choosing broken shells for the construction of its tube. This being the case, it is evident that the creature will be most abundant on those shores where shells are most plentiful, and that it may usually be found on those parts of our coast where the sand is almost wholly composed of shells.

The tube resembles in texture that of the preceding species, but it is rather firmer, owing to the superior size of the material with which it is built. Generally the fragments of shell are very small, no larger, indeed, than those of which the well-known shell-sand is composed. But, in many instances, the Terebella has been known to choose shells in an entire state, and thus to give its tube a very extraordinary aspect.

The process of tube-making is conducted with much rapidity. Supposing that, as is often the case, a storm has arisen, and washed away the projecting portion of the tube, the Terebella sets to work to make a new portion in lieu of that which has been destroyed. Spreading out its long and delicate tentacles in all directions, it brings the materials towards its mouth, and always apportions the number of tentacles employed to the load which is to be carried. Having brought the materials within reach, it arranges them in regular circles, agglutinating them together with some secretion which has the property of hardening under water into a gelatinous substance. The union of these two materials—namely, the gelatinous secretion and the shells which are fastened to it—gives to the tube the two needful qualities of strength and flexibility.

Small particles of shell are generally affixed by the whole of their surface, but the large pieces are only fastened by one edge, so that a tube that is made wholly of large fragments looks as if it were covered with scales, like those of a serpent.

It is a remarkable fact that the Terebella does not form tubes during the early portions of its life, but swims about freely, like the nereis and other marine annelids. It has a head, eyes, feet, and antennæ, and roams about at will; whereas, in its perfect state, it has neither head, nor eyes, nor antennæ, nor true feet, the last-mentioned organs being modified into the tufts of hooks, and bristles, by means of which it moves up and down its tube. The reader may perhaps remember that the barnacles and many other stationary marine animals are free during their preliminary epochs, and only become fixed when they attain the perfect form. To our minds, the former seems the more perfect, and certainly the more agreeable state of existence; but we cannot measure the feelings of such an animal by our own, and may be sure that the creature enjoys existence as much while shut up in a tube, as when roaming the ocean at liberty.

ANOTHER species, *Terebella figulus*, sometimes called the POTTER, prefers mud as the material for its dwelling, and contrives to make the dark sea-mud so adhesive that it is capable of being formed into a tube.

As may be easily imagined, this tube is extremely fragile, and cannot be removed entire from the water without the exercise of much care, its own weight being mostly sufficient to tear it asunder. The walls of the tube are tolerably thick, and the tube itself is of some size, measuring nearly half an inch across, and is always found to be protected by the earth upon which it is placed. It is a rather curious fact that the tentacles of this species are of extraordinary length, extending for some eight or nine inches beyond the entrance of the tube, the animal itself measuring little more than four inches in length.

THE last species of Terebella that will be mentioned, is a very small and very remarkable species. It has been appropriately termed the Weaver Terebella (*Terebella textrix*) from the curious submarine home which it makes.

Not content with using the glutinous secretion as a means for binding together the muddy particles of which the tube is made, it spins a kind of web, bearing some resemblance to that of the spider, and being quite a complicated piece of work. This web is composed of many threads, which are very strong, but are also very fine, and in consequence are almost invisible when in the water, and as their substance is quite translucent, like the threads of isinglass. The threads encircle the body, and as it is only made in the month of May, when the eggs are deposited, it is in all probability employed more for the sake of guarding the eggs than protecting the body.

The tube of the Weaver Terebella is very small, not sufficing to cover more than half the body. The worm seems to be more independent of its tube than is usually the case, frequently vacating and returning to it, and sometimes making two or three tubes near each other, and living in any of them which it may happen to prefer at the time.

We now come to a group of tube-building annelids which are called Sabellæ, because they live in the sand, and in most cases form their tubes of that material. The general appearance of the tube is extremely variable. In some cases it bears so great a resemblance to the dwelling of the serpula, that a practised eye is needed to discover the distinction.

One very conspicuous species is the Trumpet Sabella (*Sabella tubularia*) which is generally found attached to stones or shells. The material of which it is made, is that hard, calcareous matter which is employed by the serpula, and at first the two tubes seem to be exactly alike. A more detailed examination will, however, show that it is not twisted like that of the serpula, but is nearly straight, looking very much like the military trumpet, or "tuba," of the ancient Romans. In some cases this tube attains considerable length, measuring eight or nine inches from tip to mouth. It is a solitary animal, and as far as is yet known, is never found grouped in masses, like many allied species.

The gill-fan of this species is exceedingly beautiful, being white, dotted profusely with scarlet, and expanding into a graceful feathery coronet. Although the resemblance to the serpula is very close, the animal may easily be distinguished by the absence of the beautiful operculum or stopper, which forms so conspicuous a feature in the serpula.

PERHAPS the most plentiful species of this genus is the common SABELLA (*Sabella alveolaria*), which may be found in countless myriads on many of our coasts. On several sandy shores, especially those of the southern coast, the wanderer by the sea may perceive masses of hard, agglutinated sand, pierced with innumerable holes. These masses are of great size, and in some places are strong enough to bear the pressure of a foot, though in others a slight push with the hand is sufficient to detach a portion.

If this perforated sand be closely examined, it will be seen to consist of a vast number of tubes, which are fixed together, and are further consolidated by sand which has washed over them, and lodged between them. When the water covers the sand mass, a delicate feathery tuft is seen to protrude from each hole, so that the general aspect is full of beauty. These tufts are the tentacles of the Sabella, and when examined with a microscope of moderate power, each tentacle is seen to be composed of a central shaft, with projecting teeth or fringes on both sides. There are about eighty of these tentacles, and as they are extremely flexible and always in motion, their appearance is peculiarly elegant.

Nothing is easier than to examine the structure of this Sabella, though the task of isolating a single tube is not an easy one. A penknife will soon break up the tube, and a pair of forceps will easily pull out the inhabitant, in spite of the array of bristles and hooks wherewith it clings to its habitation. It is but a little creature in point of length, but in point of width it nearly fills the diameter of the tube. The extremity of the body, however, is very small and slender, and is doubled back upon itself, with its tip pointing to the mouth of the tube.

The structure of the tube is extremely variable. Some individuals seem to give all their endeavours towards making their dwelling as long and strong as possible, while others are content with a tube which is barely long enough to shelter the whole body. They work with great rapidity, and when confined in an aquarium, will build their sandy homes nearly as well as if they were at liberty in the sea. Many interesting experiments have been made upon their modes of working, and by a judicious supply of different substances, they may be forced to build tubes of various colours and forms.

Very little care is required in keeping the Sabella in the aquarium. The beautiful little worm requires no feeding, and the only precaution that is needed, is to see that the water is pure. All dead animals should be carefully removed from the aquarium, as they are sure to putrefy very quickly, and the noxious gases evolved in that process will soon destroy all the other inhabitants.

Generally, when the Sabella feels ill, it is obliging enough to come out of its cell and die, in full sight; but now and then one of them will retreat to the bottom of its cell and die there, in which case it is as noisome and as difficult to discover as a dead rat behind the wainscot. Still, although the number of apertures in a single group of Sabella tubes renders it difficult for the possessor to identify the particular tube in which the defunct worm may be lying, its presence can generally be detected by a kind of whitish growth that appears at the mouth of the tube, and that tells its tale to an experienced eye. The best plan of getting out the inmate, is by a slender wire hooked at the end. This can be pushed down to the very bottom of the cell, when a twist of the wire will mostly secure the dead worm, and the nuisance may thus be removed.

Another species belonging to this genus, the STRAIGHT SABELLA (*Sabella unispira*) is remarkable for the form of its tube, which is nearly cylindrical and scarcely possesses any curve at all.

THERE is another group of tube-making marine annelids which are remarkable for the transparency of their newly constructed dwellings. Of these, a very singular example is found in the SILKWORM AMPHITRITE (*Amphitrite bombyx*).

The reader will remember that one, at least, of the Terebellæ can make a structure which is as transparent as isinglass, and will not, therefore, be surprised to find that another annelid possesses similar powers. The tube of the Silkworm Amphitrite is longer than the body, and is made entirely of the gelatinous secretion which in most of the species is used as a cement for fastening together the sand, shells, mud, and other materials of which the tube is formed. In this creature, however, the secretion is so plentiful, that it forms the whole of the tube.

Nor does it content itself with a single tube, but forms several, one after the other. When first made, the tube is so beautifully transparent, that the body of the inhabitant can be seen almost as plainly as through glass; but in process of time, it becomes incrusted with mud and sand, and almost looks as if it were made of very dirty leather. The average length of an adult specimen is three inches, and its beautiful gill-fan is decorated with brown and yellow. As is the case with most of the tube-inhabiting worms, it is a very timid creature, jerking itself into the tube on the least alarm, and contracting the orifice after it has retired into seclusion

ANOTHER species, the FAN AMPHITRITE (*Amphitrite ventilabrum*), forms a long and tough tube, which is apparently made of shoe-leather, but which is really formed from mud and the cement which is secreted by the animal. Mr. Rymer Jones has very lucidly described the mode of construction: "We will suppose a specimen, with its plume fully expanded, in a jar filled with its native element. In this condition, if a drop of liquid mud be dropped from above into the water so as to disturb its cleanliness, the animal immediately begins to arouse itself, and all the thousands of cilia that fringe its branchial plumules are discovered to be in vigorous activity, collecting by their incessant action the diffused muddy particles into a loose mass, which is soon perceived visibly accumulating in the bottom of the funnel. Meantime, the neck, or first segment of the body, rising unusually high above the orifice of the tube, exhibits two fleshy lobes or trowels, beating down the thin edges as they fold and clasp over the margin, like our fingers pressing a flattened cake against the palm of the hand.

"During these operations, the muddy materials are seen descending between the roots of the fans towards the trowels; while another organ, perhaps the mouth, is also occupied, it may be, in compounding the preparation with adhesive matter. As the bulk of the muddy mass diminishes, the activity of the worm abates; it is soon succeeded by repose, and then the tube is found to have received evident prolongation."

The gill-fans of this species are most lovely, forming a nearly complete funnel by their regularly radiating arrangement, and being coloured gorgeously with scarlet, green, brown, and gold.

Then there is the FUNNEL AMPHITRITE (*Amphitrite infundibulum*), so called because the tentacles are so perfect in shape, that when fully spread they form a circular funnel, in which scarcely the slightest break is perceptible. This funnel is rendered more complete by the curious fact, that in each half of the fan the tentacles are united by a delicate and transparent web, reaching nearly to their tips. The prevailing colour of these beautiful fans is purple, darkest towards the tips, and changing gradually to chestnut at the base.

This is not a plentiful species, and is mostly obtained by means of the dredge or drag, which tears up a portion of the ground and brings with it the Funnel Amphitrite and its dwelling. This is of a tubular shape, and generally black, with a slight mixture of green. The colour, however, is very variable, and depends much upon the age of the tube; for when freshly made, it is nearly as transparent as that of the previous species. There are other species of Amphitrite, all of which are interesting, but which our diminished space will not admit.

SHOULD the reader happen to be an entomologist, he will readily call to mind the tiny cylindrical cases that are made by certain lepidopteran larvæ, belonging to the great family Tineidæ, and which are found so plentifully upon the leaves of oak, hazel, and other trees. If he should happen to be something of an aquarian naturalist, and fond of looking for marine curiosities, he may find attached to submarine plants, certain little cylindrical cases which are wonderfully like those of the moths. They are very small indeed, scarcely thicker than the shaft of an ordinary pin, and measuring scarcely more than the eighth of an inch in length. Their colour is pale brown, their surface is rough, and they are stuck upon the seaweed in great confusion, without the least attempt at arrangement.

These are the habitations of a very small crustacean (*Cerapus tubularis*), popularly called the CADDIS SHRIMP, because the tube which the creature makes is analogous to that which is formed by the caddis larvæ. The animal which inhabits this case is a curious little being, very like the long-bodied, long-legged caprellæ, that are so plentiful among seaweeds, and furnished with two pairs of long and stout antennæ, and two pairs of grasping feet. As the tube is too short to contain the entire animal, the long antennæ are always protruded, and occasionally the powerful grasping feet are also thrust out of the opening.

The antennæ are continually flung forward and retracted in a manner that reminds the observer of the movements of the acorn barnacle, each grasp being evidently made for the purpose of arresting any passing substance that may serve for food. This remarkable little crustacean is generally found upon the well-known alga which produces the Carrageen, or Irish moss (*Chondrus crispus*). It will not, however, be found upon those plants which can be plucked by hand, but resides in deeper water, so that the best method of procuring it is to go out in a boat, throw the drag overboard, and then examine the algæ which are torn from their attachments.

THE BEAVER AND ITS HOME.

CHAPTER XXI.

SOCIAL HABITATIONS.

SOCIAL MAMMALIA.

The BEAVER—Its form and aquatic habits—Need for water and means used to procure it—Quadrupedal engineering—The dam of the Beaver—Erroneous ideas of the dam—How the Beaver cuts timber—The Beaver in the Zoological Gardens—Theories respecting the Beaver's dam—How the timber is fastened together—Form of the dam, and mode of its enlargement—Beaver-dams and coral-reefs—The house or lodge of the Beaver—Its locality and structure—Use of a subterranean passage—How Beavers are hunted—Curious superstition—"Les Paresseux."

WE now come to the SOCIAL HABITATIONS, and give precedence to those which are constructed by Mammalia.

Of the Social Mammalia, the BEAVER (*Castor fiber*) takes the first rank, and is the best possible type of that group. There are other social animals, such as the various marmots and others; but these creatures live independently of each other, and are only drawn together by the attraction of some favourable locality. The Beavers, on the other hand, are not only social by dwelling near each other, but by joining in a work which is intended for the benefit of the community.

The form of the Beaver is sufficiently marked to indicate that it is a water-loving creature, and that it is a better swimmer than walker. The dense, close, woolly fur, defended by a coating of long hairs, the broad, paddle-like tail, and the well-webbed feet, are characteristics which are at once intelligible. Water, indeed, seems to be an absolute necessity for the Beaver, and it is of the utmost importance to the animal that the stream near which it lives, should not grow dry. In order to avert such a misfortune, the Beaver is gifted with an instinct which teaches it how to keep the water always at or about the same level, or, at all events, to prevent it from sinking below the requisite level.

If any modern engineer were asked how to attain such an object, he would probably point to the nearest water-mill, and say that the problem had there been satisfactorily solved, a dam having been built across the stream so as to raise the water to the requisite height, and to allow the superfluous water to flow away. Now, water is as needful for the Beaver as for the miller, and it is a very curious fact, that long before millers ever invented dams, or before men ever learned to grind corn, the Beaver knew how to make a dam and insure itself a constant supply of water.

That the Beaver does make a dam is a fact that has long been familiar, but how it sets to work is not so well known. Engravings representing the Beavers and their habitations, are common enough, but they are generally untrustworthy, not having been drawn from the natural object, but from the imagination of the artist. In most cases the dam is represented as if it had been made after the fashion of our time and country, a number of stakes having been driven into the bed of the river, and smaller branches entwined among them. The projecting ends of the stakes are neatly squared off, and altogether the work looks exactly as if it had been executed by human hands. One artist seems to have copied from another, so that the error of one man has been widely perpetuated.

Now, in reality, the dam is made in a very different manner, and in order to comprehend the mode of its structure, we must watch the Beaver at work.

When the animal has fixed upon a tree which it believes to be suitable for its purpose, it begins by sitting upright, and with its chisel-like teeth, cutting a bold groove completely round the trunk. It then widens the groove, and always makes it wide in exact proportion to its depth, so that when the tree is nearly cut through, it looks something like the contracted portion of an hour-glass. When this stage has been reached, the Beaver looks anxiously at the tree, and views it on every side, as if desirous of measuring the direction in which it is to fall. Having settled this question, it goes to the opposite side of the tree, and with two or three powerful bites cuts away the wood, so that the tree becomes overbalanced and falls to the ground.

This point having been reached, the animal proceeds to cut up the fallen trunk into lengths, usually a yard or so in length, employing a similar method of severing the wood. In consequence of this mode of gnawing the timber, both ends of the logs are rounded and rather pointed, as may be seen by reference to the illustration. In the Zoological Gardens may be seen many excellent examples of timber which has been cut by the Beaver: the logs and cut stumps which are given in the illustration were sketched from those objects.

The next part of the task is, to make these logs into a dam. Now, whereas some persons have endeavoured to make the Beaver a more ingenious animal than it really is, and have accredited it with powers which only belong to mankind, others have gone to the other extreme, and have denied the existence of a regularly built dam, saying that it is entirely accidental, and caused by the logs that are washed down by the stream, after the Beavers have nibbled off all the bark.

That this position is untenable is evident from the acknowledged fact that the dam is by no means placed at random in the stream, just where a few logs may have happened to lodge, but is set exactly where it is wanted, and is made so as to suit the force of the current. In those places where the stream runs slowly, the dam is carried straight across the river, but in those where the water has much power, the barrier is made in a convex shape, so as to resist the force of the rushing water. The power of the stream can, therefore, always be inferred from the shape of the dam which the Beavers have built across it.

Some of these dams are of very great size, measuring two or three hundred yards in length, and ten or twelve feet in thickness, and their form exactly corresponds with the force of the stream, being straight in some parts, and more or less convex in others.

The dam is formed, not by forcing the ends of the logs into the bed of the river, but by laying them horizontally, and covering them with stones and earth until they can resist the force of the water. Vast numbers of logs are thus laid, and as fast as the water rises, fresh materials are added, being obtained mostly from the trunks and branches of trees which have been stripped of their bark by the Beavers.

The reader will remember that many persons have thought that the dam of the Beaver is only an accidental agglomeration of loose logs and branches, without any engineering skill on the part of the animals. There is some truth in this statement, though the assertion is too sweeping. For, after the Beavers have completed their dam, it obstructs the course of the stream so completely that it intercepts all large floating objects, and every log or branch that may happen to be thrown into the river is arrested by the dam, and aids in increasing its dimensions.

Mud and earth are also continually added by the Beavers, so that in process of time the dam becomes as firm as the land through which the river passes, and is covered with fertile alluvium. Seeds soon make their way to the congenial soil, and in a dam of long standing, forest trees have been known to grow, their roots adding to the general stability by binding together the materials. It is well known that the fertile islands formed on coral reefs are stocked in a similar manner. Originally, the dam is seldom more than a yard in width where it overtops the water, but these unintentional additions cause a continual increase.

The bark with which the logs were originally covered, is not all eaten by the animals, but stripped away, and the greater part hidden under water, to serve for food in the winter time. A further winter provision is also made by taking the smaller branches, diving with them to the foundations of the dam, and carefully fastening them among the logs. When the Beavers are hungry, they dive to their hidden stores, pull out a few branches, carry them on land, nibble away the bark, and drop the stripped logs on the water, where they are soon absorbed by the dam.

We have now seen how the Beavers keep the water to the required level, and we must next see how they make use of it. The Beaver is essentially an aquatic mammal, never walking when it can swim, and seldom appearing quite at its ease upon dry land. It therefore makes its houses close to the water, and communicating with it by means of subterranean passages, one entrance of which passes into the house or "lodge," as it is technically named, and the other into the water, so far below the surface that it cannot be closed by ice. It is, therefore, always possible for the Beaver to gain access to the provision stores, and to return to its house, without being seen from the land.

The lodges are nearly circular in form, and much resemble the well known snow houses of the Esquimaux, being domed, and about half as high as they are wide, the average height being three feet and the diameter six or seven feet. These are the interior dimensions, the exterior measurement being much greater, on account of the great thickness of the walls, which are continually strengthened with mud and branches, so that, during the severe frosts, they are nearly as hard as solid stone. Each lodge will accommodate several inhabitants, whose beds are arranged round the walls.

All these precautions are, however, useless against the practised skill of the trappers. Even in winter time the Beavers are not safe. The hunters strike the ice smartly, and judge by the sound whether they are near an aperture. As soon as they are satisfied, they cut away the ice and stop up the opening, so that if the Beavers should be alarmed, they cannot escape into the water. They then proceed to the shore, and by repeated soundings, trace the course of the Beavers' subterranean passage, which is sometimes eight or ten yards in length, and by watching the various apertures are sure to catch the Beavers. This is not a favourite task with the hunters, and is never undertaken as long as they can find any other employment, for the work is very severe, the hardships are great, and the price which they obtain for the skins is now very small.

While they are thus engaged, they must be very careful not to spill any blood, as if they do so, the rest of the Beavers take alarm, retreat to the water, and cannot be captured. They also have a curious superstitious notion, which leads them to remove a knee-cap from each Beaver and to throw it into the fire. They would expect ill-luck were they to omit this ceremony, which is wonderfully like the custom of our fishermen of spitting into the mouth of the first fish they catch, and the first money which they take in the day, "for luck."

Generally, the Beavers desert their huts in the summer time, although one or two of the houses may be occupied by a mother and her young offspring. All the old Beavers who have no domestic ties to chain them at home, take to the water, and swim up and down the stream at liberty, until the month of August, when they return to their homes. There are, also, certain individuals called by the trappers "les paresseux," or idlers, which do not live in houses, and make no dam, but abide in subterranean tunnels like those of our common water rat, to which they are closely allied. These "paresseux" are always males, and it sometimes happens that several will inhabit the same tunnel. The trapper is always pleased when he finds the habitation of an idler, as its capture is a comparatively easy task.

SOCIABLE WEAVER BIRD.

CHAPTER XXII.

SOCIAL BIRDS.

The SOCIABLE WEAVER BIRD and its country—Description of the bird—Nest of the Sociable Weaver—How begun and how carried on—Materials of the nest—The tree on which the nest is built, and its uses—Dimensions of the nest and disastrous consequences—A Hottentot and a lion—Supposed object of the Social nest—Average number of inhabitants—Analogy with Dyak houses—Enemies of the Sociable Weaver, the monkey, the snake, and the parrakeet.

WE now come to the SOCIAL BIRDS, one of which is as pre-eminent among the feathered tribes as is the beaver among mammalia. This is the SOCIABLE WEAVER BIRD, sometimes called the SOCIABLE GROSBEAK (*Philetœrus socius*).

This species is allied to the Weaver Birds, some of which have already been described, and makes a nest which is no whit inferior to those which have already been mentioned. The Sociable Weaver Bird is a native of Southern Africa, and in some places is very plentiful, its presence depending much upon the trees which clothe the country. It is not a large bird, measuring about five inches in length, and is very inconspicuous, its colour being pale buff, mottled on the back with deep brown.

The chief interest about the species is concentrated in its nest, which is a wonderful specimen of bird architecture, and attracts the attention of the most unobservant traveller. Few persons expect to see in a tree a nest which is large enough to shelter five or six men; and yet that is often the case with the nest of the Sociable Weaver Bird. Of course so enormous a structure is not the work of a single pair, but, like the dam of the beaver, is made by the united efforts of the community. How it is made will now be described.

Large as is the domicile, and capable at last of containing a vast number of parents and young, it is originally the work of a single pair, and attains its enormous dimensions by the labours of those birds which choose to associate in common. The first task of this Weaver Bird is to procure a large quantity of the herb which really seems as if made expressly for the purpose. This is a grass with a very large, very tough, and very wiry blade, which is known to the colonists as Booschmannie grass, probably because it grows plentifully in that part of Southern Africa where the Bushmen, or Bosjesmans live.

They carry this grass to some suitable tree, which is usually a species of acacia, called by the Dutch colonists Kameel-dorn (*Acacia giraffa*), because the giraffe, which the Dutch persist in calling a kameel or camel, is fond of grazing on the leaves. This is a most appropriate tree for the purpose, as the wood is extremely hard and tough, and the branches are therefore able to bear the great weight of the nests. This tree is used in Southern Africa for many purposes wherein hardness and endurance are required, such as the axle-trees of the wooden waggons, which have to withstand such rough usage, the upright timbers of houses, and the handles of tools, especially those which are intended for agricultural purposes.

The birds then hang the Booschmannie grass over a suitable

branch, and by means of weaving and plaiting it, they form a roof of some little size. Under this roof are placed a quantity of nests, increasing in number with each successive brood. The nests are set closely together, so that at last they look like a mass of grass pierced with numerous holes, and it is really wonderful that the birds should be able to find their way to their own particular homes. To human eyes, the nests are as much alike as the houses in a modern street, before the blinds, the flowers, and other additions have communicated an individuality to each dwelling; but, notwithstanding this similarity, the inmates glide in and out without any hesitation.

Although the same nest-mass is occupied for several successive seasons, the birds refuse to build in the same nests a second time, preferring to make a fresh domicile for each new brood. In consequence of this custom, when the birds have entirely filled the roofs with their nests, they do not desert it, but enlarge the roof, and build a second row of nests, just like the combs of a wasp's or hornet's habitation.

Layer after layer is thus added, until the mass becomes of so enormous a size that travellers have mistaken these nests for the houses of human beings, and been grievously disappointed when they came near enough to detect their real character. There is a story of a Hottentot and a lion, which will give an idea of the dimensions of these nests. A Hottentot, who was engaged in some task, was suddenly surprised by a lion, and instinctively made for the nearest tree, which happened to be a kameel-dorn. Up the tree he sprang, and finding one of the branches occupied by the nest of the Sociable Weaver Bird, he took refuge behind the grassy mass, and was thus concealed from the pursuer.

The lion, in the meantime, arrived at the foot of the tree, but could not see his intended prey. The unlucky Hottentot, however, peeped over the nest in order to see whether the coast was clear, and was spied by the lion, who made a dash at the tree. The man shrank back behind the nest, but his imprudent movement brought its own punishment.

Knowing that the ascent of the tree was impossible, and at the same time unwilling to leave its prey, the lion sat down at the foot of the tree, and kept watch upon the man. Hour after hour the lion mounted guard over its prisoner, until thirst overpowered hunger, and the animal was forced reluctantly to quit its post and seek for water. The man then scrambled down the tree, and made the best of his way homewards, little the worse for his imprisonment except the fright, and a skin scorched by long exposure to the sun. The artist has introduced this little episode into the illustration, because it enables the reader to judge of the enormous size of the nest.

Season after season the Weaver Birds continue to add their nests, until at last the branch is unable to endure the weight, and comes crashing to the ground. This accident does not often occur during the breeding months, but mostly takes place during the rainy season, the dried grass absorbing so much moisture, that the weight becomes too great for the branch to bear.

The nest group which is shown in the illustration is of medium size, as can be ascertained by its shape. In its early state, the nest-mass is comparatively long and narrow, spreading out by degrees as the number of nests increases, so that at last it is as wide and as shallow as an extended umbrella. The dimensions of some of these structures may be gathered from the fact, that Le Vaillant counted in one unfinished edifice, beside the deserted nests of previous seasons, no less than three hundred and twenty nests, each of which was occupied by a pair of birds engaged in bringing up a brood of young, four or five in number.

Those who are acquainted with Borneo and the customs of its inhabitants, cannot fail to perceive the analogy between these social nests of the Weaver Bird and the "long houses" of the Dyaks, each of which houses is in fact one entire village, sheltering a whole community under a single roof.

The Weaver Birds have but few enemies. First, there are the snakes, which are such determined robbers of nests, swallowing both eggs and young; and then there are the monkeys, which are capable of sad depredations whenever they can find an opportunity. Monkeys are extremely fond of eggs, and there is scarcely a better bribe to a monkey, ape, or baboon, than a fresh raw egg. The bird which laid it is almost as great a dainty, and a monkey seems to be in the height of enjoyment if a newly-killed bird be put into its paws. It always begins by eating the brain, and then tears the carcase to pieces with great deliberation. A mouse is quite as much appreciated as a bird, provided that it has been recently killed, and that the blood has not congealed.

However, the structure of the nest forms an insurmountable barrier to the snake, and the monkey can only reach a few of the cells which are near the edge. The worst enemies are certain little parrakeets, which are delighted to be able to procure nests without the trouble of building them, and which are apt to take possession of the cells and oust the rightful owners.

CHAPTER XXIII.

SOCIAL INSECTS.

Arrangement of groups—Nests of POLYBIA—Curious method of enlargement—Structure of the nests—How concealed—Various modes of attachment—A curious specimen—The HIVE BEE, and its claims to notice—General history of the hive—Form of the cells—The royal cell, its structure and use—Uses of the ordinary cells—Structure of the Bee-cell—Economy of space—How produced—Theories of different mathematicians—Measurement of angles—A logarithmic table corrected by the bee-cell—The "lozenge," a key to the cell—How to form it—Beautiful mathematic proportions of the lozenge—Method of making the cell or a model—Conjectured analogy between the cell and certain crystals—Effect of the cell upon honey—The HORNET and its nest—Its favourite localities—Difficulties of taking a hornet's nest—Habits of the insect—Mr. Stone's method of taking the nest—The SYNŒCA and its habitation—Beautiful nests in the British Museum—Description of the insect—Nest of the EUCHEIRA—Its external form—Curious discovery in dissection—A suspended colony—Conjectures respecting the structure—Nest from the Oxford Museum—Remarkable form of its doors, and material of which it is made—The SMALL ERMINE MOTH—and its ravages—Its large social habitation—General habits of the larva—why the sparrow does not eat them—The GOLD-TAILED MOTH, and its beautiful social nest—Description of a specimen from Wiltshire—Illustration of the theory of heat—The BROWN-TAILED MOTH and its nest—Social habitations of the PEACH and SMALL TORTOISESHELL BUTTERFLIES.

AFTER the Social Birds come the SOCIAL INSECTS, to which the following chapter is dedicated.

The reader will probably have noticed that several insects, especially those of the hymenopterous order, seem to have been omitted in previous chapters, although they might fairly claim admission into the ranks of Builders, Pensiles, Burrowers, and Subaquatics. The fact is, that some of them unite the characteristics of several groups, and may therefore be placed in either of them. For example, the South American wasp, which makes the nest called popularly the "Dutchman's pipe," may be ranked either as a builder, a pensile, or a social insect. In such cases, therefore, I have endeavoured to select that characteristic which seems to be marked most strongly, and have arranged the insects accordingly.

Just as the hymenoptera are chief among the pensiles and the builders, so are they chief among the Social Insects, and the species which may be placed in this group are so numerous, that it will only be possible to make a selection of a few, which seem more interesting than the others.

POLYBIA.

In the British Museum there are some very remarkable nests made by hymenopterous insects belonging to the genus *Polybia*, several of which are drawn in the accompanying illustration. As it was desirable to include more than one specimen, the figures are necessarily much reduced in size. Neither the nests nor the insects, however, are of large dimensions, and the former are so sombre in colour as well as small in size, that they would not of themselves attract any attention. Their nests, however, are extremely interesting, as may be seen from the examples which are figured in the illustration.

On the left hand may be seen a nearly spherical nest, which is evidently hollow, and has cells both on the outside and within the cover. These cells are not placed vertically, with their mouths downward, like those of the wasp and hornet, nor horizontally like those of the bee, but are set with their mouths radiating from the centre of the nest. Moreover, there is another curious circumstance connected with the nest. If it were to be opened, it would be seen to be composed of several concentric layers, very much like those ivory puzzle-balls which the Chinese make so beautifully.

The method by which the nest is formed is very simple, though not one that is usually seen among the hymenoptera. The layers of combs are made like hollow spheres, the mouths of the cells being outwards, and as soon as a layer is completed, the insects protect it from the weather by a cover of the same material as is used for the construction of the cells. When they require to make a fresh layer of cells, they do not enlarge the cover, as is the case with the wasp and hornet, but place the new cells upon the surface of the cover, and make a fresh cover as soon as the comb is completed. Thus the nest increases by the addition of concentric layers, composed alternately of comb and cover.

In the nest which is in the British Museum, the insects have commenced several patches of comb on the outside of the cover, and one such patch is shown in the illustration.

On the right of the globular nest is another curious structure, also made by insects of the same genus, and having a kind of similarity in its aspect. This nest, however, is very much longer in proportion to its width, and being fixed throughout its length to a leaf, is not so plainly visible as the last mentioned specimen. Indeed, when the leaf has withered, as is the case with the object from which the drawing was made, the dull brown of the nest coincides so completely with the colour of the faded leaf, that many persons would overlook it unless their attention were specially drawn towards it.

On the extreme right of the illustration, and in the upper corner, is seen a nest which is also the work of insects belonging to the genus Polybia, and it is pendent from a bough, like the habitation of the Chartergus and other pensile hymenoptera.

In the same collection there are many more specimens of social nests formed by insects belonging to this genus, two cases being quite filled with them. One is attached to the bark of a tree, and resembles it so closely that it seems to be made of the same substance, this similarity of aspect being evidently intended as a preservative against the attacks of birds and other insect-loving creatures, which would break up the nest, and eat the immature and tender grubs. Most of the nests are fixed to leaves, and are different forms, according to the species which made them. They are mostly fixed to the under sides of the leaf, so that the weight causes the leaf to bend and to form a natural roof above them. The shape of the nest seems to depend much on the character of the plant to which it is fixed. Those that are fastened to reeds are long and slender, and generally much narrower than the sword-shaped leaf on which they rest. Others, which are fastened to short and broad leaves, adapt themselves so closely to the shape of the leaf, that, if removed, they would enable any one to conjecture the form of the leaf upon which they had been fixed.

One such nest is very remarkable. In general form it bears a singular resemblance to the nest of the fairy martin, which is figured at page 313, though its materials are entirely different. The nest is flask-shaped, and its base is fastened to a leaf which it almost covers. The body of the nest is oval, and the entrance, which is small, is placed at the end of a well-marked neck. The shell of the nest is extremely thin, not in the least like the loose, papery structure of an ordinary wasp-nest, nor the paste-board-like material which defends the nest of the Chartergus. It is rather fragile, and in thickness is almost double that of the paper on which this account is printed.

The name of the species which builds this curious nest is *Polybia sedula*, and the specimen was brought from Brazil.

For the reasons which have been given at the beginning of this chapter, the Hive Bee has been reckoned among the Social Insects.

The Bee has always been one of the most interesting insects to mankind, on account of the direct benefit which it confers upon the human race. There are many other insects which are in reality quite as useful to us, and indeed are indispensable, but which we neglect because we are ignorant of their labours. The Bee, however, furnishes two powerful and tangible arguments in its favour—namely, honey and wax—and is sure, therefore, to enlist our sympathies in its behalf.

Independently, however, of these claims to our notice, if the Bee never made an ounce of honey—if the wax were as useless to us as wasp-comb—if the insect were a mere stinging creature, with a tetchy temper, it would still deserve our admiration, on account of the wonderful manner in which it constructs its social home, and the method by which that home is regulated.

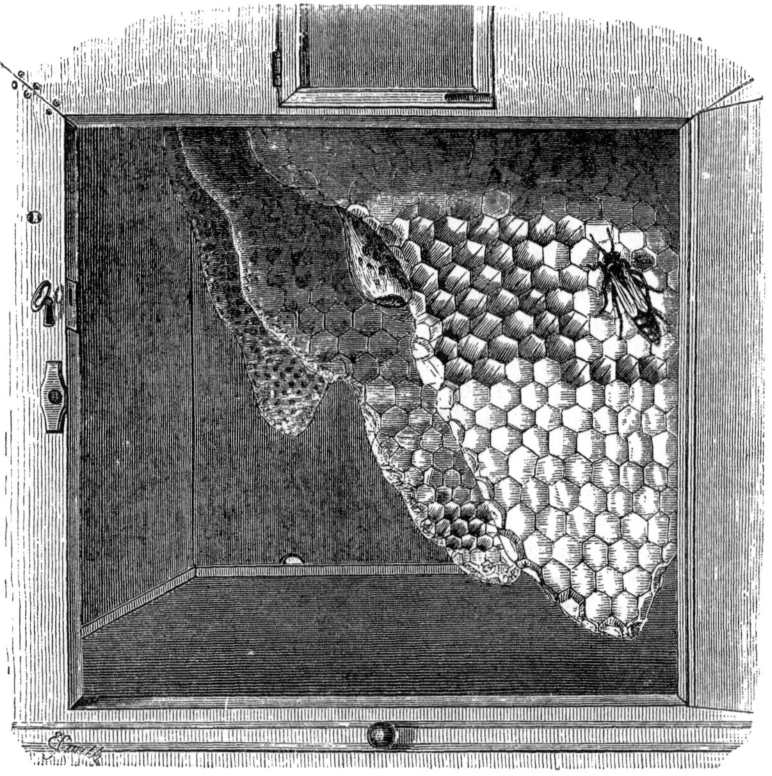

BEE-HIVE.

The accompanying illustration shows the lower part of the interior of a hive, which is supposed to be viewed through a plate of glass set in the back. When the hive is in full operation, the combs are so crowded with Bees that their structure can hardly be seen; but in the illustration the Bees are supposed to have gone away, with the exception of one individual.

I need not in this place repeat the well-known facts respecting the constitution of the Bees, nor describe the duties of the Queen, Drone, and Worker Bees. Suffice it to say, that the former is the mother as well as the queen of the hive; that the workers are undeveloped females, which are properly called neuters; and that the drones are males, which do no work, and have no stings.

In the illustration, the Queen Bee is seen walking over the combs, and in this position she exhibits the peculiarities of form which distinguish her from her subjects, and which enables an experienced eye to detect her at once amid a crowd of workers. In the Queen Bee, the abdomen is long in proportion to its width, and the wings slightly cross each other when closed; the latter being a very conspicuous badge of sovereignty. The drones are easily distinguished by their generally larger size, their larger eyes, and the wide, blunt, and rounded abdomen.

The lower part of the comb, in the foreground, is formed of cells which are closed at their mouth, and which do not show the hexagonal shape as well as those which are yet empty. Some of the empty cells are shown above, and the Queen Bee is represented as making her way towards them.

There are three kinds of cell in a hive; namely, the worker-cell, the drone-cell, and the royal-cell. Of these, the two former are hexagonal, but can easily be distinguished by the greater size of the drone-cell; while the royal-cell is totally unlike the nursery of a subject, whether drone or worker, and is almost always placed on the edge of a comb. One of these cells is shown in the illustration, and may be seen on the edge of the comb in the foreground. It is very much larger than an ordinary cell, and is built with a lavish expenditure of wax that affords a curious contrast with the rigid economy observed in the structure of the other cells. The difference of size between the worker and drone-cells is shown in the central comb, where the worker-cells are seen below, and the drone-cells above.

The little grub which is placed in the royal cell is not fed with the same food which is supplied to the other Bees, but lives upon an entirely different diet, and which is, apparently, of a more stimulating character; and it is now well known, that if a young grub which has been hatched in one of the worker-cells be removed into the royal-cell, and supplied with royal food, it becomes developed into a queen, and, in time, is qualified to rule and populate a hive. This remarkable provision of nature is intended to meet a difficulty, which sometimes occurs, when the reigning queen dies, and there is no royal larva in the cell.

Although the primary object of the bee-cell is to serve as a storehouse and a nursery, it also is made to answer other purposes. When the Bee seeks repose, it almost invariably creeps into a cell, and buries itself deep therein, the whole head, thorax, and part of the abdomen being hidden. If a hive be examined in the winter-time, every cell that happens to be empty will be tenanted by a Bee; and when the poor insects are put to death by the absurd and cruel plan of smothering them with the fumes of burning sulphur, they will be found to have vainly sought escape from the suffocating vapour by forcing themselves into the recesses of the empty cells.

As a general fact, the Bees place the honey in the coolest part of the hive, and the young brood in the warmest; so that bee-keepers are enabled to procure honeycomb of wonderful purity by affixing glass or wooden caps to their hives. These caps are necessarily cooler than the body of the hive, and therein the Bees will store large quantities of honey.

The chief point which distinguishes the comb of the Hive Bee from that of other insects, is the manner in which the cells are arranged in a double series. The combs of the wasp or the hornet are single, and are arranged horizontally, so that their cells are vertical, with the mouths downwards and the bases upwards, the united bases forming a floor on which the nurse wasps can walk while feeding the young inclosed in the row of cells immediately above them.

Such, however, is not the case with the Hive Bee. As every one knows, who has seen a bee-comb, the cells are laid nearly horizontally, and in a double series, just as if a couple of thimbles were laid on the table with the points touching each other and their mouths pointing in opposite directions. Increase the number of thimbles, and there will be a tolerable imitation of a bee-comb.

There is another point which must now be examined. If the bases of the cells were to be rounded like those of the thimbles, it is clear that they would have but little adhesion to each other, and that a large amount of space would be wasted. The simplest plan of obviating these defects is evidently to square off the rounded bases, and to fill up the ends of each cell with a hexagonal flat plate, which is actually done by the wasp. If, however, we look at a piece of bee-comb, we shall find that no such arrangement is employed, but that the bottom of each cell is formed into a kind of three-sided cup. Now, if we break away the walls of the cell, so as only to leave the bases, we shall see that each cup consists of three lozenge-shaped plates of wax, all the lozenges being exactly alike.

These lozenge-shaped plates contain the key to the bee-cell, and their properties will therefore be explained at length. Before doing so, I must acknowledge my thanks to the Rev. Walter Mitchell, Vicar and Hospitaller of St. Bartholomew's Hospital, who has long exercised his well-known mathematical powers on this subject, and has kindly supplied me with the outline of the present history.

If a single cell be isolated, it will be seen that the sides rise from the outer edges of the three lozenges above-mentioned, so that there are, of course, six sides, the transverse section of which gives a perfect hexagon. Many years ago Maraldi, being struck with the fact that the lozenge-shaped plates always had the same angles, took the trouble to measure them, and found that in each lozenge, the large angles measured 109° 28′, and the smaller, 70° 32′, the two together making 180°, the equivalent of two right angles. He also noted the fact that the apex of the three-sided cup was formed by the union of three of the greater angles. The three united lozenges are seen at fig. 1.

Some time afterwards, Reaumur, thinking that this remarkable uniformity of angle might have some connexion with the wonderful economy of space which is observable in the bee-comb, hit upon a very ingenious plan. Without mentioning his reasons for the question, he asked Kœnig, the mathematician, to make the following calculation. Given a hexagonal vessel terminated by three lozenge-shaped plates; what are the angles which would give the greatest amount of space with the least amount of material?

Kœnig made his calculations, and found that the angles were 109° 26′ and 70° 34′, almost precisely agreeing with the measurements of Maraldi. The reader is requested to remember these angles. Reaumur, on receiving the answer, concluded that the Bee had very nearly solved the difficult mathematical problem, the difference between the measurement and the calculation being so small as to be practically negatived in the actual construction of so small an object as the bee-cell.

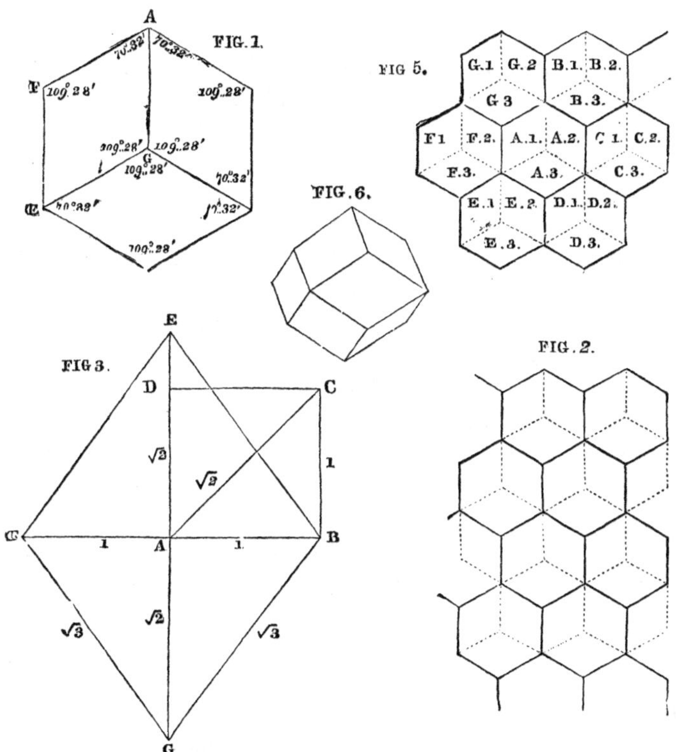

Mathematicians were naturally delighted with the result of the investigation, for it showed how beautifully practical science could be aided by theoretical knowledge, and the construction of the bee-cell became a famous problem in the economy of nature. In comparison with the honey which the cell is intended to contain, the wax is a rare and costly substance, secreted in very small quantities, and requiring much time for its production; it is therefore essential that the quantity of wax employed in making the comb should be as little, and that of the honey contained in it as great, as possible.

For a long time these statements remained uncontroverted. Any one with the proper instruments could measure the angles for himself, and the calculations of a mathematician like Kœnig would hardly be questioned. However, Maclaurin, the well-known Scotch mathematician, was not satisfied. The two results very nearly tallied with each other, but not quite, and he felt that in a mathematical question precision was a necessity. So he tried the whole question himself, and found Maraldi's measurements correct, namely, 109° 28′, and 70° 32′.

He then set to work at the problem which was worked out by Kœnig, and found that the true theoretical angles were 109° 28′, and 70° 32′, precisely corresponding with the actual measurement of the bee-cell.

Another question now arose. How did this discrepancy occur? How could so excellent a mathematician as Kœnig make so grave a mistake? On investigation, it was found that no blame attached to Kœnig, but that the error lay in the book of logarithms which he used. Thus, a mistake in a mathematical work was accidentally discovered by measuring the angles of a bee-cell—*a mistake sufficiently great to have caused the loss of a ship whose captain happened to use a copy of the same logarithmic tables for calculating his longitude.*

Now, let us see how this beautiful lozenge is made. There is not the least difficulty in drawing it. Make any square, ABCD (fig. 3) and draw the diagonal AC.

Produce BA towards F and AD, both ways to any distance.

Make AE and AG equal to AC, and make AF equal to AB. Join the points EFGB, and you have the required figure.

Now comes a beautiful point. If we take AB as 1, being one side of the square on which the lozenge is founded, AE and AG will be equal to $\sqrt{2}$, and EF, FG, GB, and BE, will be equal to $\sqrt{3}$, as can be seen at a glance by any one who has advanced as far as the 47th proposition of the first book of Euclid.

Perhaps some of my readers may say that all these figures may be very true, but that they do not show how the cell is formed. If the reader will refer to fig. 4, he will see how the theory may be reduced to practice. After he has drawn the lozenge-shaped figure which has just been described, let him draw upon cardboard nine of them, as is shown in the illustration (fig. 4.) Then let him cut out the figure, and draw his penknife half through the cardboard at all the lines of junction. He will then find that the cardboard will fold into an exact model of a bee-cell, the three lozenges which project from the sides forming the base, and the others the sides. This cell will, of course, have very short sides; but by the simple expedient of widening the lozenges, which form the sides, without altering the angles, the imitation cell can be made of any desired length.

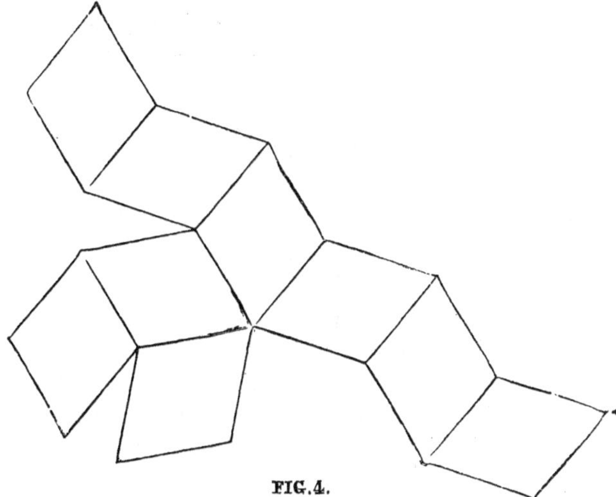

FIG. 4.

The best way of showing this beautiful structure is to make two models, one to lie flat or be folded and opened at discretion, and the other formed into a cell, and the angles written upon the cardboard. A little gummed paper will hold the sides together, so that the model can be handled without breaking. A very amusing puzzle may be formed by cutting out the nine lozenge-shaped pieces of cardboard, and by requesting that they be so put together as to form the model of a bee-cell.

We have not yet exhausted the wonders of the bee-comb.

If we take a piece of comb from which all the cells have been removed, and hold it up to the light, we shall see that the cells are not placed opposite each other, but that the three lozenges which form the base of one cell form part of the base of three other cells, as is seen in fig. 2. Thus a still further economy of material is attained, while the strength is enormously increased, each of the edges formed by the junction of two lozenges making a buttress which performs precisely the same office as the buttresses of architecture.

The same principle is observable throughout the cell, which even at its edges is supported by three cells, and gives partial support to three others. As the walls of the cells are extremely thin, the Bee always strengthens their mouths by a thick edging of harder wax than that with which the cell itself is made. The engineer who plans girders, boilers, and other objects of a similar character, acts in precisely the same manner, and strengthens the comparatively thin and yielding plates by flanges or angle-irons.

Many inquirers have asked themselves how the Bee constructs the comb, and on what principle it proceeds. To this question there have been several answers, none of which appear to be satisfactory. One ingenious entomologist remarked, that when the Bee placed the claws of its forefeet against each other, the limbs embraced a hexagonal space, of which the thorax formed one side. Another, a very popular solution of the question, is that which may be called the "sculpture" theory.

The Bee that commences the task is supposed to work a lump of wax on the stick or bar which supports a comb, and then to excavate a circular hole on one side, the interior of the hole being shaped like a concave lens. Round this hole or basin the Bee then excavates six other holes of equal diameter, so that their edges nearly touch each other. It then cuts away the wax from each basin until the material is reduced to the requisite thinness, and thus obtains the hexagonal cell. In the meantime, another Bee is working in the same manner on the opposite side of the wax, taking care, however, to make the centre of its first basin correspond with the union of three basins on the opposite side. A similar system of sculpturing is carried on, so that at last a series of hexagonal basins is formed, from which rise the walls of the future cells.

There is an amount of plausibility about this theory, which is very attractive. It must, however, be remembered that the Bee is still supposed to execute problems which are as difficult as that which they are presumed to explain.

In the first place, the Bee must strike perfect circles from centres the distance of which from each other must be accurately adjusted. Again, these centres must be so placed that the centre of the circle sculptured on one side of the comb must be equidistant from the centres of the three adjacent circles on the other side—a problem of no easy accomplishment, even with the aid of rule and compass. Then, if the circles be not perfect, or their centres wrongly placed, or the hollow of one cut deeper than that of another, or the hemispherical form of the hollow not precisely just, the whole accuracy of the angles is destroyed, and the entire comb would be as distorted as the first essays of a young carpenter.

Then there is another explanation, which may be called the "equal pressure" theory. The Bee is, according to the advocates of this theory, supposed to construct all the cells of a cylindrical shape, and the cells are supposed to assume the hexagonal form by equal pressure in all directions. Every one knows that cylinders made of a yielding substance, always become hexagonal if pressed together, and a similar process was supposed to cause the hexagonal shape of the bee-cell.

There is another theory, which I believe to be entirely original, which is suggested by the well-known mathematician and crystallographer above mentioned. Mr. Mitchell writes to me as follows: "It may not be out of place to remark that the bee-cell forms a mould, as it were, of the natural form of a crystal. There is in nature a great variety of crystals, hexagonal prisms terminated by three planes, like the bee-cell. These have many different angles. But there is one form, called the rhombic dodecahedron, see fig. 6, very frequently found in natural crystals of the garnet, which has precisely the same angles as the bee-cell.

"Certain crystals split naturally into planes precisely like the lozenges which have already been described in giving the key to the structure of the bee-cell. May it not, therefore, be possible that wax, which is a crystallizable substance, cleaves in this particular direction, and does the Bee use this property in forming its cell? Though this vague conjecture should prove to be true, we shall not less admire the marvellous instinct which combines this fact with the structure of the cell."

It would, of course, be easy to fill many pages with the

account of the Hive Bee and its habits; but as this work is restricted to the habitations of animals, we can only look upon the Bee as a maker of social habitations. It will, however, be necessary to mention the material of which the comb is made.

The other hymenoptera obtain their materials from external sources. The hornet and wasp have recourse to trees and branches, and bear home in their mouths the bundles of woody fibres which they have gnawed away. The upholsterer and leaf-cutter Bees are indebted to the petals and leaves of various plants, and various wood-boring insects make their homes of the woody particles which they have nibbled away. The Bee, however, obtains her wax in a very different manner.

If the body of a worker Bee be carefully examined, on the under sides of the abdomen will be seen six little flaps, not unlike pockets, the covers of which can be easily raised with a pin or needle. Under these flaps is secreted the wax, which is produced in tiny scales or plates, and may be seen projecting from the flaps like little semilunar white lines. Plenty of food, quiet, and warmth are necessary for the production of wax, and as it is secreted very slowly, it is so valuable that the greatest economy is needed in its use. It is, indeed, a wonderful substance; soft enough when warm to be kneaded and to be spread like mortar, and hard enough when cold to bear the weight of brood and honey. Moreover, it is of a texture so close that the honey cannot soak through the delicate walls of the cells, as would soon be the case if the comb were made of woody fibre, like that of the hornet or wasp.

Indeed, it is a most remarkable fact that the Bee should be able to produce not only the honey, but the material with which is formed the treasury wherein the honey is stored. Honey itself is again scarcely less remarkable than wax. The Bee goes to certain flowers, inserts its hair-clad proboscis into their recesses, sweeps out the sweet juice, passes the laden proboscis through its jaws, scrapes off the liquid and swallows it. The juice then passes into a little receptacle just within the abdomen called the "honey-bag," which is apparently composed of an exceedingly delicate membrane, and seems to discharge no other office than that of a vessel in which the juice can be kept while the Bee is at work.

As soon as the honey-bag is filled, the Bee flies back to the hive and disgorges the juice into one of the cells. But, during that short sojourn in the insect, the juice has undergone a change, and been converted into honey, a substance which is quite unlike that from which it was formed, and which has an odour and flavour peculiarly its own. How this change is wrought is at present unknown, for the little bag in which the transformation is made is composed of a membrane that seems incapable of exerting any influence upon the substance contained within it.

All food that is eaten by the Bee passes through the honey-bag, which is closely analogous to the crop of a bird, and it would seem that the honey ought rather to pass into the stomach than be disgorged at the will of the insect. However, it is well known that many birds feed their young by disgorging food, and the Bee is enabled to perform the same operation by means of a little valve which leads from the honey-bag into the stomach, and is plainly perceptible even with the unassisted eye. Under ordinary circumstances the valve just allows the food to pass gently and gradually into the stomach; but the violent effort, which is made in ejecting the food, closes the valve, and only allows the honey to flow upwards through the mouth.

The office of the worker and drone cells is two-fold—first, to act as nurseries for the insects while passing through their preliminary stages, and next to serve as repositories for food, whether liquid or solid. The egg of the Queen-Bee is placed nearly at the bottom of the cell, exactly on the angle where the points of the lozenges meet. It is soon hatched into a little white grub, which is assiduously fed by the nurses, and grows with wonderful rapidity. As soon as it has eaten its last larval meal, it spins a silken cover over the cell, and remains there until it has become a perfect insect. It then bites its way out, and after a day or so devoted to hardening and strengthening its limbs, it leaves the hive and joins in the labours of the community.

No sooner is the Bee fairly out of its waxen nursery, than the workers clear out the cell, and prepare it for the reception of honey. As soon as the cell is filled, the Bees close up the entrance with a waxen door, which is air-tight, and serves to preserve the honey in proper condition. Those who wish to eat honey in its pure state should always purchase it in the comb.

If it be stored in pots, however well they may be sealed, it always crystallises, and in that state is injurious to digestion. Moreover, it is so extensively adulterated, that a pot of really pure honey is not readily obtained.

Besides the honey, "bee-bread" is placed in the cells. This is a compound of honey and the pollen of flowers, and is chiefly used as food for the young grubs. We may often see the Bees hastening home with a load of yellow pollen on each of the hinder pair of legs, and this pollen is destined to be made into bee-bread.

Such, then, is a brief outline of the wonderful social habitation which is made by the Hive Bee.

We now come to an insect which is as well known by name as the bee, though not so familiar to our eyes. This is the common Hornet (*Vespa crabro*), which is tolerably plentiful in many parts of England, but seems to be almost absent from others.

The nest of the Hornet is much like that of the wasp, except that it is proportionately larger, and is almost invariably built in hollow trees, deserted outhouses, and places of a similar description. Whenever the Hornet takes up its residence in an inhabited house, as is sometimes the case, the inmates are sure to be in arms against the insect, and with good reason. The Hornet is much larger than the wasp, and its sting is proportionately venomous. It is popularly said that three Hornets can kill a man; and although in such a case the sufferer must previously have been in bad health, the poisonous properties of the Hornet are sufficiently virulent to render such a saying popular.

Moreover, the Hornet is an irascible insect, and given to assault those whom it fancies are approaching its nest with evil intentions. It is not pleasant to be chased by wasps, but to be chased by Hornets is still less agreeable, as I can personally testify. They are so persevering in their attacks that they will follow a man for a wonderfully long distance, and if they be struck away over and over again, they will return to the charge as soon as they recover from the shock. There is a deep ominous menace in their hum, which speaks volumes to those who have some acquaintance with the language of insects; and no one who has once been chased by these insects will willingly run the same risk again.

Mr. S Stone, whose interesting letter upon the wasp has already been mentioned, tells me that he has been successful in breeding Hornets as well as wasps, and forcing them to build nests much more beautiful than they would have made if they had been at liberty.

HORNET.

One nest, when of moderate size, was removed from the head of a tree, and placed in a large glazed box similar to those which have been mentioned in connexion with the wasp. Within the box the Hornets continued their labours, and a most beautiful nest was produced, symmetrical in shape, and variegated with wonderfully rich colours. "Such a nest as that," writes Mr. Stone, "is not produced by Hornets in a general way. They do not trouble themselves to form much of a covering, especially when a small cavity in the head of a tree is selected, which is often the case. The walls of the chamber they consider a sufficient protection for the combs.

"If you expect them to form a substantial covering, the combs must be so placed as to have ample space around them, and if you expect them to fabricate a covering of great beauty, you must select the richest coloured woods, and such as form the most striking contrasts, and place them so that the insects shall be induced, nay, almost compelled, to use them in the construction of their nest. This is exactly what I did with reference to the nest in question."

Knowing from experience the difficulty of assaulting a Hornet's nest, I asked Mr. Stone how he performed the task, and was told that his chief reliance was placed on chloroform. Approaching very cautiously to the nest, he twists some cotton wool round the end of a stick, soaks it in chloroform, and pushes it into the aperture. A mighty buzzing immediately arises, but is soon silenced by the chloroform, and as soon as this result has happened, mallet, chisel, and saw are at work, until the renewed buzzing tells that the warlike insects are recovering their senses, and will soon be able to use their formidable weapons. The chloroform is then re-applied until they are quieted, and the tools are again taken up.

The extrication of a nest from a hollow tree is necessarily a long and tedious process, on account of the frequent interruptions. Even if the insects did not interfere with the work, the labour of cutting a nest out of a tree is much harder than could be imagined by those who have not tried it.

Moreover, the habits of Hornets are not quite like those of the wasps. At night, all the wasps retire into their nest, and in the dead of night the nest may be approached with perfect safety, the last stragglers having come home. Hornets are apt to continue their work through the greater part of the night, and if the moon be up, they are nearly sure to do so. Therefore, the nest-hunters are obliged to detail one of their party as a sentinel, whose sole business it is to watch for the Hornets that come dropping in at intervals, laden with building materials or food, and that would at once dash at the intruders upon their domains. Fortunately, the light from the lanterns seems to blind them, and they can be struck down as they fly to and fro in the glare.

The nest that has just been mentioned, was rather deeply imbedded in the tree, and cost no less than six hours of continuous labour, the week of excavation having been begun at eight P.M. and the nest extracted at two A.M. on the following morning.

In the illustration is seen a portion of a lately begun nest, much reduced in size, as may be conjectured from the dimensions of the insects that are crawling upon it. As the arrangement of the combs is identical with that of the wasp-nest, the interior is not disclosed. Another reason for showing the exterior of the nest is, that the reader might see how the Hornet forms the paper-like cover, and the manner in which the insects can enter at different parts, instead of having but a single entrance, as is the case with several hymenopterous nests which have been mentioned.

IN many parts of Brazil there may be seen the social nests of certain hymenopterous insects, which are very aptly termed *Synœca,* this name being derived from two Greek words, which signify sociality.

The nests of these insects have some resemblance to those of certain Polybiæ, which have already been described. They are, however, of much greater size, and as they are rather heavy, are affixed to tolerably strong branches. One such nest, which is now in the British Museum, has been built upon a post, nearly encircling it above, and sloping off to a rounded point, nearly two feet below the highest portion. Another is fixed to a rather stout, straight, and upright branch. The nests are dark brown in colour, and as they are fixed to objects of a similar hue, are very inconspicuous. The insect which makes this nest is nearly as large as the English hornet.

The walls of the nests made by insects of this genus are very thin and fragile, not unlike those of the structure built by Polybia sedula. In one nest the cover is remarkably elegant, being shaped like the half of a melon cut longitudinally, and moulded into ribs which run transversely across the nest, and have a gentle and regular curve. These ribs project about a quarter of an inch, and are nearly half an inch wide, and are as round and regular as if they were produced by cords wound upon the combs.

The insect which made this nest is of a deep steel-blue colour, looking nearly black in a dim light. The head is rather large, and the abdomen is rounded and small, being connected with the thorax by a footstalk of moderate length.

THE two remarkable nests which are figured in the large illustration come from different parts of the world, but, as they are similar in many respects, they are placed close to each other.

The long, flask-like nest was brought from Mexico, by Owen Rees, Esq. in 1834. Even before it was opened, its structure was evidently full of interest. The colour is dull white, not unlike parchment, and the texture of the materials is nearly as hard, stiff and close as that substance. When placed under the microscope, it is seen to be composed of a vast number of shining threads, crossing and recrossing each other in every direction, and producing a material like very thin, but stiff felt.

It was suspended to a branch, but could not swing in the wind, because a twig descended into the neck and prevented any lateral motion. At the bottom of the nest there is a small and nearly circular aperture, through which the insects are enabled to make their exit and entrance. The length of the nest is about eight inches.

So much for the exterior. On opening the nest, however, a most singular state of things was discovered. A great number of pupæ, evidently those of some butterfly, were suspended by their tails to the walls and to the twig which runs down the nest. In this nest they were about one hundred in number, and they were hung to the whole of the upper part of the nest, but without any particular order.

On seeing this nest, an entomologist naturally asks how and when the insects made it. That they did not form it of a small size and then add to its dimensions in order to suit their growth, is evident from the fact that no trace of enlargement can be perceived. It is most probable that as in the case of the processionary moth, the caterpillars spin their silken home when they are three parts grown, and in consequence have but a short time to spend in it before they pass into the quiescent pupal form.

It is evident that the insects make their escape from their pensile home as soon as they have broken out of the pupal skin,

NEST OF SOCIAL LEPIDOPTERA. *See* page 449.

because the aperture is so small that they could not possibly pass through it when their wings were thoroughly expanded and dried. Of what form and colour these wings might be, was for a long time a mystery. Mr. Westwood, who first opened a nest, carefully dissected some of the pupæ, and by cautiously softening the withered membranes in warm water, succeeded in spreading out the wings sufficiently to learn the general form of the nervures and the shape of the "cells," as the spaces between the nervures are named.

Specimens of the perfect insect have now been obtained, and are seen to be butterflies closely resembling in shape the lovely heliconidæ, which are so plentiful in Southern America, but of very simple colours, the general hue being blackish brown, diversified by a broad, but indistinctly marked white band across the wings. Examples of the nest have lately been sent to Vienna, but any one who wishes to see the specimen from which the above sketch and description were taken, may do so by visiting the museum at Oxford, where the perfect butterflies may also be seen. The scientific name of the butterfly is *Eucheira socialis.*

On the upper part of the same illustration may be seen a curious object, that looks something like a flattened pincushion fastened to the branches. This is the nest of a social insect, and is, I believe, an unique specimen. It was brought from Tropical Africa by Vernon Wollaston, Esq. and is so remarkable as to deserve a detailed description.

In length it measures eight inches, and in width five and a half inches, its depth being about three inches. The aspect of the exterior gives but little promise of the exceeding strength of the structure, which is as hard and elastic as the side of a silk hat, rebounding when pressed in precisely the same manner. When cut, this covering is seen to be double, the outer case being very thin, and formed of orange-brown silken threads, and the inner being made of many successive layers of dark brown silk, so that it looks very like undressed leather.

The most extraordinary part of the nest, however, is the provision which is made for the exit of the inmates. Set upon different parts of the nest are thirteen or fourteen little conical protuberances, which do not project very far from the general surface, and are quite inconspicuous. On examination, these prominences are seen to be composed of stiff silken threads, which converge to a point, precisely like those which guard the entrance of the emperor moth's cocoon, so that any inhabitant can crawl out, but no enemy can crawl in.

This nest, like the preceding, may be seen in the museum at Oxford.

SMALL ERMINE MOTH.

THERE is a very pretty, very interesting, and very destructive insect, called by entomologists the SMALL ERMINE MOTH (*Yponomeuta padella*), which is very plentiful in this country, and by gardeners is thought to be much too plentiful. It can easily be recognised by its long narrow wings, the upper pair of which are soft silvery, or satiny white, spotted with black, and the lower pair dark brown. The expanse of the spread wings is about three quarters of an inch.

In its winged and pupal states the insect is perfectly harmless, but in its larval condition it becomes a terrible pest. Most caterpillars wage war singly on the foliage, and though they do much damage, their ravages are conducted in a desultory manner. The Small Ermines, however, band themselves together in hosts, and march like disciplined armies to the attack, invading a district and completely devastating it before they proceed to another.

They live in large tents, placed among the branches of some tree, and composed of silken threads, which are loosely crossed and recrossed in various directions. From this centre the caterpillars issue in vast numbers, each individual spinning a strong silken thread as it proceeds, which acts as a guide to the nest, just as the fabled clue led through the intricacies of Rosamond's bower. When once these caterpillars have taken possession of a tree, they are sure to strip it of its leaves as completely as if the foliage had been plucked by hand. It is a very curious sight to watch the systematic manner in which these troublesome insects set about their work, how they send out pioneers which lead the way to new branches, either by crawling up to them or by lowering themselves to them by means of their silken life-lines, and how soon they are followed by their ever hungry companions.

Perhaps the reader may wonder why the little birds do not eat these caterpillars. When they have nearly stripped the branch, they are very conspicuous, especially as they make their way from bough to bough along their silken bridges. Indeed, a proprietor of a garden that was much damaged by this moth did once mention the immunity of the caterpillars as a proof that any tenderness to small birds was misplaced, saying that if the sparrows were half as insectivorous as I mentioned, they would long ago have eaten all the caterpillars.

Now, at the first glance, there seemed to be some reason in this remark; but a short look at one of the damaged trees explained the reason why the sparrows did not eat the caterpillars. The birds literally dared not approach the insects; for the silken threads which traversed the branches in all directions were an effectual barrier, striking against the wings and terrifying the poor birds. We all know that a few threads of fine cotton passed from bough to bough of a gooseberry-bush will deter any little bird from settling on it; and, in the same manner, the silken threads of the caterpillars deter the birds from settling on the branches. These threads are very elastic, and of marvellous strength, considering their tenacity, producing most uncomfortable sensations when they come across the face, and being nearly as strong as the fibres spun by the common silkworm.

The caterpillar which works all this damage is rather slender, and is covered with black dots along the back.

ANOTHER well-known British insect which constructs social habitations is the GOLD-TAILED MOTH (*Porthesia chrysorrhœa*), a familiar and beautiful insect, with wings of soft downy plumage, and snowy-white in colour, and a tuft of yellow hair at the end of the tail. The perfect insect may often be seen sticking on the trunks of trees in gardens, waiting until the evening, when it will fly off to its labours.

When the moth has laid its eggs, it plucks off the beautiful yellow tuft at the end of the tail, and with it forms a roof over the pile of eggs, laying the hairs so artificially as to make a perfect thatch. When the larvæ are hatched, they retain their sociability, and spin for themselves a common domicile. This house is very remarkable. Viewed on the exterior, it is seen to be a bag-like structure of whitish silk, rather strong and tough, but very yielding.

One of these nests, which I found in Wiltshire, is now before me. It was found in a hedge, about two feet from the ground, and is rather a complicated structure. The scaffolding, so to speak, of the nest is formed by a horizontal spray of three small twigs, and it is strengthened by the long hedge-grass which crossed the spray. Seeds of different kinds are woven into the walls, so that a comparatively small portion of the silk is exposed to view.

When cut open, it shows a singularly beautiful structure within. There are several sheets of silken tissue, each becoming more delicate, and the innermost being white, shining like satin; whereas the outer covering is dull-white, and very tough, clinging to the scissors so that a straight cut is almost impossible. Delicate walls divide the interior into several compartments, in all of which are evidences that the caterpillars must have resided for some time. The reason why the creatures make this nest is, that they are hatched towards the end of summer, and in consequence are forced to pass the winter in the larval condition, so that some warm residence is needful for them. It is well known that air is a very bad conductor of heat, and, in consequence, the successive sheets of silk which cover the nest, and which inclose layers of air between them, form a protection which is far warmer than would be obtained by a solid mass of silk measuring twice the thickness of the three walls, together with their intervening spaces.

THERE is an allied insect, popularly called the BROWN-TAILED MOTH (*Porthesia auriflua*), which spins a social nest that in many respects resembles that of the Gold-tailed Moth. The nest, however, is scarcely so elegant, nor is the silken web so beautifully delicate. Much, however, depends upon surrounding conditions, such as the disposition of the twig on which the nest is placed, and the presence or absence of leaves, whether those of the tree or of other plants that happen to grow in close proximity.

THESE nests are very firmly constructed, and the walls are solid, as is needful for insects which are obliged to pass the winter within them. There are, however, many caterpillars which live socially, and which spin a common habitation, but which leave it before the cold weather comes on, and, in consequence, do not need such thick walls. Any hedgerows where nettles are found will supply numerous examples of such nests, made by the curious caterpillars which afterwards assume the lovely and familiar forms of the PEACOCK and SMALL TORTOISE-SHELL BUTTERFLIES (*Vanessa Io* and *Vanessa urticæ*). Great black masses of these caterpillars may be seen upon the nettles, and, on examining them closely, they will be seen to reside within a common home, made of tough silken threads, very loosely spun, and forming a kind of net, with long and irregular meshes.

CHAPTER XXIV.

SOCIAL INSECTS CONTINUED.

A curious Ant from India (*Myrmica Kirbii*)—Locality of its nest—Description of the nest, its material and mode of structure—A nocturnal misadventure—The DRIVER ANT of Africa—Description of the insect—Reason for its name—Its general habits—Destructive powers of the Driver Ant—How the insects devour meat and convey it home—How they kill snakes—Native legend of the python—Their mode of march—Fatal effects of the sunbeams—An extemporised arch—Method of escaping from floods—Site of their habitation—Modes of destroying them—Living ladders and their structure—Method of crossing streams—Tenacity of life—A decapitated Ant—Mode of biting—Description of the insect—Curious nest of a Brazilian Wasp—Weight of the nest and method of attachment—Variety of Polistes nest—*Polistes aterrimus* and its singular nest—Beautiful structure of an unknown Polistes.

ALTHOUGH several species of Ants have been mentioned under the title of burrowing insects, there are many which possess very interesting habits, and which may here take their place among the creatures which build social habitations. Among them is a curious insect inhabiting India, and discovered by Colonel Sykes, the well-known naturalist, who called it *Myrmica Kirbii.*

This insect forms its nest on the branches of trees and shrubs, and Colonel Sykes mentions that he has found their curious habitation on the branches of the Kurwund shrub, *Carissa Carandas,* and on the Mango-tree, *Mangifera Indica.*

The nests are more or less spherical, and are about as large as an ordinary foot-ball. The material of which they are made is cow-dung, which is spread in flakes in a manner that reminds the observer of the outside cover of a wasp's nest. The flakes are placed upon each other like the tiles of a house, so that although the insects can creep into the nest beneath the flakes, no water can enter. On the summit of the nest is one very large flake, that acts as a general roof to the structure.

Within the nest are placed a number of cells made of the same material as the exterior, and in them may be found insects in every state of development, eggs in one, larvæ in another, and pupæ in a third. No provision seems to be laid up within the nest, so that the inhabitants must depend on their daily excursions for their food.

When Colonel Sykes brought home the first nest he discovered, he hung it to the tent-pole, preparatory to examining it in the morning. "In the night the men were awakened by repeated punctures and general irritation of the skin, but the darkness prevented them from discovering their tormentors, and they continued to toss and tumble in their beds for some hours in no very complacent state of mind. At last they got up, dressed themselves, and abandoned the tent; but the evil was rather aggravated than abated, as parts of their persons which had previously escaped had now their share of suffering. At daylight they discovered to their consternation that they were covered with minute ants, which had filled their pantaloons, penetrated the sleeves of their coats and every other part of their habiliments. On inspecting the tent, they found the interior teeming with multitudes of little angry beings, in busy progress, seeking to resent the outrage which had been committed on the community by the removal of their abode."

The insects are extremely small, barely one fifth of an inch in length, and are reddish in colour.

PERHAPS one of the most terrible of insects is that which is appropriately called the DRIVER ANT of Western Africa (*Anomma arcens*).

This insect is a truly remarkable creature. Although it is to be found in vast numbers, it has never been found in the winged condition, and neither the male nor the female have as yet been discovered. The workers are uniform in colour, but exceedingly variable in size. Their hue is deep brownish black, and their length varies from half an inch to one line, so that the largest workers nearly equal the common earwig, while the smallest are no larger than the familiar red ant of our gardens. In the British Museum are specimens of the workers, which form a regular gradation of size, from the largest to the smallest.

They are called Driver Ants, because they drive before them every living creature. There is not an animal that can withstand the Driver Ants. In their march, they carry destruction before them, and every beast knows instinctively that it must not cross their track. They have been known to destroy even the agile monkey, when their swarming host had once made a lodgment on its body, and when they enter a pigstye, they soon kill the imprisoned inhabitants, whose tough hides cannot protect them from the teeth of the Driver Ants. Fowls they destroy in numbers, killing in a single night all the inhabitants of the hen-roost, and having destroyed them, have a curious method of devouring them.

DRIVER ANTS.

The Rev. Dr. Savage, who has experimented upon these formidable insects, killed a fowl and gave it to the Ants. At first, they did not seem to pay much attention to it, but he soon found that they were in reality making their preparations. Large parties of the insects were detached for the purpose of preparing a road, and worked with the assiduity which seems to be a characteristic of these energetic insects. Numbers of them were employed in smoothing the road to the nest by removing every obstacle out of the way, until by degrees a tolerably level road was obtained. The Ants are possessed of strength which seems gigantic when compared with their size, carrying away sticks four or five times as large as themselves, and never failing to pounce upon any grub or insect that might happen to be lurking beneath their shelter. They always carried such burdens longitudinally, grasping them with their jaws and legs, and passing the load under the body. Some of these roads are more than two hundred yards in length.

Meanwhile, the other Ants were busy with the fowl. Beginning at the base of the beak, they contrived to pull out the feathers one by one, until they stripped it regularly backwards, working over the head, along the neck, and so on to the body. This was evidently a very hard task, as the insects did not possess sufficient strength to pull out the feathers by main force, and were consequently obliged to grub them up laboriously by the roots. The next business was to pull the bird to pieces, and at this work they were left. Unfortunately the experiment was spoiled by the natives, who stole the fowl, thinking that the Ants had eaten so many of their poultry that they were justified in retaliation. Others chose to excuse themselves by saying that they thought the fowl to be a fetish offering to the Ants, and accordingly took it away from them.

The large iguana lizards fall victims to the Driver Ants, and so do all reptiles, not excluding snakes. It seems, from the personal observations of Dr. Savage, that the Ants commence their attack on the snake by biting its eyes, and so blinding the poor reptile, which only flounders and writhes helplessly on one spot, instead of gliding away to a distance.

It is said by the natives, that when the great python has crushed its prey in its terrible folds, it does not devour it at once, but makes a large circuit, at least a mile in diameter, in order to see whether an army of Driver Ants is on the march. If so, it glides off, and abandons its prey, which will soon be devoured by the Ants; but if the ground is clear, it returns to the crushed animal, swallows it, and gives itself to repose until the process of digestion be completed. Whether this assertion be true or not, Dr. Savage cannot say; but it is here given in order to show the extreme awe in which the natives hold the Driver Ants.

So completely is the dread of them on every living creature, that on their approach whole villages are deserted, and in extreme cases the entire population is forced to take to the rivers, knowing that the insects will not enter water unless obliged to do so; although on occasions they do not hesitate to commit themselves to the waves, as will presently be seen.

The order of their marching is very curious, and is well described by Dr. Savage:—

"Their sallies are made in cloudy days, and in the night, chiefly in the latter. This is owing to the uncongenial influence of the sun, an exposure to the direct rays of which, especially when the power is increased by reflection, is almost instantaneously fatal. If they should be detained abroad till late in the morning of a sunny day by the quantity of their prey, they will construct arches over their path, of dirt agglutinated by a fluid excreted from their mouth. If their way should run under thick grass, sticks, &c., affording sufficient shelter, the arch is dispensed with; if not, so much dirt is added as is necessary to eke out the arch in connexion with them. In the rainy season, or in a succession of cloudy days, the arch is seldom visible; their path, however, is very distinct, presenting a beaten appearance, and freedom from everything moveable.

"They are evidently economists in time and labour; for if a crevice, fissure in the ground, passage under stones, &c., come in their way, they will adopt them as a substitute for the arch.

"In cloudy days, when on their predatory excursions, or migrating, an arch for the protection of the workers is constructed of the bodies of their largest class. Their widely-extended jaws, long slender limbs, and projecting antennæ, intertwining, form a sort of network, that seems to answer well their object. Whenever an alarm is given, the arch is instantly broken, and the ants, joining others of the same class on the outside of the line, who seem to be acting as commanders, guides, and scouts, run about in a furious manner, in pursuit of the enemy. If the alarm should prove to be without foundation, the victory won, or danger passed, the arch is quickly renewed, and the main column marches forward as before, in all the order of an intellectual military discipline."

Sometimes, as is usual in tropical countries, the rain descends like a flood, converting in a few minutes whole tracts of country into a temporary lake. The dwellings of the Driver Ant are immediately deluged, and, but for a remarkable instinct which is implanted in the insects, most of the Ants, and all the future brood, would perish. As soon as the water encroaches upon their premises, they run together and agglomerate themselves into balls, the weakest (or the "women and children," as the natives call them) being in the middle, and the large and powerful insects on the outside. These balls are much lighter than water, and consequently float on the surface, until the floods retire and the insects can resume their place on dry land.

The size of the ant-balls is various; but they are, on an average, as large as a full-sized cricket-ball. One of these curious balls was cleverly caught in a handkerchief, put in a vessel, and sent to Mr. F. Smith, of the British Museum, who has kindly presented me with several specimens of the insect.

When a colony of these insects has been established neaı a house, the inhabitants naturally endeavour to destroy it. The habitation is very simple and artless, and generally consists of a mere hole in a rock or bank, in which the creatures assemble. They are very fond of usurping the sepulchres of the dead, which are usually excavated in the sides of hills, and are about eighteen inches in depth.

The natives generally try to destroy the colony by heaping dry leaves of the palm upon the dwelling, and setting fire to the heap. When this plan was tried, it was found to be very unsatisfactory; for the greater mass of the insects contrived to make their escape, and were found upon neighbouring trees, clinging in heavy bunches and long festoons, which connected one branch with another, and formed ladders over which the insects could pass. These festoons were made in a very curious manner.

First, a single Ant clung tightly to a branch, and then a second insect crawled cautiously down its suspended body, and hung to its long, outstretched limbs. Others followed in rapid succession, until they had formed a complete chain of Ants, which swung about in the wind. One of the largest workers then took its stand immediately below the chain, held firmly to the branch with its hind limbs, and dexterously caught with its fore-legs the end of the living chain as it swung past. The ladder was thus completed, and fixed ready for the transit of insects; and, in a similar way, the whole tree was covered with festoons of Ants, until it was blackened with their sable bodies.

They can even cross streams by means of these ladders. Crawling to the end of a bough which overhangs the water, they form themselves into a living chain, and add to its length until the lowermost reaches the water. The long, wide-spread limbs of the insect can sustain it upon the water, especially when aided by its hold on the suspended comrade above.

Ant after Ant pushes forward, and the floating portion of the chain is thus lengthened, until the free end is swept by the stream against the opposite bank. The Ant which forms the extremity of the chain then clings to a stick, stone, or root, and grasps it so firmly, that the chain is held tightly, and the Ants can pass over their companions as over a suspension bridge. In the illustration a column of Driver Ants is shown on the march. The vanguard of the column has crossed the stream by means of the living ladder, which is seen suspended from a branch, and extended across the water. The fragile tube which they build is also shown, and a few of the larger architects are drawn of the natural size. The smaller specimens will not emerge from the tunnel.

There is a species of Ant in Ceylon which makes living bridges in precisely the same manner as the Driver Ant. In Mr. E. Sullivan's "Bungalow and the Tent" there is the following passage:—"I have seen Ants form a bridge from one stick to another. I even saw one leave his companions, who were clustered at the end of a stick, unable to reach another at a short distance, make a considerable circuit, ascend the stick they were aiming at from another direction, and by stretching out his body as far as possible, enable the pioneer of the main body to reach him, and thus complete the chain of communication, by which the rest immediately crossed. It would be difficult to prove that this was not reason."

Finding that the comparatively gradual action of fire permitted the active insects to escape before the heat finally reached them, Dr. Savage waited until they had settled in their home, and then poured upon them a few gallons of boiling water, which was instantaneously fatal. As for cold water, they seem to care little for it; having been immersed for twelve hours, and although they were apparently dead when removed, yet they soon recovered themselves, and ran about as lively as ever. Their tenacity of life is indeed wonderful, and injuries which would immediately kill almost any creature seem to have no immediate effect upon their vigour. Another fact, illustrating their tenacity of life, may here be stated.

"The head of one of the largest class, when dissevered from the body, grasped the finger of an attendant so furiously, as to cause an immediate flow of blood. It was left in a glass tumbler from three P. M. till the next morning at eight o'clock, when the finger was again applied, and apparently as severe a wound as before inflicted. Another individual of the same class was decapitated at seven A. M., and at half-past nine next morning, twenty-six and a half hours from the time of decapitation, a piece of newspaper was held between the jaws, which it grasped and retained with considerable force.

"I then applied the small finger of my right hand, which it bit severely; indeed, so powerful was the grasp, that the point of the mandibles met beneath the cuticle. It then partly withdrew one mandible, and, pointing it more perpendicularly, penetrated deeper than the other, and thus at every stroke giving to the mandible a direction more vertical, wounding and cutting wider and deeper, precisely in the manner of the insect in possession of all its parts and powers. The sensation at each thrust was like that of a pin, and equally painful; and when the mandibles were withdrawn, the blood flowed as freely. The head continued to give signs of life for more than thirty-six hours after decapitation. The body to which it belonged lived still longer, or more than forty-eight hours."

It is a very remarkable fact that the insect should be so tenacious of life under circumstances that would be instantly fatal to most creatures, and yet should die suddenly under conditions in which many insects live and thrive. The reader will remember that the direct action of the sun's rays will kill the Driver Ant in less than two minutes, and yet there are ants of the same country which run about freely in the blazing sunshine, traversing with impunity the heated ground, which blisters the bare hand, and being able to secrete abundant stores of the liquid which they use in making their habitation.

In Dr. Livingstone's well-known work, there are several interesting accounts of ants and their habits, and one anecdote bears so aptly on the subject, that I give it in the writer's own words.

After describing the terrible drought at Chonuane, when the river Kolobay ran dry and the fish perished, when the crocodile himself was stranded and died, and the native trees could not hold up their leaves, he proceeds as follows:—"In the midst of this dreary drought, it was wonderful to see those tiny creatures, the Ants, running about with their accustomed vivacity. I put the bulb of a thermometer three inches under the soil in the sun at mid-day, and found the mercury to stand at 132° to 134°; and if certain beetles were placed on the surface, they only ran about a few seconds and expired.

"But this boiling heat only augmented the activity of the long-legged Black Ants; they never tire; their organs of motion seem endowed with the same power as is ascribed by physiologists to the muscles of the human heart, by which that part of the frame never becomes fatigued, and which may be imparted to all our organs in that higher sphere to which we fondly hope to rise.

"Where do these Ants get their moisture? Our house was built on a hard, ferruginous conglomerate, in order to be out of the way of the White Ant, but they came despite the precaution; and not only were they in this sultry weather able individually to moisten soil to the consistency of mortar for the formation of galleries, which in their way of working is done by night (so that they are screened from the observation of birds by day in passing and repassing towards any vegetable matter they may wish to devour), but, when their inner chambers were laid open, these were also surprisingly humid; yet there was no dew, and the house being placed on a rock, they could have no subterranean passage to the bed of the river, which ran about three hundred yards below the hill. Can it be that they have the power of combining the oxygen and hydrogen of their vegetable food by vital force as to form water?"

In corroboration of this opinion, Dr. Livingstone mentions an insect found in Angola, and which is allied to the common cuckoo-spit (*Aphrophora spumaria*) of England, which has the property of pouring out great quantities of water, so that a group of seven or eight insects will produce three or four pints of water in the course of the night. After stating that he believes the water to be produced, not from the sap of the tree, but from the atmosphere, he proceeds as follows :—

"Finding a colony of these insects busily distilling on a branch of the *Ricinus communis*, or castor-oil plant, I denuded about twenty inches of the bark on the tree-side of the insects, and scraped away the inner bark, so as to destroy all the ascending vessels. I also cut a hole in the side of the branch, reaching to the middle, and then cut out the pith and internal vessels. The distillation was then going on at the rate of one drop in each sixty-seven seconds, or about two ounces five and a half drams in twenty-four hours. Next morning the distillation, so far from being affected by the attempt to stop the supplies, supposing they had come up through the branch from the tree, was increased to a drop every five seconds, or twelve drops per minute, making one pint in every twenty-four hours.

"I then cut the branch so much that during the day it broke; but they still went on at the rate of a drop every five seconds, while another colony on a branch of the same tree gave a drop every seventeen seconds only, or at the rate of about ten ounces four and one-fifth drams in every twenty-four hours. I finally cut off the branch; but this was too much for their patience, for they immediately decamped, as insects will do from either a dead branch or a dead animal. The presence of greater moisture in the air increased the power of these distillers; the period of greatest activity was in the morning, when the air and everything else was charged with dew."

Three species of Driver Ant are known, namely, the common species, which has already been described, *Anomma Burmeisteri*, and a smaller species, *Anomma rubella*.

The two first insects are deep, shining black, and resemble each other so closely that an unpractised eye could not distinguish between them, while the last may be easily known by its brownish red hue.

The specimens which have already been mentioned are now before me, and curious beings they are. The largest are black, with a slight tinge of red, and have an enormous head, almost equalling one-third of the entire length. It is deep and wide as well as long, as indeed is necessary for the attachment of the muscles which move the enormous jaws. These weapons are sharply curved, and when closed, they cross each other, so that when the insect has fairly fixed itself, its hold cannot be loosened unless the jaws are opened. It is useless, therefore, to kill the ant, for its head will retain its grasp in death as well as in life. Beside the sharp points of the mandibles, they are further armed with a central tooth, which is so formed that when the mandibles are quite closed, and the points crossed to the utmost, the tips of the central teeth meet and form another means of grasping.

There is no vestige of external eyes, and even the half-inch power of the microscope fails to show the slightest indication of visual organs. As, however, the horny coat of the head is sufficiently translucent to permit the articulation of the jaws to be seen through it, when a very powerful light is thrown upon the head and the eyes of the observer are well sheltered, it is possible that the insect may have some sense of sight, and at all events will be able to distinguish light from darkness.

The limbs are of a paler red than the body, and although they are slender and delicate, their grasping power is very great. Two of my specimens had grasped each other's limbs with such force that they could not be separated without damaging the insect, and it was not until the rigid joints were softened with moisture, and then with the aid of a magnifier, that I succeeded in disengaging the insects.

The smaller specimens are not so black as the larger, nor are their jaws so proportionately large, but they are still formidable insects, if not from their individual size, yet from their collective numbers and their reckless courage, which urges them to attack anything that opposes them. Fire will frighten almost any creature, but it has no terrors for the Driver Ant, which will dash at a glowing coal, fix its jaws in the burning mass, and straightway shrivel up in the heat.

In the collection of the British Museum may be seen a very remarkable nest, which is made by some species of wasp at present unknown.

The material of which it is formed is mud, or clay, which is worked by the insect until it has attained a wonderful tenacity and strength, and is rendered so plastic as to be worked nearly as neatly as the waxen bee-cell. It is of rather a large size, measuring about thirteen inches in length, by nine in width, and filled with combs. Unfortunately, in its passage to this country, it was broken and much damaged, but the fragments were col-

MUD WASP.

lected and skilfully put together by Mr. F. Smith, who has succeeded in restoring the nest to its original shape, with the exception of an aperture through which the interior of the nest may be seen.

The accident was in so far an advantage, that it gave opportunities of studying the construction of a nest which is at present unique, and which the officers of the Museum might be chary of cutting open, particularly as its materials are so brittle. The walls of the nest are remarkably hard and solid, but extremely variable in thickness, some parts being nearly three times as strong as others. The upper portions of the nest are the thickest, the reason for which is evident on inspecting the specimen.

The nest was found in a Guianan forest, near the river Berbice, suspended to a branch, which passed through a hole in the solid wall of the nest. In the actual specimen, the branch is wanting; but in the illustration it has been restored, in order to show the manner in which the winged artificers suspended their wonderful home. As is always the case with pensile nests, the foundation is laid at the top, thus carrying out Dean Swift's suggestion for a new patent in architecture. A large quantity of clay is worked round the chosen branch, and made very strong, in order to sustain the heavy weight which will be suspended from it. This clay foundation is wonderfully hard, though very brittle, this latter quality being probably due to the long residence in a room which is always kept warm and dry by artificial means. In the open air, and in the ever-damp, though hot atmosphere of tropical America, the clay would probably be much tougher, without losing the necessary hardness.

The combs are not flat, like those of an ordinary wasp-nest, but are very much curved, so that when the nest is laid open they almost follow the curve of the walls. This peculiar form of the comb is shown in the illustration. The cells are not very large, scarcely equalling the worker cells of the common burrowing wasp of England.

One of the most remarkable points in the construction of this nest is the entrance. In pensile nests, the insect usually forms the opening below, so that it may be sheltered from the wind and rain. Moreover, it is usually of small dimensions, evidently in order to prevent the inroads of parasitic insects and other foes, and to give the sentinels a small gateway to defend. But the particular Wasp which built this remarkable nest seems to have set every rule at defiance, and to have shown an entire contempt of foes and indifference to rain.

As may be seen by reference to the illustration, the entrance is extremely long, though not wide, and extends through nearly the length of the nest, so that the edges of the combs can be seen by looking into the aperture. The edges of the entrance are rounded, so that the outer edge is wider than the inner; but it is still sufficiently wide to allow the little finger of a man's hand to be passed into the interior; while its length is so great, that forty or fifty insects might enter or leave the nest together.

THE remarkable fact has already been mentioned, that two species of Wasp will inhabit the same nest, and amicably work at the same edifice. Entomologists have long been aware that two species of Ant will dwell in the same nest, and live upon friendly terms, although the association of the working part of the community is not voluntary, but compulsory.

The Ant which employs enforced labour is called the AMAZON ANT (*Polyergus rufescens*), and is tolerably common on the Continent. This insect is not furnished with jaws which are capable of performing the work that usually falls to the lot of the neuters; but the same length and sharpness of the mandibles which unfit the insect for work, render it eminently capable of warfare. When, therefore, a colony of the Amazon Ants is about to establish itself, the insects form themselves into an army, and set off on a slave-hunting expedition.

There are at least two species of Ant which act as servants to the Amazon Ants, the one being named *Formica fusca*, and the other *Formica cunicularia;* and to the nests of one or other of these insects the Amazons direct their march.

As soon as they reach the nest, they penetrate into all its recesses, in spite of opposition, and search every corner for their spoil. This consists solely of the pupæ which will afterwards be developed into neuters; and vast numbers of the unconscious young are carried off in the jaws of the conquerors. The rightful owners and relatives of the captured young cannot resist the enemy, as their shorter though more generally useful jaws are unable to contend with the long and sharply-pointed weapons of their foes.

After the marauding army has returned, the living spoils are carefully deposited in the nest, where they are speedily hatched into perfect insects of the worker class, and immediately take on themselves the labours of the nest, just as they would have done in their own home. The Amazon Ant seems to be utterly incapable of work; and in one notable instance, when a number of them were confined in a glass-case, together with some pupæ, they were not only unable to rear the young, but could not even feed themselves, so that the greater number died from hunger. By way of experiment, a single specimen of the slave Ant (*Formica fusca*) was introduced into the case, when the state of affairs was at once altered. The tiny creature undertook the whole care of the family, fed the still living Amazon Ants, and took charge of the pupæ until they were developed into perfect insects.

Some writers have enlarged upon the hard lot of the slave Ants, imagining their servitude to be as distasteful to them as it is sometimes made to human slaves. Mr. Westwood, however, points out very clearly that any compassion bestowed upon them is wasted, and that the lot of the "helots"—if they may be so called—is precisely that for which they were made. The labours which the little creatures undertake are not arbitrarily forced upon them by the dread of punishment, but are urged upon them by the instincts implanted within them. They would have worked in precisely the same manner, and with exactly the same assiduity, in their own nests as in that of their captors, and the labours are undertaken as willingly in the one case as in the other.

They find themselves perfectly at home, and are in every respect on a par with their so-called masters. In point of fact, however, the real masters in the nest are the slaves, for upon them the Amazons are dependent from their earliest days to the end of their life, and without them the entire community would perish. The slaves have no other home but that to which they have been brought, and are no more to be pitied than are dogs, cattle, and other domestic animals that never have freedom. Indeed, none but solitary animals can be free even in the wild state, for they are held in absolute servitude by the leaders of the herds, and, if they dare to disobey, are summarily punished.

As the slaves are always neuters, it is necessary that fresh importations should be made as fast as the demand for workers exceeds the supply; and it is really a wonderful thing that the

Amazon Ants should always select the pupæ which will afterwards be developed into neuters, and never take those from which males or females will issue.

The Amazon of the Continent is not the only Ant which enslaves the neuters of another species, for in different parts of the world several species of Ants have been observed which seize upon workers belonging to other nests, and bring them to do the work of the home. A Brazilian species (*Myrmica paleata*) has been observed to act in this manner.

In the accompanying illustration are shown two remarkable nests, made by insects of the same genus, which have been placed side by side in order to show the different manner in which cells are arranged by insects which are closely allied to each other.

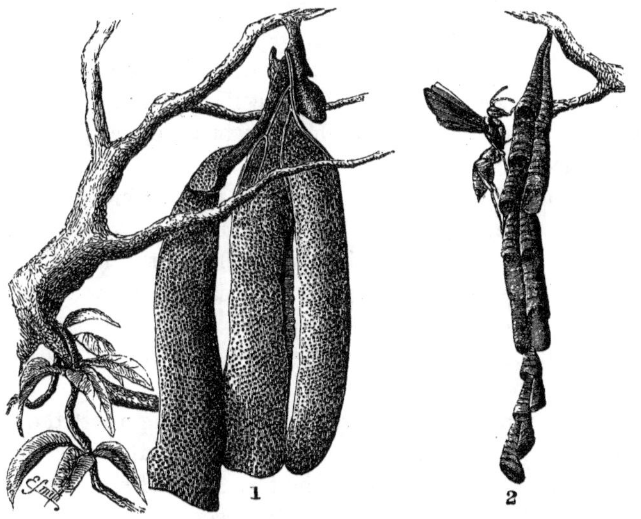

NESTS. POLISTES ATERRIMA, ETC.

The smaller, though more conspicuous nest, is formed by an insect which is called *Polistes aterrima*, a name which is very appropriate to the creature, inasmuch as the generic title signifies a builder, or the founder of a city, and the specific name signifies intense blackness, and is given to the insect on account of its colour. In general shape the insect resembles other species of its genus, but is rather larger than the generality of its kin, and is conspicuous for its deep black colour.

The method of nidification which this species adopts is very peculiar. The cells are formed with beautiful regularity, but are arranged in a very curious fashion. They are placed with their mouths downwards, as is frequently the custom with the cells of hymenoptera, but are not quite perpendicular, inclining alternately to either side. Each cell is set rather lower than its predecessor, so that the general effect is very peculiar, and gives to the cell-group a character which renders it at once recognisable.

THE second nest which is represented in the same illustration is scarcely so striking in appearance as the preceding, but is equally interesting, and to many minds more so, because the architect is at present unknown, and there is some little mystery about the edifice. That it is the work of a Polistes is evident enough from the character of the cells, but it is not made by any of the numerous species whose nests are already in European collections.

The colour of the combs is a rather sombre brown, so that at a little distance the whole group would easily escape notice, especially if it were buried in the depths of the forest, as is the case with many similar nests. The cells are about the same size as those of the curious clay nest which has already been described, but they are undefended by any covering, and exposed to the weather.

It has been suggested by some observers that the combs might have been originally protected by an outer case, and that the case itself has been lost. The formation of the branch, however, from which the combs are hung, serves to militate against any such theory, as the twigs project so far that they must have been enveloped by the covering if it ever existed, while upon them there is no trace of any such material as that of which the nest is made. The inference is, therefore, that they were never intended to be protected by a cover, but that they were intentionally exposed to the air, as is the case with the habitations of Polistes and several allied insects, whose homes will shortly be described.

One of the most curious points in the construction of this nest is the manner in which it is suspended to the branch. As is shown in the illustration, the combs are comparatively narrow at the point of attachment, and gradually increase in width, so that their weight, when filled with the young brood, must be considerable, and the strain on the upper part of the comb very great. The manner in which the insect has met this difficulty is really wonderful. It has not made the upper part of the comb to consist of a solid mass, as is the case with the clay nest which has just been described, but has utilized almost every portion of the comb from the top to the bottom. But, in order to obtain the needful strength, the upper part is constructed after a manner that is widely different from that which prevails upon the lower and wider portion of the comb.

If one of the combs were broken across, the lower half would much resemble, except in colour, the nest of an ordinary wasp, except that the cells are smaller, and the material stronger. But, towards the top, the partitions between the cells become thicker, and in consequence the cells are fewer. This increased strength is chiefly found in the partitions which run perpendicularly, and which are so thick, that the hexagonal form of the cells becomes obscured, the great object being, not the exact shape of the cells, but their ability to bear the weight of the comb below.

The general effect of this modification can be easily imitated by taking an oblong piece of linen, rounding the corners, and plaiting one end, just as ladies gather in the upper part of an apron. The longitudinal folds will then represent the perpendicular partitions of the cells, and will show how strength is gained without needless expenditure of material. The strengthened partitions do not run quite perpendicularly, but are slightly irregular, just as would be the case with the folds of the linen if it were fastened to a branch by the plaited end, and suffered to hang loosely.

The history of Social Insects would be incomplete without the mention of several British insects, which are plentiful enough, but which are scarcely known as well as they deserve. These are the creatures which are popularly known as Carder Bees, because they prepare the materials for their nest in a manner similar to that which is employed in carding cotton wool or heckling flax.

Several species of Carder Bee are known, all belonging to that familiar group of insects called Humble Bees. Among these, as among Humble Bees in general, there is a great variety of colour, so that the same species has been called by different names, even by skilled entomologists. For example, in Kirby's admirable monograph of British Bees, no less than seven varieties of the commonest species of Carder Bee (*Bombus muscorum*) are given as separate species.

That such mistakes should be made is no matter of surprise when we take into consideration the capriciousness with which the colours of this species are distributed among its members. Among the queen Bees, the abdomen is sometimes marked with rings of yellow, black and red, and is sometimes red at the base and tip and black in the middle. The worker has usually a yellowish abdomen, with one or two blackish bands, but in some cases the whole abdomen is black, except a small patch on the base and another at the tip. The male Bee has generally the abdomen coloured like the first-mentioned example of the worker, but sometimes it is wholly black, and in many cases it is black except the tip, which is dun. Indeed, these insects are so extremely variable, that the only method of determining their true arrangement is by taking a great number of nests, breeding the inmates, and subjecting them not only to careful external examination, but also to dissection of their internal anatomy.

The specific title "muscorum," *i. e.* "of the mosses," which is given to this Bee, is due to the material of which the nest is usually made. It was generally thought to be made exclusively of moss, but is in fact constructed of various substances, according to the locality. Mr. F. Smith mentions several instances where the Bees had made use of very singular and unexpected materials.

In one case, Bees were seen flying into a stable through the latticed window, collecting the little hairs that had fallen from the horses during the process of currying, making them up into bundles, and flying off with them. On being watched carefully, one of the Bees was seen to alight on some grass, not very far from the stable, and among the grass was found the nest, which was composed entirely of horsehair. Unfortunately this remarkable nest was destroyed before it was completed.

"Another very interesting deviation from the usual economy of the moss-building Bees was observed by Dr. William Bell. During the summer of 1854, a robin built its nest in the porch of his cottage at Putney. Some time after this had been observed, a Humble Bee took possession of the nest, and adapted it to her own purpose. He was unfortunately unable to identify the species by capturing a specimen, the nest having been destroyed; but Dr. Bell saw the Bee on one occasion, and observed that it was black, with yellow bands, probably *Bombus pratorum.*"

Moss, however, is the favourite material of the Carder Bees, and wherever it can be obtained, they will use no other substance, though in places where it is scarce, or not to be found, they employ leaves, grass, or any other suitable material. Whatever may be the material, the Bee always takes great pains to disentangle the fibres, in order to be able to weave them in a systematic manner into the nest. This process is conducted by means of the legs, the Bee seizing the fibre with her fore-feet, and passing it under her body by means of the remaining pairs of legs, forming it, as she does so, into a small bundle which can be easily carried off.

The object of the moss and other substances is very simple. The Carder Bees do not build their nests, like those of many Humble Bees, beneath the surface of the ground, but upon it choosing a spot where there is a slight hollow of an inch or two in depth. The moss is then woven so as to form a domed cover to the cells, this dome being of variable dimensions, according to the number of cells which it covers, but seldom reaching more than three or four inches in height above the ground. As in very rainy weather this mossy dome would not be waterproof, the insects line it with a very coarse, dark-coloured wax, similar to that of which the breeding cells are made.

The entrance to the nest is always at the bottom; for although the insects will sometimes make an opening at the top, they seem to do so merely for the purpose of admitting air and warmth, and never enter or leave the nest through it, closing it at night or in rainy weather. Generally, a kind of tunnel or arched entrance leads into the nest, like the passage into an

Esquimaux snow-house, an edifice to which the moss-covered dome of the Carder Bee bears no small resemblance.

The best time to search for these Bees is in the hay-making season, when the scythemen often come upon them during their work; and a promise of some small reward for this or any other structure will probably produce a tolerable barvest of nests, as well as of hay.

CHAPTER XXV.

PARASITIC NESTS.

Various Parasites—Parasitic Birds—The CUCKOO and its kin—The COW BIRD and its nest—Size of its egg—Comparison between the Cuckoo and the Apteryx—The ÆPYORNIS—The BLUE-FACED HONEY-EATER or BATIKIN—General habits of the bird—Singular mode of nesting—The SPARROW-HAWK and its parasitic habits—The KESTREL, its quarrel with a Magpie—The PURPLE GRAKLE or CROW BLACKBIRD—Its curious alliance with the Osprey—Wilson's account of the two birds—The SPARROW as a parasite—Curious behaviour of the STORK—Parasitic Insects—The ICHNEUMON FLIES—The parasite of the CABBAGE CATERPILLAR—Its numbers and mode of making its habitation—Trap-doors of the cells—The Australian Cocoon and its parasites—The OAK-EGGER MOTH, its cocoons and enemies—The PUSS MOTH—Its remarkable cocoon—Powerful jaws of the parasite—RUBY-TAILED FLIES and their victims—Modes of usurpation—The CUCKOO FLIES or Tachinæ—Parasites within pupæ—Parasites on vegetables—The GALL FLIES and their home—British Galls, their shapes, structures, and authors—Foreign Galls, and their uses.

WE now pass to another branch of this inexhaustible subject, and come to those creatures that are indebted to other beings for their homes. In some cases, the habitation is simply usurped from the rightful proprietors, who are either driven out by main force or are ousted by gradual encroachment. In other cases, the deserted tenement of one animal is seized upon by another, which either inhabits it at once, or makes a few alterations, and so converts it to its own purposes. In many instances, however, the habitation of the parasite is found within the animal itself; and in some cases the entire body forms the habitation of the parasite.

Several examples of the first description of parasites have already been given under other headings. For instance, where the puffin invades the rabbit-burrows, and drives out the rabbits by dint of courage and a powerful beak; or where the Coquimbo owl and rattlesnake take possession of the homes which had been excavated by the prairie dog. Examples of the second description of parasites have also been given. The kingfisher for instance, usurps the deserted hole of a water-shrew; and the humble-bee and wasp usually take advantage of the deserted burrow of some rat or mouse. In the account of the sociable weaver-bird, mention is also made of certain little green parrots, which are apt to take possession of the great nest, and use it for their own purpose. And in the last chapter an example was mentioned where a carder-bee established herself in the deserted nest of a wren, and so saved herself the trouble of fetching materials and building a dome.

BIRDS of various kinds are notorious parasites, the Cuckoos ranking as chief among them, inasmuch as they make no nest at all, but simply lay their eggs in the nests of other birds, and foist upon them a supposititious offspring, which occupies the entire nest and monopolises all the care of its foster-parents.

All Cuckoos, however, do not possess this habit; for some of the group build nests which are remarkable for their beauty, and tend their young as carefully as do any birds. The celebrated Honey-finders, for example, which are found in most hot portions of the globe, are notable for their skill in architecture. The nests of these birds are pensile, and not unlike those of the African weaver-birds, which have already been described. They are made of tough bark, torn into filaments, and are flask-like in shape, hung from the branches of trees, and having their entrance from below.

Then there is the well-known COW-BIRD of America (*Coccygus Americanus*), which is closely allied to the common cuckoo, and yet which builds its own nest, and rears its own young. "Early in May," writes Wilson, "they begin to pair, when obstinate battles take place among the males. About the 10th of that month they commence building. The nest is usually fixed among the horizontal branches of an apple-tree; sometimes in a solitary thorn, crab, or cedar, in some retired part of the woods. It is constructed with little art, and scarcely any concavity, of small sticks and twigs, intermixed with green weeds and blossoms of the common maple. On this almost flat bed the eggs, usually three or four in number, are placed; these are of an uniform greenish blue colour, and of a size proportionate to that of the bird.

"While the female is sitting, the male is generally not very far distant, and gives the alarm by his notes, when any person is approaching. The female sits so close, that you may almost reach her with your hand, and then precipitates herself to the ground, feigning lameness, fluttering, trailing her wings, and tumbling over, in the manner of the partridge, woodcock, and many other species. Both parents unite in providing food for the young."

In this narrative, two points are especially worthy of notice. In the first place, the egg of the Cow-bird is proportionate in size to the bird which laid it. Now, one of the most remarkable facts connected with the history of the common cuckoo is, that although the bird is as large as a small hawk, its egg is scarcely half as large as that of a thrush or blackbird, as indeed is needful for its admission into the nest of a hedge-sparrow or redstart.

Here, then, we have an example of a bird laying an egg which is extremely small in proportion to its own size, while in the apteryx or kiwi-kiwi of New Zealand, we have an example of a bird laying an egg which is absolutely gigantic in proportion to its own size. The apteryx is not a large bird, certainly not larger than a guinea fowl, and yet its egg looks like that of a swan, and weighs just one quarter as much as the bird which produced it. Thus it is evident that the dimensions of an egg afford no certain criterion respecting the size of the bird that laid it, and although a large bird usually lays a large egg, and a small bird lays a little one, the cases *may* be reversed, as in the instance just mentioned.

All naturalists are familiar with the gigantic egg laid by some bird unknown, and called by the provisional name of *Æpyornis*, or "tall-bird." This egg makes that of the ostrich itself shrink into insignificance, for its lineal measurement is precisely double that of a large ostrich egg, and its cubic bulk is eight times as great. In fact, the æpyornis egg looks as gigantic by the side of an ostrich egg, as does an ostrich egg near that of a duck. It was therefore imagined that the æpyornis must be at least eight times as large as the ostrich, and a height of sixteen feet was attributed to the unknown bird.

Now, it is easy to work out this problem by the rule of three, and to give the result in figures; but when that result is compared with existing facts, it becomes startling. On paper, a height of sixteen feet for an ostrich-like bird seems rather gigantic, but does not appear to carry with it any idea of its real magnitude. The height of a very fine ostrich being about seven or eight feet, we say that the æpyornis must be twice as tall as an ostrich, and so dismiss the subject from our minds. But, when we come to compare the imaginary bird with actually existing beings, we shall better understand the dimensions of a bird that measured sixteen feet in height. Sixteen feet is the average height of the adult giraffe, the females varying from thirteen to sixteen feet, and the males from fifteen to eighteen.

It is impossible to say that there never was a bird as large as a giraffe, but all our present knowledge controverts such an idea. If, however, we keep in mind the comparative dimensions of the apteryx and its egg, we must be prepared to find that the æpyornis, although necessarily a large bird, may not be larger than an ostrich, and need not be so large.

Thus, then, the comparative size of an egg is by no means an unimportant fact in natural history, and the comparison of two such birds as the apteryx and the cuckoo may at least save us from the danger of generalizing too hastily.

The second point in the history of the Cow-bird is its love for its young, which is quite equal to the affection that is manifested by the lapwing and other birds that endanger themselves in order to draw attention away from their offspring, and directly opposed to the indifference towards the young which seems to actuate the ordinary cuckoo.

In Australia there is a large group of rather pretty birds, popularly called Honey-eaters, because they feed largely on the sweet juices of many flowers, although the staple of their diet consists of insects. They seem indeed to occupy in Australia the position which is taken in America by the humming-birds, and by the sun birds of the old world. To this group belong many familiar and interesting species, such as that which produces a sound like the tinkling of a bell, and is in consequence called the Bell-bird; the different species of Wattle Birds; the odd, bald-headed Friar Birds, and the splendidly decorated Poe Birds.

One species of it, which comes in the present section, is the Blue-faced Honey-eater of New South Wales, called by the natives Batikin (*Entomyza cyanotis*). It is a pretty bird, the plumage being marked boldly with black and white, and a patch of bare skin round the eyes being bright azure. This peculiarity has earned for the bird the specific title of *cyanotis*, or "blue-eared."

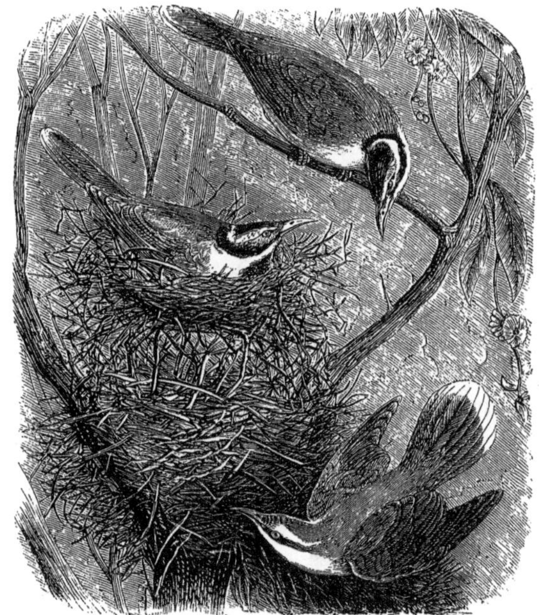

HONEY-EATER IN NEST.

Like all the Honey-eaters, it is a most lively and interesting bird, and to the careful observer affords an endless fund of amusement. It is never still, but traverses the branches with astonishing celerity, skipping from one to another, probing every crevice with its needle-like tongue, hanging with its head downwards, and even suspending itself by a single claw, while it secures a tempting insect. It is generally to be found on the eucalypti, or gum-trees, and is one of the stationary birds, remaining in the same locality throughout the year.

The generality of the Honey-eaters are skilful architects, but the Batikin seems not to share the ability of its relatives, or, at all events, not to exercise it. Mr. Gould thinks that the bird can hardly depart so far from usual custom as to be incapable of building a nest, but he has never found such a nest, nor heard of one. The Batikin is one of the parasitic group, usurping the nest of another bird, and taking possession of it in a very curious fashion.

In Australia there is a bird belonging to the genus *Pomatorhinus*, which somewhat resembles the bee-eater, except in plumage, which is quite dull and sober. This bird builds a large, domed edifice, and appears to make a new nest every year. The deserted nests are always usurped by the Batikin, which establishes herself without any trouble. The reader would naturally imagine that when the bird finds herself in possession of so large and warm a nest, she will pass into the interior, and hatch her young under the protection of the roof. This plan, however, she does not follow, preferring to take up her abode on the very top of the nest, exposed to all the elements. She takes very little trouble about preparing her home, but merely works a suitable depression upon the soft dome, lays her eggs in it, and there hatches them.

The reader will remember that there are several birds which form a supplementary nest upon the exterior of the original domicile, and the parasitic nest of the Batikin is evidently an extension of the same principle.

In England we have many parasite birds, one of which is the common Sparrow-Hawk (*Accipiter Nisus*), which is in the habit of usurping the nest of the common crow, magpie, or other bird, and laying its handsome eggs therein.

Whether it forcibly drives away the rightful owner, or whether it contents itself with a nest which has already been abandoned, is not precisely known, different naturalists inclining to opposite opinions. In all probability, therefore, both disputants are right, and the Sparrow-Hawk takes a deserted nest when it can find one, and when it cannot do so, attacks birds which are in actual possession of a suitable nest, and takes possession of their home In such a case, the combat must be a sharp one, for both crow and magpie are courageous birds, nothing inferior in determination to their assailant, and armed with bills which are much larger, and quite as formidable as that of the Sparrow-Hawk.

The KESTREL (*Tinnunculus alaudarius*) is also in the habit of laying its eggs in the nests of other birds, and may possibly eject the rightful owner by main force. This opinion is rendered probable by a fact mentioned by Mr. Peachey, in the "Zoologist." A man was passing a tree, and hearing a loud screaming proceeding from a nest at the summit, he had the curiosity to climb the tree. The screams still continued, and on putting his hand into the nest, he found two birds struggling, the uppermost of which he caught. This proved to be a Kestrel, and as soon as it was secured, the other bird, which was a magpie, flew out, evidently having been worsted by its antagonist.

Then there is the well-known STARLING (*Sturnus vulgaris*), which is a notably parasitic bird, delighting to take the nests of the jackdaw, pigeon, and other birds, and to use them as its own. Every one who has a dovecote knows how apt are the Starlings to usurp the boxes intended for the pigeons, and how in consequence it is accused of killing the young of the pigeons, and sucking their eggs, two accusations which I believe to be wholly false. Were the Starlings to be thus predaceous, the pigeons would be quite aware of their depredations, and would appear greatly disturbed whenever the robbers were seen. As, however, the pigeons in one box live in perfect amity with the Starlings in the next, it is very unlikely that the latter birds prey in any way upon the former.

THERE is a group of birds which are popularly called Grakles, and are scientifically known as Quiscalinæ. They are also called Boat-tails, because their tail-feathers are formed so as to take the shape of a canoe. One species, the PURPLE GRAKLE, or CROW-BLACKBIRD (*Quiscalus versicolor*), is conspicuous as a parasitic bird, and selects a most extraordinary spot for its nest.

Generally, the predaceous birds are avoided and feared by the rest of the feathered tribes, and if a hawk or eagle show itself, the smaller birds either hide themselves, or try to drive away the intruder by force of numbers or swiftness of wing. The Purple Grakle, however, is devoid of such fears, at all events as far as one species of predaceous bird is concerned, and boldly takes up its abode with the osprey or fish-hawk (*Pandion haliaëtus*).

The nest of the osprey is a very large edifice, made of sticks, grass, seaweed, leaves, and similar materials. The foundations are made by sticks almost as thick as broom-handles, and some two or three feet in length, on which are piled smaller sticks, until a heap some four or five feet in height is made. Interwoven with the sticks are stalks of corn and various herbs, the larger sea-weeds and large pieces of grass, the whole mass being a good load for an ordinary cart, and as much as a horse can be reasonably expected to draw. The bird retains the nest year after year, and, as has been shown from actual observation, the same spot has been occupied for so long a term that the branches of the tree became rotten, and the nest fell to the ground. In this case it is evident that a succession of birds must have occupied the same nest.

It has been observed that whenever a tree is occupied by the osprey, it dies in a short time, though no one is aware of the precise nature of the injury which kills it. Some persons say that the fish-oil which is spilled by the birds is the cause of death; but when we remember that there is no better manure than fish, we can hardly believe that the alleged cause is the real one. Other persons think that the real cause of death is the huge mass of decaying vegetable and animal substances which is placed on the branches, and that the drippings from the nest fall into casual interstices of the branches, and gradually kill it from above downwards. So firmly are the materials fastened together, that when a tree falls on which an osprey nest is built, large masses of the nest hold together in spite of the shock.

The construction of the osprey nest has been described somewhat at length, because the manner in which the Purple Grakle becomes a parasite could not be understood unless the structure of the nest were comprehended.

As the sticks of which the foundation of the nest are made are very large, and not regular in form, considerable interstices are left between them, and in such spots the Grakle chooses to nidificate.

In writing of the osprey, Wilson remarks as follows: "There is one singular trait in the character of this bird which is mentioned in treating of the Purple Grakle, and which I have had many opportunities of witnessing. The Grakles, or Crow-Blackbirds, are permitted by the fish-hawk to build their nests among the interstices of the sticks of which its own is constructed,—

several pairs of Grakles taking up their abode there, like humble vassals around the castle of their chief,—laying, hatching their young, and living together in mutual harmony. I have found no less than four of these nests clustered round the sides of the former, and a fifth fixed on the nearest branch of the adjoining tree, as if the proprietor of this last, unable to find an unoccupied corner on the premises, had been anxious to share, as much as possible, the company and protection of this generous bird." In another place, the same writer remarks that the curious allies "mutually watch and protect each other's property from depredators."

These Grakles exist in great numbers, and sweep over the land in vast flocks, like our own starlings, their wings sounding like the blast of a tempest as they rise from the ground, and their bodies darkening the air. "A few miles from the banks of the Roanoke, on the 20th of January, I met with one of these prodigious armies of Grakles. They rose from the surrounding fields with a noise like thunder, and, descending on the length of road before me, covered it and all the fences completely with black; and when they again rose, and, after a few evolutions, descended on the skirts of the high timbered woods, they produced a most singular and striking effect, the whole trees, for a considerable extent, seeming as if hung in mourning; their notes and screaming the meanwhile resembling the distant sound of a great cataract, but in more musical cadence, swelling and dying away on the ear, according to the fluctuations of the breeze."

It is evident that such vast multitudes of birds cannot all have been nurtured in the interstices of osprey nests. Indeed, the generality of the birds build in tall trees, usually associating together, so that fifteen or twenty nests are made in the same tree. The nests are well and carefully made of mud, roots, and grasses, about four inches in depth, and warmly lined with horsehair and very fine grasses. The fact that the bird possesses this capability of nest-building, gives more interest to the occasional habit of sharing its home with the osprey—a privilege of which it seems to avail itself whenever an osprey's nest is within reach.

The colour of this bird appears at a little distance to be black, but is in reality a very deep purple, changing in different lights to green, violet, and copper, and having a glossy sheen like that of satin.

OUR little friend the SPARROW (*Passer domesticus*) is occasionally a parasite, following to some extent the custom of the purple grakle, though it does not select a bird of prey for its companion.

On the Continent, the common stork builds largely, and in several countries is protected by general consent, the slaughter of a stork, or the destruction of its nest and eggs, being visited with a heavy fine. In consequence of this immunity, the stork is very tame, building upon houses as freely as does the martin, and being considered as a bringer of good luck when it does so.

Any disused chimney is sure to have a stork's nest upon the top, and so is a pillar, or any ruin. The nest of the stork bears a general resemblance to that of the osprey, and, with the exception of the sea-weed, is made of similar materials. It is of huge dimensions, and chiefly consists of sticks and reeds, heaped together without much arrangement, and having on the top a slight depression, in which the eggs are laid. As is the case with the osprey nest, considerable interstices are left between the sticks, and in these spots the Sparrow loves to place its nest. Mr. F. Keyl has told me that he has repeatedly seen the storks and Sparrows thus living in amity together, the stork appearing to extend to a weaker bird that protection which it receives from mankind.

WE now pass to the Parasitic Insects. As this work is intended to describe dwellings which are in some way formed by the creatures, it is necessary to exclude all the parasite insects that may exist upon the animal, and make no habitation, such as the ticks, as well as those which are merely parasitic within the animal, such as the various entozoa.

Of Parasitic Insects, the greater number belong to that group of hymenoptera which is called Ichneumonidæ, and which embraces a number of species equal to all the other groups of the same order. Being desirous of producing, as far as possible, those examples of insects which have not been figured, I have selected for illustration several specimens which are now in the

British Museum, one or two of which have only been recently placed in that collection.

THE best known of all the Ichneumonidæ is that tiny creature called *Microgaster glomeratus*, of which a casual mention has already been made in page 270.

A group of these insects and their cells is now before me, and will be briefly described.

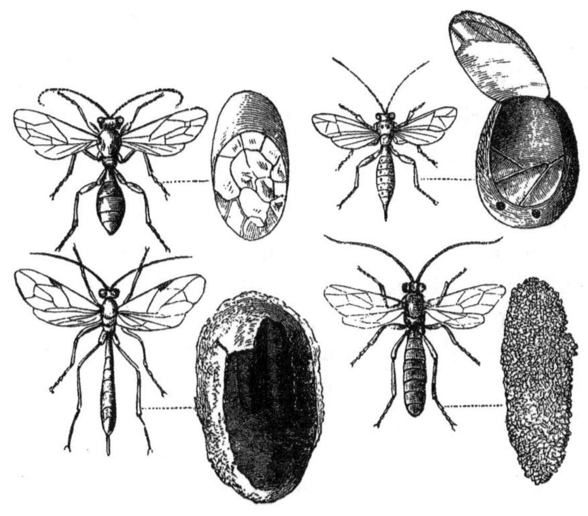

PARASITIC INSECTS.

COCOON OF OAK-EGGER MOTH. (*Cryptus fumipennis.*)

COCOON FROM NEW SOUTH WALES. (*Pimpla.*)

COCOON OF PUSS MOTH. (*Paniscus glaucopterus.*)

COCOON OF GOAT MOTH. *Lamprosa setosa.*)

The insects themselves much resemble in general form the Burnet ichneumon which is mentioned in pp. 270, 273, but are smaller, blacker, and not nearly so beautiful, although their wings gleam with an iridescence nearly as brilliant. Small as it is, this tiny insect is extremely valuable to us, and to the gardener is beyond all value, though, as a general rule, the gardener knows nothing about it. Were it not for this ichneumon, we should scarcely have a cabbage or a cauliflower in the garden; for the noisome cabbage caterpillars would destroy every leaf of the present plant, and nip the growth of every bud which gave promise for the future.

Every one knows the peculiarly-offensive caterpillars which eat the cabbages, and which are the offspring of the common large white butterfly. In the spring, the butterflies may be seen flitting about the gardens, settling on the cabbages for a few moments, and then flying off again. They look very pretty, harmless creatures, but, in fact, they are doing all the harm that lies in their power. Forty or fifty eggs are thus laid on a plant, and if only one quarter of the number are hatched, they are quite capable of eating every leaf. In process of time, they burst from the egg-shell, and commence their business of eating, which is carried on without cessation throughout the whole time of the larval existence, with a few short intervals, while they change their skins.

When they are full grown, they crawl away from the plant to some retired spot, and there suspend themselves, preparatory to changing into the pupal condition. A few of them succeed in this task, but the greater number never achieve the feat, having been the unwilling nourishers of the ichneumon flies. Just before the larva is about to pass into the pupal state, a number of whitish grubs burst from its sides, and each immediately sets to work at spinning a little yellow, oval cocoon. The walls of the cocoon are hard and smooth, especially in the interior; but the outside is covered with loose floss-silk, which serves to bind all the cocoons together. Generally, they are very loosely connected; but a group of these little objects is now before me, where the cocoons are formed into a flattish oval mass, about the size and shape of a scarlet-runner bean, split longitudinally, and are bound so tightly together, that their shape can barely be distinguished through the enveloping threads.

As is the case with the cells of the Burnet ichneumon, each cell is furnished with a little circular door, which exactly resembles in shape and dimensions the circular pieces of paper that are punched out of the edges of postage-stamps. On the average, about sixty or seventy ichneumon flies are produced from a single cabbage caterpillar.

The groups of yellow cells are very plentiful towards the middle of summer and the beginning of autumn, and may be found on walls, palings, the trunks of trees, in outhouses, and, in fact, in every place which affords shelter to the caterpillar. Nothing is easier than to procure the insects from the cocoons, as the yellow mass needs only to be put into a box, with a piece of gauze tied over it by way of a cover. Nearly every cocoon will produce its ichneumon, and as the little creatures are not strong-jawed enough to bite through the gauze, they can all be secured.

There are many species of Microgaster; but those which have been mentioned are the most important, and make the most interesting habitations.

THE large oval cocoon was brought from New South Wales, and is evidently the produce of some lepidopterous insect, probably a moth allied to the silkworm. Upon the larva which constructed the cocoon an ichneumon has laid her eggs, and the consequence has been that the caterpillar has been unable to change into the pupal condition, but has succumbed to the parasites which infested it. These insects are not of minute dimensions, like the Microgaster, but are tolerably large, and in consequence can be but few in number. The cells are very irregular in shape, and are not rounded like those of many Ichneumonidæ, but have angular edges.

In this, and in one or two other examples which are shown in the illustration, the reader will note a peculiarity in the development of the parasite. The Microgaster larvæ emerge from the caterpillar just before it undergoes its change into the pupal condition, and effectually prevent that change by killing the creature in which they had been nurtured. But, in many instances the ichneumon larva delays its escape until the caterpillar has completed its cocoon, and in some cases waits until the change into the pupal state has been achieved. In the present example, the larva has permitted the cocoon to be made, and then killed the caterpillar, the reason of this delay being that the cocoon is very firm and strong, and affords an impregnable shelter to the parasite. The names of the Parasites are placed beneath the cocoons.

WITHIN the same case there are several cocoons in which a similar calamity has befallen the caterpillars which made them. There is, for example, a cocoon of the OAK-EGGER MOTH (*Lasiocampa quercus*), the interior of which resembles that of the insect which has just been described, except that the cells of the parasite are more numerous. This species of caterpillar is peculiarly subject to the attacks of the ichneumon flies, as is well known to all practical entomologists, who lose many of their carefully-bred specimens by means of these insects.

There is also one of the winter cocoons of the GOAT MOTH caterpillar, the inmate of which has been pierced by the ichneumon fly, and killed by its young. As the species of ichneumon is a large one, only a single individual was produced, and as may be seen from the cell of the parasite which is placed by the side of its victim, the habitation of the ichneumon is so large that it must have occupied nearly the entire cocoon of the dead caterpillar.

IN another room, placed among the series of British moths, is a cocoon of a PUSS MOTH (*Cerura vinula*), which has been occupied by two ichneumon larvæ.

If the reader should happen to know the cocoon of this moth, he will remember that it is made of wood-scrapings, glued together with a cement secreted by the insect, and that its walls are so hard that a tolerably strong knife is required in order to cut it open. That the eggs of a parasite should be introduced into the body of the larva is not an extraordinary circumstance, but that the perfect insects should be able to make their way out of such a cocoon is really wonderful. The interior of this cell is hard and smooth as if made of polished ebony, and its concavity renders it more difficult of penetration. Yet these singular insects contrive to make their way through the sturdy walls.

The ichneumons which usually attack the Puss Moth are rather large insects, belonging to the genus *Ophion*, and have long, slender, curved abdomens, and long antennæ slightly twisted at the ends. The colour is orange, diversified with black. Those which have made their cells in the above-mentioned cocoon belong to the species called *Paniscus glaucopterus*, and are of a yellowish hue. It sometimes happens that the insects fail in making their way through the cell-walls and die in the interior. This accident, however, seems chiefly to befall the ichneumons produced in cocoons which are kept in houses for the purpose of breeding the Puss Moth, and which are in consequence harder and more dry than those which remain in the open air, adhering to the trunks of trees.

THOSE splendid insects which are popularly called RUBY-TAILED FLIES, or FIRETAILS, and scientifically are termed *Chrysididæ*, are also to be numbered among the parasitic insects.

They make no nests for themselves, but intrude upon those of various mason and mining bees, and several other insects. The Firetail does not, however, lay its eggs in the body of the larva, but makes its way into the nest while the rightful owner is absent, and places an egg near that of the bee. The egg of the parasite is sometimes hatched at the same time with that of the bee, but generally later. In the first instance, the larva feeds on the provisions which were supplied for the bee, and so starves the poor creature to death; and in the latter case, it is not hatched until the young bee is large and fat, and capable of affording ample subsistence to the parasite, which fastens upon it and devours all the after portions.

Then there are the CUCKOO FLIES (*Tachinæ*), which bear some resemblance to the common house-fly, but which are parasitic, feeding on the larvæ of other insects, and selecting the same species which are persecuted by the firetails. When the Tachina larva has eaten that of the mason bee, it forms an oval cocoon, and there remains until the time for becoming a perfect insect. A single larva of the mason bee seems to be sufficient for the Tachina grub, as Mr. Rennie has recorded an instance where two larvæ of the mason bee were in a nest into which a single egg of the Tachinæ had been introduced. The parasitic larva devoured one of the rightful inhabitants, but did not touch the other, and the cocoons of the bee and the Tachina were formed side by side.

SOMETIMES, as has already been mentioned, the chrysalis itself of a lepidopterous insect becomes the home of the parasite. I have found the pupæ of various butterflies absolutely filled with tiny ichneumon flies of the most brilliant colours; and in the British Museum there is an excellent example of a chrysalis, which has been filled by a single ichneumon fly, of such a size that the little chrysalis from which it was taken seems scarcely capable of holding it and its cocoon.

WE now pass to a remarkable series of insects belonging to the same order as the ichneumons, but parasitic upon vegetables and not on animals. Their scientific name is *Cynipidæ*, and they are popularly known as GALL FLIES, because they cause those singular excrescences which are so familiar to us under the name of Galls. This group comprises a vast number of species, all of which have a strong family resemblance, though they greatly differ from each other in size, form, and colour.

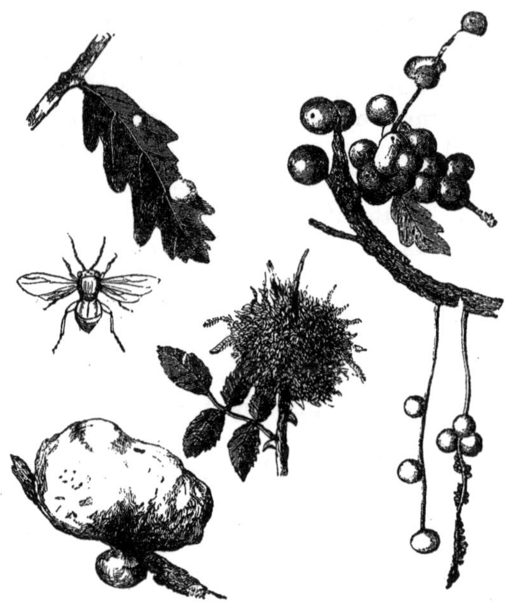

BRITISH GALLS.

Leaf Galls of Oak. Bedeguar of Rose. Galls of Cynips Kollari.
Cynips Kollari (Slightly magnified) Currant Galls of Oak.
Oak Apple.

In the accompanying illustration are given several examples of British Galls, most of which are tolerably common in this country, and some of which can be found in plenty.

IN the left hand upper corner of the illustration is a figure of an oak-leaf, upon which are two globular projections. These are the well-known "cherry-galls," which are made by a little insect called *Cynips quercus-folii*. They are beautifully coloured, some being entirely scarlet, while others are white, orange, and red, in various gradations, something like the colour of a nearly ripe peach, or those of a Newtown pippin. Perhaps they bear more resemblance to the apple than to the peach, because their surface is highly polished and shining, much like that of the American apple.

These galls may be found in profusion upon the oak-leaves, and are most plentiful upon pollard oaks, upon the youngest trees, or upon the oak underwood that sprouts around a felled trunk. In such cases the leaves are much larger, and fuller of juice than those which spring from adult trees, and the development of the gall is proportionately increased. Wherever there is a thick growth of oaken underwood, the numberless galls which stud the leaves have a remarkably beautiful effect, provided that the observer lies on the ground, or stoops sufficiently low to perceive the under-surface of the leaves, to which the galls are attached.

If one of the galls be cut open with a knife, it will be found to consist of a soft, pulpy substance, fuller of juice than an apple, and somewhat resembling the consistence of a hothouse grape. In the very centre of the soft mass the knife will meet with resistance in the shape of a globular cell of hard, woody texture, and in the middle of the cell will be found a tiny grub, perfectly white, very fat, somewhat resembling the grub of the humble bee, and curved so as to fit the globular cell in which it lies. This is the little being for whose benefit the gall was formed, and the little white grub feeds on the juices of the gall, precisely as the larva of the ichneumon fly feeds on the soft portions of the insect in which it temporarily resides.

On seeing the little creature thus snugly ensconced in the receptacle which serves it at once for board and lodging, a question naturally arises as to the manner in which it was placed there. No aperture is perceptible in the gall, not a hole through which air can reach the enclosed larva, which must, therefore, be capable of existing without more air than can pass through the minute pores of the vegetable substance in which it lies, or must be able to respire by means of the oxygen which is given out by living plants.

The question, indeed, is very like the well-known query as to the manner in which a model of a waggon and four horses can find its way into a bottle, the neck of which is so small as to prevent even the head of the waggoner from passing. The answer is similar in both cases. The bottle was ingeniously blown over the waggon and horses, and the gall was formed around the grub.

When the leaf is in its full juiciness, and the sap is coursing freely through its textures, a little black insect comes and settles upon the leaf. She is scarcely as large as a garden ant, but has four powerful and handsome wings, which can be used with much agility. An entomologist, on seeing her, would at once pronounce her to belong to the order hymenoptera, and to be closely allied to the ichneumon flies which have just been described.

Running to and fro upon the leaf, she fixes upon one of the nervures, and there remains for a short time, evidently busy about some task, which is very important to her, but which her minute size renders impossible to be observed with the naked eye. If, however, a magnifying glass be applied very carefully to the leaf, the following process will be seen.

From the abdomen there projects a tiny hair-like ovipositor, which is coiled in such a manner that it can be protruded to a considerable length. This ovipositor is thrust into the leaf, so as to produce a hole, which is widened by the action of the boring instrument. Presently, the blades of the ovipositor separate, and a single egg is seen to pass between them, so that it is lodged at the bottom of the hole. Into the same aperture is then poured a slight quantity of an irritating fluid, and the insect flies away, having completed her task. The whole proceeding, indeed, is, with the exception of the deposition of the egg, precisely the same as that which takes place when a wasp uses its sting, the ovipositor and sting being but two slightly different forms of the same organ, and the irritating fluid of the cynips being analogous to the poison of the wasp.

The effect of the wound is very remarkable. The irritating fluid which has been projected into the leaf has a singular effect upon its tissues, altering their nature, and developing them into cells filled with fluid. As long as the leaf continues to grow, the gall continues to swell, until it reaches its full size, which is necessarily variable, being dependent on that of the leaf. I have, for example, many specimens of these galls, of different sizes, from which the insects have escaped, showing that they had attained their full size. On the juices of the gall the enclosed insect lives, until it reaches its full term of imprisonment, when it eats its way through the gall and emerges into the world. In some cases, it undergoes the whole of its change within the gall, but in others, it makes its way out while still in the larval state, burrows into the earth, and there changes into the pupal and perfect forms.

To the unassisted eye, the insect which forms the leaf-gall presents no especial attraction, as it is simply, to all appearance, a little black fly. When placed under the microscope, however, it soon proves to be a really beautiful creature, though not possessing the brilliant and gem-like hues which distinguish many of its relatives. The body still retains its blackness, but has a soft tint on account of the white and shining hairs with which it is thickly studded. The eyes are large, stand boldly from the head, and the many lenses of which these organs are composed are so boldly defined, that even in so small an insect they can be distinguished with a very low power of the microscope. Indeed, the inch and a half object glass is quite powerful enough to define them, while the half-inch glass makes them look like the pits in a lady's thimble.

The chief beauty of the insect, however, lies in the wings, which are very large in proportion to the size of the owner, are traversed by a few, but strong nervures, and glow with a changeful radiant lustre, like mother-of-pearl illuminated with living light. In order to see these wings properly, the insect should be laid on some black substance, and the light concentrated upon them by the various means which a microscopist can always employ.

THE oak is a tree that seems to be especially loved by gall-insects, which deposit their eggs in its leaves, its twigs, its flowers, and even in its roots. One of the most familiar examples of oak-galls is that which is called the oak-apple, and which is produced by a species of insect called Cynips terminalis. Although the insect is not of very great size, the gall which it produces is sometimes enormous, being as large as a common golden pippin or nonpareil apple, and therefore very conspicuous upon the tree. It is coloured in the same manner as the cherry-gall, but seldom so brilliantly, and the exterior is not so smooth and polished.

The resemblance to a veritable fruit is much closer at the beginning of the season than in the autumn, as a number of small leaf-like projections surround its base, just as if they were a half-withered calyx. These, however, fall off as the summer advances, and are no more seen.

If the oak-apple be cut with a knife, the first touch of the steel betrays a marked difference between its substance and that of the cherry-gall. Its texture is neither so firm nor so juicy, but is of a softer, drier, and more woolly character. Moreover, the knife passes through several resisting substances, which, when the gall is quite severed, prove to be separate cells, each containing a grub. From each of these cells, which are extremely variable in number, a kind of fibre runs toward the base of the gall, and it is the opinion of some naturalists that these fibres are in fact the nervures of leaves which would have sprung from the bud in which the gall-fly has deposited her eggs, and which, in consequence of the irritating fluid injected into the tree, are obliged to develop themselves in a new manner.

To procure the insects of this and many other galls is no very difficult task. The branch to which they adhere should be cut off, and placed in a bottle of water, and a piece of very fine gauze tied net-wise over it. The insects, although they can eat their way out of the gall in which they have been bred, never seem to think of subjecting the gauze to the same process, and therefore can be always secured. It is needful, however, to procure galls which are tolerably near their full age, as a branch can only be kept alive for a limited time, and if the supply of nourishment be cut off by the death of the branch, the enclosed insect becomes stunted, if not deformed.

The galls produced by Cynips terminalis are those which are so greatly in request upon the twenty-ninth of May, and which, when covered with gold-leaf, are the standards under which the country boys are in the habit of levying contributions. A figure of this gall is seen in the illustration.

SOME years ago, when I was calling at the office of the *Field* newspaper, then recently started in its race for popularity, I was shown some oak-branches containing a vast number of hard, woody, spherical galls, and asked if I could tell the name of the insect which had produced them. They had recently made their appearance in the country, and no one knew anything about them. A branch beset with these galls is shown in the right hand upper corner of the illustration, the figures being necessarily much reduced.

I was totally unacquainted with them, but, in the following year, found many of them on Shooter's Hill, in Kent, where the growth of oaks is very dense. At the present day they have increased so rapidly that they outnumber almost every species, if we except the tiny spangle-galls, and I have bred great quantities of the insect. The creature which made them is named *Cynips Kollari,* in honour of the celebrated entomologist, and is plentiful on the Continent. I believe that it has long been known in Devonshire, though in Kent it has only recently made its appearance.

The galls produced by this insect are wonderfully spherical, of a brown colour, smooth on the exterior, and about as large as white-heart cherries. Each contains a single insect, which undergoes all its changes within the gall, and eats its way out when it has attained the perfect form. Occasionally two galls become fused together, and in my collection there is a very curious example of these twin galls. They form a figure like that of a rude hour-glass, and each portion has contained an insect. The inhabitant of one portion has eaten its way out and escaped, but the other has met with a singular fate. By some untoward error, it has taken a wrong direction, and instead of issuing into the world in the ordinary way, has hit upon the neck which connects the two galls, so that, instead of merely piercing half the diameter of the gall, it would have been forced to gnaw a passage equal to three half diameters.

Natural powers are always adjusted to the work which their possessors have to perform. The insect was gifted with the capability of eating her way through the walls of her own habitation, but not with the power of making a passage through another gall afterwards. As a natural consequence, she has died from exhaustion before she could emerge into the air; and when I cut the double gall, in order to see how the inmates had fared, I found the dead insect lying near the middle of the second gall, so that she was even farther from the outer air than when she started on her course.

The Cynips Kollari is larger than the generality of the family, equalling a small house-fly in dimensions. Its colour is pale brown. A figure of the insect may be seen in the illustration.

NEARLY in the centre of the illustration is seen a figure of the well-known gall that is so common on the rose, whether wild or cultivated, and which is popularly known by the name of BEDEGUAR. This gall is caused by a very tiny and very brilliantly-coloured insect, named *Cynips rosæ*, which selects the tender twigs of roses, and deposits its eggs upon them.

I have now before me quite a collection of these galls, some of which are so variable in shape that they scarcely seem to have been made by the same species of insect. When the Cynips rosæ deposits her eggs upon the rose, the effects are rather remarkable. Each egg becomes surrounded with its own cell or gall, and the whole of them become fused into one mass. The exterior of these galls is not smooth, like that of the specimens which have been described, but is covered with long, many-branched hairs, which stand out so thickly that they entirely conceal the form of the gall itself.

Reaumur, who gave much attention to galls, thought that the hairs were formed by the exudation of sap through little orifices in the growing gall, just as the web of the spider is formed by the exudation of a glutinous liquid from minute pores. This theory, however, is scarcely tenable, because sap has no power of hardening into threads when exposed to the air, and, besides, a well-defined vegetable structure is seen in the hairs, which would not be the case if they were merely hardened sap. Moreover, if the hairs were formed in this manner, they could not have the power of throwing out the tiny branchlets with which they are studded, or of ramifying like the bough of a tree, as is often the case with them.

The number of galls in a single Bedeguar is mostly very great. A specimen of average size, taken at random from the drawer in which the galls are kept, was. when fully clothed, as large as a golden pippin. When the hairy clothing was removed, its size notably diminished, and it was then seen to be composed of a large number of woody tubercles, varying much in size and shape. Their average dimensions, however, are about equal to those of an ordinary pea. The tubercles in question are fused together more or less strongly, some falling off at a slight touch, while others cannot be separated without the use of the knife. There are about thirty-five of these wooden knobs.

On selecting one of the knobs, and examining it, a few very small circular holes are seen, showing that the insects have made their escape from the cells. Indeed, one or two of the insects were found entangled amid the dry and crisp hairs that surrounded the gall, and which formed a second barrier, which they could not penetrate. When, however, a sharp knife is carefully used, the woody tubercle can be laid open in several directions, and then proves to be a congeries of cells fused together into one mass, and varying from four to twenty in number, according to the size of the insect. Perhaps, on an average, ten cells may be reckoned in each knob.

In many of the cells the perfect insect may be found, the death of the rose-branch, and the consequent deprivation of sap, having so hardened the walls of the cells that the inmates have been unable to make their way out. In other cells may be seen certain odd little objects, amber-coloured, hard, shining, and appearing to the unaided eye to be nearly spherical. They are about as large as dust-shot. For a long time I could not satisfy myself about them, not being able clearly to ascertain whether they were deceased insects or merely hardened sap. That they were probably of insect origin was evident from the fact that they were always found in cells which had no opening, and from which the insect had not escaped.

At last, however, one of them happened to lie on the paper so that it could be well illuminated, and then the whole mystery was unfolded. These strange little objects were the pupæ of the insects, which had died in the cells, and shrivelled up into the singular forms which have been described.

The cells are of different sizes, some being more than ten times as large as others. The superior dimensions of the cell seem to be obtained at the expense of the walls, so that the large cells can be broken by the finger and thumb, while the small cells cannot be opened without the knife.

The insects themselves are equally variable, some being mere dots of shining blue and green, while others are about as large as the common red ant of the gardens, but with plumper bodies. In consequence of these two facts, the large, strong-jawed insect can easily make its way through the comparatively thin walls of the large cell in which it was enclosed, while the small and necessarily weak-jawed specimens are utterly unable to pierce the walls of their cells, which are so thick that they must bore a hole equal in length to that of their whole body before they can escape into the air. Consequently, the great mass of the insects that are found in the cells are the small specimens, the larger having made their escape. I find that on an average twenty small insects are thus found in proportion to one of the larger kind.

Nothing is easier than the rearing of insects from this as well as other galls, but to decide upon the species which make them is by no means so easy a task as appears on the surface. Even should the experimenter find the right species of insect in the gauze bag, he has to go through the wearisome task of searching through the family of Cynipidæ, and identifying the species—a process which every entomologist is rather apt to postpone until the visionary period when he shall have leisure.

But it is very probable that the required insect does not make its appearance at all, and that the little hymenoptera which make their way out of the cells, or are found dead within them, are not the rightful occupants of the galls. For the Cynipidæ are as liable to parasites as other insects, and it frequently happens that from a single many-chambered gall will issue insects that sadly puzzle an amateur, as they seem to belong to at least two distinct species. The very gall which has just been described affords a good example of this fact, for in some of the chambers are specimens of the true *Cynips rosæ*, and in others are insects which belong to another family, the Ichneumonidæ, which, as the reader may remember, are parasites upon other insects. They have evidently introduced their eggs into the cells occupied by the larvæ of *Cynips rosæ*, so that the larvæ which have been hatched from these eggs have fed upon the legitimate occupants, and come to maturity in the cells that were designed for others.

Insects of totally different orders sometimes make their appearance. When I began to take to pieces the gall which has been described, I was rather surprised to find among the long hairs an empty cocoon of the Galleria moth, whose ravages have been mentioned in an earlier part of the volume. On further dissecting the gall, no less than twelve other cocoons were found, all buried so deeply in the hairs and among the woody cells that they could not be seen until the hairy clothing was removed. A person who was entirely ignorant of entomology might naturally fancy that the moths were the architects of the gall from which they had apparently issued. How they obtained access to the galls, and on what food they lived, are two problems that I can by no means solve. The drawer in which the galls were placed is tightly closed, and all bee, wasp, and hornet combs have been so treated with corrosive sublimate, that they have not been touched by the caterpillars from which the moths had been developed.

THERE is another gall, very common in England, which is found upon the oak, and which is generally thought, by persons who are unacquainted with botany or entomology, to be the buds which naturally grow upon the tree.

In these curious galls, the excrescences with which they are covered take the form of leaves instead of hairs, as is the case with the bedeguar and many other galls. These bud-like objects may be found on the young twigs, and may be easily recognised by their shape, which somewhat resembles that of a pine-apple, and the curious manner in which their leafy covering lies regularly over them, like the tiles upon an ornamental roof. The size of the gall is rather variable, but it is, on an average, about as large as an ordinary hazel-nut.

The gall is so wonderfully bud-like that I have known the two objects to be confounded—the immature acorns in their cups to be carried off as galls, while the real galls were left on the tree. The incipient naturalist who made the mistake kept the buds for some eighteen months, and was sadly disappointed to find that no insects were produced from them.

The insect whose acrid injection produces this curious effect upon the tree is rather larger than the leaf-gall insect, and is of more slender proportions. It has been suggested that the object of the leafy or hairy covering is, that the insect, which remains in the gall throughout the winter, should have a warm house by which it may be protected from the chilling frost as well as from the wind and rain.

IF the reader will again refer to the illustration, he will see that from the same branch on which the *Cynips Kollari* has formed so many galls, depend two slender threads supporting one or two globular objects. These are popularly called CURRANT-GALLS, because they look very much like bunches of currants from which the greater part of the fruit has been removed. Their colour, too, is another reason for giving them this name, as they are sometimes scarlet, resembling red currants, and sometimes pale cream colour, thus imitating the white variety.

These galls are placed upon the catkins of the oak, which are forced to give all their juices to the increase of the gall, instead of employing them on their own development. Some authors think that the insect which forms them is a distinct species, while others think that the galls are the production of the same insect which forms the leaf-gall, the punctures being made in the stalk of the catkin and not in the nervure of the leaf.

That this supposition may be correct is evident from the fact that the same insect which forms the oak-apples does also deposit its eggs in the root of the same tree, causing large excrescences to spring therefrom, each excrescence being filled with insects. I have often obtained these root-galls, several of which are now before me, some having been cut open, in order to show the numerous cells with which they are filled, and others left untouched, in order to exhibit the form of the exterior. Being nourished by the juices of the root, they partake of the sombre hues which characterise the part of the tree from which they spring, and do not display any of the colours which are seen on the oak-apples which spring from the twigs.

There are, however, distinct species of gall insects which pierce the roots of the oak-tree. One of them is termed *Cynips aptera*, and makes a pear-shaped gall about one-third of an inch in diameter. Each gall contains a single insect, and a number of the galls are often found attached by their narrow end to the root-twigs of the tree, something like a bunch of nuts on a branch. There is another insect which is termed *Cynips quercus-radicis*, which forms a many-chambered gall of enormous size, containing a small army of insects. Mr. Westwood mentions that one of these galls in his possession was five inches long, one inch and a quarter wide, and produced eleven hundred insects, so that the entire number was probably fourteen or fifteen hundred.

No one who is accustomed to notice the objects which immediately surround him can have failed to observe the curious little galls which stud the leaves of several trees, and which are appropriately called SPANGLE-GALLS, because they are as circular, and nearly as flat, as metallic spangles.

These objects had been observed for many years, but no one knew precisely whether their growth was due to animal or vegetable agency. That their substance was vegetable was a fact easily settled, but some botanists thought that they were merely a kind of fungus or lichen, while others supposed that they were the work of some parasitic insect.

When closely examined, these "spangles" are seen to be discs, very nearly but not quite flat, fastened to the leaf by a very small and short central footstalk. Reaumur set at rest the question of their origin by discovering beneath each of them the larva of some minute insect, but he could not ascertain the insect into which the larva would in process of time be developed. The task of rearing the perfect insect from the gall is exceedingly difficult, the minuteness of the species and the peculiar manner in which the development takes place, being two obstacles which require a vast expenditure of care and patience before they can be overcome.

Supposing a branch containing a number of infested leaves to be placed in water and surrounded with gauze, it will die in a week or two, and yet there will be no sign of an insect. If the branch be kept until the winter has fully set in, the desired insects will still be absent, and the experimenter will probably think that his trouble has been thrown away. The real fact is, that the little insects are not developed until the spring of the following year, and that they pass through their stages of the pupal and perfect forms after the leaves have fallen, and while they are still lying on the ground.

Mr. F. Smith, who has given so much time and research to the history of the hymenoptera, has discovered the insect that inhabited the galls to be *Cynips longipennis*, and has remarked that the perfect insects do not make their appearance until the month of March.

WE now pass from the British galls to those which are found in various other countries. A few of the more interesting examples are figured in the accompanying illustration.

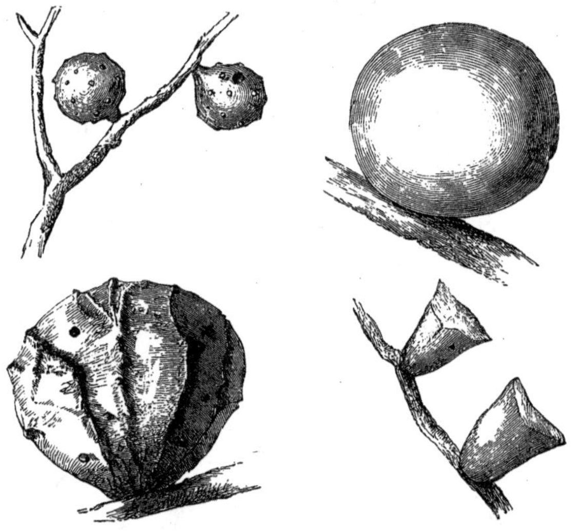

FOREIGN GALLS.

Ink Galls. Dead Sea Apple.
Hungarian Gall. Galls of Cynips polycera.

Should the reader have the curiosity to examine for himself the structure of the British galls (as I trust he will do) he will find that when he cuts a juicy specimen, such as that of the leaf-gall, his fingers will presently be stained with purple-black. He may wash his hands as much as he likes, but he will not wash away the stain, which soon looks as if it had been produced by spilling ink on the hands. There is reason for such an appearance, inasmuch as the staining liquid is really ink, though of a paler hue than that which is used for writing. A little lemon-juice will soon discharge the colour, and then the soap and water will remove the last remnants of the stain.

Ink is made by mixing a solution of the sulphate of iron (properly called green vitriol or copperas) with a decoction of certain oak-galls. Perhaps I may mention that a "decoction" signifies water in which any substance has been submitted to boiling heat, but not dissolved. Tea, for example is, when properly made, a decoction of the leaf, though when made with hot, but not boiling water, it has no right to the name. The solution of copperas is only pale green, and that of the gall is nearly colourless, although when mixed, they become deeply black. The old practical joke of forcing a dupe to stain his hands and face black, depended on the knowledge of these properties.

Before the victim went to wash his hands, some of the decoction of galls was poured into the water, while the towel with which he was supplied had been damped with the copperas solution and then dried. The consequence of this combination was, that although the hands and face might be washed perfectly clean, yet as soon as they were dried with the prepared towel the union of the two substances produced ink, and both hands and face were deeply stained.

Now when a gall is cut with a knife, the slightly acid juice acts upon the steel, and so a kind of ink is produced, which is pale, but still a veritable ink. There is a well-known method of secret writing which depends on this property of iron and tannin, the principle contained in the galls.

A quill pen is dipped in the solution of copperas, and the required message is written, usually between the lines or among the words of a letter on unimportant subjects, so as to avoid the suspicions which would be aroused by a sheet of blank paper. The almost colourless solution leaves no mark, and the letter passes without comment, until it reaches the person who is in the secret. He pours some decoction of galls into a wide and flat vessel, and warily dips the letter into it, so as to wet it; or he saturates a cloth with the decoction, and lays the letter upon it. The tannin then acts upon the solution of iron, ink is formed by their combination, and the formerly invisible words immediately become plain and legible.

A decoction of oak-bark would make ink, though of inferior quality, and so would tea, inasmuch as the tea-leaf contains a large amount of tannin. In fact, whenever the ink in the bottle become thick, it can always be restored by adding to it a little strong tea, which not only gives the requisite liquidity, but does so without affecting the blackness, which would probably be the case if simple water were added.

The two principal ingredients of the ink which is in common use are sulphate of iron and the galls of a species of oak called *Quercus infectoria,* which grows in large quantities in the Levant. They are technically termed Aleppo galls, and are divided into several classes, according to their value. Besides these two ingredients a little gum is added, in order to give consistency, and a very little corrosive sublimate or creosote, to prevent the growth of mould. The proportions are generally six ounces of pounded galls, four ounces of copperas, and four ounces of gum arabic to six pints of water.

In the upper left hand corner of the illustration two of these galls are seen upon a branch of the oak.

They are necessarily much reduced in size, their ordinary dimensions being about equal to those of Cynips Kollari. For the purposes of trade they are divided into black, blue, green, and white galls. The last mentioned class of galls includes those from which the insects have escaped, and which are consequently weakened in astringency. They are so called because they assume a paler hue than the three first classes, in which the insect still remains. In shape, the ink-gall is nearly spherical, with a slight tendency to a pear-like form, and their exterior is defended by a few short, stout, and rather sharp prickles.

I cannot but think that the gall-insect affords a proof that the most insignificant objects of creation have their uses, provided that we could only discover them. Nature is a vast treasure-house, or rather a city of treasure-houses, very few of which have been unlocked because no one has found the keys. No one indeed is likely to do so, as long as he chooses to despise "little things," and if the only acknowledged benefit conferred on mankind by the insect tribes had been the ink-gall, it is a boon so great that every insect ought to deserve our respect as the possible donor of some similar aid to civilization at present unknown.

IN the right hand upper corner of the illustration is seen a gall of some size, and nearly spherical. This is the celebrated DEAD SEA APPLE, of which such strange stories have been told.

This so-called fruit was said to be lovely and beautiful to the eye, but, instead of containing sweet juice, to be filled with bitter ashes, which filled the mouth as soon as it was bitten. Of course, the ashes were supposed to be drawn by the tree from the sunken remnants of the three evil cities beneath the bituminous waves of the Dead Sea, and to present tangible evidence of their existence.

This story, which was implicitly believed for many centuries, was at length as decidedly discredited, and the whole narrative of the ash-filled fruit denounced as a mere fable. However, recent researches have proved, as is often the case, that the main facts of the story are true, though the inference to be derived from them has been entirely mistaken. In the first place, these seeming fruits are not produced by any of those trees which are known to gardeners as fruit-bearers, but are found only upon a species of oak, which is in fact the same tree that furnishes the ink-galls of commerce. At the proper season of the year, the oaks, which are of low stature, and more like scrubby bushes than the stately trees which are suggested by the name of oak, are seen to be covered with round, fruit-looking objects, beautifully coloured, and closely resembling ripe apples. If, however, they are cut open, they will be found to be the habitation of a species of gall-fly, which has been named by Mr. Westwood, *Cynips insana.*

It is evident that if any one were to bite a gall, especially one that was produced from the oak, the exceedingly astringent properties of the excrescence would produce a very rough and ash-like sensation to the palate, which would be increased by the dryness of its substance. Except in size, they much resemble the gall of commerce, and many persons have thought that they are produced by the same insect. These galls are also known as Mala Sodomitica, and Mad Apples, the latter term being the origin of the specific title "*insana*," applied by Mr. Westwood to the insect.

IMMEDIATELY below the Dead Sea Apple, and in the right hand lower corner, may be seen two remarkable objects, which would scarcely be recognised as galls except by an experienced eye. They are, however, the production of an insect called by entomologists *Cynips polycera.*

These galls are found in many parts of Germany, upon the oak-tree, and are at once recognised by their remarkable form. As may be seen by reference to the illustration, they are shaped something like miniature sugar-loaves, and stand boldly from the branch with their broad end uppermost. The body of the gall is slightly conical, so that if cut transversely, it would present a circular section. The end, however, is constructed after a peculiar fashion.

It is nearly flat, "abruptly truncated" according to scientific language, and throws out several projections like horns or spines. The reader will remember that the ink-gall also possesses short and sharp projections, but they start from all parts of the surface, whereas in the present species they belong wholly to the flattened end. Their number is variable, so that the end of the gall is sometimes triangular, and sometimes squared, beside assuming other forms according to the number of projections. This remarkable form has earned for the insect the name of *polycera,* this term being derived from two Greek words which signify "many-horned." The insect which forms this curious gall is about half as large as *Cynips Kollari.*

THE last example which is represented in the illustration is also found in Germany upon the oak, and is made by an insect which is called *Cynips Hungarica.*

This gall is represented of the natural size, whereas all the others are much diminished, in order to be inserted in so limited a space. It is a very remarkable object, and cannot be mistaken for any other species. Its surface is traversed by a variable number of irregular ridges, which all radiate from the stem, and so pass longitudinally over the gall. The whole of the ridges are rough and sharp-edged, but at intervals they shoot out into hard-pointed horns, much like those which arm the preceding species. Indeed, the whole substance of the gall is remarkable for its hard texture, for when cut with a knife it offers as stubborn a resistance as if it were seasoned oak or elm.

That a hymenopterous insect should be able to bore its way through so hard a substance, and to make a tunnel barely wide enough for the passage of its body, and nearly three-quarters of an inch in length, is really surprising. The insect is not a large one, and resembles *Cynips Kollari* so closely that an inexperienced observer would certainly mistake it for that insect, the distinctions being so trifling that they can only be detected by means of the microscope.

CHAPTER XXVI.

PARASITIC NESTS (CONCLUDED.)

The Oak-tree, and its aptitude for nourishing Galls—COMPOUND GALLS, or one Gall within another—The SENSITIVE GALL of Carolina—The Fungus of wine-vaults—Galls and the Insects which caused them—Colours of Galls—Whence derived—The Galls of various trees and plants—The Cynips parasites upon an insect—Galls produced by other insects—Mr. Rennie's account of the BEETLE GALL of the Hawthorn—The BEETLE GALL of the Thistle—DIPTEROUS GALL-MAKERS—Leaf-Miners and Galls—Size of the larvæ of Leaf-Miners—The perfect insect and their beauty—Method of displaying the insect—SOCIAL LEAF-MINERS—DIPTEROUS LEAF-MINERS—Animal Galls—The CHIGOE and its habits—Its curious egg-sac—Difficulty of extirpating it—The penalty of negligence—The BREEZE FLIES and their habitations—WURBLES and their origin—Their influence upon cattle—The CLERUS and its ravages among the hives—The DRILUS, its remarkable form and the difference between the sexes—The curious habitation which it makes.

THE reader cannot but notice the singular aptitude possessed by the oak-tree for nourishing galls. No part of the tree seems to escape the presence of a gall of some sort, diverting its vital powers into other channels. The tree, however, does not appear to suffer from them, and it is just possible that they may be useful to it. The leaves are studded with galls, and so are their stems. The branches are covered with galls of various shapes, sizes, and colours, some bright, smooth, and softly coloured, like ripe fruit, others hard, harsh, spiny, and rough, as if the very essence of the gnarled branches had been concentrated in them. There are galls upon the flowers, galls upon the trunk, and even galls upon the root.

Some oak-galls may be called compound galls. M. Bosc mentions a small gall which is found upon the American oak. It is not larger than a pea, and if shaken is found to contain some hard substance loosely lodged in its interior. When the gall is cut open, a very curious state of things is seen. The walls are very thin, so that in spite of the small dimensions, the cell is larger than that of many cynipidæ. Within the cell, no insect is discovered, but in its place a little spherical object, about as large as a No. 5 shot, which is very hard, and rolls about freely in the interior. If this be opened, the larva is found within it, reminding the adept in fairy lore of the white cat whose gifts were enclosed in a succession of nuts, each within the other. How these singular little cellules are made is not known, though their discoverer expended great trouble and patience upon them.

The same naturalist mentions another species of gall, also found upon the oak in Carolina. It is spherical, covered with prickles like a thistle, and beset with a thick downy covering of rather long hair. Many other galls possess these characteristics, but the most curious point connected with this species is, that the hairs are as mobile as those of the sensitive plant, and as soon as they are touched, sink down, and never afterwards regain their former position.

There is a kind of fungus which is found in wine-vaults, and which exhibits a similar phenomenon. When newly grown, it hangs in great masses, like tufts of pure cotton-wool. But to carry a specimen away is impossible, for, as soon as it is touched, it begins to contract, and in a minute or two shrivels up into a flat membraneous mass, that looks like the web of the house-spider. M. Bosc was unable to rear any of the inmates of these galls.

THE size of a gall is no criterion of the dimensions or numbers of the insect which made it. Even in the galls which infest the oak, the smallest galls often furnish the largest insects, and in some specimens brought from Greece, the gall is as large as an ordinary black-currant, while the cell would contain a red-currant, showing that the inhabitant of the cell must be a large one in order to fill it. Again, although the oak-apple and rose-bedeguar do contain a great number of insects, there are many examples where galls scarcely so large as a pea contain from ten to fifteen insects, while the ink-gall and the large Hungarian gall are inhabited by a single insect.

One of the most curious problems is, to my mind, that of the brilliant colours with which many of these galls are decorated. That the rose-bedeguar should be so beautifully adorned with scarlet and green is a fact which does not seem to excite any astonishment, inasmuch as it may be said that the colours which ought to have been developed in the petals and the leaves have been diverted from their proper course, and forced to exhibit themselves in the gall.

Botanists and physiologists will see that this idea is quite groundless, but to the uninstructed and popular mind it has a sort of plausibility that often commands assent. But when we come to the oak-tree the case is at once altered, and some other cause must be found for the lovely colours of its galls. The cherry-galls are as brightly coloured as any apple, and the soft hues of the oak-apple are nearly as beautiful though not so brilliant. Yet the oak possesses no such store-house of colour as is popularly attributed to the rose. Its leaves are simple green, and its flowerets are so colourless as scarcely to be distinguished by the unassisted eye.

Whence then are derived these beautiful colours? Some hasty observers, who have neglected the first rule of logic, and drawn an universal conclusion from particular premises, have said that the colours of the gall are derived from the insect; adducing, as a proof of their assertion, the brilliant colours which equally deck the rose-bedeguar and the *Cynips rosæ* from which it sprang. But if they had only followed the example of careful naturalists, who, like Dr. Hammerschmidt, have examined and drawn between two and three hundred species of galls, so hasty a generalization would never have been made. The cherry or leaf-gall of the oak is every whit as gorgeously coloured as the bedeguar of the rose, while the insect that made it is quite black. It is true that the diaphanous wings glitter as if they were made of polished gems; but this appearance is due, not to the wings themselves, but to the myriad hairs with which they are regularly studded, each hair acting as a miniature prism by which the light is refracted and broken into the resplendent hues of the rainbow.

Many other trees beside the oak are chosen by certain species of gall-fly, and even the herbs and flowers do not escape the ravages of these remarkable insects. The white poppy, from which is obtained the opium of commerce, is attacked by a species of gall-fly, which lays its eggs in the large head, or pod, and sometimes does much damage to the plant, the delicate divisions between the seed vessels being rendered quite hard and solid, and the pod itself deformed. Mr. Westwood has described a species of gall-fly which infests the turnips, and another species is known to lay its eggs upon wheat.

As if to show that the family of Cynipidæ is really related to the ichneumons, it has been discovered that some species of this family are actually parasitic upon other insects. In treating of this remarkable fact, Mr. Westwood writes as follows:—"The relations of these insects with the following families (*i. e.* Evanidæ and Ichneumonidæ) have been already noticed. It had always appeared to me contrary to nature that a tribe of vegetable-feeding insects should be arranged in the midst of parasites; nor was it until I had an opportunity of ascertaining the parasitic habits of some of the species of the family, that I was enabled to form a just notion as to the true value of the parasitic or herbivorous nature of these insects. In June, 1833, I detected a minute species, *Allotria victrix*, in the act of ovipositing in the body of a rose-aphis, and I subsequently succeeded in hatching specimens of the perfect insect from infested aphides."

A figure of the tiny insect is given, as it appeared while in the act of depositing its eggs, and has a rather remarkable effect from the fact that the very minute dimensions of the parasite make the aphis look quite a large insect. Other species of this family are also known to be parasitic. The rose-aphis is certainly infested by two species of gall-fly, and probably by more, while the aphides which are found on the willow, the cow-parsnip, and other plants, also fall victims to the Cynipidæ. There is one genus of this family, called *Figites,* which is parasitic on the larva or pupa of certain dipterous insects.

THE Cynipidæ are not the only insects that produce galls upon different plants. For example, several species of beetle are known to pass their earlier stages in swellings produced by the puncture of the parent insect. There is a little weevil of a greyish brown, which is mentioned by Mr. Rennie as forming a gall upon the hawthorn.

"In May, 1829, we found on a hawthorn at Lee, in Kent, the leaves at the extremity of a branch neatly folded up in a bundle, but not quite so closely as is usual in the case of leaf-rolling caterpillars. On opening them up, there was no caterpillar to be seen, the centre being occupied with a roundish, brown-coloured, woody substance, similar to some excrescences made by gall insects (*Cynips*).

"Had we been aware of its real nature, we should have put it immediately under a glass, or in a box, till the contained insect had developed itself; but instead of this, we opened the ball, where we found a small yellow grub coiled up, and feeding on the exuding juices of the tree. As we could not replace the grub in its cell, part of the wall of which we had unfortunately broken, we put it in a small pasteboard-box with a fresh shoot of hawthorn, expecting that it might construct a fresh cell. This, however, it was probably incompetent to perform; it did not, at least, make the attempt, and neither did it seem to feed on the fresh branch, keeping in preference to the ruins of its former cell.

"To our great surprise, although it was thus exposed to the air, and deprived of a considerable portion of its nourishment, both from the fact of the cell having been broken off, and from the juices of the branch having been dried up, the insect went through its regular changes, and appeared in the form of a small greyish brown beetle of the weevil family.

"The most remarkable circumstance in the case in question, was the apparent inability of the grub to construct a fresh cell after the first was injured,—proving, we think, beyond a doubt, that it is the puncture made by the parent insect when the egg is deposited that causes the exudation and subsequent concretion of the juices forming the gall." Although the insect in question succeeded in attaining the perfect state, it would probably be of stunted growth in consequence of the deprivation of food. Such, at all events, is the case with insects of other orders, when their supply of food is at all checked while they are in the larval state.

THERE is another weevil, scientifically called *Cleonus sulcirostris*, which is one of the gall-makers. It is one of the largest of the British weevils, being more than half an inch in length, and is very simply clad in grey and black.

If the reader desires to discover the larva of the beetle he may probably be successful by going to any waste spot where thistles are allowed to grow, and examining them carefully about the stems and roots. Nothing is more common than to find the stems of thistles swollen in parts, and in many cases the root is affected as well as the stem. Fortunately for the gardener, who hates thistles, even though he should be a Scotchman, as is so often the case with skilled gardeners, the larva of the Cleonus feeds on the juices of the plant at the expense of its life, so that the thistle dies just before the seed is developed, and a further extension of the plant is thereby prevented.

THERE are also gall-making insects among the Diptera. Such, for example, is the THISTLE-GALL FLY (*Urophora Cardui*), which produces large and hard woody galls upon the thistle, as well as several species of the larger genus Tephritis, some species of which live in the parts of fructification of several flowers, the common dandelion being infested by them.

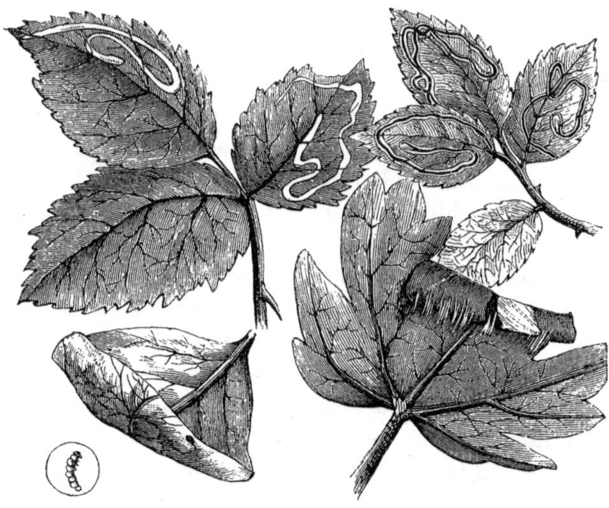

LEAF MINERS AND ROLLERS.

WE may now describe, at fuller length than has hitherto been done, another group of insects, which live between the membranes of leaves, and which belong to different orders.

If the reader will carefully examine the leaves of any rose-tree which grows in the open air, he will certainly remark that many of them are notable for certain curious markings, which look something like the rivers in a map, and which traverse the leaf in various ways. They all, however, agree in one point, namely, their gradual and regular increase in diameter. At their origin, they are so small that the finest thread could hardly pass through them, but in proportion as they increase in length they increase in width, so that at their termination they are sometimes the twelfth of an inch in width.

These marks are the tracks made by very small larvæ, which live between the membranes of the leaves, and feed upon the parenchyma, or soft substance which lies between the two membranes. They follow no rule in their meanderings, but traverse the leaf in a variety of ways. Sometimes they never leave the edge, but follow every little serration of the leaf with perfect accuracy. Sometimes they form a kind of spiral, and sometimes they wander irregularly over the whole leaf. Generally, the insect does not cross the track which it has once made, being diverted from doing so by some wonderful instinct. There are instances, however, where the insect has crossed its own track, not only once, but several times.

If the little gallery be opened at the widest extremity, one of three things will be found. Sometimes there is a tiny white grub, very much resembling the larva of certain beetles, and having the rings which represent the thorax rather wider than those which will afterwards be developed into the abdomen. As the little creature is able to live between the membranes of a leaf so thin as that of the rose or oak, it is evident, to the most superficial observer, that the insect which will be developed from it must be of very minute dimensions.

The larva of all winged insects is very large in proportion to the same insects when they have obtained their perfect form, much of the substance being taken up by the wings. As a natural consequence, it follows that the larger the wings, the larger must be the grub, the size of the body being quite a secondary consideration. In the present case, the larvæ which we are supposed to examine belong to the lepidopterous order, in all of which insects the wings, when present at all, are of great comparative size. If, then, the full-grown larva is so small that it can lie concealed between the membranes of a leaf without causing any conspicuous alteration in its outline, it is evident that the perfect insect must be of almost microscopical minuteness. Accordingly, it has been found that the little moths which have been bred from such caterpillars are so small that they have almost escaped observation until comparatively late years.

How small these insects are may be imagined from the fact that many species of the *Microlepidoptera,* as they are fitly named, do not occupy, even with their wings spread, a space larger than is taken up by the capital letter at the beginning of this sentence. To "set" these tiny creatures is necessarily an extremely difficult task, and cannot be accomplished by the ordinary plan of running a pin through the thorax, and extending the wings on the "setting-board." The only method of displaying them is to set them on white cardboard by means of gum, which is strengthened by many entomologists with various substances. A sheet of cardboard covered with specimens of Microlepidoptera neatly set is a very pretty sight, but needs the aid of the microscope before it can be perfectly seen.

Even to the unaided eye, the tiny moths are seen to be beautifully decorated, their wings gleaming in favourable lights like the throat of the humming-bird. But when placed under the microscope, especially if it be furnished with a binocular tube, and illuminated by a suitable light, the wings are positively dazzling in their brightness, and hues that formerly seemed to be but dun and bronze or brown, suddenly flash out into gold and emerald, each scale distinct and shining as if of burnished metal.

Sometimes, when opening the extremity of the leafy tunnel, we find a tiny chrysalis lying in the little chamber, and awaiting the time for the shell to burst and the perfect insect to emerge. Later in the year, we shall find neither larva nor pupa, but shall see a little hole in the leafy chamber, from which issues the shattered end of an empty chrysalis-shell, showing that the moth has made its escape into the outer air.

Two examples of other mined leaves may be seen upon the illustration, both drawn from the actual object. The specimen in the right-hand upper corner was taken from the bramble, and has been mined by the larva of a little moth called *Nepticula anomella.* It is a very pretty little creature, though its hues are not brilliant without the aid of the microscope. The upper wings are brown, but their tips are beautifully coloured with bright chestnut. The lower wings are pale grey, without any of the brilliancy that distinguishes the upper pair. They possess, however, a compensating beauty in the long, feathery fringe with which they are edged, and which, when subjected to the microscope, is seen to consist of the ordinary scales of the wings exceedingly developed both in length and width.

The leaves on the left hand were taken from the garden-rose, and have been mined by the larva of another species of the same genus, *Nepticula aurella*.

This beautiful little moth derives its specific name from the peculiar colouring of the upper wings, which are bright chestnut, relieved by a broad band of gold across their centre. The tips of these wings are fringed, and the lower pair are nearly white, and edged with a fringe similar to that which has already been described.

As a general rule, the leaf-mining caterpillars are solitary, and if even two or three are found in the same leaf, each leads an isolated life, and does not inhabit the same burrow as its neighbour. There are, however, exceptions to this rule, as to most others, and certain species of leaf-miners inhabiting the henbane, live harmoniously together between the membranes of the same leaf. They are larger than the ordinary species, and are remarkable for their power of burrowing into a fresh leaf when ejected from their former habitation, a power which does not seem to belong to the caterpillars of the Microlepidoptera.

As the meandering tracks of the Microlepidoptera upon the leaves of various plants are very similar in general aspect, I have caused figures of common leaf-rollers to be inserted in the illustration. If the reader will turn to pages 294—298, he will see an account of certain moths whose larvæ roll up the leaves in which they reside. The lilac-leaf on the left hand of the illustration has been rolled up by the larva of a moth named *Gracillaria syringella,* the generic title being given to it on account of its graceful form, and the specific name because it is fond of frequenting the syringa shrub. The larva is shown just below, of its natural size, and the hole through which the perfect insect has escaped can be seen upon the rolled portion of the leaf.

On page 247 may be seen an account of the manner in which the larva performs a task so apparently impossible as rolling up a leaf of such great comparative size. When the little caterpillar is contrasted with the leaf on which it is at work, the contrast is almost ludicrous, for it seems nearly as impossible for so little a creature to roll up so large a leaf as for a man to roll up one of the armour-plates of an iron-clad ship. The manner by which this task is achieved is described on the above-mentioned page, and the reader will be the better able to understand the description if he compares it with the illustration.

The moth which has inhabited this leaf is called *Gracillaria semifascia.*

To return to our leaf-miners.

Although the greater number of these insects belong to the lepidoptera, the rule is by no means an universal one. Many beetles are thus parasitic within the leaves of plants, and, as a general rule, they belong to the family of Curculionidæ, or weevils. There are also several species of dipterous insects which have this habit, among which may be named the CHRYSANTHEMUM FLY (*Tephritis artemisiæ*), which burrows into the leaves of the flower. There is also a genus of flies called Phytomyza, *i. e.* Plant-sucker, the different species of which select particular plants and burrow between the membranes of their leaves. The holly, for example, is infested by one species, the honeysuckle by another, and the common hart's tongue by a third.

WE must now glance at a few of the insects that are parasitic upon other animals. Their numbers are very great, but we must restrict ourselves to those which construct some sort of a habitation.

The only insect which can be said to be parasitic on man, and at the same time to form a habitation, is the celebrated CHIGOE (*Pulex penetrans*), otherwise called the JIGGER, or EARTH FLY. This terrible pest is a native of Southern America and the West Indian islands, and is too well known, especially by the negroes and natives.

This insect, which is closely allied to the common flea, and much resembles it in general appearance, contrives to hide itself under the nails of the fingers or toes, usually the latter. Having gained this point of vantage, it proceeds very gradually to make its way under the skin, and, strange to say, does so without causing any pain. There is a slight irritation, rather pleasing than otherwise, to which a novice pays no attention, but which puts an experienced person on his guard at once.

The male Chigoe is innocent of causing any direct injury to man, the female being the cause of all the mischief. As soon as she is settled, her abdomen begins to swell until it becomes quite globular, and of great comparative size, and containing a vast quantity of tiny eggs. Pain is now felt by the victim, who generally has recourse to the skilful old dames, who have a kind of monopoly of extracting Chigoe "nests." With a needle, they carefully work round the globular body of the buried insect, taking great care not to break it, as if a single egg remains in the wound, all the trouble is wasted. By degrees they gently eject the intruder, and exhibit the unbroken sac of eggs with great glee. To prevent accidents, however, the wound is filled with a little Scotch snuff, which certainly causes rather a sharp smarting sensation, but effectually destroys any egg or young insect that may perchance have escaped notice.

Europeans and natives of the better caste escape easily enough, because they always take warning by the first intimation of a Chigoe's attack, and generally succeed in killing her before she has succeeded in burying herself. Moreover, the shoes and stockings of civilized man protect his feet, and the gloves guard his hands, so that the insect does not find many opportunities of attacking the white man.

But the negroes, and especially the children, suffer terribly from the Chigoe. Children never are very apt at sacrificing the present to the future, and the negro child is perhaps in this particular the least apt of all humanity. The Chigoe is in consequence seldom disturbed until it has made good its entrance, and even then would not be mentioned by the child, on account of the pain which he knows is in store for him. But the experienced eyes of the matrons are constantly directed to the feet of their children, and if one of them is seen to hold his toes off the ground as he walks, he is immediately captured and carried off to the operator, uttering dismal yells of apprehension.

He certainly has good reason for his fears. The Chigoe nest is duly removed, and then, partly to prevent the hatching of any egg that may have escaped during the operation, and partly to punish the delinquent for his disobedience, the hollow is filled, not with snuff (which is too valuable a substance to be wasted), but with pounded capsicum. The discipline is certainly severe, but it is necessary. After a child has once paid the penalty of negligence, he seldom chooses to bring such a punishment on himself a second time, and as soon as he feels the first movements of a Chigoe, away he goes to have it removed before it can burrow under the skin.

If the Chigoe be allowed to remain, the results are disastrous. Swellings make their appearance along the limbs, the glands become affected, and if the cause is permitted to remain undisturbed, mortification takes place, and the sufferer dies. So the red-pepper discipline, severe as it may be, is an absolute necessity with those who are unable to reason rightly, or to exercise forethought for the future. Every evening the negro quarter of the villages is rendered inharmonious by the outcries of the children who have neglected to report themselves in proper time, and who in consequence are suffering the penalty of their negligence.

THERE are some insects which produce upon animals certain swellings which are analogous to the galls upon trees. Such, for example, is the well-known BREEZE FLY (*Œstrus bovis*), which is so troublesome to cattle. The larvæ of this insect live under the skin of the animal, and in some manner raise a large swelling, that is always filled with a secretion on which they live. In fact, the swelling is a gall produced on an animal instead of a plant, and the enclosed insect feeds in a similar manner upon the abnormal secretion which is induced by the irritation of its presence.

The larvæ are fat, soft, oval-bodied creatures, and are notable for the flattened end of the tail, on which are placed two large spiracles or breathing-holes.

Although the larva which inhabits the vegetable gall seems to have but small need of air, and to all appearance can exist without any apparent channel of communication with the external atmosphere, such is not the case with the inhabitant of the animal gall. An opening is always preserved in the upper part of the swelling, and the tail of the grub is tightly pressed against the aperture so as to ensure a constant supply of air.

In the months of May and June, these swellings may be found in great plenty. They are mostly seen upon young cattle, and as a general rule are situated close to the spine. So common indeed are they, that out of a whole farm-stock of cattle I have seen almost every cow under the age of four years attacked by the Breeze Fly, and counted from two or three to twelve or fourteen upon a single animal. It is said that as many as forty have been detected upon a single cow, but such an event has not come within my own observation.

The swellings caused by the Breeze Fly are called Wurbles, or Wornils, and can be easily detected by passing the hand along the back. Strangely enough, the cow does not appear to feel any pain from the presence of these large parasites, nor does she suffer in condition from them, although it would seem that they must keep up a continual drain upon the system. Indeed, some experienced persons have thought that, instead of being injurious, they are absolutely beneficial.

When the grub has reached its full development, it pushes itself backwards out of the gall, and falls to the ground, into which it burrows. Presently, the skin of the pupa becomes separated from that of the larva, and the latter dies, and becomes the habitation in which the pupa lives. The head portion of the skin is so formed that it flattens when dry, and can easily be pushed off, like the lid of a box, permitting the perfect fly to escape. Even when the insect is still in its pupal condition this lid can be removed, so that the pupa can be seen within its curious habitation. I may mention here that insects which are thus covered while in their pupal state, so as to show no traces of the creature within, are said to undergo a "coarctate" metamorphosis. Nearly all the diptera are examples of the coarctate insects.

Before we close the subject of parasites, it will be needful to give a brief account of one or two parasitic insects which possess points of peculiar interest in the habitations which they make, or in the places wherein they find their abode.

One of these insects is a rather pretty beetle, termed *Clerus alvearius.* In its perfect state it is innocent enough, but in its larval state it is so destructive among the hives, that all bee-keepers will do well to destroy every Clerus that they can catch.

It is generally to be found on flowers, licking up their sweet juices by means of a brush-like apparatus attached to the mouth. The wing-cases of most of the species are bright red, barred or spotted with purple.

The larva is of a beautiful red, and is hatched from an egg placed in the cell occupied by the bee-grub. As soon as it is hatched, it proceeds to feed upon the bee-grub, and devours it. Unlike many insects with similar habits, it is not content with a single grub, but proceeds from cell to cell, devouring all their inhabitants. When it has eaten to the full, it conceals itself in the cell, and spins a cocoon of rather small dimensions in comparison with its own size. In process of time, it is developed into a perfect insect, and then breaks out of its cocoon and leaves the hive, secure from the bees, whose stings cannot penetrate the horny mail in which it is encased.

THERE is another beetle which is parasitic upon snails, and which, in its larval and pupal states, is only to be found within those molluscs. Its scientific name is *Drilus flavescens*, the latter name being given to it in honour of its yellow-tinted wing-cases, which present a pretty contrast with the black thorax. It is a little beetle, scarcely exceeding a quarter of an inch in length, and is remarkable for the beautiful comb-like antennæ of the male. As for the female, she is so unlike her mate that she has been described as a different insect. She has no pretensions to beauty, and can scarcely be recognised as a beetle, her form being that of a mere soft-bodied grub. Moreover, the size of the two sexes is notably different. The male is, as has already been observed, only about a quarter of an inch long, while the female is not far from an inch in length, and is broader than the length of her mate, antennæ included.

This curious insect lives in the body of snails, the common banded snail of our gardens being its usual prey. When it is about to change into the perfect state, it makes a curious cocoon, of a fibrous substance, which has been well likened to common tobacco, the scent as well as the form increasing the resemblance. The grub or larva of this beetle bears a very great resemblance to the perfect female, and indeed is so similar that none but an entomologist could distinguish the two creatures. It is furnished with a number of false legs, as well as with a forked appendage at the end of the tail, by which it is enabled to force its way into the body of its victims. The head is pointed, and the jaws are very powerful.

CHAPTER XXVII.

BRANCH-BUILDING MAMMALIA.

The DORMOUSE in Confinement, and at Liberty—Nest of the Dormouse—Its position, materials, and dimensions—Entrance to the nest—The winter treasury—The LOIRE and the LEROT—Man as a Branch-builder—Moselekatze—His conquests—Effects upon the people—Branch-houses—Their approaches.

WE now come to another division of the subject, namely, the nests that are built in branches, and adhering to the system which has been followed through the progress of the work, we shall take first the branch-building mammalia.

There are but few mammals which can be reckoned in this division, but our little island produces two of them, namely, the squirrel and the DORMOUSE (*Muscardinus avellanarius*). The former of these animals has been already described at page 196.

The pretty little brown-coated, white-bellied Dormouse is familiar to all who have been fond of keeping pets. There is no difficulty in preserving the animal in health, and, therefore, it is a favourite among those who like to keep animals and do not like the trouble of looking after them. It is, however, rather an uninteresting animal when kept in a cage, as it sleeps during the greater part of the day, and the sight of a round ball of brown fur is not particularly amusing.

When kept in confinement, it is obliged to make for itself a very inartificial nest, because it is deprived of proper materials and a suitable locality. It does its best with the soft hay and cotton wool which are usually provided for it, but it cannot do much with such materials. But when in a state of liberty, and able to work in its own manner, it is an admirable nest-maker. As it passes the day in sleep, it must needs have some retired domicile in which it can be hidden from the many enemies which might attack a sleeping animal.

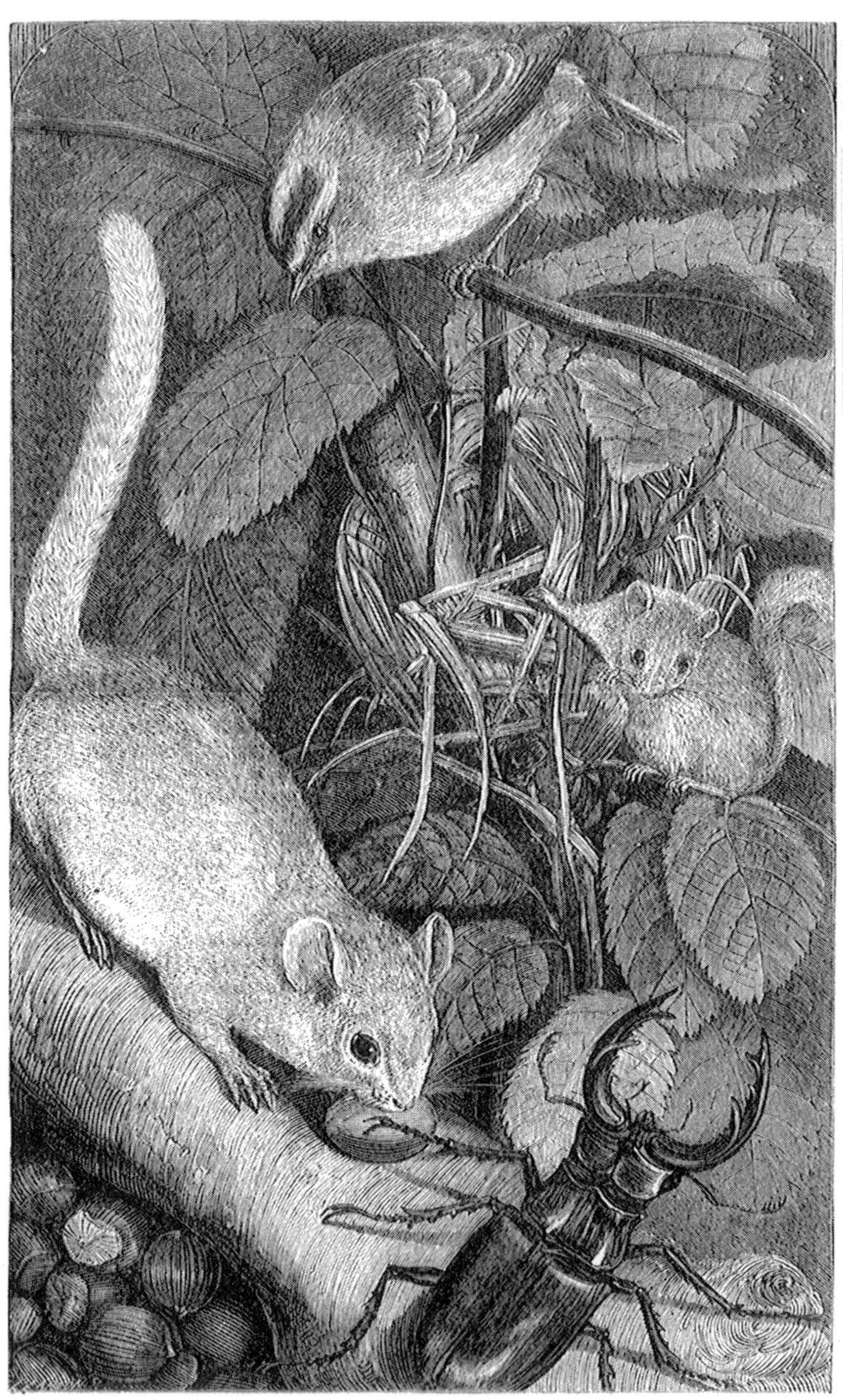

DORMOUSE AND NEST.

One of these nests is depicted in the illustration, and the specimen from which it is drawn forms a part of my collection. It was situated in a hedge about four feet from the ground, and, as may be seen by reference to the illustration, is placed in the forking of a hazel branch, the smaller twigs of which form a kind of palisade round it. The substances of which it is composed are of two kinds, namely, grass-blades and leaves of trees, the former being the chief material. It is exactly six inches in length by three inches in width, and is constructed in a very ingenious manner, reminding the observer of the pensile nests made by the weaver birds, which have already been described at page 199.

Two or three kinds of grass are used, the greater part being the well-known sword-grass, whose sharp edges cut the fingers of a careless handler. The blades are twisted round the twigs and through the interstices, until they form a hollow nest, rather oval in shape. Towards the bottom the finer sorts of grass are used, as well as some stems of delicate climbing weeds, which are no larger than ordinary thread, and which serve to bind the mass together. Interwoven with the grass are several leaves, none of which belong to the branch, and which are indeed of two kinds, namely, hazel and maple, and have evidently been picked up from the ditch which bounded the hedge. Their probable use is to shield the inmate from the wind, which would penetrate through the interstices of the loosely woven grass-blades

The entrance to the nest is so ingeniously concealed, that to find it is not a very easy matter, even when its precise position is known, and in order to show the manner in which it is constructed, one of the Dormice is represented in the act of drawing aside the grass-blades that conceal it. The pendent pieces of grass that are being held aside by the little paw are so fixed, that when released from pressure, they spring back over the aperture and conceal it in a very effectual manner.

Although the Dormouse uses this aerial house as a residence, it does not make use of it as a treasury. Like many other hibernating animals, it collects a store of winter food, which generally consists of nuts, grain, and similar substances. These treasures are carefully hidden away in the vicinity of the nest, and in the illustration the animal is shown as eating a nut which it has taken from one of its storehouses beneath the thick branch.

During the winter the animal does not feed much upon its stores, inasmuch as it is buried in the curious state of hibernation during the cold months. At the beginning of spring, however, the hibernation passes off, and is replaced by ordinary sleep, with intervals of wakefulness.

Now, while the animal hibernates, the tissues of the body undergo scarcely any change, even though no nutriment be taken. But, as soon as the creature resumes its ordinary life, waste goes on, and the creature soon feels the pangs of hunger. As the food of the Dormouse consists chiefly of seeds and fruits, it could not find enough nourishment to support the body, and would therefore perish of hunger but for the stores which instinct had taught it to gather in the preceding autumn.

In the illustration, the stag-beetle and the golden-crested wren have been introduced to show the comparative size of the animals. The old Dormouse does not fear the beetle, and tranquilly pursues his meal, but the young one is rather discomposed at the intrusion of the big black insect, and meditates a retreat into the nest.

There are several species of Dormouse, which have similar habits, and at least two of them are found in Europe. These are the LOIRE (*Myoxus glis*) and the LEROT (*Myoxus quercinus*). The former of these animals is sometimes called the Fat Dormouse, because it was in ancient days considered as a great delicacy, and carefully fattened in places called *gliraria*. This animal is found in France.

In many parts of the same country the Lerot is a great pest to the gardens, because it is fond of fruit, and has a special liking for the ripest peaches, nectarines, and similar choice fruit.

BESIDES those which have been mentioned, several other species of mammalia make aerial nests in the branches of trees, though such nests are only to be considered as exceptions from the general rule. Perhaps the most singular of these exceptions is that which has been discovered in Africa, where human beings systematically build their houses in trees. This curious fact was discovered by Mr. Moffatt, the well-known missionary, in the course of his travels.

Two traders had been in the country which was ruled by the ferocious chief, Moselekatze, who has derived an unenviable name for his relentless cruelty, which will always eclipse his well deserved reputation as a man of commanding genius and of subtle intellect. He was, in fact, a savage Napoleon, and, if possible, even a more wonderful man, inasmuch as he had no education, and created the terrible power which he so skilfully wielded.

King of the Zulu Kafirs, he had organized a vast military establishment, and had invented a system of warfare so ingenious, as to entitle him to the name of a born general. All able-bodied men were forced to serve as soldiers, drafted into different regiments according to their capacities of strength, swiftness, or cunning, and when they went into action had the alternative of victory or death, a fugitive being invariably killed by the executioner.

When his white visitors were about to return to their homes, Moselekatze thought that it would be a good opportunity of extending his knowledge, and consequently his influence, by learning the manners and customs of white men, and therefore sent two of his councillors to accompany the visitors to their homes, to inspect their proceedings, and then to return and report what they had seen. The envoys carried out the instructions of their master, though their brains could scarcely retain the vast stores of new facts which were continually poured into them, and in due time they wished to return to their own country.

Here, however, was a difficulty. In order to reach Zulu-land, they must pass through tracts inhabited by other nations, all of which had been invaded and harried by the conquering troops of Moselekatze, and they knew full well that if their identity were recognised, they would be murdered in retaliation by the incensed owners of the land.

In this strait Mr. Moffatt offered to accompany them until they had reached the boundaries of their land, and set off with them. When he had fulfilled his promise, he was about to return, but his guests begged so earnestly that he should go on and visit their king, that he yielded to their request. Contrary to the usual habits of the Kafir, Moselekatze was grateful to Mr. Moffatt, saying that "the kindness which had been done to his servants had been done to him, Moselekatze the son of Machobane."

On this journey Mr. Moffatt's attention was taken by a magnificent tree, under whose shadow were a number of human beings moving about. On approaching nearer, he found that the tree was close to water, and on looking upwards he saw that a number of little huts were among the branches. Seventeen of these huts were completed, and three more were in course of erection. These were the dwellings of the natives who had been seen under the tree, and were constructed in a very ingenious manner.

Where two or three branches spread their forked boughs horizontally, a number of sticks were laid so as to form a platform about seven or eight feet in diameter. Upon this platform was erected the hut, a necessarily small edifice, consisting of sticks fastened together so as to make a conical-shaped hut, about six feet in diameter at the bottom, and barely as much in height, so that a tall man could hardly lie at full length even when occupying the very centre of it. The roof of the hut was made of grass, and the sides were wattled with the same substance. As the hut was always placed at one edge of the scaffold, the opposite edge afforded a small landing or platform, about a foot or eighteen inches in width. The only method of approaching these curious huts was by means of notches cut in the trunk of the tree, the owners not daring to trust to any less difficult means of ascent.

We now ask ourselves why the natives chose to live in such small and inconvenient dwellings, when there was ample space on the fertile ground for a village. Moselekatze was the cause. His armed hordes, with their wonderful discipline, had swept over the country, destroyed all military power, carried off the cattle, in which consists the wealth of the South African, killed many warriors, and disarmed the rest. Under these circumstances, the wild beasts began to increase in number and audacity, and the enfeebled members of the tribe were, perforce, obliged to abandon their ordinary mode of life, and to reside among the branches where the lions could not reach them. During the day they were tolerably safe, but at night they retired to the trees.

In one of these aerial huts Mr. Moffatt passed the night, having previously shot a rhinoceros, and put the hump into a deserted ant-hill which was used as an oven. During the night the lions came and did their best to devour the meat, the savoury smell of which attracted them on all sides. Fortunately for the travellers, the oven was too hot for the lions, and although they growled and snarled over it all night, they dared not attack it, and retired in the morning. The chief food of the people who inhabit these huts consisted of locusts and roots, for their cattle were gone, they could not make fences wherein to inclose a patch of cultivated ground, the lions had driven away the smaller game, and the few weapons which had escaped Moselekatze were insufficient for the slaughter of the larger and more powerful animals.

CHAPTER XXVIII.

FEATHERED BRANCH-BUILDERS.

The Rook and its nesting-place—Materials and structure of the nest—Some habits of the Rook—The Crow—Difference between the nest of the Rook and the Crow—The Heron and its mode of nidification—The Heronry at Walton Hall—Rustic ideas respecting the Heron's nest—The Chaffinch—Locality and structure of its beautiful nest — Mode of obtaining materials — The Goldfinch and its home — Distinction between the nests of the Goldfinch and Chaffinch—The Bullfinch—Locality and form of its nest—Variability of structure—The Blue-eyed Yellow Warbler—Curious materials of its nest—Its remarkable habits—The Bald-headed Eagle—Why so called—Wilson and Audubon's account of its nest — The Golden Oriole and its beautiful nest — Mode of catching the bird — The Red-winged Starling, its value and demerits—Its gregaricus habits—Locality and structure of its nest—The Yellow-breasted Chat and its odd ways—Its courage and affection for its nest and young—Structure of its nest—The Ringdove and its curious nest—The Whitethroat—Description of the locality and structure of the nest—Reasons for its various popular names—The Mocking Bird—The Water-hen and its nesting—Its habit of covering the eggs.

We pass now to the many birds which build their nests on branches of trees or shrubs, and which may therefrom be termed Aerial Builders. A vast proportion of the feathered tribes select branches as a site for their habitation, so that only the remarkable examples will be mentioned or figured.

Perhaps the most conspicuous of all ordinary branch-nests are those which are made by the Rooks and the Crows.

Every one has seen the nests of the former of these two birds. They are large, dark, and are placed upon the topmost boughs of the tree, so that they can be seen at a considerable distance. Their position is evidently intended as a safeguard against the attacks of various enemies, among which the bird-nesting boy is pre-eminently the most dangerous. Scarcely would the boughs endure the weight of a cat or monkey, and so slender are they in many cases, that the spectator wonders how they can support the nest with its living contents of a parent and three or four young.

The foundation of the nest is composed of sticks of various sizes and lengths, all, however, being tolerably light and dry, the Rook generally carrying up the dead branches that have been blown down by the winds of the preceding winter. These are usually interlaced among the spreading branches of a convenient spray, and thus form a rude basket-work, in which will lie the softer materials on which the eggs and young are to repose. The lining is composed almost entirely of long and delicate fibrous roots, which are intertwined, so as to make an interior basket very similar in general construction to the twig basket of the exterior, and being so independent of it that, with a little care, it can be lifted out entire.

On this soft bed are laid the eggs, which are four or five in number, and are rather variable in colour, the usual tint being greenish grey, largely spotted, mottled, and splashed with dark brown, in which a shade of green is visible. They vary in size as well as in hue, and from the same nest I have taken eggs of so different an aspect that a casual observer would probably think them to be the production of distinct birds.

The principal labours of nest-building fall on the young birds, inasmuch as the elders mostly return to the same domicile every successive season, and are seldom obliged to make an entirely new nest. The young builders are sometimes aggrieved at this distribution of labour, and try to equalize it by helping themselves to the sticks belonging to other proprietors. The general community, however, never suffer theft to be perpetrated, and are sure in such a case to scatter the ill-gotten materials, and force the dishonest birds to begin their labours anew.

When the young are launched upon the world and able to get their own living, the nest is used no more, but is abandoned both by parents and young, not to be again used until repaired in the spring of the following year. It is a curious point in the economy of the Rook, that, when it has abandoned its temporary home, it does not choose to repose among the trees on which the nest was made. Mr. Waterton, who possesses invaluable opportunities for studying the habits of this bird, and has developed them to the utmost, makes the following remarks upon the roosting of this bird :—

"There is no wild bird in England so completely gregarious as the Rook, or so regular in its daily movements. The ring-doves will assemble in countless multitudes, the finches will unite in vast assemblies, and waterfowl will flock in thousands to the protected lakes, during the weary months of winter; but when the returning sun spreads joy and consolation over the face of nature, these congregated numbers are dissolved, and the individuals retire in pairs to propagate their respective species. The Rook, however, remains in society the year throughout. In flocks it builds its nest, in flocks it seeks for food, and in flocks it retires to roost.

"About two miles to the eastward of this place are the woods of Nostell Priory, where from time immemorial the Rooks have retired to pass the night. I suspect, by the observations which I have been able to make on the morning and evening transit of these birds, that there is not another roosting-place for at least thirty miles to the westward of Nostell Priory. Every morning, from within a few days of the autumnal to about a week before the vernal equinox, the Rooks, in congregated thousands upon thousands, fly over the valley in a westerly direction, and return in undiminished numbers to the nest, an hour or so before the night sets in.

"In their morning passage, some stop here; others in other favourite places, farther and farther on; some repairing to the trees for pastime, some resorting to the fields for food, till the declining sun warns those which have gone farthest that it is time they should return. They rise in a mass, receiving additions to their numbers from every intervening place, till they reach this neighbourhood in an amazing flock. Sometimes they pass on without stopping, and are joined by those which have spent the day here. At other times they make my park their place of rendezvous, and cover the ground in vast profusion, or perch upon the surrounding trees. After tarrying here for a certain time, every Rook takes wing. They linger in the air for awhile, in slow revolving circles, and then they all proceed to Nostell Priory, which is their last resting-place for the night.

"In their morning and evening passage, the loftiness or lowliness of their flight seems to be regulated by the state of the weather. When it blows a hard gale of wind, they descend the valley with astonishing rapidity, and just skim over the tops of the intervening hills, a few feet above the trees: but when the sky is calm and clear, they pass through the heavens at a great height, in regular and easy flight."

This custom of the Rooks is the more curious because it is hardly possible to conceive any roosting-place which would be more acceptable to a sensible bird than the woods within the confines of Walton Hall. As has already been mentioned, the birds will occasionally rest for a while in those pleasant woods, though they ultimately take wing for the accustomed roosting-place. There is plenty of space for them; they have their choice of trees on which to settle, and the lofty wall which surrounds them ensures their freedom from all disturbance.

VERY similar in general aspect to the rook, the CROW (*Corvus corone*) builds a nest which resembles that of the rook in outward form, but is easily distinguished by an experienced eye. The lining of the nest is made of animal instead of vegetable substances, hair and wool taking the place of fibrous roots.

Viewed from the foot of the tree, the nest of the Crow is nothing but a large and nearly shapeless bundle of sticks, but when the enterprising naturalist has climbed to the summit of the tree in which it is placed, and can look into the nest, he is always gratified by the peculiarly neat and smooth workmanship of the aerial home. The outside of the nest is rough and rugged enough, but the inner nest, which is made of rabbit's-fur, wool, and hair, is woven into a basin-like form, beautifully smooth, soft, and elastic. On this bed repose the eggs, which are somewhat like those of the rook, but darker and greener, and more thickly spotted, though they are extremely variable in size and colour, and sometimes resemble so closely those of the rook that the distinction can hardly be detected.

The Crow always builds at the tops of trees, and has a wonderful knack of choosing those which are most difficult of ascent. The nests are plentiful enough, but the proportion of eggs taken is very small in comparison. There are some nests which baffle almost any one to rob successfully. An experienced nest-hunter is always endowed with a strong head, and ought to be perfectly at his ease on the summit of the loftiest trees, even though he should be obliged to crawl in fly-fashion under a branch, to hang by one hand while he takes the eggs with the other, or to suspend himself by his legs in order to get at a nest below him That a nest should escape a properly qualified hunter is simply impossible, but to secure the eggs is quite another matter.

In many cases the nest of the Crow is placed on branches so long and so slender that they will not endure the weight of a small boy, much less of a man, and the only method of getting at it is by bending down the branches. But, when the branches are bent, the nest is tilted over, and out fall the eggs, so that the disappointed hunter loses all his time and trouble.

Possibly this extreme caution may be the result of sad experience, for, although the generality of Crows' nests are placed in the most inaccessible positions, I have seen and taken many which were so easy of attainment that in a very few minutes I had ascended the tree and returned with the eggs. There are generally four or five eggs, although in some exceptional cases six eggs are said to be laid in a single nest. I never saw more than five, though I have examined very many nests. High as the nest of a Crow may be, it is always worthy of an ascent, for, even should it be an old nest and deserted by the original inhabitant, there is always a possibility that it may have been usurped by some hawk, whose beautiful eggs are always considered as prizes.

THERE is a splendid British bird, which is becoming scarcer almost yearly, which makes a nest something like that of the crow and rook, but much larger. This is the HERON (*Ardea cinerea*), one of the very few large birds which still linger among us.

On account of its own great size, the Heron makes a very large and very conspicuous nest, built chiefly of sticks and twigs, and placed on the summit of a tree.

Like the rook, the Heron is gregarious in its nesting, so that a solitary Heron's nest is very seldom seen, though now and then an exception to the general rule is discovered. To watch the manners and customs of this bird is not a very easy task, because the number of heronries in England is very small, and the shy nature of the birds renders them difficult of approach. At Walton Hall, however, the Herons are so fearless, through long-continued impunity, that they will allow themselves to be watched closely, provided that the observer is quiet, and does not make a noise, or alarm the birds by abrupt movements.

It is a very pretty sight to watch the great birds as they go to and from their nests, bringing food to their young, or flying to the lake in search of more fish. Numbers of the Heron may be seen at the water's edge, sometimes standing on one foot, with their long necks completely hidden, and their bayonet-like beaks projecting from their shoulders. For hours the birds will retain this attitude, which to a human being would be the essence of discomfort, and it is really wonderful how they can keep up for so long a time the muscular energy which is expended in holding up the spare leg and keeping it tucked under their belly.

Now and then, one of the Herons seems to wake up, and after a stretch of the neck and a flap of the wings, walks statelily and deliberately into the water, through which it stalks, examining every inch of bank and every cluster of weeds as it passes along. Presently the bird pauses, and remains quite still for some time, when the long neck is suddenly darted forwards, the beak disappears for a second among the reeds, and presently emerges, with a fish, frog, or water-rat in its gripe. The real beauty of the Heron can never be appreciated until it is seen at liberty, and in the enjoyment of its natural life. It suits the locality so well that, when it flies away, the spot has lost somewhat of its charms. As it stands in the water, intent upon catching prey, the drooping feathers of its breast wave gracefully in the breeze, and the ripples of the sunlit water are reflected in mimic waves upon its grey plumed wings.

Generally it cares little for exerting itself until towards the evening, but then it becomes impatient and restless, and is not quieted until it has obtained some food.

Some anglers have an idea that the Heron is one of the birds that ought to be ranked as "vermin," thinking that it destroys so many fish, that it ruins an angler's sport. Consequently, they kill the bird whenever they can manage to do so, and flatter themselves that they are doing good service in preserving the breed of fish. Now, even were the entire diet of the Heron to consist of fish, the bird would really do but little harm, because it can only take food in shallow water, and is seldom to be seen more than a yard or two from the bank. But the diet of the Heron is by no means exclusively of a fishy nature, inasmuch as the bird eats plenty of frogs and newts, and will often secure a water-rat even when fully grown. It is seldom that fish which are of any value to the angler come into water in which the Heron could catch them, and even if they did so, their size would prevent the bird from taking them.

At Walton Hall, where the Herons breed largely, and where they procure nearly all the food for themselves and young out of the lake, there is no lack of fish, as may be practically proved by any one who is permitted to cast a line into the water. I am a very poor fisherman, and yet I never found any difficulty in taking in the course of the morning quite as many fish as could easily be carried home.

So far indeed is the Heron from injuring the interests of the angler, that it is a positive benefactor. Mr. Waterton, who was obliged by the continual burrowing of water-rats to drain and fill up a series of large ponds, makes the following remarks on the bird:—"Had I known then as much as I do now of the valuable services of the Heron, and had there been a good heronry near the place, I should not have made the change. The draining of the ponds did not seem to lessen the number of rats in the brook; but soon after the Herons had settled here to breed, the rats became exceedingly scarce, and now I rarely see one in the place where formerly I could observe numbers sitting on the stones at the mouth of their holes, as soon as the sun had gone below the horizon."

When the Heron flies to its nest from any great distance, it generally ascends to a considerable height, and is in the habit of uttering a curious and very harsh cry, which at once tells the naturalist that a Heron is on the wing. When a Heron passes immediately over the observer, the effect is very remarkable, the long, stretched-out legs and neck and slender body looking like a large knitting-needle supported on enormous wings.

To see the Heron alight on its nest or on a branch is rather a curious sight. The bird descends, drops its long legs, places its feet on the branch, and then flaps its huge wings as if to get its balance before it settles down. The rustics have an idea that a Heron is obliged to allow its legs to dangle on either side of the nest while it sits on its eggs, and some will aver that a hole is made in the nest through which the legs can be thrust. It is scarcely necessary to say that the construction of a bird's legs prevents it from assuming such an attitude, and that the long Heron can sit as easily upon its pale green eggs as the short-limbed domestic fowl on her white eggs.

Some of our common British birds build nests that can vie, in point of beauty and delicacy, with any nest made by birds of other lands. It is scarcely possible to conceive a nest which is more worthy of admiration than that of the Long-tailed Titmouse, which has already been described; and in their own way, the

NEST OF THE CHAFFINCH.

houses erected by the Chaffinch and Goldfinch are quite as beautiful. As there are some points of similarity in the two nests, they will be mentioned in connexion with each other.

First, we will take the nest of the Chaffinch (*Fringilla cœlebs*).

Although the beautifully-spotted eggs are plentiful in the collection of every nest-hunting schoolboy, they do not come into his little museum for some time. The eggs of the blackbird, thrush, and hedge-warbler are generally the first to be found, because the nests in which they are contained are so conspicuous. But the nest of the Chaffinch is never easily seen, and its discovery requires a special training of the eye.

An experienced nest-hunter will always detect it, and it is amusing to watch the bewildered expression of a novice to whom a Chaffinch-nest is pointed out, and who cannot see it in spite of all the indications of his instructor. The bird likes to find the fork of a tree or bush, where several branches are thrown out from one spot, and so as to form a kind of cup in which the nest can lie. Tall hawthorns, or even sloe or crab-trees, especially if they grow in thick hedges, are favourite trees with the Chaffinch, and a luxuriant and untrimmed hedgerow is always prolific in Chaffinch-nests.

Within the forked branches, the bird constructs its nest, and does so in rather a singular manner. The chief material is wool, which is matted together so as to form a kind of loose felt, and with this felt are woven delicate mosses, spider-webs, cottony down, and lichens. The last-mentioned materials are stuck most ingeniously upon the outside of the nest, and have the effect of making it look exactly like a natural excrescence from the tree in which it is placed.

This pretty nest is generally deep in proportion to its width, and is lined with hairs, arranged in a most methodical manner, so as to form a cup for the eggs. The hair of the cow is much used by the Chaffinch, which may be seen collecting its stock of hairs from the fields wherein cows are pastured, not plucking them directly from the body of the animal, but searching for them in the crevices of the trees and posts against which the cattle are accustomed to rub themselves. Mostly, the bird can only procure single hairs; but when it is fortunate enough to find a tuft or bunch of hairs, it pulls them out, and works them separately into the nest, so as to ensure the needful uniformity The hair of the horse is largely used by the Chaffinch, as is the fur of several other animals; but in the generality of nests the hairs of the cow predominate.

The texture of the nest is very strong, and, owing to the

nature of the materials, is very elastic, returning to its original shape even after severe pressure. Boys seldom take the eggs of the Chaffinch, because they are so plentiful; but they are too apt to take the nest itself, knowing that it makes a safe and convenient basket for the eggs of rarer birds, and forgetting that they cause much sorrow to the poor birds that have spent so much trouble in preparing their home.

As I have already mentioned, there is some resemblance between the nest of the chaffinch and that of the GOLDFINCH (*Fringilla carduelis*).

NEST OF GOLDFINCH.

In point of beauty, neither yields to the other, for the materials are much the same, and the mode of structure is nearly identical. The nest of the Goldfinch, however, is shallower than that of the chaffinch, and the lichens and moss of which it is partly made are not stuck on the outside, but are woven so deeply into the walls that the whole surface is quite smooth.

The position of the two nests, however, is very different. Instead of choosing the forks of a bough, the Goldfinch likes to make its nest near the end of a horizontal branch, so that it waves about and dances up and down as the branch is swayed by the wind. It might be thought that the eggs would be shaken out by a tolerably sharp breeze, and such would indeed be the case, were they not kept in their place by the form of the nest. If one of the best examples be examined, it will be seen to have the edge thickened and slightly turned inwards, so that, when the nest is tilted on one side by the swaying of the bough, the eggs are still retained within. I have seen the branches of a tree violently agitated by ropes and sticks, and noticed that the eggs in a Goldfinch-nest retained their position until the branch was struck upwards close to the spot on which the nest was made, all the previous agitation having failed to dislodge them.

The lining of the Goldfinch's nest is unlike that which is used by the chaffinch. The latter bird mostly employs hair, while the former makes great use of vegetable-down, such as can be obtained from the willow, the coltsfoot, and other plants. Like other birds, the Goldfinch will not take needless trouble, and if it can find a stray tuft of cotton-wool, will carry it off, and work it into the nest. Sheep-wool is also used for the same purpose; but the bird likes nothing so well as down, and will use it in preference to any other material. On this soft bed repose the five pretty eggs, white, tinged with blue, and diversified with small greyish-purple spots. Now and then a small streak is seen; but the spots are the rule, and the streaks the exception.

Altogether, it is hardly possible to find a more beautiful group than is made by a pair of Goldfinches, their nest, and eggs.

THE nest of the BULLFINCH (*Pyrrhula vulgaris*) is unlike that of the goldfinch, though it is sometimes found in similar localities. This bird seems to be rather capricious in its ideas of nest-making, sometimes preferring trees, and sometimes building in shrubs.

There was a little spinney which I once knew, in which were any number of Bullfinch-nests, the underwood being very attractive to the birds. All the nests were built very low, seldom more than four feet from the ground, and, to the best of my recollection, were placed among the branches of hazel and dogwood. The nest of the Bullfinch is by no means so neat and smooth as that of the goldfinch, but is made in a much looser manner; the foundation being formed of slender twigs, usually those of the birch, and the inner wall of the nest woven of delicate fibrous roots. This wall is flimsy in structure, rather shallow, and neither so deep nor so round as that of the goldfinch. The lining is made of similar materials, but of a finer kind.

The quantity of sticks used as the foundation for this nest varies according to the kind of branch on which it is placed; for when the bird selects a forked twig, such as that of the hazel or dogwood, it uses a considerable quantity of sticks; but when it places its nest on the nearly horizontal spray of the fir, it finds a sufficient foundation ready made, and only just lays a few twigs to fill up a blank space. The egg of the Bullfinch is something like that of the goldfinch, but larger and more conspicuously spotted.

In some works upon the eggs and nests of birds, the Bullfinch is said to build in bushes of considerable height and size. Now, this is not necessarily the case, inasmuch as the spinney which has just been mentioned was composed entirely of trees and low brushwood, and the Bullfinches always preferred the latter. I certainly have often found their nests in tall bushes, and sometimes in trees; but they were always placed at so low an elevation, that the height of the tree or bush had no effect on that of the nest.

IF the reader will refer to page 181, he will see a short account of the Hoop-shaver Bee, which strips off the down that clothes the stem of the common bladder-campion, or white-bottle (*Silene inflata*), and uses it for the lining of her nest.

There is a bird found in North America, which has a similar habit, peeling off the downy hairs of plants, and using them in the structure of its nest. This is the pretty little BLUE-EYED YELLOW WARBLER (*Sylvia citrinella*), remarkable for the contrast

M M 2

afforded by its blue bill and eyelids with the golden yellow of its head and breast.

When the bird builds her nest, she places it in the foot of a shrub, either among briars or under them, and weaves its outer walls of vegetable fibre, using flax or tow whenever she can find it. The walls of the nest are strongly made, and woven firmly into the twigs that support it. When she has finished the outer nest, the bird goes to the fields, and carries off the hair of cattle and other animals, and weaves them into a lining, which is made softer and warmer by the downy hairs which grow on the stems of certain ferns, and which the bird plucks off with great address.

It is not only a pretty, but an useful and interesting bird. It is useful, because it is one of the insect-eaters, and may be constantly seen at work among the leaves, picking up the little green caterpillars which destroy the trees, and which form its chief food. Moreover, it brings up two broods of young during the year, each brood being four or five in number, so that the havoc which it makes among the caterpillars may be imagined. It is interesting, on account of the love which it bears towards its young, and its undaunted courage in defending them. When it is free from the cares of a family, it is as timid as any other bird, and makes the best of its way from the danger. But if its nest be approached while the eggs or young are still in it, the little bird seems to lose all fear, and devotes itself to the task of decoying away the approaching foe. It pretends to be very ill or lame, stretches out its neck, trails its wings, drops its tail, and flutters feebly along the branches, in order to delude the enemy into an idea that it is so lame, that it can easily be caught. The male is even a greater adept than the female at this practice, and, if he thinks that he has not decoyed the intruder far enough, he will slip through the branches, fly round, and repeat the process.

ANOTHER North American bird is a mighty nest-maker, trusting for safety to the inaccessible nature of the tree on which its home is placed. This is the well-known BALD-HEADED EAGLE, sometimes called the BIRD OF WASHINGTON (*Falco leucocephalus*), which has been accepted as the emblem of the United States of America. The nest of this bird has been admirably described by the two great masters of American ornithology, Audubon and Wilson; and as it is not easy to improve upon the language of those who were at the same time good observers and practised writers, their accounts will be given in their own words. The reader will perceive that the two histories are placed side by side, because the points that are omitted by one are supplied by the other.

I may mention that the term "bald-headed," as applied to this splendid bird, is by no means correct, because the head is feathered as densely as any other part of the body; but as the head of the adult bird is white, it produces an effect, when viewed at a distance, as if it were deprived altogether of feathers, and covered with a white skin. The following account is by Wilson:—

"The White-headed Eagle is seldom seen alone, the mutual attachment which two individuals form when they first pair seeming to continue until one of them dies or is destroyed. They hunt for the support of each other, and seldom feed apart, but usually drive off other birds of the same species. They commence their amatory intercourse at an earlier period than any other land bird with which I am acquainted, generally in the month of December.

"At this time, along the Mississippi, or by the margin of some lake not far in the interior of the forest, the male and female birds are observed making a great bustle, flying about and circling in various ways, uttering a loud cackling noise, alighting on the dead branches of the tree on which their nest is already preparing, or in the act of being repaired, and caressing each other. In the beginning of January, incubation commences. I shot a female, on the 17th of that month, as she sat on her eggs, on which the chicks had made considerable progress.

"The nest, which in some instances is of great size, is usually placed on a very tall tree, destitute of branches to a considerable height, but by no means always a dead one. It is never seen on rocks. It is composed of sticks, from three to five feet in length, large pieces of turf, rank weeds, and Spanish moss in abundance, whenever that substance happens to be near. When finished, it measures from five to six feet in diameter; and so great is the accumulation of materials, that it sometimes measures the same in depth, it being occupied for a great number of years in succession, and receiving some augmentation each season. When placed in a naked tree, between the forks of the branches, it is conspicuously seen at a great distance.

"The eggs, which are from two to four, more commonly two or three, are of a dull white colour, and equally rounded at both ends, some of them being occasionally granulated. Incubation lasts for more than three weeks; but I have not been able to ascertain its precise duration, as I have observed the female, on different occasions, sit for a few days on the nest before laying the first egg. Of this I assured myself by climbing to the nest every day in succession, during her temporary absence—a rather perilous undertaking when the bird is sitting.

"I have seen the young birds not larger than middle-sized pullets. At this time, they are covered with a soft cottony kind of down, their bills and legs appearing disproportionately large. Their first plumage is of a greyish colour, tinted with brown of different depths of tint; and before the parents drive them off from the nest they are fully fledged.

"I once caught three young eagles of this species, when fully fledged, by having the tree on which their nest was cut down. It caused great trouble to secure them, as they could fly and scramble much faster than any of our party could run. They, however, gradually became fatigued, and, at length, were so exhausted, as to offer no resistance when we were securing them with cords. This happened on the border of the Lake Pontchartrain, in the month of April. The parents did not think fit to come within gunshot of the tree while the axe was at work."

We will now turn to the second of these celebrated ornithologists, and see what he has to say on the nesting of this splendid bird:—

"The nest of this species is generally fixed on a very large and lofty tree, often in a swamp or morass, and difficult to be ascended. On some noted tree of this description, often a pine or cypress, the Bald Eagle builds, year after year, for a long series of years. When both male and female have been shot from the nest, another pair has soon after taken possession. The nest is large, being added to and repaired every season, until it becomes a black prominent mass, observable at a considerable distance. It is formed of large sticks, sods, earthy rubbish, hay, moss, &c.

"Many have stated to me, that the female lays first a single egg, and that after having sat on it for some time, she lays another; when the first is hatched, the warmth of that, it is pretended, hatched the other. Whether this be correct or not, I cannot determine; but a very respectable gentleman of Virginia assured me that he saw a large tree cut down, containing the nest of a Bald Eagle, in which were two young, one of which appeared nearly three times as large as the other.

"As a proof of their attachment to their young, a person near Norfolk informed me, that in clearing a piece of wood on his place, they met with a large dead pine-tree, on which was a Bald Eagle's nest and young. The tree being on fire more than half-way up, and the flames rapidly ascending, the parent eagle darted round and among the flames, until her plumage was so much injured, that it was with difficulty she could make her escape; and even then she several times attempted to return, to relieve her offspring."

THE bird next on our list is rather variable in its nesting.

The GOLDEN ORIOLE (*Oriolus galbula*) is seldom seen in England, and its nest even more seldom. Every year, however, a few stray nests are built in this country, as there are few years in which the journals devoted to natural history do not contain a notice of the bird being seen, and occasionally of its nest being found. In the warmer parts of the Continent it is plentiful, and in Italy is regularly exposed in the markets towards the middle of autumn, when it has indulged in fruit for some time and has become very plump and fat.

In this condition it is well known to epicures under the name of Becquafiga, corrupted into Beccafico. It is not easily procured, as it is a very wary bird, and does not like to venture far from covert. In the autumn, however, its love of fruit conquers its fear of man, and it haunts the orchards in numbers, making no small havoc among the fruit. Even under such circumstances it is not easy of approach, and the gunner will seldom manage to secure his prey except by imitating its peculiar and flute-like notes. He must, however, be very careful in his mimicry, for the bird has a critical ear, and if it detects the imitator, is sure to slip through the foliage and fly off to its forest stronghold.

The nest of the Golden Oriole is always placed near the

extremity of a branch, and in some cases is so constructed that it almost deserves to be ranked among the pensiles. It is always a pretty nest, and the accompanying illustration conveys a good idea of its general form. It is always more or less cup-like in shape, but the comparative depth of the cup is very variable, as in some cases it is scarcely deeper in proportion than that of the goldfinch, and rather saucer-shaped, while in others the depth

GOLDEN ORIOLES AND NEST.

even exceeds the width. Perhaps the nest may be altered in shape after the female begins to deposit her eggs, as is known to be the case with many birds, the additions being always made to the margin.

It is a remarkable fact that this enlargement of the nest should be common both to birds and insects. The reader may perhaps remember that the wasp, as well as other hymenoptera, lays an egg in the cell while it is yet shallow, and adds to the cell in proportion to the growth of the grub. The time of year, therefore, at which the nest of the Golden Oriole is found will have an influence on its shape, as the nest which is taken in the early spring, before the eggs are laid, will probably be shallower than that which is found in autumn, after the eggs have been hatched and the young reared.

The object for deepening the nest may probably be traced to the weather which happens to prevail. If the winds be light, the nest may remain in its flat and saucer-like form without endangering the safety of the eggs, but if the season should be inclement and tempestuous, a deeper nest is needed in order to prevent the eggs or young from being flung out of their home.

The body of the nest is formed chiefly of vegetable substances, usually the stems of different grasses, which are interwoven with wool, and thus made into a tolerably strong fabric. The female bird is said to be very affectionate, and to sit so closely on her nest that she will almost suffer the hand to be laid upon her before she will leave her post. In the illustration, the female bird is standing upright on the branch, and looking upwards, while the male is bending over the bough, and peering downwards, as if at some fancied foe. He can always be distinguished from his mate by the brighter gold of his plumage, the black spot between the eye and the beak, and the deeper black of his wings; whereas in the female, a tinge of blue invades the yellow, changing it to yellowish green, the wings are brown, edged with grey, and the black spot in front of the eye is altogether absent. Moreover, the breast and belly are marked with many longitudinal dashes of greyish brown.

ONE of the most variable of birds in its nesting is the well-known RED-WINGED STARLING of Northern America (*Icterus phœniceus*).

This beautiful bird derives its popular name from the fiery scarlet of the lesser wing-coverts, contrasting so boldly with the jet-black of the remaining plumage. It is known by several opprobrious names in its own country, such as Corn-thief, Maize-thief, &c. because it is popularly thought to live upon corn, whereas, like our starling, it is a most insatiable eater of the grubs, caterpillars, and other creatures which infest the corn-fields, and only eats corn at a certain time. There is, however, one season in the year, in which the Red-winged Starling becomes an arrant thief.

It is said that every living creature can be bribed, if the right bribe can only be found, and in the case of this bird, the newly-developed maize-grains present a temptation which it cannot resist. It is not alone in this predilection, for there are many other birds and some quadrupeds, the bear being the most conspicuous, which revel in the sweet, pulpy, succulent Indian corn. Even mankind is overcome with this delicacy. The white man fills his pockets with the plump ears, and munches them as he goes on with his business, and the copper-skinned native half-stupefies himself by gorging the cream-like grains. Small blame therefore to the bird for following an example which is set by its superiors. But before the maize is developed, and after it is hardened, the Red-winged Starling depends chiefly on insect-food for its subsistence, and is, therefore, a truly useful bird, deserving to be protected rather than destroyed, and only requiring to be driven out of the maize plantations for a week or two in the course of the year.

The nest of this bird is almost always built in morasses where reeds are plentiful, and in such places it almost invariably roosts, flocking to them towards nightfall in vast masses that absolutely blacken the air, now appearing as a vast dull cloud, and now suddenly flecked with blood-red patches, as the black bodies and scarlet wings are alternately turned to the spectator.

Somewhere about the end of April, the Red-winged Starling begins to make its nest. This is sometimes placed on the ground, sometimes on a grass-tussock, and sometimes in a branch, thus being more variable in position than is the nest of any other bird. The mode of structure and materials of the nest differ with the locality.

When the nest is placed on the ground, it is composed of a few rushes and grass stems, the chief care of the bird being to secure a soft lining of grass-blades. When it is built in a grass or rush-tussock, the stems are drawn together and held in their places by long grasses, so as to make a hollow wherein the nest may repose.

But when it is placed on a branch it is much more complicated. The bushes which are found in swampy places are always so slender and flexible, that much care is required in order to render them capable of bearing the nest. The bird, therefore, takes a quantity of wet rushes and long grasses, and twists them round a number of twigs, intertwining them so as to bring these twigs into a rudely-shaped hollow cylinder. From the same materials the body of the nest is formed, and the lining is made from dry grass blades. Little pains are taken to hide the nest, because the swampy nature of the ground prevents the intrusion of many foes, and in some cases three or four nests are seen close to each other on a single bush.

ONE of the common American birds, the YELLOW-BREASTED CHAT (*Icteria viridis*) is not only remarkable for its really pretty nest, but for the manner in which it defends its home.

Although so chary of being seen that an experienced ornithologist may follow it for an hour by its voice, and never catch a glimpse of the bird, it is full of talk, and as soon as a human being approaches, it begins to vociferate reproaches in an odd series of syllabic sounds, which can be easily imitated. Mocking the bird is an unfailing method of doubling its anger, and will cause it to follow the imitator for a long distance, although it will even under these circumstances keep itself hidden in the foliage. Wilson's account of the curious sounds which it utters is very graphic and interesting. "On these occasions his responses are constant and rapid, strongly expressive of anger and anxiety, and while the bird itself remains unseen, the voice shifts from place to place among the bushes, as if it proceeded from a spirit. First is heard a repetition of short notes, resembling the whistling of the wings of a duck or teal, beginning loud and rapid, and falling lower and slower, till they end in detached notes. Then a succession of others, something like the barking of young puppies, is followed by a variety of hollow, guttural sounds, each eight or ten times repeated, more like those proceeding from the throat of a quadruped than that of a bird; which are succeeded by others not unlike the mewing of a cat, but considerably hoarser.

"All these are uttered with great vehemence, in such different keys and with such peculiar modulation of voice as sometimes to seem at a considerable distance, and instantly as if just beside you; now on this side and now on that: so that, from these manœuvres of ventriloquism, you are utterly at a loss to ascertain from what particular spot or quarter they proceed. If the weather be mild and serene, with clear moonlight, he continues gabbling in the same strange dialect, with very little intermission, during the whole night, as if disputing with his own echoes.

"While the female is sitting, the cries of the male are still more loud and incessant. When once aware that you have seen him, he is less solicitous to conceal himself, and will sometimes mount up into the air, almost perpendicularly, with his legs hanging, descending, as he rose, by repeated jerks, as if highly irritated, or, as is vulgarly said, 'dancing mad.' All this noise and gesticulation we must attribute to his extreme affection for his mate and young; and when we consider the great distance from which in all probability he comes, the few young produced at a time, and that seldom more than once in the season, we can see the wisdom of Providence very manifestly in the ardency of his passions."

The nest which the bird defends with such skill and courage is very well concealed in a dense thicket, and the bird is always best pleased if it can find a bramble-bush thick in foliage and well beset with thorns. Sometimes it is forced to content itself with a vine or a cedar, and in any case it is seldom more than four or five feet from the ground. The outer wall is made of leaves, within which is a layer formed of the thin bark of the grape-vine, and the lining is formed of dried grasses and fibrous roots of plants.

An allied bird, the YELLOW-THROATED CHAT (*Vireo flavifrons*, makes a nest somewhat similar in materials, though not in locality, to that of the preceding bird. It is usually fixed in the horizontal branch of a tree or bush, and is made from the bark of the grape-vine, moss, and lichens, and is lined with fine vegetable fibres.

OF our four British pigeons, two are branch-builders. The Stockdove places its nest in holes in trees, in holes in the ground, or on the tops of pollard oaks, willows, and similarly crippled trees. The Rockdove makes its rude nest in the crevices of the rocks which it frequents. But the Ringdove and the Turtledove are true branch-builders, and are therefore noticed in this place.

We will first take the RINGDOVE (*Columba palumbus*) sometimes called the Wood-pigeon, the Woodquest or queest, and the Cushat.

The nest of the Ringdove is placed in a variety of localities, for the bird is not in the least particular in this respect. Sometimes it is situated near the top of a lofty tree, and sometimes it is found in a hedge only a few feet from the ground. I have seen nests in both localities.

Mr. Waterton mentions a curious circumstance connected with this bird. In a spruce fir-tree there was the nest of a

RINGDOVE AND NEST.

magpie, containing seven eggs, which were removed and those of the jackdaw substituted. Below this nest a Ringdove had chosen to fix her abode, and so the curious fact was seen, that on the same tree, in close proximity to each other, were magpies, jackdaws, and Ringdoves, and all living in perfect amity. It might have been supposed that the magpies and jackdaws would have robbed the nest of the Ringdove, but such was not the case. Moreover, the bird knew instinctively that she would not be endangered by her neighbours, for she came to the tree after the magpie had settled in it.

The nest of the Ringdove is of so simple a character as scarcely to deserve the name. The bird chooses a suitable spray, and lays upon it a number of sticks, which cross each other so as to make a nearly flat platform. Many birds make a similar platform as the foundation of their nest, but with the Ringdove it constitutes the entire nest. So slight is the texture of the platform, that when the two white eggs are laid upon it they can be discerned from below by a practised eye, and it really seems wonderful that they can retain their position on such a structure.

Moreover, the open meshes of the nest allow the wind to blow freely between the sticks, so that nothing would seem to be more uncomfortable for the young. Above, they can certainly be sheltered by the warm body and protecting wings, but below they seem to be exposed to every blast. Yet they find shelter enough, and not only find it, but make it. With the generality of birds, the droppings are conveyed away by the parents, but with the Ringdove they are allowed to remain, when they rapidly fill up all the open interstices, and form a dry scentless plaster, which effectually defends the tender bodies of the young from the wind, and has the further effect of consolidating and strengthening the nest.

Although the nests are plentiful enough, and the eggs are common in the cabinet of oologists, it is not very easy to find a nest that is furnished with this curious plaster, probably because some one of the many foes which persecute the Ringdove has discovered the nest, stolen the eggs, or killed the parent before the young birds were hatched.

It has already been mentioned that, with many branch-building birds, the thickness of the nest, or of the platform on which it is placed, is regulated by the exposed or sheltered position of the branch, and such is the case with the Ringdove. Although in some instances, the platform is so flimsy that the eggs can be seen through the interstices, in other cases it is from half an inch to an entire inch in thickness. In all cases, the longest twigs are first laid, and followed by those of smaller size; and, although the whole structure is very rude, it is always made with sufficient care to assume a tolerably circular shape.

The Turtledove (*Columba turtur*) builds a nest of very similar form, and, if possible, even slighter in construction.

EVERY one knows the common catchweed so plentiful in waste ground. The long trailing stems of this plant are used by a pretty little bird in making its nest, and are most ingeniously twined among the branches into the needful shape. The bird which uses this plant is the WHITETHROAT (*Curruca cinerea*) sometimes called the Haychat and Nettle-creeper. Its ordinary name is due to the white feathers of the throat, and it is called Nettle-creeper because it is so active among the weeds that fringe the hedgerows. The nest is always placed low, and I have mostly found it towards the top of some stubby bush or shrub, about three feet from the ground. Although placed in such localities, it is not very easy of discovery, as it is well hidden by the foliage, and in most instances the boughs must be pressed aside before the nest can be made clearly visible. Although the catchweed is used by the bird in making the framework of the nest, it does not consider itself bound to employ no other substance, but uses grass blades and vegetable fibres. The lining of the nest is simply made of fine hay, among which are twined a variable number of horsehairs, sometimes only twenty or thirty, and sometimes in such a quantity as almost to conceal the hay. It is in allusion to the lining of the nest that the bird is called Haychat. The nest varies much in thickness, probably in proportion to the density of the bush in which it is placed.

THE celebrated MOCKING-BIRD of America (*Turdus* [or *Mimus*] *polyglottus*), is also one of the Branch-builders.

The situation chosen by the bird is always variable, depending almost entirely on the nature of the district and the character of the inhabitants. Should the bird be resident in some wild part of the country, it takes some pains to conceal its nest, choosing the most impenetrable thicket that can be found. A thick thorn-bush is a favourite spot, because the sharp points serve to deter intruders from forcing their way to the nest; and the cedar is sometimes chosen, because its dark masses of foliage afford such a cover for the nest that it can scarcely be detected even by one who is looking for it.

But, should the bird build in some inhabited locality, where it is taught by instinct that it will not be molested, it makes its nest close to the house, and cares not to hide it. Six or seven feet from the ground is the usual height at which the nest is placed, and the bird has so little anxiety about its nest that it often builds upon the branches of a pear or apple-tree. The nest itself is rather a pleasing specimen of bird architecture, and is mostly built upon a slight foundation of delicate twigs, intermixed with dry weeds of the preceding year. The body of the nest is formed of straw, grass, wool, and vegetable fibres, and the lining is almost wholly composed of very fine fibrous roots.

Although the bird is so careless about concealing its nest, it is jealously anxious about intruders, and attacks indiscriminately any beast, reptile, or bird that approaches the favoured spot. Dogs are forced to run away from the sharp beak and buffeting wings of the angry bird; the cat finds that the ascent of a tree while a pair of infuriated birds are pecking her nose and blinding her eyes is an impracticable task, and even man himself is attacked by the fearless defenders of the home.

The worst and most treacherous foe however, is the black snake (*Coryphodon constrictor*), a harmless reptile, but one that is much dreaded by uninstructed pedestrians, because it imitates the manners of the rattlesnake with such fidelity that it is generally reckoned among the poisonous serpents. This snake lives mostly on rats, mice, young birds, and eggs, and in pursuit of the last-mentioned dainties will ascend trees and traverse any branch which holds a nest.

The very sight of the black snake inflames the Mocking Bird with fury, and he instantly darts at it, avoiding its stroke with admirable quickness, and dealing a rapid succession of blows on the reptile's head. The black snake is peculiarly vulnerable about the head, and even tries to retreat, but is prevented from escaping by the Mocking-Bird, which redoubles his efforts and easily beats the reptile down. As soon as he sees his advantage he seizes the snake by the neck, lifts it from the ground, buffets it with his wings, pecks it again as it drops, and ceases not until the hated enemy is left dead on the ground.

THE well-known WATER HEN or MOOR HEN (*Gallinula Chloropus*) is nearly, though not quite, as variable in its nesting as the red-winged starling lately described. The nest is always placed near the water, but the bird seems to be very indifferent about the precise locality.

Sometimes it is made on the ground, and in that case is laid among sedges and rushes where the water cannot reach it. The Water Hen, however, is not averse to nesting in a warm and comfortable place, for Mr. Waterton mentions that on one occasion, when he had built a neat little brick house for a duck, and furnished it with dry hay for a nest, a Water Hen took possession of it, and the duck had to find a home elsewhere.

WATER HEN AND NEST.

Sometimes the nest is made on a branch, and in that case the bird selects a very low bough which overhangs the water. I have found several nests thus placed, and in one case the only method of getting at the nest was to enter the water and swim round to it. It is a large and rudely made nest, and from its size appears to be more conspicuous than is really the case.

When it is placed on a bough, the twigs of the same branch often dip into the water, and the nest looks like a bunch of weeds and other *débris* that have floated down the stream and been arrested by the branch.

The similitude is increased by a curious habit of the bird. When she leaves her nest, she pulls over her eggs a quantity of the same substances as those which form the materials of the nest, so that they are completely hidden from sight, and the form of the nest is quite obscured. It is true that the nest is not unfrequently found with the eggs exposed, but this apparent negligence is always caused by the frightened bird dashing off at the approach of the intruder, and having no time to cover her eggs properly. The object of covering the eggs was once thought to be the retention of heat, the neighbourhood of water being imagined to be injurious. As, however, many birds build as close to the water as does the Water Hen, and do not cover the eggs, it is evident that concealment and not warmth is the object to be attained.

I may mention that the illustration was sketched from a nest before it was removed, and that most of the nests have been drawn in the same manner from actual objects.

The eggs are many in number, seldom less than six, and often eight, and their united weight is far from inconsiderable, as they are fully proportioned to the size of the bird. The young are the oddest little beings imaginable, looking like spherical puffs of black down, rather than birds. They take to the water at once, and if the reader can manage to watch the mother and her little family, he will see one of the quaintest and prettiest groups that our country can supply. The little black balls swim about quite at their ease, keeping within a short distance of their parent, and traversing the water with a rapidity that reminds the observer of the *gyrini*, or whirligig beetles. In spite of the prolific nature of the bird, it is not so numerous as it might be, having many enemies in its youth, the worst of which is the pike, which comes up silently from below, opens its terrible jaws, and absorbs the unsuspecting bird.

CHAPTER XXIX.

FEATHERED BRANCH-BUILDERS CONCLUDED.

The SEDGE-WARBLER—Its nest and loquacity—The REED-WARBLER—Use of its peculiar tail—Localities haunted by the bird—Song of the Reed-Warbler—Its deep and beautifully balanced nest—Colour of the eggs—The INDIGO BIRD—The CAPOCIER—Familiarity of the bird—Le Vaillant's experiments—How the nest is made—Division of labour—Lover's quarrels—Structure of the nest—Humming-birds again—The FIERY TOPAZ—Its nocturnal habits—Appearance of the nest—Its shape and the materials of which it is made—The HERMIT HUMMING-BIRDS and their nests—The RUBY-THROATED HUMMING BIRD—Variable dimensions of the nest—Concealment—Mr. Webber and his discoveries—Variable form and positions of the nest—Materials of which it is made—Its deceptive exterior—Feeding of the young—The VERVAIN HUMMING BIRD—How the nest assumes its shape—The RED-BACKED SHRIKE—Use of the Shrike in falconry—Their singular mode of feeding—Impaled prey—Conspicuous character of the nest—Popular ideas concerning the Red-backed Shrike—Structure of the nest—The HEDGE SPARROW—Its proper title—Carelessness about its nest—Foes of the Hedge Sparrow—Its fecundity.

ANOTHER bird that loves to build near water is the pretty little SEDGE WARBLER (*Salicaria phragmitis*).

The nest of this bird is placed at a very low elevation, usually within a foot or so from the ground, and raised upon rushes, reeds, or other coarse herbage, which is found abundantly in such places. There is more material in the nest than might be supposed from the size of the bird and the slender stems by which it is supported. Viewed from the exterior, it seems to have the ordinary cup-shaped form which is so prevalent among small birds, but looked at from above, the apparent depth is seen to be owing to the mass of material, the hollow being singularly small and shallow. It is a well-made nest, the general framework being formed of leaves of grass-blades, while strength, warmth, and density are attained by the quantity of wool and hair which are woven into the fabric.

The Sedge Warbler is well known for its loquacity, and its ceaseless chatter. Should it be silent, a stone flung among the reeds and sedges will always induce it to recommence its little song.

THE remarkably beautiful nest which is here shown is built by one of the British birds, but is not often found, on account of the localities where it is placed.

NEST OF THE THE REED WARBLER.

The architect of this nest is the REED WARBLER (*Salicaria* [or *Curruca*] *arundinacea*). It is a pretty little bird, bright brown above, yellow-brown below. In some respects it resembles the sedge warbler, but does not possess the remarkable wedge-shaped tail of that bird. R. Mudie, in his History of British Birds, offers the following suggestion respecting this difference of form. When treating of the sedge warbler, he remarks that the slender head, pointed bill, and wedge-shaped tail are useful to the bird by enabling it to glide between the tall aquatic plants among which it resides and finds its food. Of the Reed Warbler he writes as follows:—

"That the bird is not adapted for so many situations as the sedge bird, might be inferred from the different form of the tail, which is more produced and not wedge-shaped, so that while it answers better as a balance on the bending reeds or other flexible aquatic plants, it would not be so convenient among the unyielding sprays of a hedge or brake. The bird rarely, if ever, perches upon the tops of reeds, even on its first arrival, and when the song of invitation to a mate is given, its place is on a leaf or a leaning stem, though upon an emergency it can cling to an upright one, the stiff feathers of the tail acting as a sort of prop.

"It is not easily raised, and remains but a very short time upon the wing, but it is by no means timid on its perch, upon which, if it be very flexible, it sits with its wings not quite closed, but *recovered*, so as to have a little hold on the air, and thereby either prevent its fall or be ready when a gust comes, to bear it to a more secure footing. Its food is found wholly over the stagnant waters. The Reed Warbler does not come until the reeds are considerably advanced, and it departs before they are cut; so that it dwells in peace, and especially in the mornings about the end of May and the beginning of June it may be observed with the greatest ease."

Still, although the bird be common, and although it is bold enough to admit of approach, it is not generally familiar, simply because none but professed naturalists are likely to look for it in the spots which it frequents. The Reed Warbler loves a large patch of marshy land almost wholly covered with stagnant water, and full of the reeds among which its home is made. Such a place is not agreeable to the pedestrian, for although an hour spent in wading through water knee-deep is no difficult or even unpleasant task, yet no one likes to meet also with mud of various and unknown depths, as is the case in the great reed swamps where the birds most love to build. Even the song of the Reed Warbler does not attract attention. Though musical in tone, it is very feeble in power and monotonous in character, consisting of several hurried notes in a low warble, which can only be heard at a little distance.

The nest of this bird is supported between three or four reeds, as is shown in the illustration, and is remarkably deep in proportion to its width. The object of this depth is evident. To bend as a reed before the wind is a proverbial saying, and any one who has seen a large mass of reeds on a stormy day must have been impressed with their graceful curves. As the blasts of the wind pass over them, they bend in successive waves like the billows of the sea, and are sometimes bowed so low that their tips nearly reach the water.

A nest, therefore, which rests on such pliant supports must be thrown out of its perpendicular by every breath of wind, and unless it were very deep the eggs would be flung out. The great depth, however, of the nest counteracts the deflection of the reeds; and, however fiercely the storm may rage, the Reed Warbler sits securely in her nest, even though it be sometimes nearly bowed to the surface of the water. The materials of the nest are generally taken from the immediate neighbourhood, the body of the nest being composed of broken rushes and moss bound together with reed leaves, and the lining made almost wholly of cows' hair.

In the illustration the nest is represented as it appears during a rather smart breeze. The reeds are all bowed down by the force of the wind, and the nest is leaning so much to one side, that its contents would be flung into the water were it of the ordinary cup-shaped form. The tiny inmates, however, are perfectly secure in their home, and crouch in the bottom of the nest, so that there is no fear that they may be thrown out. The parent birds are busily attending on their little family, one having just brought an insect which all the gaping mouths are eager to devour, while the other is setting off in its turn to perform the like office. The little eggs are rather pleasing in colour, being very pale green, almost fading to whitish grey in parts, and being mottled and speckled with brown or green darker than the ground hue of the shell. As is usually the case with similar birds, they are four or five in number.

MANY foreign birds are excellent branch-builders.

One of the best known is a lovely little bird, which is familiar to us through the mediumship of taxidermists, who are always glad to insert a few specimens in their glass cases of brightly plumaged birds. This is the INDIGO BIRD, or BLUE LINNET of America (*Spiza cyanea*), which derives its name from the hue of its feathers. Viewed in some lights, the plumage is a rich, deep azure, shining with a satiny lustre in the direct light of the sunbeams, and deepening into indigo in the shadows. But the most remarkable point in the hue is, that in certain lights it changes to that peculiar green which is known to artists as "verditer," so that the bird seems absolutely to change its colour if its position be shifted for only a few inches. Consequently, a well arranged group will have two specimens placed in such a manner that one glows in all the glory of its azure dress, while another is vivid green. The wings are black, and retain their colour in all lights.

The nest of the Indigo Bird is set in a bush, and, according to Wilson, is upheld by two twigs, one passing up each side, so as to preserve the balance. To the twigs it is firmly bound with the strong flaxen fibres of which the walls are formed, and its lining is made of fine grasses.

IN Southern Africa there is a small, simply coloured, but interesting bird, called by Le Vaillant the CAPOCIER (*Drymoica maculosa*) because it builds in a cotton-yielding tree, called by the Dutch colonist Capoc-bosche.

The attention of that naturalist was directed to the bird in the following manner.

Being, in common with all true naturalists, a lover of birds in their living state, and being in no wise disposed to kill them without necessity, he had contrived to tame a pair of little brown birds, which at last became so familiar that they would enter his tent. On these terms they remained until the beginning of the breeding season, when they began to come less regularly, and then to absent themselves for several successive days. About this time they became thieves. M. Le Vaillant was accustomed to keep on his table a quantity of tow and cotton-wool, which he used in stuffing and otherwise preparing the skins which he had procured for his collection. The birds seemed suddenly to take a wonderful fancy to the tow and cotton-wool, and were continually flying off with them, sometimes stealing a piece that was nearly as large as both the birds together.

Struck with this sudden fancy of the birds, Le Vaillant determined to watch them, and soon traced them to a capoc-bosche tree which grew at some distance, and in a remarkably retired spot. Among the branches of this tree they had already begun their nest, which consisted of a quantity of moss pressed tightly into the forks of a bough, and which was at the time only in a rudimentary condition. The moss, in fact, was the foundation of the nest, upon which the beautiful walls were intended to be built, just as in the habitation of many other birds there is a foundation of substances more solid than the materials of which the walls are made.

Into this nest the Capociers were weaving the stolen stores of cotton-wool, working it in a manner that will be presently described. Le Vaillant soon discovered that the legitimate substance of the nest-walls was the soft, white down produced by certain plants, and that the birds used an enormous amount of materials in comparison with their own size. As, however, they found that upon the naturalist's table was always a plentiful supply of vegetable down and fibres ready plucked, they ingeniously saved themselves the trouble of collecting, and simply resorted to the hospitable tent.

The male was the principal collector of materials, and the female the chief architect. He used to fly off, and return with a mass of cotton-wool, moss, or tow, and deposit it close to the spot where his mate was at work. Then she would take the materials, arrange them, press them into form, and only ask his assistance in carrying out her plans. He pressed, and pecked and pulled the cotton-wool so as to reduce it to a kind of felt, but did not seem to originate any architectural ideas, leaving them to his more ingenious mate.

Le Vaillant's account of the mode of working is so interesting and elegant that in justice to himself it must be given in his own words. After describing the process of fetching materials and laying them in their places, he proceeds as follows :—

"This agreeable occupation was often interrupted by innocent and playful gambols, though the female appeared to be so actively and anxiously employed about her building as to have less relish for trifling than the male, and she even punished him for his frolics by pecking him well with her beak. He, on the other hand, fought in his turn, pecked, pulled down the work which they had done, prevented the female from continuing her labours, and, in a word, seemed to tell her, 'On account of this work you refuse to be my playmate, therefore you shall not do it.'

"It will scarcely be credited that, entirely from what I saw and knew respecting these little altercations, I was both surprised and angry at the female. In order, however, to save the fabric from spoliation, she left off working, and fled from bush tc bush, for the express purpose of teasing him. Soon afterwards, having made matters up again, the female returned to her labour, and the male sang for several minutes in the most animated strains. After his song was concluded, he began again to occupy himself with the work, and with fresh ardour carried such materials as his companion required, till the spirit of frolic again became buoyant, and a scene similar to that which I have described occurred. I have witnessed eight interruptions of this kind in one morning. How happy birds are! They are certainly the privileged creatures of nature, thus to work and sport alternately, as fancy prompts them.

"On the third day the birds begun to rear the side walls of the nest, after having rendered the bottom compact by repeatedly pressing the materials with their breasts, and turning themselves round upon them in all directions. They first formed a plain border, which they afterwards trimmed, and upon this they piled up tufts of cotton, which was fitted into the structure by beating and pressing it with their breasts and the shoulders of their wings, taking care to arrange any projecting corner with their beaks, so as to interlace it into the tissue, and to render it more firm. As the work proceeded, the contiguous branches of the bush were enveloped in the side walls, but without damaging the circular cavity of the interior. This part of the nest required many materials, so that I was quite astonished at the quantity which they used.

"On the seventh day their task was finished, and, being anxious to examine the interior, I determined to introduce my finger, when I felt an egg that had been probably laid that morning, for on the previous evening I could see that there was no egg in it, as it was not quite covered in.

"This beautiful edifice, which was as white as snow, was nine inches in height on the outside, whilst in the inside it was not more than five. Its external form was very irregular, on account of the branches which it had been found necessary to enclose; but the inside exactly resembled a pullet's egg placed with the smaller end upwards. Its greatest diameter was five inches, and the smallest four. The entrance was two-thirds or more of the whole height as seen on the outside, but within it almost reached the arch of the ceiling above."

One of the most remarkable points of this singularly beautiful nest is the firm texture of the walls. Externally, the nest looks as if it were a mere large hollow bunch of cotton-wool with a hole near the top, and seems to be so fragile that the eggs would fall through the fabric. But when the inside of the nest is viewed, it is seen to be composed of a kind of felt, as firm and close as if it had been formed by human art, so that neither wind nor wet can penetrate; and it is capable of upholding a much greater weight than would be introduced into it. To pull out a tuft of the cotton-wool is impossible without tearing a hole in the fabric, so closely are the delicate fibres interwoven with each other.

In the accompanying illustration are shown the nests of two species of Humming Bird.

The oddly-shaped nest which occupies the upper part of the drawing is made by the Fiery Topaz (*Topaza pyra*), one of the most magnificent of these lovely birds. Indeed, Prince Lucien Buonaparte calls it the most beautiful of the Trochilidæ, and it is hardly possible to imagine a bird that can surpass it in brilliancy. The body is fiery scarlet, the head velvet-black, the throat glittering emerald, with a patch of crimson in the centre; the lower part of the back is also green, and the long, slender, crossed feathers of the tail are purple with a green gloss. So magnificent a bird can have but few rivals, and there is only one species which even approaches it in beauty. This is the Crimson Topaz (*Topaza pella*), a bird which is nearly allied to it, and which much resembles it in general colouring. It may, however, be distinguished by the colour of the body, which is crimson instead of scarlet.

Curiously enough, although it is bedecked with resplendent hues, which seem to need the presence of daylight, and to be made expressly for the purpose of reflecting the brightest beams of the sun, yet the lovely bird is one of the night wanderers, being seldom seen as long as the sun is above the horizon, and preferring to seek its food while the world is shrouded in darkness. Perhaps the reader may remember that the sea-mouse, whose iridescent garment possesses all the tints of the rainbow, is also a darkness lover, and passes its life sunk in the black mud of the sea-shore.

The nest which is built by the Fiery Topaz is really a wonderful structure.

Its shape is remarkable, and is well shown in the illustration. It is fastened to the branch with extreme care, as is clearly necessary from its general form. The most curious point about the nest is, however, the material of which it is made. When it was first discovered no one knew how the bird could have built

FIERY TOPAZ AND HERMIT.

so strange a structure. It looked as if it were made of very coarse buff leather, and was so similar in hue to the branches that surrounded it, that it seemed more like a natural excrescence than a bird's nest. The reason for this similitude was simple enough. It was made of a natural excrescence, and therefore resembled one.

When the Fiery Topaz wishes to build a nest, it goes off to the trees, and searches for a kind of fungus belonging to the genus boletus, and with this singular material it makes its home. It is tough, leathery, thick and soft, and in some curious manner the bird contrives to mould the apparently intractable substance into the shape which is represented in the illustration. The non-botanical reader may form an idea of the appearance of the nest, by supposing it to be made of German tinder, which is, in fact, a kind of boletus which has been pressed, dried, and steeped in a weak solution of nitre.

The lower figure in the same illustration represents the nest of another Humming Bird (*Phaëthornis eurynome*), belonging to the pretty little group which are popularly called Hermits, and which may be recognised by the peculiar shape of the tail, which is regularly graduated, the two central feathers being, however, much longer than the others. They are inhabitants of Venezuela.

All the Hermits are remarkable for the beauty of their homes, and the present species is mentioned as affording a good example of nest-making. The nest is always long and funnel-shaped, and is hung either to a leaf or the delicate twig of a tree, according to circumstances. The materials of which the nest is made are rather various, consisting of vegetable fibres, especially those downy, cotton-like filaments which are furnished by so many plants, of small herbs, and spider webs. The last mentioned substance is employed for the purpose of binding the materials together, and is used also in fastening the nest to the support on which it hangs.

THERE is another species of this beautiful group, called the RUBY-THROATED HUMMING BIRD (*Trochilus colubris*), which is generally accepted as the typical species. This lovely bird is plentiful in many parts of America, and is sometimes seen as far north as Canada. It derives its popular name from the feathers of the throat, which glitter as if made of burnished metal, and glow with alternate tints of ruby and orange. The general colour of the body is green, and the wings are purple-brown. The two sexes are coloured after the same manner, with the exception of the ruby gorget, which only belongs to the male, and which is not attained until the second year. There is no species more common in museums and ornamental cases than this, because it is as plentiful as it is lovely. That it should be plentiful, or indeed that any species of Humming Bird should be anything but scarce, is matter of wonder, inasmuch as they never lay more than two eggs, and in all probability do not rear more than three, or perhaps four, young in the course of a season.

The general habits of this tiny bird are well worthy of notice, but at present we must content ourselves with it as it appears in its nest-making capacity. Being a very small bird, only three inches and a half in total length, and very slenderly made, the uest is necessarily small. But, although we so often find that little birds build large nests, we cannot but notice that the nest of this Humming Bird is even smaller than the size of its occupant seems to require. It is round, neatly made, and has thick walls and a small hollow.

The bird has a wonderful power of concealing the nest, which cannot be discovered except by a practised nest-hunter, so closely does it resemble a knob upon a branch. So careful, too, is the female of her home, that she does not fly straight to it, but rises high in the air, and then darts down among the branches with such rapidity that the eye cannot follow her movements, and she is fairly seated in her nest before the spectator knows exactly in which direction she has gone.

This curious trait seems to have been discovered by Mr. C. W. Webber. He had successfully tamed some Ruby-throats, and determined to find a nest, so that he might obtain the young. After finding that a pair of Humming Birds had been seen near a certain spot on a river, he set himself determinately to discover the nest. By degrees they were watched to a point of the river, but there they always disappeared, as they had a habit of shooting perpendicularly into the air until their tiny bodies were lost to sight. At last, however, the patient watchfulness of the observer was rewarded by catching a glimpse of the female bird, as she descended perpendicularly from the height to which she had risen, and in this manner was the nest discovered.

The same agreeable writer relates an anecdote respecting the discovery of a nest belonging to the Emerald-throated Humming Bird, an edifice which is very similar to that which is made by the Ruby-throat. He had been in vain looking for a nest, when "chance favoured me somewhat strangely about this time. I had been out squirrel-shooting early one sweltering hot morning, and on my return had thrown myself beneath the shade of a thick hickory, near the bank of a creek. I lay on my back, looking listlessly out over the stream, when the chirp of the Humming

Bird and its darting form reached my senses at the same instant. I was sure I saw it light upon the limb of a small iron-wood tree, that happened to be exactly in the line of my vision at that instant.

"In about five minutes another chirp and another bird darting in. I saw this one drop upon what seemed to be a knob or an angle of the limb. I heard the soft chirping of greeting and love. I could scarcely contain myself for joy. I would have given anything in the world to have dared to scream, 'I've got you, I've got you at last!' By a great struggle I choked down my ecstasy and kept still. One of them now flew away, and after waiting fifteen minutes, that seemed a week, I rose, and with my eyes steadily fixed on that important limb, I walked slowly down the bank, without, of course, seeing where I placed my feet.

"But the highest hopes are sometimes doomed to a fall, and a fall mine took with a vengeance! I caught my foot in a root, and tumbled head foremost down the bank into the river! I suppose that such a ducking would have cooled the enthusiasm of most bird-nesters, but it only exasperated mine. I shook off the water, and vowed that I would find that nest if it took me a week. But how to begin was the question, for I had lost the limb, and how was I to find it among a hundred others just like it?

"The knot that I had seen was so exactly like other knots upon other limbs all round it, that the prospect of finding it seemed a hopeless one; but, 'I'll try, sir,' is my favourite motto. I laid myself down as nearly as possible in the position which I had originally occupied, but, after some twenty minutes' experiment, came to the conclusion that my head had been too much confused by the shock of my fall and ducking for me to hope to make much out of this method. Then I went under the tree, and commencing at the trunk, with the lowest limb which leaned over the water, I followed it slowly and carefully with my eye out to the extremest twig, noting carefully everything that seemed like a knot. This produced no satisfactory result after half an hour's trial, and with an aching neck I gave it up in despair, for I saw half a dozen knots, either one of which seemed as likely to be the right one as the other.

"I now changed my tactics again, and, ascending the tree, I

stopped with my feet upon each one of those limbs and looked down along it. It was a very tedious proceeding, but I persevered. Knot after knot deceived me, but, at last, when just above the middle of the tree, I caught a sharp gleam of gold and purple among the leaves, and, looking down upon the last limb to which I had climbed, almost lost my footing for joy, as I saw, about three feet out from where I stood, the glistening back and wings of the little bird just covering the top of one of these mysterious knots that was about half the size of a hen's egg.

"The glancing head, long bill, and keen eyes were turned upwards, and perfectly still, except the latter, which surveyed me from head to foot with the most dauntless expression. It seemed not to have the slightest intention of moving, and I would not have disturbed it for the world. It was sufficient to me to gaze on my long lost treasure. Its pure white breast—or throat rather, for the breast was sunk in the nest—formed such a sweet and innocent contrast with the splendour of its back, head, and wings." The capture of the little birds which were afterwards hatched in that nest served to set at rest the question of the Humming Bird's food. They lived mostly on syrup, but were obliged to fly off and eat the tiny garden spiders as they lay in the middle of their radiating webs.

The nest of the Ruby-throated Humming Bird seems to be rather variable in form and material and situation, but has always a peculiar character which enables the experienced observer to recognise it. According to Wilson, it is sometimes fixed on the upper part of a horizontal branch, as was the case with the nest so graphically described by Mr. Webber. Sometimes it is seen actually upon the trunk of a tree, attached to the bark by its side; and in a few rare instances it has been found in a garden, attached to some strong-stalked herb. Generally, however, the bird selects a white oak sapling if it builds in the woods, and a pear-tree if it prefers the garden.

The tiny nest is scarcely more than one inch in width and the same in depth, so that its size is very small when compared with that of its occupants, which, when full grown, are more than three inches in total length. The materials of which the nest is made are principally the delicate cotton-like fibres which form the "wings" of certain seeds, such as those of the thistle, and which are so carefully woven together that they form a tolerably stout wall. Upon this wall are stuck quantities of a ligh grey lichen which is found on old fences and trees, so that the external appearance of the nest is rendered very similar to that of the branch on which it is placed. The lining is composed of the fine hairs which clothe the stalks of mullein and ferns and other pubescent plants, and forms a thick, soft bed on which repose the two minute pearly eggs.

The nest is not merely placed upon the branch, because in that case it would present a decided outline, and be comparatively easy of recognition. On the contrary, the base of the nest is partly continued round the branch, so that the whole fabric rises gradually from the bough, as if it were a natural excrescence.

When the young are hatched they are fed by thrusting their beaks into the opened mouths of their parents, and extracting the supply of liquid sweets which have been collected from the flowers.

THERE is another species of this group that builds a very pretty branch-nest. This is the VERVAIN HUMMING BIRD (*Mellisuga minima*), one of the minutest of the feathered race. From the point of its beak to the end of its tail it only measures two inches and three-quarters, so that when stripped of its feathers it seems more like an insect than a bird.

Its popular name is derived from its fondness for the West Indian vervain (*Stachyarpheta*) a very common weed in neglected pastures, with a slender stem, a blue flower, and averaging a foot in height. Wherever the vervain is plentiful, the Humming Bird is sure to be found, darting here and there, now poising itself before a flower, and probing its recesses with the long, slender tongue, and now shooting for hundreds of feet into the air, and then descending diagonally, as if shot from a gun, towards the flower from which it started, and balancing itself before its blue petals as if it had not moved.

The nest of this bird is proportionately small, and is beautifully made of vegetable fibres, such as the silk-cotton of the bombax, and, when the eggs are laid, is only just large enough to contain them and to retain the body of the mother bird. When, however, the young are hatched, the parents add to the walls of the nest, which, by degrees, alters its shape completely. At first it very much resembles an immature acorn-cup, but when the young are ready for flight, it is deep like an ordinary coffee-cup. This pretty little bird is common in Jamaica.

In the hedgerows of our own country may often be found a nest which is not only pretty in itself, but remarkable for its accessories. This is the home of the Red-backed Shrike (*Enneoctonus collurio*).

The predatory habits of the Shrikes are well known, and one species, the Great Grey Shrike (*Lanius excubitor*), was formerly used as a falcon for the purpose of catching winged game. True, the bird was not considered as a veritable hawk, and in the old days of sumptuary laws, when each degree of rank had its own particular species of hawk, this was a fact of some significance, showing that those who thus employed the Shrike were not of gentle blood.

The popular notion of the time supplied another reason why the Shrike was looked upon with disdain as a bird-catcher. It was supposed to use guile in securing its prey, instead of openly conquering in fair chase. "Sometimes," writes an old sporting author, "upon certain birds she doth use to prey, whome she doth entrappe and deceive by flight, for this is her desire. She will stand at pearch upon some tree or poste, and there make an exceeding lamentable crye and exclamation, such as birds are wonte to do, being wronged or in hazard of mischiefe, and all to make other fowles believe and thinke that she is very much distressed and stands in need of ayde; whereupon the credulous sellie birds do flocke together presently at her call and voice, at what time if any happen to approach neare her she out of hand ceazeth on them, and devoureth them (ungrateful subtill fowle!) in requital for their simplicity and pains.

"Heere I end of this hawke, because I neither accompte her worthy the name of a hawke, in whom there resteth no valour or hardiness, nor yet deserving to have any more written upon her propertie and nature. For truly it is not the property of any other hawke, by such devise and cowardly will to come by their prey, but they love to winne it by main force of wings at random, as the round winged hawkes doe, or by free stooping, as the hawkes of the tower doe most commonly use as the falcon, gerfalcon, sacre, merlyn, and such like."

The Shrikes have a peculiarity which is not shared by any other predacious bird. When they have slain their prey, no matter whether it be bird, beast, reptile, or insect, they take it to some thorn tree, and there impale it, pressing a long and sharp thorn into the body, so as to hold it firmly. The Great Grey Shrike will thus impale the smaller birds, frogs, field-mice, and other creatures which are nearly as large as itself, and in some instances it has been known to kill and impale the thrush. It does not always employ thorns for this purpose, but will use sharply-pointed splinters of wood, or even an iron spike if no better instrument can be found.

Why it should have recourse to such a singular mode of holding its prey is quite a mystery. Some have said that the digestive organs of the Shrike are incapable of dissolving fresh meat, and that the bird is obliged to render its prey semi-putrid by exposure before it can venture to make a meal. But, as the Shrike frequently eats a little bird or insect as soon as caught, this theory falls to the ground.

Whatever theory may be right or wrong, the fact remains that the Shrikes impale the creatures which they have killed, and prefer to hang them near their nests. The Red-backed Shrike is not so formidable a foe to birds as its larger relative, but makes insects its chief prey. The nest of this Shrike always affords a curious sight, and as the bird is plentiful it may easily be seen.

There is not the least difficulty in finding a Shrike's nest, for the owner really seems to use every means which can attract attention. In the first place, it is a bird of insatiable curiosity. It will follow, or rather precede, a human being for half an hour at a time, keeping always some thirty or forty yards in front, settling near the top of a hedge, and wagging its long tail up and down as if to make itself more conspicuous. Last year I amused myself by making a Shrike move up and down a long hedge for a very long time, while I was insect-hunting among the flowers. Whenever the Shrike begins to act in this manner, it may generally be presumed that a nest is at no great distance.

Then, if perchance the careful observer should note these signs and approach the spot where the nest is placed, the bird sets up a hideous squall, just as if it intended to inform the searcher that he was right at last. The alarm cry of the blackbird is quite enough to draw attention as the bird flies through the underwood; but at all events it is only a short cry, and the bird is soon out of sight; but the Shrike remains on or near the nest while it continues to utter its harsh screams, and flies away noisily when the intruder is close at hand.

The nest itself is large, and not concealed with any care, while around it are stuck humble bees, cocktail beetles, ground beetles, and a variety of other insects, each impaled upon a thorn, and forming admirable indications to the nest-hunter. Sometimes, but seldom, young birds are impaled instead of insects, and in such cases they are always callow nestlings, and are fixed by a thorn run between the skin and the flesh, instead of being pierced through the body, as is the method employed with insects.

There is a popular idea that the bird always has *nine* impaled creatures at hand, and that when it eats one it catches another, and with it replaces the one which has been eaten. In consequence of this notion, which prevails through several counties, the bird is called Nine-killer. The generic name, Enneoctonus, is composed of two Greek words which have a similar signification. So strongly is this idea held by some persons, that I have seen a treatise upon instinct, where the Shrike was gravely produced as an example of arithmetical powers possessed by birds. These theories generally fail when confronted by facts. I have seen numberless Shrikes' nests; and, though in some cases there may been nine impaled animals, in some there were more and in others less.

The nest itself is neatly, though loosely, built of roots, moss, wool, and vegetable fibres, and is lined with hair. I have mostly noticed it about five feet from the ground; and, although it is said to be closely hidden, have always found it a peculiarly conspicuous nest.

THE last branch-building bird which will be mentioned in these pages is the well-known HEDGE SPARROW, or HEDGE ACCENTOR, as it ought rightly to be called (*Accentor modularius*). The bird derives its popular name from two peculiarities, one of person and the other of habits. As its general tints are brown and black, the name of Sparrow has been given to it, although it rightly belongs to the warblers. It may easily be distinguished from the sparrow by its slender form, its blue-grey colour, and the absence of the black patches that mark the head and throat of the true sparrow.

It is very plentiful in England, and that it should be so is rather remarkable on account of the exposed situation and conspicuous form of its nest. The red-backed shrike is remiss enough in placing its nest; but the Hedge Sparrow seems to be utterly heedless on the subject, and appears absolutely to invite the attention of its foes, which are many.

First and foremost comes the bird-nesting boy, whose eyes are generally so sharp that to conceal a nest from him is no easy matter. Then the Hedge Sparrow is one of the earliest builders, and so hasty is it in its proceedings that I have seen the half-finished nest filled with the snows of early spring. The bird had been in such a hurry to set up housekeeping that she would not even wait until the leaves were sufficiently large to shelter the nest; and, as might be expected, the snow found an easy entrance into the unprotected edifice. In consequence, moreover, of this passion for early building, the nest is so conspicuous an object in the leafless hedge, that the most casual glance cannot fail to detect it, while the natural foes of the bird, namely, the boy, the stoat, the cat, cuckoo, and others, find it the easiest of their prey.

The boy, for example, who might not be able to reach the nest of the shrike, which is placed some five or six feet from the ground, has no difficulty whatever in harrying that of the Hedge Sparrow, which is seldom more than two feet from the ground. Moreover, although the older nest-hunters will not trouble themselves about eggs so common as those of the Hedge Sparrow, the novices, and even many who ought to know better, can never resist the round, shining blue shells, as they lie snugly packed away in their basket-like receptacle.

Then there is the cuckoo, that flies about the hedgerows, peering into every nest and looking out for a foster home for her eggs, which she cannot hatch, and for the young which she cannot cherish. There is, perhaps, no nest which is easier to be seen or more accessible when discovered than that of the Hedge Sparrow, and the consequence is, that the cuckoo's egg is oftener to be found in the nest of the Hedge Sparrow than in that of any other bird. This circumstance is certainly unfortunate to the Hedge Sparrow, who is obliged to give up the whole of her nest to a supposititious offspring, and to bestow upon a single intruder all the care and attention which would otherwise have been lavished upon the five rightful occupants.

Besides the cats, rats, and weasels, there are direful feathered foes, such as the shrike, which steals away the young and carries them to its home, where it hangs them up in its natural larder, and the magpie, which will sometimes work great havoc among the young or eggs. Now and then the owl makes a meal of a young bird, as I can testify from personal experience, and the viper is always ready to glide up the stems of the shrubs amid which the bird has built, to insert its baleful head into the nest, and to swallow the callow young.

Still, as the Hedge Sparrow generally produces two broods of young in a year, and sometimes three, all her offspring are not destroyed by these foes, and she may have the satisfaction of rearing some of her own. The nest is nicely, substantially, but not elegantly made, as, indeed, might be inferred from its lowly position. Nests upon or near the ground are very seldom made with much attention to elegance of architecture, the greatest trimness of nest-building skill being displayed in those homes which are placed upon lofty branches or suspended from slender twigs. It is a rather large nest, and is made of moss, wool, hair, and similar materials. As is generally the case with the group of birds to which the Hedge Sparrow belongs, the eggs are five in number, and on an average, three young in each brood attain maturity.

PENSILE SPIDER'S NEST.

CHAPTER XXX.

SPIDERS AND INSECTS.

Remarkable Spider Nests in the British Museum—Seed-nests and Leaf-nests—Nests of the TUFTED SPIDER—Form and colouring of the Spider—Its curious limbs—Nests illustrative of the hexagonal principle—Nest of the ICARIA—The equal pressure and excavation theories—Nest of MISCHOCYTTARUS and its remarkable form—Nest of the RAPHIGASTER—Summary of the Argument—The PROCESSIONARY MOTH—Reasons for its name—How the larvæ march—Damage done by them to trees—A natural remedy—The CALOSOMA and its habits—The GIPSY MOTH—Its ravages upon trees and mode of destroying it—The social principle among Caterpillars—Mr. Rennie's experiments—The LACKEY MOTH—Supposed derivations of its popular name—The eggs, larvæ, and perfect insects—Habits of the Moth—The BROWN-TAILED MOTH—Locality where it is found—Its ravages abroad—Nests of the ICARIA as they appear in branches—The APOICA and its remarkable nests—Moth Nests from Monte Video.

WE have already seen several nests built by SPIDERS, some of which are made in the earth, others are strictly pensile, and others may fairly come into the present group. The specimens from which the drawings were made are in the collection of the British Museum, some in the upper and others in the lower rooms. Of the architects, the manner in which the nests were made, and the reasons why they were so singularly constructed, I can say nothing, because no record is attached to the specimens. Still, they are so curious that they have found a place in this work, and it is to be hoped that the very fact of their publicity will induce travellers to search for more specimens and to describe their history.

Differing as they do in shape, colour, and material, they have one object in common, namely, the rearing of the young. They are clearly nests in the true sense of the word, being devoted not to the parents, but to the offspring. At the upper part of the illustration may be seen a number of long, spindle-shaped bodies, suspended from a branch. These are drawn about half the full size, in order to allow other specimens to be introduced into the same illustration for the purpose of comparison. In colour they are nearly white, with a slight yellowish tinge, and are very soft and delicate of texture, so that when viewed in a good light they form a very striking group of objects.

In the opposite upper corner of the illustration may be seen a remarkable nest, which few would recognise as the work of a spider. Such, however, is the case, the creature being urged by instinct to take several concave seed-pods, and to fix them together as seen in the drawing. The seed-pods are fastened firmly together with the silken thread of which webs are made, and in the interior the eggs are placed. The drawing is reduced about one-third in proportion to the actual object. Several of these singular nests are in the collection at the British Museum.

Occupying the lower part of the illustration is seen a leaf upon which are piled a number of fragments of leaves, so as to form a rudely conical heap. This is also the work of a spider, and is made with even more ingenuity than the two preceding specimens. In the first instance, the spider has spun a hollow case of silk, similar in principle of construction, though not in form, to the spherical egg cases made by several British spiders. In the second instance, the creature has chosen a number of concave seed-pods, and, by adjusting their edges together and fastening them with silk, made a hollow nest, which only requires to be lined in order to make it a fit nursery for the young. But, in the present example, the work of nest-making has been much more elaborate, for the structure has been regularly built up of a great number of pieces, each being arranged methodically upon the other, very much as children in the streets build their oyster-shell grottoes. The labour must have been considerable, even if the spider had nothing to do but to arrange and fasten together pieces of leaves which had already been selected.

TUFTED SPIDER. SPHERICAL SPIDER NESTS.

THE large, oval, cocoon-like nests which are seen in the accompanying illustration are made by the TUFTED SPIDER of the West Indies, a creature which derives its name from the remarkable tufts of stiff, bristle-like hairs which decorate the limbs. A very fine specimen of this remarkable Spider is now before me, having been taken out of its bottle with extreme difficulty, owing to the great length of the limbs, and the weight of the prolonged abdomen.

The length of the body is one inch and a half, of which measurement the abdomen alone occupies two-thirds. The average circumference of the abdomen is five-sixths of an inch ; and, as it varies very little thoughout its entire length, that portion of the body is very solid and heavy. The colour is deep chocolate-brown, curiously marked with circular dots of bright yellow, and further diversified with stripes of the same colour, especially over the fore-part of the abdomen. Two bold yellow bars are also drawn transversely across the under surface of the abdomen. The thorax is deep brown, and clothed with short hairs of greyish yellow, set so densely that the dark colour of the thorax cannot be seen without close inspection. There are, however, three black squared spots on each side, and a black spot occupies the centre. The animal is armed with a formidable pair of poison-jaws, of a deep shining black, at the ends of which the curved fangs are bent inwards like the venomous teeth of the rattlesnake. On the front of the thorax, and looking directly forward, are the eight eyes, the four smallest being arranged closely together in the centre, in the form of a square, and the four largest being set on bold prominences so as to form a large oblong, in the centre of which is the square.

The limbs are of considerable length. The first pair of legs, which are the longest, measure two inches and a half in length, and the expanded second pair measure four inches and a half. The most remarkable point about the spiders is the peculiarity from which it derives its name. The first, second, and fourth pairs of legs are adorned with dense hairy tufts, the first pair having two tufts each, and the others only one. The third pair of legs are much shorter and smaller than the others, and are destitute of tufts. As the legs themselves are bright yellow-brown and the tufts are deep black, the contrast of colour is very bold and agreeable to the eye. The entomological reader may perhaps remember that social exotic beetles are also decorated with tufts upon their antennæ and limbs. Of the curious spherical spider nests, with their black cross-bars, nothing is known except the mere fact of their existence. They are about as large as full-sized black currants.

In the accompanying illustration three most remarkable nests are given, all of them the work of hymenopterous insects, and all serving in some degree to illustrate the hexagonal system of cell-building, so common among the hymenoptera.

MISCHOCYTTARUS ICARIA. RAPHIGASTER.

Of these, perhaps, the central figure is the most interesting, because it entirely sets at rest a question which is periodically agitated. It is made by an insect belonging to the genus *Icaria.* Perhaps my readers may remember that on page 429, the celebrated bee-cell problem is described, and that mention is made of the many theories which have been invented to solve the riddle. Among them the two most conspicuous are those which are known as the equal pressure theory and the excavation theory. Differing as they do in many respects—one attempting to prove that each cell is forced into the hexagonal shape by the pressure of six cells surrounding it, and the other that the cell is made hexagonal by the cutting away of material from six surrounding cells—they both agree in one point, namely, that the normal shape of the cell is cylindrical, and that it only assumes the hexagonal form by mechanical means.

These questions were briefly mentioned, because an entire omission of them would appear negligent, but they were not followed up because the nests that would set them at rest belonged to another group. We will first take the central nest.

The specimen from which this was drawn was fortunately in an unfinished state, only eight cells being made, and some of these but partly finished. As the reader may see by reference to the illustration, all the cells are hexagonal, whether finished or incomplete, and moreover, that the edges of the hexagon are quite sharp and well defined.

Now, if either of the two theories were true, the cells would not have assumed this shape. Where are the six surrounding cells that are needed to compress the outermost cell into an hexagonal? Or where are the six surrounding cells from which the hexagon was excavated? There are none. The outermost cell, for example, is perfectly free on five of its sides, being only attached to the neighbouring cell by the sixth side. Compression, therefore, has not been employed, because there is nothing that can compress it; neither has excavation been used, because there is no material to be excavated. No one, on looking at this group of cells, can deny that the hexagonal form is produced by the direct labours of the insect, and not by any secondary mechanical means.

Perhaps some one who has not examined the actual object might say that the materials of which the cells are made are sufficiently stiff to need no support of contiguous cells. Now the substance of this remarkable nest is singularly slight, the walls being not thicker than the paper on which this account is printed, and the material is quite soft, as may be seen by the curvature produced by the mere weight of the structure. Yet none of the cells are united by more than three sides, the greater number by two only, and the external cells merely by a single side, leaving five sides and four angles perfectly free.

In this particular specimen the material has evidently been

varied, the insect having been forced to employ different substances in forming its home, as is seen by the pale and dark rings alternately surrounding the cells. The insect which makes this curious home is of moderate size, and is greyish-black, banded with yellowish-white. The abdomen is tolerably stout and sharp-pointed, and is attached to the thorax by a short brownish footstalk. This insect is a native of Natal. Other species of the same group will be mentioned in the course of the following pages.

In the left-hand upper corner may be seen a very remarkable triple nest depending from a branch. This is the work of an insect called *Mischocyttarus labiatus*, which belongs to the family Polistidæ. Like the nest of the preceding insect, it is attached to the bough by a slender and tolerably long footstalk, and the mouths of the cells are downwards, as is always the case with these insects.

Generally, the group of cells is single, but occasionally a more perfect nest is found, which, like the specimen figured in the illustration, has three distinct cell groups, each pendent from the centre of the group above. This may seem rather a dangerous method of suspending the nest, but it is not more so than that which is employed by the common wasp, which builds tier under tier of cells, hanging each tier from its immediate predecessor by little pillars of the same paper-like material as that of which the cells are constructed; or very much, indeed, as the roadway of a suspension bridge is hung from its arch instead of being placed upon it. The insect itself is smaller than the preceding, and is almost uniformly brown.

The last of these three groups is particularly entitled to notice, on account of its bearing upon the hexagonal principle, which has been so often mentioned. The name of the insect is Raphigaster Guiniensis, and, as its name implies, it is a native of Western Africa.

The nest consists of a group of long cells, and suspended from a footstalk. The material of which the nest is composed is peculiarly soft and flimsy, reminding the observer of the worst and most porous French paper. The cells are so thin that the light shines through their delicate walls, and they are so soft that they yield to the least pressure. Each cell is small at the base, and increases regularly towards the mouth, like a reversed sugar-loaf.

Now, if the real cause of the hexagonal form were to be found in the equal pressure of surrounding cells, the central cells of this group ought to be hexagons, for they are soft, pliable, and their conical form renders them peculiarly liable to be squeezed out of shape. Yet, on examining the nest, we find that all the cells retain their conical form, the central cells being as rounded as those on the exterior, and their mouths being as circular.

These examples entirely destroy both theories.

In the first instance we have nests of which the cells are perfectly hexagonal throughout, although some of them are only attached by one side, and are not pressed upon at either of the five remaining sides. We find that the external angles are as sharp, and their internal measurement as true, as those which occupy the very centre of the bee-comb; so that pressure is clearly not the cause of the hexagon. That excavation is not the cause is also evident, from the fact that the external cells cannot have been excavated, and yet are hexagonal.

These examples, therefore, show that the hexagonal form can exist without pressure. But, as if to show that pressure can exist without producing the hexagonal form, we have the nest of the Mischocyttarus, whose long, delicate, soft-walled cells are grouped round each other, and yet retain their conical form, so that at any part of them a transverse section would show a circular edge.

The insect which makes this nest is rather long, measuring perhaps an inch in length. The colour is pale yellow, and the abdomen is much elongated, and attached to a slender footstalk or peduncle nearly as long as itself. Several of the cells have been occupied by larvæ which have begun to assume the pupal condition, as is shown by the white covers over their mouths.

One of the most remarkable of these branch-building insecet is that which has been appropriately named the Processionary Moth (*Cneethocampa processionea*). This curious moth lays a number of eggs, mostly upon the oak, and as soon as they are hatched the little creatures begin to form their home.

Externally it is not unlike that of the brown-tailed moth, but it differs in one respect, namely, that it is not divided into separate chambers, and has only one aperture. When the larvæ sally out for the purpose of procuring food, they spin guide lines, as is the case with many other caterpillars. But, instead of going out singly into the world, each to find its own food in its own way, they march out in regular order, like a military party on a foraging expedition.

PROCESSIONARY MOTH AND CALOSOMA.

A single caterpillar is always the leader, and often is followed by one or two others in Indian file. Presently, however, the caterpillars march two deep, and, if a large number should be on the move, the line is sometimes from five to six deep. They are all very close to each other, so that the procession flows on in one unbroken line, and until the observer is close to it, he cannot see that its component parts are moving at all.

On referring to the illustration, the reader will see that the artist has represented a nest of the Processionary moth, part of which has been torn open so as to show the absence of partitions in the interior. A number of the caterpillars are also shown, but in the middle of the nest is one grub of very great size, being, in fact, six or seven time as large as the caterpillars. This creature has been introduced because it is generally to be found in the nest of the Processionary moth, and because it is one of the most useful insects that a careful agriculturist can protect.

It is the larvæ of a beautiful beetle, called scientifically *Calosoma sycophanta,* which is represented below in the act of ascending the tree. The beetle is a lovely blue-green, but the larva is as unsightly a being as can well be conceived, its body being fat, flattish, and scaly, and its colour black. This creature feeds entirely upon the various caterpillars and other larvæ, even eating those of the destructive sawflies. At the end of the tail are two horny spines, and the head is furnished with a pair of curved, sharp, and powerful jaws, by means of which it seizes its prey.

Instinct teaches these grubs to find their prey, and it may easily be imagined that when they approach a nest of the Processionary moth they are not slow to avail themselves of the opportunity. Indeed, so sure are they of discovering their prey, that Réaumur asserts that he never opened a nest of the Processionary moth without finding one or more specimens of their rapacious enemy, as many as five or six having been seen in a single nest. They are most voracious creatures, as indeed is evident from their structure; and, as each grub will eat several large caterpillars in a day, the havoc which is made in the nest may easily be imagined. The caterpillars have no means of defence or escape. They cannot leave their home, and they cannot kill or expel the intruder. All that they can do is to go out and eat, and come back and be eaten, their numbers ever diminishing, like the companions of Ulysses in the Cyclops' cave.

But for the exertions of this most useful insect, the ravages of the Processionary caterpillars would be greatly increased, for the creature does not only eat them while in the larval condition, but feeds upon them after they have become pupæ. Sometimes, however, this extreme voracity defeats its own purpose. It occasionally happens that a grub of the Calosoma habitually gorges itself to such an extent with Processionary caterpillars that it becomes fat, unwieldy, and scarce able to move. If, when it is in this condition, leaner and hungrier grubs should come across it, they are too apt to seize upon it and devour it in sheer wantonness, even though the nest be full of their legitimate prey.

Knowing the habits of this grub, a French entomologist, M. Boisgerard, managed very ingeniously to avail himself of its devouring capacities. There is a well-known insect, *Bombyx dispar*, popularly called the Gipsy Moth, which is very common in France, though scarce in most parts of England. The larva of this moth is destructive to trees, feeding on their leaves, and then retreating to a cunning little hiding-place, in some crevice of the bark. Finding that his trees were infested with these caterpillars, M. Boisgerard procured a number of female Calosomas, and placed them on the trees. They laid their eggs, and in due season the larvæ were hatched. In process of time the destructive grubs increased so much that they ate all the noxious caterpillars, and at the end of the third year the trees were cleared, and the Calosoma beetles had to go elsewhere for a living.

In England the Calosoma is exceedingly rare, all specimens hitherto captured having been apparently blown over the sea from the Continent or brought in ships. Towards the South of France it is plentiful enough, as is needed from the enormous multitudes of crop-destroying caterpillars on which it feeds. There is, however, a closely allied species, *Calosoma inquisitor*, which is not so scarce, and, although comparatively seldom seen, may be captured by those who know where to look for it. I have taken it in Bayley Wood, near Oxford.

The reader may remember that two species of wasp, namely Vespa vulgaris and Vespa germanica, will work harmoniously at the same nest. This curious sociability, which is contrary to the usual custom of nature, is shared by moths as well as wasps. When experimenting upon the nests of this species, M. Réaumur found that several distinct broods of caterpillars would spin a common web and live in peace together, just as if they had been the offspring of one mother. Mr. Rennie, however, carried the experiments still farther, and found that two different species would act in the same social manner.

"We ourselves ascertained during the present summer (1829) that this principle of sociality is not confined to the same species, nor even to the same genus. The experiment which we tried was, to confine two broods of different species to the same branch, by placing it in a glass of water to prevent their escape. The caterpillars which we experimented upon were several broods of the brown-tail moth (*Porthesia auriflua*) and the lackey (*Clisiocampa neustria*). These we found to work with as much industry and harmony in constructing their common tent as if they had been at liberty in their native trees; and when the lackeys encountered the brown-tails they manifested no alarm nor uneasiness, but passed over the backs of one another as if they had made only a portion of the branch.

"In none of their operations did they seem to be subject to any discipline, each individual appearing to work in perfecting the structure from individual instinct, in the same manner as was remarked by M. Huber in the case of the hive bees. In making such experiments, it is obvious that the species of caterpillars experimented with must feed upon the same sort of plant."

One remark ought to be made on this interesting narrative. The author lays some stress on the fact that the two insects belonged not only to different species, but to different genera. It must, however, be remembered that although the distinction of insects into species is easy enough, their grouping into genera is quite arbitrary, depending entirely on the classifier. Linnæus, for example, divided all the butterflies into two genera, while the modern classification admits some thirty genera. While, therefore, we may lay every stress on the species, we need not trouble ourselves much about the genus.

The two moths mentioned in this history are very different in appearance, and the larvæ are still more unlike. They have, however, this point of similarity, that they construct large dwellings upon branches, spinning them of silk, and making them large enough to contain a whole brood at once. The Lackey moths are so called on account of the bright colours of the caterpillars, which are striped and decorated like modern footmen. Some species, however, derive the name from a different source.

When the mother insect lays her eggs, she deposits them on a small branch or twig, disposing them in a ring that completely encircles the twig, as a bracelet surrounds a lady's wrist. When she has completed the circle, she covers the eggs with a kind of varnish, which soon hardens, and forms a perfect defence from the rain. The varnish is so hard, and binds the eggs so firmly together, that, if the twig be carefully severed, the whole mass of eggs can be slipped off entire. As this varnish produces the same effect on eggs as lacquer does upon polished metal, preserving the surface and defending it from moisture, the insect is called the Lacquer, a word which has been corrupted into Lackey.

In wet weather the Lackey caterpillars prefer to remain in their silken home, leaving it only for the purpose of feeding. They never lose their way, because, like the larvæ of the little ermine moth, which has been already described, they continually spin a single silken thread as they go along, and are, therefore, provided with an infallible guide to the track. Before they change to the pupal state they leave the nest.

The larva of this species is a very prettily marked creature, the body being striped with blue and yellow and white. The moth itself is yellow, with a slight tinge of orange, and across the upper pair of wings runs a dark band edged on either side by a paler streak. As there is another allied species, which lives on various seaside plants, the present insect ought more properly to be called the Tree Lackey. The moth seems to be rather periodical and local; for, although specimens are found annually in most years, they swarm to such an extent in certain places, that whole rows of fruit trees are denuded of their leaves, and covered with the silken webs of the pretty but destructive caterpillars.

THE BROWN-TAILED MOTH is another of the arboreal insects, and spins a web very like that of the gold-tailed moth, which has already been described. In some seasons it is more numerous than in others, and occasionally seen in vast multitudes. This phenomenon is often observable among insects, as is well known to all practical entomologists, and in more than one instance the caterpillars of the Brown-tailed Moth have been so plentiful as to become a positive pest.

They are social larvæ, and, as they are hatched late in the autumn, they spin a joint web, in which they can be secure throughout the winter months. As the brood is mostly numerous, and as two or more broods may unite in forming a common dwelling, their habitation is extremely large, often enveloping several branches together with their twigs and leaves. Like the nest of the gold-tailed moth, it is divided into chambers, and is externally irregular in form, depending entirely for its shape upon the locality in which it is constructed.

Even in this country it is sometimes plentiful enough to annoy the farmer, who does not like to see his hedgerows disfigured by the silken tents spun by these caterpillars; but in France it has occurred in such hosts as to entail a serious loss upon the agriculturist, whole rows of trees having been stripped of their leaves, and the denuded branches covered with the sheets of web in which lay the destroying armies.

ICARIAS.

On the accompaying illustration may be seen a number of curious nests, composed of long hexagonal cells, set side by side. These are made by several species of a hymenopterous insect belonging to the genus *Icaria,* and may be advantageously compared with the central figure in the illustration on page 570.

These nests, or rather these series of cells, are made after a singular fashion. First, the insect attaches to the branch a footstalk composed of the same material as that with which the cells are formed. This footstalk, although slender, is very hard, solid, and tough, and can uphold a considerable weight, as is necessary from the manner of constructing the nest. She then makes a cell after the ordinary wasp-fashion, attaching it to the footstalk with its mouth downwards, and at first making it comparatively short. When the cell has nearly attained its due length, a second is placed alongside the first, and a third is added in like manner, each being lengthened as required. As the cells at the base of the series are finished first, it is evident that they gradually diminish towards the end, those at the extremity being often not one quarter so long as those at the base.

The material employed in making these cells is woody fibre, like that which is used by our common British wasps, and the colour is rather dark yellowish brown, so that, in spite of the curious method in which the nest-groups project from the branches, they are not seen so readily as might be imagined from their eccentric form.

In these, as in many other forms of cells made by hymenopterous insects, is to be found an enigma which as yet is unsolved, and for the mention of which I am indebted to Mr. F. Smith, of the British Museum: *all the cells are of equal size.*

Now this point, which would not particularly strike an ordinary observer, is of the greatest importance to those who have studied the economy of insects, and have bestowed much thought upon them. If we examine the nest of a hive bee, and take any single comb, we shall find that the cells are extremely variable in size—the largest being those which are occupied by the future queens, the smallest those which are the nurseries of the worker bees, and the intermediate cells those in which the drones or males are hatched.

If we examine the nest of the common wasp or hornet, we still find the cells of various sizes, corresponding with the sexes and uses of the occupants; and if we look at that of any species of humble bee, the same fact is clearly perceptible. But in the nests of the Icarias and similar insects, no such variation is discoverable, and no distinction can be found between the male and female cells. The natural question therefore arises, whether all the members of each brood, or rather each cell-group, are of the same sex; or whether one nest produces males and another females, just as one portion of the comb is given to males, another to females, and another to neuters, in the case of the hive bee. No matter how large may be the nest of an Icaria, or how full of cells it may be, the cells are all so alike in shape and size that they must apparently be the cradles of insects belonging to the same sex.

This fact is curiously brought out in a remarkable series of cell-groups which have been placed on a single leaf, some of which are shown in the lower part of the illustration. The leaf is rather long, and, being dry, is now curved by its own force. This leaf seems to have possessed some fascination for the Icarias, as upon the upper surface no less than fifteen nests have been established, none of great length, and all nearly or quite completed. In none are the cells perfectly straight, all having a slight curve downwards on account of the delicate material of which they are made.

The insect which builds these curious cells is a commonplace-looking creature, of a soft, greyish-brown colour, with a moderately large head and a little rounded abdomen, not very unlike the *Cynips Kollari*, which has already been described. It is a native of India, and the nests which have been mentioned were sent from Bareilly.

THERE is also a nest made by an allied insect belonging to the genus Sphex, and coming, as far as is known, from the same locality. As in the preceding case, the habitation is placed in the hollow of a leaf, which has either curled in the process of withering, or has been bent by the insect and retained in its cylindrical form by silken threads. On inspecting the nest it is seen to consist of a mass of delicate threads that are not unlike those of the common silkworm, which cross and recross each other so as to fill up the hollow of the leaf, and to form a receptacle for the future young.

The insect is a large one, being a full inch in length, and of rather a formidable appearance. In colour it is black, with the exception of a few narrow transverse streaks of bright yellow upon the fore part of the thorax, and the abdomen is placed at the end of a long footstalk.

IN connection with this branch of the subject, I must call the attention of the reader to the curious cell-group of *Polistes aterrima*, a figure of which is given on page 461.

At first sight it looks as if the winged architect had intended to make a cell-group like that of the Icaria, and had been brought abruptly to a conclusion. A close inspection, however, shows that the structure is intentional, and not merely the result of accident. The cells are all placed with their mouths downwards, and are set in a very peculiar manner.

In the Icaria nests the cells are arranged so as to form a more or less elongated mass, widest in the middle and tapering to the end, the variation in width being caused by the different arrangement of the cells, which are two deep at the commencement, reach three or four deep at the middle, and are reduced at the end to a single row of cells. But in the nest of the Polistes all the cells are two deep, being arranged in two regular rows.

On reference to the illustration, which is exactly one-half the length of the original specimen, the reader will see that the cells are of different sizes, and might, therefore, fancy that the upper or larger cells are those which nurture the larvæ that are destined to be perfect males and females, and that the lower or smaller cells are intended for the workers or neuters. But a careful examination shows that such a supposition would not be correct. It is true that the upper cells are larger than the lower, but this increase of size is simply owing to the fact that they are completed, while the five lower cells are still unfinished.

The attention of the reader is particularly called to the dark bars which are at regular intervals across the cells. These are not merely intended to represent effects of light and shade, nor to show that the little architect has used materials of different colours so as to form alternate bands ; but each dark line represents a distinct and well-marked ridge, and each ridge is evidently the result of an addition made to the cell.

Here, then, we have a clue to the manner in which the insect builds its curious home. The cells at the upper part of the nest were originally in the same unfinished state as those at the lower extremity. The insect made a tolerably deep cell, laid an egg therein, and then proceeded to lay the foundation of other cells, placing an egg in each.

When, however, she has formed the third or fourth cell, the eggs in the first and second chambers have been hatched into larvæ, and need to be fed. So, the mother insect has now to suspend her architectural labours and to divide her time between the erection of fresh cells and the feeding of the larvæ in those which were first made. Soon, however, her labours become more onerous and more complicated. The larvæ or grubs, which inhabited the first two or three cells, have grown so rapidly that they are fast becoming too large for their pendent cradles. The cells must therefore be enlarged to suit the increased dimensions of the inhabitants. This also is done, each addition being marked by the ridge which has been already mentioned.

Thus, then, the hard-worked mother insect is forced to engage in three distinct labours, namely, the building of new cells, the enlargement of existing cells, and the nurture of the larvæ. A short reference to the illustration will now give the reader a clear idea of the cell-group. Above is a series of completed cells, each occupied by a full-grown larva, one of which is being fed by the mother insect, and below is a series of incomplete cells, each of which has received an egg, but neither of which is fit to sustain the weight of the larvæ. If, however, the nest had been allowed to remain in the forest, instead of being carried off to the British Museum, the five lower cells would have been completed like their more perfect predecessors above.

The observant reader will probably have noticed that from the mouths of several cells a scoop-like projection is seen to issue. These projections have been faithfully rendered by the draughtsman, although they denote a certain imperfection in the specimen. They are evidently the result of hard usage, and show that part of a completed cell has been broken away. As is often the case, the fracture has its value, inasmuch as it shows that the normal form of the cell is hexagonal, and that the angles are quite sharp and firm without needing either pressure or excavation to make them so.

It is much to be regretted that in England there is no representative of this interesting group of insects by which the above-mentioned problem might be solved. We might then know whether males and females belong to the same brood, are nurtured in the same cells, and are of the same size. We might learn whether or not the males are bred in separate establishments divided from the other sex like diœcious plants. As it is, however, we are in ignorance respecting these points, and no one resident in tropical countries seems to have the energy to conduct a series of experiments which would involve expenditure of time and labour. Travellers and residents in tropical countries are often admirable collectors, but, with few exceptions, are poor observers, except of facts that pass immediately under their observation. They make valuable collections, and record many useful isolated facts; but, unfortunately, they seldom appear able to carry on a series of experiments that would occupy several successive years, and thus we lose much valuable as well as interesting knowledge, and waste much time in trying to discover by inference that which we ought to know from observation.

The last point which will be noticed in connexion with this remarkable cell-group is, that it is perfectly protected from rain. Slight and delicate as is the structure, appearing scarcely thicker than the silver paper with which engravings are guarded, it may be deluged with water without being wetted. Over the whole of the cells the insect contrives to lay a thick coat of some varnish-like substance, which at the same time gives the exterior of the cells a polish, binds them more firmly together, and renders them waterproof. The varnish is nearly transparent, but has a blackish hue, which gives to the whole cell-group a uniformity of aspect which would be wanting if the protecting substance were itself colourless.

The insect is, at first sight, black in colour, as is expressed by its specific title. A closer examination in a more favourable light shows that the true colour is a green so deep as to appear black, but having a perceptible bronze gloss in certain lights. The wings are equally sombre in aspect, looking as if they had been held above the flame of a badly-trimmed lamp, and received all the soot upon their translucent membranes. Indeed, their peculiar colour can only be expressed by the word "smoky."

IT is hardly possible to overrate the wonderful varieties of form that are assumed by the nests of insects,—varieties so bold and so startling that few would believe in the possibility of their existence without ocular demonstration. No rule seems to be observed in them; at all events no rule has, as yet, been discovered by which their formation is guided; neither has any conjecture been formed as to the reason for the remarkable forms which they assume.

Perhaps, of all the nests in the splendid collection of the British Museum, there are none that cause so much surprise as the wonderful group which is represented in this illustration.

APOICA.

Many persons pass through the room, and even take some notice of the various nests with which they are surrounded, but they seldom notice the peculiarities of this group until pointed out to them. When, however, their attention is directed towards it, they never fail to express their surprise at so curious a structure, and their admiration of the manner in which these natural homes are constructed.

In a recent portion of this work the hexagonal principle of

construction has been mentioned as applied to separate cells of certain hymenoptera, and that no explanation has been given of the mode by which the six-sided cells are made. It is always easier to explode previous theories than to supplant them with one that really explains the enigma; and such is certainly the case with these cells. But when we come to examine the group of nests which are made by an insect called *Apoica pallida*, the subject takes a wider range, and we are even more hopelessly bewildered than before. In the nests already mentioned the cells are hexagonal, but in these specimens the entire nests assume a form more or less hexagonal, as may be seen by reference to the illustration.

In order to prevent misunderstanding, I must here remark, that the seven nests or cell-groups were not all found adhering to a single branch, as seems to be the case, but that they have been placed near each other in such a manner as to allow of easy comparison, and to show their peculiar form. The large mushroom-shaped nest in the centre, and the small cell-group which occupies the extremity of the bough, appear as they were formed by the insects, but the others have only been arranged for the convenience of comparison.

Even their position has been necessarily altered. Nests of this kind are always placed with the mouths of the cells downward; but, as their peculiar form could not easily be seen if they were allowed to retain their natural position, some of them have been set on their edges, so as to exhibit their outline to the spectator. This is notably the case with that nest which occupies the left hand of the illustration, and which is the most striking of all the specimens.

If the reader will refer to the illustration, he will see that the nests are by no means uniform in size or shape. The larger one, for example, which occupies the centre, rather exceeds ten inches in diameter, while the small nest at the end of the same branch is scarcely half as wide, and the others are of all the intermediate sizes. In shape, too, they differ, some being perfectly hexagonal, others partly so, while others again are nearly circular, though on a careful inspection they show faint traces of the hexagonal form.

We will now examine these nests, and see where they agree with and differ from each other.

In the first place, their upper surfaces are more or less convex, according to their size; and whether they are circular or hexagonal, the convexity remains the same. This form is evidently intended for the purpose of making them weather-proof; for the rain torrents that occasionally deluge the country would soon wash to pieces any nest whereon the falling drops could make a lodgment. The surface is therefore as smooth as that of the various pasteboard wasps which build in the forests of tropical America.

The upper surface being convex, it naturally follows that the under surface is concave, inasmuch as the cells are of tolerably equal length. In fact, the nests somewhat resemble very shallow basins with very thick sides, and bear an almost startling resemblance to the cap of a very large and very well-shaped mushroom, the central specimen being so fungus-like in form that, if it were laid on the ground in a waste and moist spot, it would soon be picked up as a veritable mushroom. The colour, too, is yellowish brown, and the surface has a kind of semi-polish that increases the resemblance.

In the nests of our common wasp, or hornet, the sheets of paper which form the exterior show plainly where each successive flake has been deposited, and the sweep of the insect's jaws is marked distinctly upon the yielding material. Even in the case of the few British species which build pensile nests in the open air, the separate flakes can be distinguished, though they are not so clearly marked as in those homes which are defended from the weather by earth or wood. Our temperate region knows no such sudden vicissitudes of weather as take place near the equator, and there is no need for insect habitations to possess very great strength or powers of resisting water. But in these nests the cover is so beautifully uniform, that no trace of a jaw can be detected upon it.

Agreeing in general appearance, the nests vary somewhat in colour. Of the eight specimens, the generality are of the mushroom-like hue which has already been mentioned. Others, however, rather vary in this respect, and the uniform yellowish brown is pleasingly diversified by patches of red. One of the nests, however, boldly departs from the general uniformity, the surface being not only reddish brown over its whole extent, but as rough as if made of sand-paper, or from the skin of a dogfish.

One or two, again, are much darker than the others; while one is almost white, with only a tinge of grey.

Another point in these nests is, that although they vary so much in diameter, their thickness is almost uniform. The reason is evident enough. As the young larvæ attain a tolerably uniform size, and are not boldly divided into large males, larger queens, and little workers or neuters, the cells are of equal length. Therefore, whether the number be great or small, the thickness of the cell-group remains unchanged, though the diameter may increase to any reasonable amount.

All the nests are fixed in the same manner, a branch or twig passing through the upper surface. When the nest increases in size, the original support is often found to be too slight; and in that case, others are added. The smaller nests are upheld by a single twig only, but the largest is supported at no less than three points, two tolerably stout branches passing through the side of the cover, and a smaller twig supporting the top.

Another point to be noticed is, that the size of the nest is no criterion of its shape. It is not necessarily circular because it is large, nor horizontal because it is small. The eight examples in the British Museum show every gradation of shape between the hexagon and the circle, without the least reference to size.

How the insect forms these wonderful cell-groups is an enigma to which not the least clue can be found. In proportion to the size of the architect, they are simply enormous, and yet the sides and angles are as true and just as if they were single cells. It is very clear that neither the theory of excavation or of equal pressure can apply to these nests, and an additional reason is afforded why these theories should be abandoned. It is to be regretted that the only reasoning is of the destructive kind; but at present we have no data on which to found a theory that seems in the least tenable.

In the nest to which reference has been made, the insects have carried out the hexagonal principle in a curious manner. A number of cells whose mouths are closed with a white silken cover prove that the inmates are undergoing their metamorphosis, and are in a transitional state between the larva and the perfect insect. Instead, however, of being scattered at random throughout the nest, the inhabited cells are arranged in the most systematic manner, a group occupying the centre, and being surrounded at a little distance with a row of covered cells which follow the shape of the exterior outline, and therefore take the shape of a hexagon.

The insect well deserves its scientific title. The generic name *Apoica* is formed from two Greek words, which signifies a colony, and the specific title *pallida* is given in reference to the hue of the body. It is not a handsome nor even a striking insect, being long, slender, and very pale yellow, looking as if it had once been decorated with a brighter covering. It has altogether a faded and semi-bleached look, suggesting to a practical entomologist that it had been subjected to sulphur-fumes, and thereby lost its colouring. Even the wings have the same pallid hue as the body, but with a white cast, and altogether the insect seems far too purposeless of aspect to construct houses which demand so much energy as those which we have just examined.

THE last example of insect pensile nests is, I believe, one that has not yet been described, owing to its recent arrival in this country.

Whilst I was examining some specimens in the insect-room of the British Museum, two gentlemen brought for examination a box full of insect habitations, which they could not identify with those of any known species. At first sight they appeared to be specimens of galls, but a more careful inspection soon showed their real character. They were formed very much like those of the Housebuilder Moth (see page 283), but with a singular addition. Several specimens are now before me, which will be briefly described.

The foundation of the nest is a structure of leaf-stems and fragments of leaves, varying much in size, some being thicker than crowquills, and others as fine as ordinary needles. These are arranged cross-wise upon each other, so that the nest might easily be mistaken for that of a large caddis-worm. The nests, however, differ much in form, size, and material,—some being half as large again as others, and some being made almost entirely of large pieces of leaf, and others chiefly of stems, among which the leaf-fragments are closely pressed.

We will now proceed to cut open one of these nests in order to view its structure.

The outer covering is remarkably close, stiff, and tough, although very thin, and crackles like parchment as the scissors pass through it. When cut, it is found to be almost distinct from the nest which it covers, being only attached to the projecting ends of the leaf-stems, and so slightly fastened to them that it can be lifted off without injury, only leaving a few threads adherent to the stem.

We now turn back the severed flap, and the body of the nest comes to view. In the dry state the leaf-stems are so hard that they require a strong and sharp pair of scissors to penetrate them. I nearly broke a moderately fine pair of scissors in a vain endeavour to open the nest. Even in their fresh state the stems must have been tolerably strong, and the architect must have possessed a powerful pair of jaws for their severance. The stems are crossed upon each other, much as confectioners cross sticks of chocolate, so that the ends slightly protrude, and a hollow space is left in the centre. Pressed tightly among the sticks are fragments of leaves, not torn from the small delicate portions, but cut completely through the largest nervures, and seeming, indeed, as if the strongest parts of the leaves were intentionally selected. In the specimens now before me the upper surface of the leaf is always towards the exterior of the nest.

We now take a very strong and sharp pair of scissors, push one point into the nest, and carefully cut a flap corresponding with the severed portion of the silken cover. The flap is easily turned back, and discloses a smooth and silken lining, much resembling that which forms the cover. The lining, however, is softer than the cover, and does not crackle when bent. Thus we see that the nest consists of four distinct layers: first, the soft silken lining, then a cover of leaf-fragments, then a protecting *chevaux-de-frise* of stems, and lastly a cover of silk, so that the inhabitant is as well protected from weather and foes as can be imagined.

The next proceeding is to discover the architects of the nests. This is easily done, for some of the architects have assumed their perfect state during the voyage home, while others are preserved by spirits, in which their discoverer has thoughtfully placed some specimens.

Here I may be allowed to mention that the example set by Mr. W. J. Tomkinson, who sent over these interesting objects, is one which is well worthy of imitation. Residents in other countries are too apt to forget the interests of their own, and they soon become familiar with the objects which at first are new and strange to them, and at last become entirely indifferent. Even when they do take the trouble to collect and send home a few objects, they do so in such a manner that they are almost useless, no description being given of them, and no clue afforded which can help the home-staying student.

Here, however, proper pains have been taken, and the value of the objects is in consequence multiplied a hundred-fold. A number of nests were sent as they were collected from the branches, and, in order to show that the architect is not confined to one species of tree, they have been carefully selected from several trees, such as the oak, acacia, and alder. My specimens are taken from the last-mentioned tree. Knowing that the pupæ would become moths in the course of the voyage, Mr. Tomkinson placed a number of them in the box, so that a perfect series of the insect has been obtained, namely, the male and female, pupa and larva, some in the dried state and others in spirits, in order that the internal anatomy might be examined.

Before the male caterpillar changes into a chrysalis it reverses its position, so that the head is close to the orifice which was previously occupied by the tail. When it has completed its change, and is about to issue into the world, it forces itself out of the nest as far as the base of the abdomen. The female never leaves her home, and never changes her attitude, and scarcely changes her form. After she has emerged from the pupal states, she seems to return to her former condition, and would be taken by any ordinary observer for a caterpillar of more than ordinary fatness. She has no wings, and no legs to speak of, these members being needless in a creature that never changes her position. It is rather curious that the males should ever be able to find their spouses, but they are probably led by an instinct which we cannot comprehend, as is the case with several of the larger British moths.

The male is a rather small though stoutly made insect, and is not at all attractive in colour, being simple brown, with a few black markings on the wings. The antennæ, however, are very beautiful, being doubly feathered, like those of the Housebuilder Moth, the feathering being widest at the base, and narrowing gradually to the tip. The whole of the body is clothed with long, dense, and soft hair, of a pale brown, and having a silken lustre. These beautiful nests were brought to the Museum by E. H. Armitage, Esq., who kindly presented me with the specimens which have been described.

A SOMEWHAT similar nest, but of a much more formidable aspect, was discovered by W. B. Lord, Esq. R.A., and has been figured in the *Boys' Own Magazine* for August, 1864. The shape of the nest is very remarkable, and is exactly that of a soda-water-bottle, suspended by its neck. A very tolerable imitation of this curious nest could be made by coating a soda-water bottle with clay, and sticking it full of porcupine quills, with the points radiating on every side. The following is Mr. Lord's own description:—

"On looking closely at the thorny, sinuous branches, we shall see a number of little pendent prickly things, each hanging to its own silken cord, like juvenile hedgehogs 'lynched' by the fairies of the spring.

"These are a peculiar species of 'tree-caddis,' which, as far as I know, are as yet undescribed by any one. Their cases are curiously armed with thorns, nipped from the tree on which they hang. The thorns are all disposed with their points outwards, and are stuck into a strong, glutinous material of which the body of the case is composed, and they look for all the world like the spikes of *chevaux-de-frise.* A web-like skein of singularly strong material serves as a rope whereby to swing the caddis-case from the branch to which it is attached. And a nest more difficult to swallow, and hard to digest, its enemies would be rather puzzled to find."

As is frequently the case with such nests, the peculiar form serves a double purpose, namely, protection and concealment, the sharp points of the thorns performing the former duty, and their similarity to surrounding objects the latter. Acacias are conspicuous for the thorns with which their branches and sometimes their trunks are studded, and in several species the wooden bayonets are several inches in length, and as large and sharp as porcupine quills. These thorns are crowded thickly on the branches, and always diverge from each other, so that the hand can scarcely be insinuated among the boughs without suffering several wounds. The nest being surrounded with these thorns, it is evident that all ordinary foes would be baffled by such an array of points, no matter how anxious they might be to get at the creature within.

The thorns are equally efficient as a means of concealment. for, as they are taken from the tree itself, they cause the nest to harmonize so perfectly with surrounding objects, that it is not very easily perceived.

As long as the caterpillar remains in its larval state, and is obliged to feed, it traverses the branches freely, carrying with it the prickly home, and bearing the whole of its weight as it moves. But when the pupal stage has nearly arrived, the nest is suspended to the branch by strong silken threads, and thenceforth remains immoveable.

CHAPTER XXXI.

MISCELLANEA.

The POLYZOA and their varied and beautiful forms—The RAFT SPIDER—Why so called—Mode of obtaining prey—Mice and their homes—The CAMPAGNOL or Harvest Mouse—Its general habits—Its winter and summer nest—Its storehouse and provisions—Entrance to the nest—The WOOD MOUSE and its nest—Uses of the Field Mice—The DOMESTIC MOUSE—Various nests—Rapidity of nest-building—A nest in a bottle—The cell of the QUEEN TERMITE—Its entrances and exits—Size of the inmates—The FUNGUS ANT and its singular home—Material, structure, and size of the nest—The CLOTHES MOTHS and their various species—Habitations of the Clothes Moth, and the method of formation and enlargement—The ELK and its winter home—The snow fortress and its leaguers—Its use, advantages, and dangers—The ALBATROS and its mode of nesting—Strange scenes—The EDIBLE SWALLOW—Its mode of nesting—Origin of its name—Description of the nest—Curious legend respecting the bird—The EAGLE and its mode of nesting—Difficulty of reaching the eyrie—The NIGHTINGALE and its nest—Other ground-building birds and their temporary homes—The NODDY—Perilous position of the eggs, and young—The COOT, and its semi-aquatic nest.

IN this, the concluding chapter, are described sundry habitations which cannot well be classed in any of the previously mentioned groups, and which present some peculiarities which render them worthy of a separate notice.

We will begin with two aquatic habitations, one fixed, and set below the surface of the water, and the other moveable and floating upon it.

ANY one who is in the habit of frequenting the sea-shore must have observed certain rough, leaf-like objects, which are popularly called Sea Mats, one of which is shown, of its natural size, at Fig. 13. These objects are popularly supposed to be sea-weeds, and are therefore called Lemon-weeds, because they give forth an odour which somewhat resembles the fragrant oil

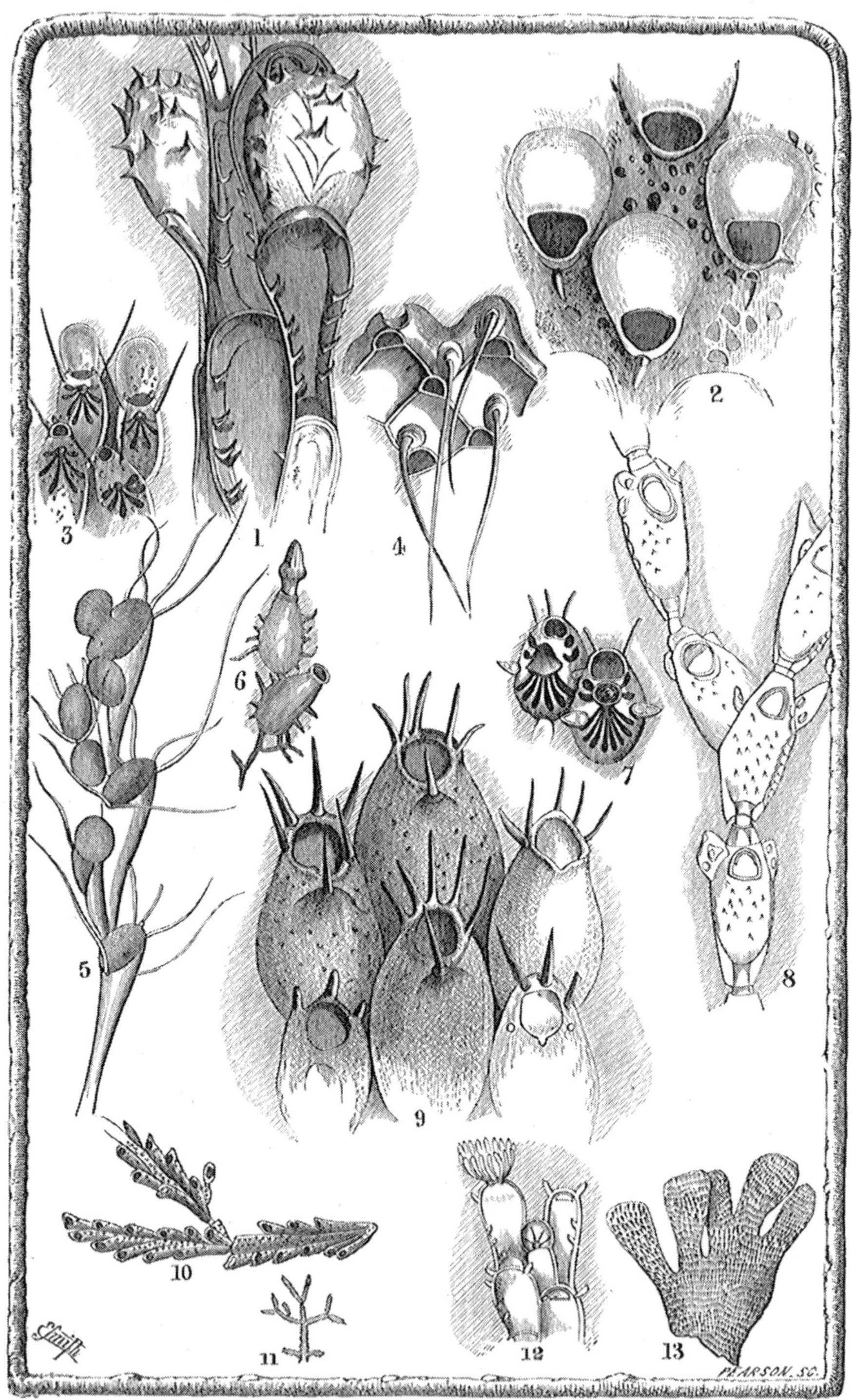

POLYZOA.

of the lemon-peel. Their origin, however, is of an animal and not a vegetable nature, for they are the submarine houses of a vast class of animals technically called POLYZOA, because many distinct animals are associated together in one community.

For many years their place in the scale of creation was extremely dubious, but they are now acknowledged to be allied to the molluscs, and to be really of a higher nature than the bright and active insects. In general form they so resemble the zoophytes, that the two great groups have been confused together, although they are in reality farther apart than the monkey and the snail. They assume various forms, which may be reduced to three, namely, a flat, leaf-like shape, a form as of delicate branching twigs, and a flat series of cells spreading over other substances, such as the stem or frond of a sea-weed, an empty shell, or submerged plant, and similar objects.

The most familiar of these Polyzoa is the common Sea Mat, which has already been mentioned, and which is called by the scientific name of *Flustra Foliacea.* If the finger be passed over one of these leaf-like objects, a decided roughness is felt, like the surface of a file, and if it be drawn through the hand from base to tip, the roughness becomes so marked that the leaf can hardly pass. The reason for this roughness may be seen by reference to Fig. 12, which is a magnified representation of a part of the same object. It will be seen that the Flustra is composed of a vast number of cells, each cell being furnished with little tooth-like projections, which produce the roughness already mentioned.

The cells are formed by their inhabitants, and very much resemble in outward form the polypi which fill the cells of the zoophytes, each of which possesses a beautiful plume of tentacles, as is seen in the uppermost cell. When the animals are at rest they are withdrawn so closely into their cells that they cannot be seen; but when they are hungry and desirous of feeding, they push themselves out of their houses, like so many snails out of their shells, expand their beautiful plumes, and await the food which is brought to them by the action of the water. Under the microscopes they are lovely objects, and even when they are dead their empty habitations are full of beauty.

A detailed description of all the species which are represented in the illustration would occupy too much space, even if it were desirable. Suffice it to say, that they have been selected in order to show the wonderful variety that exists among them, and the exceeding beauty of their forms. Perhaps the most elegant forms are to be found among the Lepralia, plenty of which are to be found on any sea-coast, spreading over the fronds and stems of sea-weeds, and sometimes entirely hiding their surface. Further information may be obtained by reference to works on Natural History, and especially to the admirable monograph on this subject, written by Mr. Bush, which forms part of the catalogue of the British Museum.

The names of the different objects are as follows :—

1. Farciminaria aculeata.
2. Lepralia reticulata.
3. ——— Gattyæ.
4. Cupularia Lowei.
5. Bicellaria gracilis.
6. Buskia nitens.
7. Lepralia alata.
8. Catenicella perforata.
9. Lepralia spinifera.
10. Crisia eburnia (magnified).
11. ——— (Natural size).
12. Flustra foliacea (magnified).
13. ——— (Nat. size).

RAFT SPIDER

THE second aquatic habitation is of a very curious character, and is made by a spider. The reader will remember that the water spider is in the habit of constructing beneath the water a permanent home, to which it retires with the prey which it has caught, and in which it brings up its young. There is another spider which frequents the water, but which only makes a temporary and moveable residence. This is the RAFT SPIDER (*Dolomedes fimbriatus*) which is represented in the illustration of its natural size.

As may be seen by reference to the figure, it is a large species, being, indeed, one of the largest British spiders, its size depending more upon the dimensions of the body than the length of the limbs. It is a remarkably handsome spider, its general colour being chocolate-brown, and a broad orange band being drawn so as to mark the outline of the abdomen and thorax. There is a double row of small white spots upon the surface of the abdomen, and a number of short dark transverse bars give variety to the colouring. The limbs are pale red.

This creature belongs to that group of spiders which do not live in a web, and wait for casual insects, but which chase their prey after the manner of carnivorous vertebrates. Indeed, it may fairly be said to belong to the large group of wolf spiders, and is nearly allied to them.

The Raft Spider is only to be found in fenny or marshy places, and is mostly seen in the fens of Cambridgeshire, where its remarkable habits have long been known. Not content with chasing insects on land, it follows them in the water, on the surface of which it can run freely. It needs, however, a resting-place, and forms one by getting together a quantity of dry leaves and similar substances, which it gathers into a rough ball, and fastens with silken threads. On this ball the Spider sits, and allows itself to be blown about the water by the wind. Apparently, it has no means of directing its course, but suffers its raft to traverse the surface as the wind or current may carry it.

There is no lack of prey, for the aquatic insects are constantly coming up to breathe the air; and although they may only remain on the surface for a second or two, the Spider can seize them before they can gain the safe refuge of the deeper water. Then there are insects, such as the gnat, which attain their wings on the surface of the water, and can be taken by the Spider before they have gained strength for flight. Also, there are insects which habitually traverse the water in search of prey, and which are themselves seized by the more powerful and equally voracious Spider. More than this, moths, flies, beetles, and other insects, are continually falling into the water, and these afford the easiest prey to the Raft Spider, who pounced upon them as they vainly struggle to regain the air, and then carries them back to its raft, there to devour them in peace.

The Spider does not merely sit upon the raft, and there capture any prey that may happen to come within reach, but when it sees an insect upon the surface, it leaves the raft, runs swiftly over the water, secures its prey, and brings it back to the raft. It can even descend below the surface of the water, and will often crawl several inches in depth. This feat it does not perform by diving, as is the case with the water spider, but by means of the aquatic plants, down whose stems it crawls. Its capability of existing for some time beneath the surface of the water is often the means of saving its life; for, when it sees an enemy approaching, it quietly slips under the raft, and there lies in perfect security until the danger has passed away.

There is, living in the same localities, a closely-allied species, the PIRATE SPIDER (*Lycosa piratica*), which has similar habits, chasing its prey on the water, and descending as well below the surface. It does not, however, possess the power of making a raft.

IN a previous chapter of this work, the beautiful pensile nest of the Harvest Mouse has been described and figured, and the burrows of other species of mouse have been cursorily mentioned. I shall now proceed to describe the nests of the common Field Mice, together with the habitation of the little brown-coated, long-tailed, sharp-nosed rodent, that is so familiar in houses unguarded by cats or traps.

We will first take the nest of the SHORT-TAILED FIELD MOUSE, otherwise termed CAMPAGNOL, or FIELD VOLE (*Arvicola arvensis*). This pretty little creature, whose red back, grey belly, short ears, and blunt nose, might be seen daily if human eyes were more accustomed to observation, is extremely plentiful in the fields, especially those of a low-lying and marshy character, such as water meadows and hay-fields near rivers.

Though more nocturnal than diurnal in their habits, the little creatures are not afraid of daylight, and I have often captured them when the sun was at its meridian height. But they are so smooth and easy in their movements, harmonise so well with the colour of the soil, and glide so deftly between the grass, that they can scarcely be distinguished even when the blades are only a few inches in length. I have known them to traverse the ground while a game at cricket was proceeding, and to cross the closely-mown space between the wickets, as if serenely conscious of their invisibility.

They seem to glide rather than to walk, and thread their way silently and without noise. Even when the grass is short, a little patch of reddish earth attracts no attention, and the red-brown fur of the mouse is so similar to such earth, that few would notice it. But if a more attentive observer finds that in a few seconds the ruddy patch has changed its place, his suspicions are at once aroused, and he examines the moving tint more curiously. He must, however, keep his eye upon it as he moves towards it, for if he once loses sight of it, he will in all probability miss it altogether, and think that his eye must have deceived him.

Towards the evening, however, the Campagnol is less fearful, and not only traverses the fields, but ascends the shrubs and plants in search of food. It climbs nearly as well as a squirrel, its sharp nails hooking themselves into every irregularity of the bark, and its long, finger-like toes clasping round the grass stems and little twigs like the paws of a monkey. An autumnal evening is the best time for watching the Campagnol, and if the observer will only remain perfectly quiet, and keep a good opera-glass in readiness, he will be greatly interested by the little animal. A hedge in which are plenty of dog-roses is a likely place for the Campagnol, as the animal is very fond of the ripe hips, and ascends the shrub in search of its daily food. When it reaches the branch, bending with the scarlet load, the mouse runs swiftly and sure-footed as a rope-dancer, and carries off a store of the fruit, partly for present consumption and partly for a stock of winter food.

For the little creature is not one of the hibernating animals, or, at all events, the semi-sleep is of so light a character that the mouse comes often abroad, even in the depth of winter. It is undeterred by severe frost, and takes little heed of snow, as is proved by its tiny footmarks being tracked in the white and yielding substance.

This little mouse makes two kinds of nest, one for the winter, and another for the summer. The winter nest is below ground, and is approached by a hole varying much in length. As the cavity in which the nest reposes is larger than the tunnel, and of a globular form, it is mostly usurped by the wasp when the Mouse deserts it for summer quarters. Sometimes it is placed at some depth in the ground; but usually is only a few inches from the surface. This is the nest to which Burns refers in his well-known poem upon the Field Mouse whose nest he had inadvertently ploughed up.

Besides the winter nest itself, the animal has a storehouse or cellar in which are placed the provisions intended for winter use, when the weather prohibits the Mouse from leaving its home, or when the surrounding shrubs and bushes are plundered of their fruits and denuded of their bark. In this storehouse the animal conceals quantities of hips and other provisions, among which are found numbers of cherry-stones.

The summer nest is of entirely a different construction, being placed above ground, though tolerably well concealed. The following account of it, by Mr. J. J. Briggs, appeared originally in the *Field* newspaper. "No wonder that in districts where they are difficult to keep down they increase with rapidity, for, like the common Mouse, they are prolific breeders. I have found nests of this Mouse in almost every week from the end of May to the middle of August, and each containing from one to ten young, usually from five to seven. The young look poor helpless creatures, being both blind and naked. They leave the nest in about a month, but remain with their parents for some time afterwards.

"The nest is placed on the ground in a pasture or meadow: a field of mowing grass is preferred, but I have found it among corn, where the long herbage affords the coveted quiet and concealment; but when the crop is cut, the nest is laid bare, and the young frequently fall a prey to hawks and other depredators. The nest is built in a little hollow on the surface of the earth, just concealed at the bottom of the stems of grass. If you pull it out it looks like a lump of herds or flax, being composed of numerous small pieces of grass nibbled to a fine texture with care by the parent animals.

"I have taken up dozens of nests to examine, but in no single instance could I ever find an entrance to the interior. How the parents gain admittance to it seems extraordinary. This remark applies to the nest of the White-bellied Field Mouse, and White, of Selborne, notices the same fact with reference to the harvest-mouse. How the young are suckled seems marvellous, unless the conjecture be correct that the female opens a fresh aperture in the nest each time she visits her young, and closes it again when she departs.

"The parents show considerable affection for their young. If a nest be exposed by the mower they do not desert it, but on the contrary endeavour to conceal it from observation as well as they can, by drawing round it the neighbouring grasses and plants."

The same writer remarks that he has several times caught the Short-tailed Field Mouse in the hedges while "bat-fowling" at night for small birds. He has also found that when the Mouse eats hips, it nibbles off one end and extracts the seeds, rejecting the husks as uneatable. Man, however, acts in just the reverse manner, rejecting the seeds with their cottony envelopes, and eating the sweet husk, or sometimes boiling it up with sugar and making it into a conserve.

The cherry-stones are mostly obtained through the agency of blackbirds, thrushes, and other feathered fruit lovers. These birds pluck the cherries, often leaving the stones adhering slightly to the stalks, or dropping them on the ground. In the fomer case the stones are sure to be flung down when the legitimate owner gathers the fruit, so that the Mouse who is fortunate enough to live in a cherry-growing district is sure of a winter stock of food. Several hundred cherry-stones are sometimes placed in a single storehouse, affording sustenance to several mice.

The animal eats them in a peculiar manner. Instead of splitting them open by using the chisel-edged teeth or wedges, after the manner of schoolboys opening nuts and peach-stones with their pocket-knives, the Mouse nibbles off one end of the stone so as to make a little hole, and through this small aperture it contrives to extract the solid kernel.

The LONG-TAILED FIELD MOUSE or WOOD MOUSE (*Mus sylvaticus*) also makes a winter nest, in which it lives, but to which it does not absolutely confine itself, making several nests in the course of a season, and selecting such spots as appear to please its fancy at the time. Mr. Briggs remarks that he has known one of these mice to make a nest in three days.

One species of Field Mouse sometimes does good service to

mankind, through its habits of storing up its winter stock of provisions. Lately in the country about Odessa vast armies of mice were seen, and evidently did much damage. Not only did they eat the crops, but they swarmed into the houses in such numbers that traps could hardly be set fast enough, twenty or thirty being often taken in a single day.

Hurtful though they were in some senses, they nevertheless had their uses. The country is liable to the attacks of locusts, which in that year happened to be particularly numerous. These destructive insects, as is the case with many of their order, lay their eggs enclosed in capsules, something like the well-known egg-cases of our too common cockroach. The mice were very fond of the egg-capsules, and not only devoured them as part of their daily food, but carried them away, laid them up in their treasuries for a winter store, thus thinning the locust armies far more effectually than man could have done.

WE now come to the COMMON MOUSE of our houses (*Mus musculus*).

This little animal is a notable house-builder, making nests out of various materials, and placing them in various situations. There seems to be hardly any place in which a Mouse will not establish itself, and scarcely any materials of which it will not make its nest. Hay, leaves, straw, bitten into suitable lengths, roots, and dried herbage, are the usual materials employed by this animal when it is in the country.

When it becomes a town mouse and lives in houses, it accommodates itself to circumstances, and is never in want of a situation for a nest or materials wherewithal to make a comfortable house. It will use up old rags, tow, bits of rejected cord, paper, and any such materials as can be found straggling about a house; and if it can find no fragments, it helps itself very unceremoniously to, and cuts to pieces, books, newspapers, curtains, or garments.

Many instances of remarkable Mouse nests are recorded, among which the following are worthy of mention.

As is usual, at the end of autumn, a number of flower-pots had been set aside in a shed, in waiting for the coming spring. Towards the middle of winter, the shed was cleared out, and the flower-pots removed. While carrying them out of the shed the owner was rather surprised to find a round hole in the mould, and therefore examined it more closely. In the hole was seen, not a plant, but the tail of a mouse, which leaped from the pot as soon as it was set down. Presently another mouse followed from the same aperture, showing that a nest lay beneath the soil. On removing the earth, a neat and comfortable nest was found, made chiefly of straw and paper, the entrance to which was the hole through which the inmates had fled.

The most curious point in connexion with this nest was, that although the earth in the pot seemed to be intact except for the round hole, which might have been made by a stick, none was found within it. The ingenious little architects had been clever enough to scoop out the whole of the earth and to carry it away, so as to form a cavity for the reception of their nest. They did not completely empty the pot, as if knowing by instinct that their habitation would be betrayed. Accordingly, they allowed a slight covering of earth to remain upon their nest, and had laboriously carried out the whole of the mould through the little aperture which has been mentioned. The flower-pot was placed on a shelf in the shed, and the earth was quite hard, so that in the process of excavation there was little danger that it would fall upon the architects.

Another nest was discovered in rather an ingenious position. A bird had built a nest upon a shrub in a garden, and, as is usual in such cases, had placed its home near the ground. A Mouse of original genius saw the nest, and perceived its value. Accordingly, she built her own nest immediately below that of the bird, so that she and her young were sheltered as by a roof. So closely had she fixed her habitation, that, as her young ran in and out of their home, their bodies pressed against the floor of the bird's nest above them. No less than six young were discovered in this ingenious nest.

Another very remarkable nest of the Common Mouse has been chronicled in the same journal to which reference has repeatedly been made. "Early in March we set a hen; and, as her nest was a basket, a sack was placed under and around it, so as to keep in the heat. When the hen was set, she was in good feather, wearing an ample tail, according to her kind (the Brahma); but as the three weeks went on, her tail seemed much broken, assumed a dilapidated appearance, and finally became a mere stump. This excited notice and surprise, as there was nothing near her against which she was likely to spoil her tail.

"When the chickens were hatched, and they and their mother were taken to a fresh nest, and the old one removed, it was found that a Mouse had constructed a beautiful nest under the basket. The body of the nest was made of tow scraped from the sack, and chopped or gnawed hay from the hen's nest; while the lining was made of the feathers of her tail, which had evidently been removed, a small bit at a time, as wanted, until all the feathers were reduced to stumps, showing marks of the Mouse's teeth. We should have liked to have heard the hen's remarks on the transaction, when the Mouse was nibbling her tail."

In this case the Mouse improved on the conduct of her relative that built in the garden; for, by placing her nest in such a position, she not only secured the very best materials for her home, but enjoyed the advantage of the regular and high temperature which proceeded from the body of the sitting hen, and which was admirably adapted for the well-being of her young family.

The last example of a remarkable Mouse-nest is that which is figured in the accompanying illustration, and which was drawn from the actual object.

A number of empty bottles had been stowed away upon a shelf, and among them was found one which was tenanted by a Mouse. The little creature had considered that the bottle would afford a suitable home for her young, and had therefore conveyed into it a quantity of bedding, which she made into a nest. The bottle was filled with the nest, and the eccentric architect had taken the precaution to leave a round hole corresponding to the neck of the bottle. In this remarkable domicile the young were placed; and it is a fact worthy of notice, that no attempt had been made to shut out the light. Nothing would have been easier than to have formed the cavity at the underside, so that the soft materials of the nest would exclude the light; but the Mouse had simply formed a comfortable hollow for her young, and therein she had placed her offspring. It is therefore evident that the Mouse has no fear of light, but that it only chooses darkness as a means of safety for its young.

The rapidity with which the Mouse can make a nest is somewhat surprising. One of the Cambridge journals mentioned, some few years ago, that in a farmer's house a loaf of newly-baked bread was placed upon a shelf, according to custom. Next day, a hole was observed in the loaf; and when it was cut open, a Mouse and her nest were discovered within, the latter having been made of paper. On examination, the material of the habitation was found to have been obtained from a copy-book, which had been torn into shreds, and arranged into the form of a nest.

MOUSE NEST IN BOTTLE.

Within this curious home were nine young mice, pink, transparent, and newly born. Thus, in the space of thirty-six hours at the most, the loaf must have cooled, the interior been excavated, the copy-book found and cut into suitable pieces, the nest made, and the young brought into the world. Surely it is no wonder that mice are so plentiful, or that their many enemies fail to exterminate them.

A GENERAL account of the TERMITES, or WHITE ANTS as they are popularly but erroneously called, has been given under the head of Building Insects, and it has been mentioned that the female, or queen, has a cell distinct from the habitation of her subjects, and that she never leaves it until her death. In order that the reader should understand more fully the structure of the royal cell, an illustration of it is here introduced.

When viewed from the outside, it would hardly be recognized for the habitation of an insect, for it looks like a large lump of hardened clay, about as large as an ordinary French roll, and not very unlike it in shape. On a closer inspection, a number of little holes may be seen, and these apertures afford an unfailing indication as to the real nature of the clay lump. Fig. 2 represents the external appearance of one of these cells.

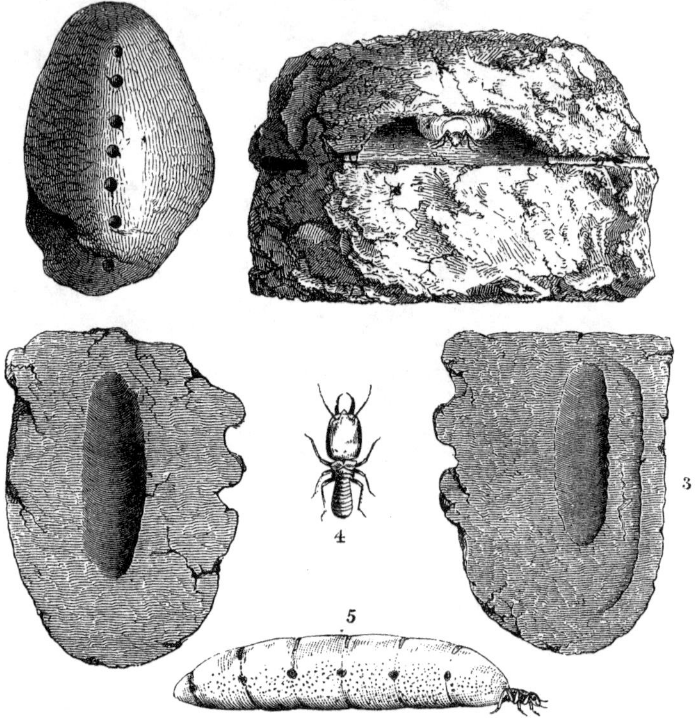

TERMITE CELL.

Supposing that a queen Termite cell be cut vertically, so that the knife passes through either of the little round holes, it will present an appearance which is shown at Fig. 1. The large hollow of the cell is nearly filled by the body of the female, whose head and thorax are seen in the cavity. On either side is a section of the little holes, which are shown to be cylindrical passages communicating with the interior of the cell. The worker Termites, being very small, can traverse these passages with perfect ease, while the enormous body of the female is utterly unable to pass.

Through these passages the workers are continually passing, some entering with empty jaws, and others emerging, each holding between its mandibles an egg, which it is conveying to the nurseries. So rapidly are the eggs laid, that the workers are fully employed in carrying them out and placing them under the charge of the nurses.

The contrast in size between the workers and the queen can easily be seen by reference to the illustration. At Fig. 5 is shown the queen, and in the right hand of Fig. 1 is seen one of the workers passing through the tunnel. None but the workers can pass through so small an aperture, for the fighters or soldiers are of very much greater size than the workers, as may be seen at Fig. 4.

The queen, however, is necessarily very much reduced in size, as, if she had been drawn of her full dimensions, she would have occupied the whole length of the drawing. Before she is immured in the royal cell, she is by no means a large insect, the abdomen being in ordinary proportion to the thorax and head. But, when she has been fairly installed in her office her abdomen begins to enlarge, until it becomes so enormous that she is totally unable to move, and therefore, her enforced prison is so far from being a hardship, that it is a necessary protection for her huge and soft body, which is several hundred times larger than that of her mate. Large indeed she must be, as she is calculated to produce, on the average, rather more than thirty million eggs.

Figs. 3, 3 show the appearance of the royal cell when split open longitudinally, the recess which contains the queen being seen nearly in the centre. All the drawings were taken from specimens in the British Museum, and in the cell which is here figured, the outline of the queen is quite perceptible, having been impressed on the interior of the cell. The mode by which it is enlarged is also shown, a further enlargement having been begun, but cut short by the demolition of the nest. The cells vary very much in size, probably in accordance with the dimensions of the enclosed queen. I have seen them as large as cocoa-nuts, and of an extraordinary weight, the greater portion of the mass being solid clay.

FUNGUS ANT.

THE accompanying illustration represents a most singular structure, which is in the collection of the British Museum, and may be seen in the nest room.

It is hardly possible to imagine any object which less resembles an insect's nest. It might very well be taken for a sponge, did sponge grow in the forests of Cayenne. It much resembles a fungus, and might be mistaken for an overgrown and partially decayed puff-ball. If inspected closely it seems to be made of the same fungus that furnishes the German tinder, and is indeed, like that substance, very useful as a means of stopping the flow of blood.

The real material, however, of which the nest is made, is formed of the short cottony fibres which fill the seed pods of the cotton tree (*Bombax ceiba*). The fibre is so short that it is incapable of being woven into fabrics, though it might possibly be made useful in the manufacture of paper, its texture being peculiarly soft and silky. The only uses to which it is at present applied are stuffing mattrasses and pillows, and packing delicate goods, and heading the tiny poisoned arrows which are projected through the blow-gun of the native.

These Ants, however, find the fibre useful for their work, and contrive to weave it so dexterously, that the individuality of the fibres is lost, and they are all made into a compact and uniform mass. The size of the nests varies, but is sometimes very considerable, a full-sized nest being often as large as a man's head.

The Ant itself is rather a curious little creature, dark in colour, covered with many angular protuberances, and being remarkable for a couple of long, sharp spines that project from the thorax, one on either side. Its scientific names, *Polyrachis bispinosa*, are given in consequence of these projections, the first name being composed of two Greek words, signifying many-peaked, and the second being formed from two Latin words, signifying two-spined.

THERE are many insects whose habitations are peculiarly annoying to mankind, and yet are extremely interesting to those who take an interest in the workings of instinct. Chief among these insects is the well-known CLOTHES MOTH. There are several allied species which popularly go by this name, but the most plentiful is that which bears the scientific title of *Tinea vestianella*. These destructive little creatures are proverbially injurious to clothes, especially if the garments be made of wool or furs, vegetable fabrics being not to their taste. Some species affect dried insects, and are in consequence extremely hateful to the entomologist; while their ravages on furs and feathers. and even on leather itself, render them the dread of those who, like myself, possess collections of natural history or ethnology.

In their winged state, the moths themselves do no direct harm; but their young are doubly mischievous, firstly, because they devour the fabrics in which they live, and secondly, because they cut up the cloth, fur, or feathers, in order to obtain material for their home. Possibly for the sake of concealment as well as protection, the larva instinctively forms a habitation which entirely covers its white body, and, which is almost imperceptible to the eye, because it is formed of the same materials as the fabric on which it lies.

The habitation is tubular in form, though not exactly cylindrical, being rather larger in the middle than at the ends, and open so as to allow the extremities of the caterpillar to protrude. One object in this structure is, to enable the inmate to turn in its cell, an operation which must necessarily be performed whenever the tubular home is enlarged. The process of enlargement is continually going on, and it is in consequence of this proceeding that so much material is used.

The manner in which the little creature enlarges its home is as follows :—

Without quitting its tubular home, it cuts a longitudinal slit throughout half its length or so, and opens the case to the required width. It then proceeds to weave a triangular piece of webbing, with which it fills up the opened slit, and joins the edges with perfect accuracy. As one end of the case is now larger than the other, the caterpillar turns its attention to the other end, cuts it open, widens it, and fills up the gap precisely as it had done to the first part. When the soft tube is sufficiently widened, it is lengthened by the addition of rings to each extremity.

By taking advantage of this peculiar method of house-making, observant persons have forced the Clothes Moth to make their tubular homes of any colour and almost of any pattern. By shifting the caterpillar from one coloured cloth to another, the required tints are produced, and the pattern is gained by watching the creature at work, and transferring it at the proper season. For example, a very pretty specimen can be produced by turning out of its original home a half-grown caterpillar, and putting it on a piece of bright green cloth. After it has made its tube, it can be shifted to a black cloth, and when it has cut the longitudinal slit, and has half filled it up, it can be transferred to a piece of scarlet cloth, so that the complementary colours of green and scarlet are brought into juxtaposition, and "thrown out" by the contrast with the black.

The caterpillar is not very particular as to the kind of material which it employs, and on which it feeds. Mr. Rennie makes the following observations on one of these creatures, whose proceedings he had watched. "The caterpillar first took up its abode in a specimen of the ghost-moth (*Hepialus humuli*), where, finding few suitable materials for building, it had recourse to the cork of the drawer, with the chips of which it made a structure, almost as warm as it would have done from wool. Whether it took offence at our disturbing it one day, or whether it did not find sufficient food in the body of the ghost-moth, we know not; but it left its cork house, and travelled about eighteen inches, selected the 'old lady' moth (*Mormo maura*), one of the largest insects in the drawer, and built a new apartment, composed partly of cork as before, and partly of bits clipped out of the moth's wings.

"We have seen these caterpillars form their habitations of every sort of insect, from a butterfly to a beetle, and the soft, feathery wings of moths answer their purpose very well; but when they fall in with such hard materials as the musk-beetle, or the large scolopendra of the West Indies, they find some difficulty in the building.

"When the structure is finished, the insect deems itself secure to feed on the materials of the cloth, or other animal matter within its reach, provided it is dry and free from fat or grease, which Réaumur found it would not touch. For building, it always selects the straightest and loosest pieces of wool; but for food it prefers the shortest and most compact; and to procure these, it eats into the body of the stuff, rejecting the pile or nap, which it necessarily cuts across at the origin and permits to fall, leaving it threadbare, as if it had been much worn."

From the account which has just been given, it is evident that the caterpillar must be able to turn completely round in its case, and in order to enable it to perform this evolution, the tube is much wider in the middle than at the ends.

The instinct of the parent moth enables it to discover with astonishing certainty any substance which may afford food to its future young. Stuffed birds suffer terribly from the moth, because the arsenical soap with which the skins are preserved does not extend its poisonous influence to the feathers. I have known whole cases of birds to be destroyed by the moth, all the feathers being eaten, and nothing left but the bare skins.

Even the most deadly poison, corrosive sublimate, is not effectual, unless it settles on every feather. There is now before me a stuffed golden-eye duck, preserved by myself, the close plumage of which has partially thrown off the poisoned solution, and has consequently admitted the moth in small patches of feathers, especially about the neck. There is also in my collection a Kaffir shield, made of an ox-hide, which has been washed with the solution, and is almost entirely secure from the depredations of the moth. Yet there are one or two spots where a thong has protected the hair, and in those very spots the pertinacious moths have laid their eggs, and, in one or two instances, the caterpillars have succeeded in attaining their perfect state.

If the reader will refer to the large illustration, he will see a representation of that curious temporary habitation which is popularly termed an Elk-yard.

The Elk, or Moose (*Alces malchis*), inhabits the northern parts of America and Europe, and is, consequently, an animal which is formed to endure severe cold. Although a very large and powerful animal, measuring sometimes seven feet in height at the shoulders—a height which is very little less than that of an average elephant—it has many foes and is much persecuted both by man and beast. During the summer-time it is tolerably safe, but in the winter it is beset by many perils.

In its native country the snow falls so thickly, that the inhabitants of a more temperate climate can hardly imagine the result of a heavy storm. The face of the earth is wholly changed—well-known pits and declivities have vanished—white hills stand where was formerly a level plain—tier upon tier of mimic fortifications rise above each other, the walls being scarped and cut by the wind in weird resemblance of human architecture.

During the sharp frosts, the Elk runs but little risk, because it can traverse the hard, frozen surface of the snow with considerable speed, although with a strange, awkward gait. Its usual pace is a swinging trot; but so light is its action, and so long are its legs, that it quietly trots over obstacles which a horse could not easily leap, because the frozen surface of the snow, although competent to withstand the regular trotting force, could not endure the sudden impact of a horse when leaping. As an example of the curious trot of this animal, I may mention that on one occasion an Elk was seen to trot uninterruptedly over a number of fallen tree-trunks, some of which were nearly five feet in diameter.

It is a remarkable fact that the split hoofs of the Elk spread widely when the foot is placed on the ground, coming together again with a loud snap when it is raised. In consequence of

THE ELK.

this peculiarity, the Elk's progress is rather noisy, the crackling sounds of the hoofs following each other in quick succession.

Want of food is sometimes a danger to the Elk; but the animal is taught by instinct to clear away the snow, and to discover the lichens on which it chiefly lives. The carnivorous animals, however, are always fiercely hungry in the winter-time, and gain from necessity a factitious courage which they do not possess at other times. As long, however, as the frost lasts, the Elk cares little for such foes, as it can distance them if they chase it ever so fiercely, or oppose them if by chance it should find itself in a place where there is no retreat. They do not like to attack an animal whose skin is so thick and tough that, when tanned, it will resist an ordinary pistol-bullet, and which has, besides, an awkward knack of striking with its fore-feet like a skilful boxer, knocking its foes over, and then pounding them with its hoofs until they are dead.

But when the milder weather begins to set in, the Moose is in constant danger. The warm sun falling on the snow produces a rather curious effect. The frozen surface only partially melts, and the water, mixing with the snow beneath, causes it to sink away from the icy surface, leaving a considerable space between them. The "crust," as the frozen surface is technically named, is quite strong enough to bear the weight of comparatively small animals, such as wolves, especially when they run swiftly over it; but it yields to the enormous weight of the Elk, which plunges to its belly at every step.

The wolves have now the Elk at an advantage. They can overtake it without the least difficulty; and if they can bring it to bay in the snow, its fate is sealed. They care little for the branching horns, but leap boldly at the throat of the hampered animal, whose terrible fore-feet are now powerless, and, by dint of numbers, soon worry it to death. Man, too, takes advantage of this state of the snow, equips himself with snow-shoes, and skims over the slight and brittle crust with perfect security. An Elk, therefore, whenever abroad in the snow, is liable to many dangers, and, in order to avoid them, it makes the curious habitation which is called the Elk-yard.

This winter home is very simple in construction, consisting of a large space of ground on which the snow is trampled down by continually treading it so as to form both a hard surface, on which the animal can walk, and a kind of fortress in which it can dwell securely. The whole of the space is not trodden down to one uniform level, but consists of a network of roads or passages through which the animal can pass at ease. So confident is the Elk in the security of the "yard," that it can scarcely ever be induced to leave its snowy fortification, and pass into the open ground.

This habit renders it quite secure from the attacks of wolves, which prowl about the outside of the yard, but dare not venture within; but, unfortunately for the Elk, the very means which preserve it from one danger only lead it into another. If the hunter can come upon one of these Elk-yards, he is sure of his quarry; for the animal will seldom leave the precincts of the snowy inclosure, and the rifle-ball soon lays low the helpless victims.

The Elk is not the only animal that makes these curious fortifications, for a herd of Wapiti deer will frequently unite in forming a common home.

One of these "yards" has been known to measure between four or five miles in diameter, and to be a perfect network of paths sunk in the snow. So deep indeed is the snow when untrodden, that when the deer traverse the paths, their backs cannot be seen above the level of the white surface. Although of such giant size, the "yard" is not by any means a conspicuous object, and at a distance of a quarter of a mile or so, a novice may look directly at the spot without perceiving the numerous paths. This curious fact can easily be understood by those of my readers who have visited one of our modern fortifications, and have seen the slopes of turf apparently unbroken, although filled with deep trenches.

THERE are many other animals which form temporary habitations in which they can remain concealed, because they are taught by instinct how to make their domicile harmonize with the surrounding objects.

One very familiar instance may be found in the common HARE, whose "form" is large enough to shelter the owner, and yet is so inconspicuous that the animal often lies undiscovered, though a human being has passed within a couple of paces of its home. The Hare is never at a loss for a home, and will often hide itself very effectually in a tuft of grass that seems scarcely large enough to conceal a rat. But it is by no means insensible of the value of a denser cover, and seems to have a peculiar affection for a thick, though small, clump of furze.

Within a mile or two of my house there is a heath which is partly studded with furze bushes, and which is a very paradise for various field animals. The field-mice have covered it with their "runs," which are often so slightly below the surface, that if the finger be inserted in the entrance it can be pushed along the whole length of the burrow, the only cover being a slight layer of still living moss. As to the Hares, a "form" can be found every few yards, and if a little thick stubbly furze-bush should be seen standing alone, it is nearly certain to be the home of a Hare, which has made its warm soft couch within the mass of needle-like prickles.

The TIGER has a very similar habit, and takes advantage of a certain drooping shrub, called the *Korinda*, which is of low growth, making its lair underneath the boughs, which afford at once a shelter from the sun and a concealment from enemies.

WE now pass to the Birds, the first of which is that remarkable species called the EDIBLE or ESCULENT SWALLOW (*Collocalia nidifica*). The popular name is given to it, not because itself is edible, but because its nest is eaten in some countries.

We have all heard of birds'-nest soup, and some of us may possibly have imagined that the nests in question are made of the ordinary vegetable substances, such as moss, leaves, and twigs. In reality, the nests are formed of some gelatinous substance, though its true nature is still uncertain, no one precisely knowing whether it is of animal or vegetable origin. Some persons have thought that the material is fish-spawn, which the bird fetches from the sea; others have supposed it to be a kind of seaweed, which is dissolved in the bird's crop and then disgorged; while others believe that it is secreted by certain glands in the throat, and proceeds entirely from the body of the architect.

When first made, these nests are very white and delicate in their aspect, and in that condition are extremely valuable, being sold at an extravagant price to the Chinese. They soon darken by use and exposure, and are not fit for the purposes of the table until they have been cleaned and bleached.

These nests are found in Borneo, Java, &c. and are extremely local, being confined to certain spots. The birds always choose the sides of deep cavernous precipices, so that the task of obtaining the nests is extremely dangerous. They are attached to the perpendicular rocks much as the ordinary mud-built swallow nests, and are generally arranged in horizontal layers. The caverns in which the nests are placed are extremely valuable, and are preserved with jealous care from any intruder.

EDIBLE SWALLOW.

One of these nests in my own collection is shaped much like one of the halves of a bivalve shell, and is thick at the base where it was attached to the rock, diminishing towards the extremity. On the outside it has a very shelly appearance, being made in regular layers, whose edges are as distinct as those of the oyster-shell, but which have a double and not a single curve. In shape it is somewhat oval, but the base is necessarily flat, on account of its attachment to the rock.

The material is so translucent, that when placed on printed paper and held to the light, the capital letters can be plainly read through its substance. A glance at the interior shows at once the mode of its construction. It is made of innumerable glutinous threads, which have been drawn irregularly across each other, and have hardened by exposure to the air into a material which much resembles isinglass. The natives say, that the construction of a single nest occupies a pair of birds for two full months; so that there is some probability that the material may really be secreted by the birds themselves.

The nests are only used for one purpose. They are steeped in hot water for a considerable time, when they soften into a gelatinous mass, which forms the basis of a fashionable soup, and is not unlike the green fat of the ordinary turtle. Indeed, those who have partaken of birds'-nest soup say, that if it were seasoned in a similar manner, it might easily be taken for turtle soup. The Chinese value this soup highly, thinking that it possesses great power of restoring lost strength. It is, however, far too costly to be obtained by any but the rich, the best quality fetching rather more than sixty shillings per pound.

The natives of Borneo have some curious traditions about these birds. They reverence all kinds of birds, believing their ancestors to have been born from a native woman who married a spirit; but they have some singular legends respecting the Edible Swallow. One of these wild legends has been kindly furnished to me by C. T. C. Grant, Esq. formerly of the Sarawak Government Service.

A DYAK LEGEND.

"It was many, many years ago, that a Dyak of *Semăbang* (in *Sadong*) and his young son arrived, after a long journey through the jungle, at a village called *Si-Lĕbor*. The village was extensive, the Dyaks very numerous. On arriving, the chief of the tribe placed food before the older visitor, but to his young son they offered nothing. The little fellow seeing this, and being very hungry after his journey, felt much hurt, and began to cry. 'To my father,' said he, 'you have given food, the *prĭok* of rice is before him, the fatted pig has been killed—everything you have given him. Why do you give me nothing?' But the child's appeal was useless. These strange Dyaks had hearts of stone; not a morsel was handed to the fatigued and hungry little wayfarer; so he wept on, and wept in vain.

"After a while the boy looked more cheerful; he had dried his tears, and was now engaged in catching a dog and a cat. These he put together on the mat, round which all the people were seated. The cat and the dog played, or more likely, as these animals will do, fought together; but whatever it was, there was something so ludicrous in it all, while the boy sat over them and set them at each other, that the whole assemblage burst into immoderate laughter. The boy, it would seem, was working some spell—there was an object in what he had been doing. Perhaps he was in communication with evil spirits, or under their influence; there was something ominous about it, we know not what. But to proceed: presently the sky became overcast, and gradually great volumes of black clouds came sailing up, propelled by sudden gusts of wind. One by one they rolled along, and were heaped up one on top of another, or got all broken up, as it were, in their collision. The sky appeared one mass of confusion, looking blacker and more angry as the sun gradually disappeared in the darkness. At last the storm burst forth with a fury never known before; sharp flashes of lightning, followed by awful peals of thunder, succeeded one another, fast and furious; the very ground below shook as the palm-leaf quivered in the breeze—it seemed as if the great end of all things was at hand.

"Now commenced a gradual but awful change. Amidst the rolling thunder, and the dazzling lightning, which only served to make the awful darkness visible, the village, the houses, all began to dissolve, to melt away, as it were, into burning lava, and, with his works, man perishing likewise. There you might see the grey-headed chief starting up, with his grandson in his arms, but, ere reaching the door, being gradually hardened into stone. There mothers would be seen flying with their little ones, to escape the same dreadful fate, but all in vain. There a young and helpless maiden would be clinging to her brave warrior, to that arm which had always been the first to help her, which could surely save her now. Alas! that cruel transformation! The living light in those bright eyes is gone, the tender grasp of that warm hand is cold; from flesh and blood they, too, pass away into senseless petrifactions; whilst, mingling with the shrieks and yells and invocations of the men and the *Borich*, would still be heard the boom of the thunder and the crackling of the houses. Not a man, woman, or child—no, nor even visitor—at that fatal village, save only the neglected boy, was left alive to mourn the loss of his all. One after another, they all melted, and were changed, when the heat of the storm was over, into solid rock. Houses, and all in them, succumbed beneath the fiery elements; and when the storm ceased, all lay, not a heap of charred ruins, but huge masses of smoking stone.

"A hill with precipices now marks the spot where this tragedy occurred; on the hill (itself the transformed village) are still pointed out, if people speak truth, the traces of the petrified houses. An upright rock is shown as the transformed figure of a Malay, an unhappy visitor on that awful day. There he stands with his hand still fixed on his sword hilt, once a living soul, now a lifeless stone. The whole scene, indeed, is a standing monument at once of the crime of inhospitality and its fearful punishment. Gazing on his revenge, the youth retreated. He returned to his native village, *Semăbang;* and time flew on, and ere he died he was the chief of his tribe, the grey-headed patriarch appealed to by the new and rising generation. Years, hundreds of years rolled away, fathers and mothers passed off the stage, and young children grew up to take their places, to attain manhood, to work, to become old, to die too; and so time went on, and children danced and played over the same ground that their ancestors had danced and played on for centuries before.

"At last, no great time ago, the tribe of *Semăbang* having flourished, become populous and rich, a young chief, the lineal descendant of the little hungry boy, dreamed that great riches were in store for him and his tribe if they went to Mount *Si-Lĕbor*, the petrified village. The next day a party was organized, and they went there and searched. They at last discovered a magnificent cave. With lighted torches they entered, and found it to be very extensive, and full of the celebrated edible birds'-nests. 'Ah,' said they, 'this is our portion, instead of that which was denied to our ancestor; his due was refused then, it has now been given to us, his descendants; this is our '*balas*' (revenge). Thousands and thousands of birds'-nests they brought out of the cave, which realized many reals to the discoverers. The *Si-Lĕbor* caves are now said to be the richest, and the tribes possessing them (the *Semābang* youth's descendants) the wealthiest and most prosperous in Sadong."

There are at least four species of swallow that make these curious nests, and the natives say that the entrance to the caves is always occupied by another kind of swallow, which makes a nest of mixed moss and gelatine, and which fights the valuable birds and drives them away. They therefore always attack the intruders, and endeavour to knock down their nests with stones. The nests are very small and shallow, and seem scarcely capable of accommodating either eggs or young birds. My own specimen is exactly two inches in length, one inch and three-quarters in breadth at its widest point, and scarcely more than half-an-inch deep. Its internal shape is exactly that of a spoon-bowl, one-third of which has been cut off abruptly near the handle.

None of the purely predacious birds are remarkable for their skill in architecure, and the EAGLE (*Aquila chrysaëtos*) is no exception to the general rule. The nest of this magnificent bird is nothing more than a huge mass of sticks flung at random on some rocky ledge, and having a shallow depression in which the young can lie. In general shape, or rather in shapelessness, it is not unlike the nest of the osprey, which has already been described, and it is so rudely put together that the sticks seem to afford even a less commodious bed than the bare rock.

The portion that is occupied by the young is comparatively small, and the general platform of the nest serves as a sort of larder, on which are deposited the birds, hares, lambs, and other animals which the parents have killed and brought home. Sometimes the nest will be amply supplied with food, but sometimes the parent birds are obliged to hunt daily. Young eagles are voracious beings, and if there be no sheep flocks within reach, the task of supplying them with food is a very heavy one, especially when they have nearly attained maturity. In feeding its young for the first few weeks of their life, the eagle tears the prey into little pieces, and impartially distributes the bleeding morsels to the gaping and screaming offspring. Afterwards, however, when the young eagles have gained a stout beak, the prey is merely dropped near them, and they tear it to pieces for themselves.

Generally the nest of the Eagle is placed in some inaccessible spot, and the bird seems never to be so pleased as when it can find a rocky ledge situated about half way down a precipice, and sheltered from above by a large projecting piece of rock. This projection answers two purposes. It prevents the nest from being seen from above, and also guards it from being harried by persons let down by ropes. To take an Eagle's nest is always a task of extreme difficulty, and one which tries to

EAGLE.

the utmost the nerves and endurance of the climber. It also makes considerable demands on his courage, for if the parent birds should discover the intruder, they are sure to attack him, and may very probably dash him to the ground.

Should the bold cragsman succeed in reaching the nest, he does not find it a very pleasant locality. The nostrils of the

Eagle are very useful for the purpose of respiration, but the bird has apparently little or no olfactory sensibilities. The stench that arises from an inhabited Eagle's nest is quite beyond the power of description, for the young Eagles themselves are not the sweetest beings in the world, and their evil odour is supplemented by that which arises from the refuse food that is suffered to putrefy in the very nest.

THERE are very many sea-birds which hatch their young on the shelves of precipitous rocks, and of them I have chosen for an example the bird which is called the NODDY (*Anous stolidus*). It is a species of Tern, and has long been celebrated among sailors for the ease with which it can be captured, especially if the daylight has departed.

THE NODDY.

The Noddy mostly chooses for its nesting-place some lofty precipice, and generally lays its eggs upon a shelf of the rock. Sometimes but rarely, it takes a fancy to some low and thick bush, and in any case is but an indifferent architect. Often the nest is nothing more than a heap of seaweed, on the top of which is excavated a very slight hollow; and in no case does the bird seem to exercise any skill in the disposition of materials. As it returns year after year to the same spot, and never clears away the old nest, it manages in time to accumulate a heap of seaweed that is sometimes more than two feet in thickness, and of considerable width. The bird is gregarious in its nesting, the rocky ledges being crowded with the rude nests, and the odour that proceeds from them being absolutely intolerable to human nostrils. The eggs are rather pretty, being of an orange colour, spotted and splashed with red and purple of different shades.

It is rare in England, but there are many British birds that build in a similar manner, such as the Solan goose, or gannet, the cormorant, the guillemot, and various gulls.

THE nest of the NIGHTINGALE (*Luscinia Philomela*) could hardly be classed in any of the preceding groups, and therefore takes its place among the miscellaneous habitations.

It is not built in the branches, nor in a hole, nor suspended from a bough, nor absolutely on the ground. It is always set very near the ground, and in most cases it is scarcely raised more than a few inches above the soil. In one sense it is not a pretty nest. It is certainly not a neat one, and its apparent roughness of construction is probably intended to make it less conspicuous. The discovery of a Nightingale's nest is not an easy task, unless the eye be directed to the spot by watching the movements of the bird. It is always most carefully hidden under growing foliage, and so well is it concealed that even in places where Nightingales abound, the detection of a nest is always welcome to the egg-hunter.

The materials of the nest are equally calculated for concealment, consisting of straw, grass, little sticks, and dried leaves, all being jumbled together with such "artless art," that even when a nest is seen its real nature often escapes the discoverer. If the same materials were seen in a branch at any height from the ground they would at once attract attention, but in the position which they occupy they look like a mass of loose débris that has been blown by the wind and arrested by the foliage among which it has lodged.

The eggs are equally inconspicuous, being dull olive-brown, without a spot or streak. After they are laid, the lively song of the Nightingale becomes less and less frequent, while after the young are hatched, the bird is silent until the next season. The Nightingale is as anxious to conceal itself as its nest, and never intentionally shows its brown plumage, though it will sing

THE NIGHTINGALE.

within six feet of a listener who will remain quiet. In the spring the bird seems as if it *must* sing, no matter who may be near, and its spirit of rivalry is so great, that the "jug-jug" of one Nightingale is sure to set singing all the others within hearing.

THE WANDERING ALBATROS (*Diomedea exulans*), the giant of the petrel tribe, makes it nest after a peculiar fashion

THE ALBATROS.

It chooses the summit of lofty precipices near the sea, and its nest may be found most plentifully in Tristan d'Acunha and the Marion Islands. The Albatros is lord of the country, and no other living being seems to intrude upon its nesting place. So completely do the birds feel themselves masters of the situation, that if a human being penetrates to their haunts, they quietly move about as if he were non-existent, and do not appear to take the least notice of him. On such elevated positions the cold is necessarily intense, but the Albatros cares not for the cold, and brings up its white-coated young in a temperature that few human beings like to endure longer than needful.

No particular bed seems necessary for the egg, for the mother bird simply deposits it on the bare ground, and then scrapes earth round it so as to form a small circular wall, as may be seen by reference to the illustration. If their nest be approached very closely, the alarmed parents snap their bills like angry owls, and if they wish to be very aggressive they discharge from their bills a quantity of oil; but they seem to have no ideas of actual fight. The Albatros lays only one egg.

OUR last sample of "Homes without Hands" is the ingenious structure that is made by the COOT (*Fulica atra*), the BALD COOT as it is sometimes called, on account of the horny plate on the forehead, which is pink during the breeding season, and white during the rest of the year. Although the general colour of the Coot is black, it is a pretty bird when in the water, and if the day be calm, the reflection on the surface has a very curious effect, the white patch appearing as if it rose to the surface of the water every time that the bird nods its head in the act of swimming.

The favourite nesting places of the Coot are little islands on which the grass grows rankly. Failing them it will make its nest among reeds and rushes, binding and twisting them together until they are firm enough to support the weight of the nest, the bird, and the many eggs. Should it not find either of these localities, it will build on the edge of the water, and almost invariably contrives to make its nest in such a manner that it cannot be reached from the land. The quantity of reeds, bulrushes, sedges, grass, and other materials used in the nest is very surprising; and yet, in spite of its large dimensions it is not a conspicuous object. The nest contains a great number of eggs, seldom less than seven, and sometimes twelve or fourteen. They are whitish, and profusely spotted with irregular brown marks.

In the illustration, the haunts of the Coot are well represented. In the foreground is seen one of the grass tussocks, of which a pair of Coots have taken possession, and in which the young are seen under the protection of their parents. Similar tussocks

THE COOT.

protrude from the shallow water, and from one of them the mother Coot is issuing, followed by her young brood. In the background are seen a pair of swans, one of which is bearing her young on her back, according to the custom of her kind.

As is the case with many of the illustrations to this work, the sketch was taken from nature.

INDEX.

PRINTED BY SPOTTISWOODE AND CO., NEW-STREET SQUARE, LONDON

www.ingramcontent.com/pod-product-compliance
Lightning Source LLC
LaVergne TN
LVHW021048110826
845150LV00001B/10

* 9 7 8 1 4 2 5 5 6 8 5 2 8 *